NEW PLANE AND SPHERICAL

TRIGONOMETRY, SURVEYING

AND

NAVIGATION

BY

G. A. WENTWORTH, A.M.

AUTHOR OF A SERIES OF TEXT-BOOKS IN MATHEMATICS

Teachers' Edition

BOSTON, U.S.A., AND LONDON

GINN & COMPANY, PUBLISHERS

1896

PREFACE.

THIS edition is intended for teachers, *and for them only.* The publishers will under no circumstances sell the book except to teachers of Wentworth's Trigonometry; and every teacher must consider himself in honor bound not to leave his copy where pupils can have access to it, and not to sell his copy except to the publishers, Messrs. Ginn & Co.

It is hoped that young teachers will derive great advantage from studying the systematic arrangement of the work, and that all teachers who are pressed for time will find great relief by not being obliged to work out every problem.

G. A. WENTWORTH.

TRIGONOMETRY.

TEACHERS' EDITION.

Exercise I. Page 2.

1. Reduce the following angles to circular measure, expressing the results as fractions of π: 60°, 45°, 150°, 195°, 11° 15′, 123° 45′, 37° 30′.

$$60° = \tfrac{60}{180}\pi = \tfrac{1}{3}\pi.$$

$$45° = \tfrac{45}{180}\pi = \tfrac{1}{4}\pi.$$

$$150° = \tfrac{150}{180}\pi = \tfrac{5}{6}\pi.$$

$$195° = \tfrac{195}{180}\pi = \tfrac{13}{12}\pi.$$

$$11° \ 15' = \frac{11\frac{1}{4}}{180}\pi = \tfrac{1}{16}\pi.$$

$$123° \ 45' = \frac{123\frac{3}{4}}{180}\pi = \tfrac{11}{16}\pi.$$

$$37° \ 30' = \frac{37\frac{1}{2}}{180}\pi = \tfrac{5}{24}\pi.$$

2. How many degrees are there in $\frac{2}{3}\pi$ radians? $\frac{3}{4}\pi$ radians? $\frac{5}{8}\pi$ radians? $\frac{15}{16}\pi$ radians? $\frac{7}{15}\pi$ radians?

$\frac{2}{3}\pi = 120°$. $\quad \frac{5}{8}\pi = 112° \ 30'$.

$\frac{3}{4}\pi = 135°$. $\quad \frac{15}{16}\pi = 168° \ 45'$.

$\frac{7}{15}\pi = 84°$.

3. What decimal part of a radian is 1°? 1′?

$$1° = \frac{\pi}{180} = \frac{3.1416}{180} = 0.0174533 \text{ radian.}$$

$$1' = \frac{0.0174533}{60} = 0.00029089 \text{ radian.}$$

4. How many seconds in a radian?

$$1 \text{ radian} = 57° \ 17' \ 45''$$
$$= 206265 \text{ seconds.}$$

5. Express in radians one of the interior angles of a regular octagon; dodecagon.

The sum of all the interior angles of a regular octagon is $8 \times 180° - 360° = 8\pi - 2\pi = 6\pi$. Hence each interior angle

$$= \frac{6\pi}{8} = \frac{3\pi}{4} \text{ radians.}$$

The sum of all the interior angles of a regular dodecagon is $12 \times 180° - 360° = 12\pi - 2\pi = 10\pi$. Hence each interior angle

$$= \frac{10\pi}{12} = \frac{5\pi}{6} \text{ radians.}$$

6. On a circle of 50 ft. radius an arc of 10 ft. is laid off. How many degrees does it subtend at the centre?

It subtends $\frac{10}{50} = \frac{1}{5}$ radian, or $\frac{57° \ 17' \ 45''}{5} = 11° \ 27' \ 33''$.

7. The earth's equatorial radius is approximately 3963 miles. If two points on the equator are 1000 miles apart, what is their difference in longitude?

Their difference in longitude is $\frac{1000}{3963}$ radian, or $\frac{1000}{3963} \times 57° 17' 45''$ $= 14° 27' 27''$.

8. If the difference in longitude of two points on the earth's equator is 1°, what is the distance between them in miles?

By Ex. 3, $1° = 0.0174533$ radian. Hence 1° on the earth's equator is equal to 0.0174533×3963 miles $\doteq 69.167$ miles.

9. What is the radius of a circle if an arc of 1 ft. subtends an angle of 1° at the centre?

Since 1° of arc = 1 ft., 1 radian, $57° 17' 45''$, $= 57\frac{17\frac{3}{4}}{60}$ ft. = 57 ft. 3.55 in. = the radius.

10. In how many hours is a point on the equator carried by the earth's rotation through a distance equal to the earth's radius?

The earth turns through $360° = 2\pi$ radians in 24 hours. Hence it turns through 1 radian in $\frac{24}{2\pi} = \frac{12}{3.1416}$ hours = 3 hrs. 49 min. 11 sec.

11. The minute-hand of a clock is 3 ft. 6 in. long. How far does its extremity move in 25 minutes? $(\pi = \frac{22}{7}.)$

The circumference which is passed over in 60 minutes is $2\pi \times 3\frac{1}{2}$ ft. Hence the arc passed over in 25 minutes is $\frac{25}{60} \times 2\pi \times 3\frac{1}{2} = \frac{35}{12}\pi$ $= \frac{35}{12} \times \frac{22}{7} = 9$ ft. 2 in.

12. A wheel makes 15 revolutions a second. How long does it take to turn through 4 radians? $(\pi = \frac{22}{7}.)$

The wheel turns through 2π radians in $\frac{1}{15}$ of a second. Hence it turns through 4 radians in

$$\frac{4}{2\pi} \times \frac{1}{15} = \frac{7}{11} \times \frac{1}{15} = \frac{7}{165} \text{ second.}$$

Exercise II. Page 5.

1. What are the functions of the other acute angle B of the triangle ABC (Fig. 2)?

$$\sin B = \frac{b}{c}, \qquad \cos B = \frac{a}{c},$$

$$\tan B = \frac{b}{a}, \qquad \cot B = \frac{a}{b},$$

$$\sec B = \frac{c}{a}, \qquad \csc B = \frac{c}{b}.$$

2. If $A + B = 90°$, prove that

$$\sin A = \cos B,$$
$$\tan A = \cot B,$$
$$\sec A = \csc B,$$
$$\cos A = \sin B,$$
$$\cot A = \tan B,$$
$$\csc A = \sec B,$$
$$\text{vers } A = \text{covers } B,$$
$$\text{covers } A = \text{vers } B.$$

$\sin A = \frac{a}{c}$, $\cos B = \frac{a}{c}$,

$\cos A = \frac{b}{c}$, $\sin B = \frac{b}{c}$,

$\tan A = \frac{a}{b}$, $\cot B = \frac{a}{b}$,

$\cot A = \frac{b}{a}$, $\tan B = \frac{b}{a}$,

$\sec A = \frac{c}{b}$, $\csc B = \frac{c}{b}$,

$\csc A = \frac{c}{a}$, $\sec B = \frac{c}{a}$,

$\text{vers} A = \frac{c-b}{c}$, $\text{covers} B = \frac{c-b}{c}$,

$\text{covers} A = \frac{c-a}{c}$, $\text{vers} B = \frac{c-a}{c}$.

3. Find the values of the functions of A, if a, b, c respectively have the following values:

(i.) 3, 4, 5. (iv.) 9, 40, 41.
(ii.) 5, 12, 13. (v.) 3.9, 8, 8.9.
(iii.) 8, 15, 17. (vi.) 1.19, 1.20, 1.69.

(i.) $\sin A = \frac{3}{5}$, $\cos A = \frac{4}{5}$, $\tan A = \frac{3}{4}$, $\cot A = \frac{4}{3}$, $\sec A = \frac{5}{4}$, $\csc A = \frac{5}{3}$.

(ii.) $\sin A = \frac{5}{13}$, $\cos A = \frac{12}{13}$, $\tan A = \frac{5}{12}$, $\cot A = \frac{12}{5}$, $\sec A = \frac{13}{12}$, $\csc A = \frac{13}{5}$.

(iii.) $\sin A = \frac{8}{17}$, $\cos A = \frac{15}{17}$, $\tan A = \frac{8}{15}$, $\cot A = \frac{15}{8}$, $\sec A = \frac{17}{15}$, $\csc A = \frac{17}{8}$.

(iv.) $\sin A = \frac{9}{41}$, $\cos A = \frac{40}{41}$, $\tan A = \frac{9}{40}$, $\cot A = \frac{40}{9}$, $\sec A = \frac{41}{40}$, $\csc A = \frac{41}{9}$.

(v.) $\sin A = \frac{39}{89}$, $\tan A = \frac{39}{80}$, $\sec A = \frac{89}{80}$, $\cos A = \frac{80}{89}$, $\cot A = \frac{80}{39}$, $\csc A = \frac{89}{39}$.

(vi.) $\sin A = \frac{119}{169}$, $\cos A = \frac{120}{169}$, $\tan A = \frac{119}{120}$, $\cot A = \frac{120}{119}$, $\sec A = \frac{169}{120}$, $\csc A = \frac{169}{119}$.

4. What condition must be fulfilled by the lengths of the three lines a, b, c (Fig. 2) in order to make them the sides of a right triangle? Is this condition fulfilled in Example 3?

It is $a^2 + b^2 = c^2$.

5. Find the values of the functions of A, if a, b, c respectively have the following values:

(i.) $2mn$, $m^2 - n^2$, $m^2 + n^2$.

(ii.) $\frac{2xy}{x-y}$, $x + y$, $\frac{x^2 + y^2}{x-y}$.

(iii.) pqr, qrs, rsp.

(iv.) $\frac{mn}{pq}$, $\frac{mv}{sq}$, $\frac{nr}{ps}$.

(i.)

$$\sin A = \frac{a}{c} = \frac{2mn}{m^2 + n^2},$$

$$\cos A = \frac{b}{c} = \frac{m^2 - n^2}{m^2 + n^2},$$

$$\tan A = \frac{a}{b} = \frac{2\,mn}{m^2 - n^2},$$

$$\cot A = \frac{b}{a} = \frac{m^2 - n^2}{2\,mn},$$

$$\sec A = \frac{c}{b} = \frac{m^2 + n^2}{m^2 - n^2},$$

$$\csc A = \frac{c}{a} = \frac{m^2 + n^2}{2\,mn}.$$

(ii.)

$$\sin A = \frac{2\,xy}{x-y} \times \frac{x-y}{x^2+y^2} = \frac{2\,xy}{x^2+y^2},$$

$$\cos A = (x+y) \times \frac{x-y}{x^2+y^2} = \frac{x^2-y^2}{x^2+y^2},$$

$$\tan A = \frac{2\,xy}{x-y} \times \frac{1}{x+y} = \frac{2\,xy}{x^2-y^2},$$

$$\cot A = \frac{x-y}{2\,xy} \times (x+y) = \frac{x^2-y^2}{2\,xy},$$

$$\sec A = \frac{1}{x+y} \times \frac{x^2+y^2}{x-y} = \frac{x^2+y^2}{x^2-y^2},$$

$$\csc A = \frac{x-y}{2\,xy} \times \frac{x^2+y^2}{x-y} = \frac{x^2+y^2}{2\,xy}.$$

(iii.)

$$\sin A = \frac{pqr}{rsp} = \frac{q}{s},\quad \cos A = \frac{qrs}{rsp} = \frac{q}{p},$$

$$\tan A = \frac{pqr}{qrs} = \frac{p}{s},\quad \cot A = \frac{qrs}{pqr} = \frac{s}{p},$$

$$\sec A = \frac{rsp}{qrs} = \frac{p}{q},\quad \csc A = \frac{rsp}{pqr} = \frac{s}{q}.$$

(iv.)

$$\sin A = \frac{mn}{pq} \times \frac{ps}{nr} = \frac{ms}{qr},$$

$$\cos A = \frac{mv}{sq} \times \frac{ps}{nr} = \frac{mpv}{nqr},$$

$$\tan A = \frac{mn}{pq} \times \frac{sq}{mv} = \frac{ns}{pv},$$

$$\cot A = \frac{pq}{mn} \times \frac{mv}{sq} = \frac{pv}{ns},$$

$$\sec A = \frac{sq}{mv} \times \frac{nr}{ps} = \frac{nqr}{mpv},$$

$$\csc A = \frac{pq}{mn} \times \frac{nr}{ps} = \frac{qr}{ms}.$$

6. Prove that the values of a, b, c, in (i.) and (ii.), Example 5, satisfy the condition necessary to make them the sides of a right triangle.

(i.)

$$a^2 + b^2 = c^2,$$

$$(2\,mn)^2 + (m^2 - n^2)^2 = (m^2 + n^2)^2,$$

$$4\,m^2n^2 + m^4 - 2\,m^2n^2 + n^4 = m^4 + 2\,m^2n^2 + n^4,$$

$$m^4 + 2\,m^2n^2 + n^4 = m^4 + 2\,m^2n^2 + n^4.$$

(ii.)

$$\left(\frac{2\,xy}{x-y}\right)^2 + (x+y)^2 = \left(\frac{x^2+y^2}{x-y}\right)^2,$$

$$\frac{4\,x^2y^2}{x^2 - 2\,xy + y^2} + x^2 + 2\,xy + y^2 = \frac{x^4 + 2\,x^2y^2 + y^4}{x^2 - 2\,xy + y^2},$$

$$4\,x^2y^2 + x^4 - 2\,x^2y^2 + y^4 = x^4 + 2\,x^2y^2 + y^4,$$

$$x^4 + 2\,x^2y^2 + y^4 = x^4 + 2\,x^2y^2 + y^4.$$

7. What equations of condition must be satisfied by the values of a, b, c, in (iii.) and (iv.) Example 5, in order that the values may represent the sides of a right triangle?

(iii.)

$$p^2q^2r^2 + q^2r^2s^2 = r^2s^2p^2,$$

or $$p^2q^2 + q^2s^2 = p^2s^2.$$

(iv.)

$$\frac{m^2n^2}{p^2q^2} + \frac{m^2v^2}{s^2q^2} = \frac{n^2r^2}{p^2s^2},$$

or $$m^2n^2s^2 + m^2p^2v^2 = n^2q^2r^2.$$

8. Compute the functions of A and B when $a = 24$, $b = 143$.

$$c = \sqrt{(24)^2 + (143)^2}$$

$$= \sqrt{21025}$$

$$= 145.$$

$$\sin A = \frac{24}{145} = \cos B,$$

$$\cos A = \frac{143}{145} = \sin B,$$

$$\tan A = \frac{24}{143} = \cot B,$$

$$\cot A = \frac{143}{24} = \tan B,$$

$$\sec A = \frac{145}{143} = \csc B,$$

$$\csc A = \frac{145}{24} = \sec B.$$

9. Compute the functions of A and B when $a = 0.264$, $c = 0.265$.

$$\begin{aligned} b^2 &= c^2 - a^2 \\ &= 0.070225 - 0.069696 \\ &= 0.000529. \\ \therefore b &= 0.023. \end{aligned}$$

$$\sin A = \frac{a}{c} = \frac{264}{265} = \cos B,$$

$$\cos A = \frac{b}{c} = \frac{23}{265} = \sin B,$$

$$\tan A = \frac{a}{b} = \frac{264}{23} = \cot B,$$

$$\cot A = \frac{b}{a} = \frac{23}{264} = \tan B,$$

$$\sec A = \frac{c}{b} = \frac{265}{23} = \csc B,$$

$$\csc A = \frac{c}{a} = \frac{265}{264} = \sec B.$$

10. Compute the functions of A and B when $b = 9.5$, $c = 19.3$.

$$\begin{aligned} a^2 &= c^2 - b^2 \\ &= 372.49 - 90.25 \\ &= 282.24. \\ \therefore a &= 16.8. \end{aligned}$$

$$\sin A = \frac{a}{c} = \frac{168}{193} = \cos B,$$

$$\cos A = \frac{b}{c} = \frac{95}{193} = \sin B,$$

$$\tan A = \frac{a}{b} = \frac{168}{95} = \cot B,$$

$$\cot A = \frac{b}{a} = \frac{95}{168} = \tan B,$$

$$\sec A = \frac{c}{b} = \frac{193}{95} = \csc B,$$

$$\csc A = \frac{c}{a} = \frac{193}{168} = \sec B.$$

11. Compute the functions of A and B when

$$a = \sqrt{p^2 + q^2},\ b = \sqrt{2pq}.$$

$$\begin{aligned} a^2 + b^2 &= c^2, \\ p^2 + 2pq + q^2 &= c^2. \\ \therefore p + q &= c. \end{aligned}$$

$$\sin A = \frac{a}{c} = \frac{\sqrt{p^2 + q^2}}{p + q} = \cos B,$$

$$\cos A = \frac{b}{c} = \frac{\sqrt{2pq}}{p + q} = \sin B,$$

$$\tan A = \frac{a}{b} = \frac{\sqrt{p^2 + q^2}}{\sqrt{2pq}} = \cot B,$$

$$\cot A = \frac{b}{a} = \frac{\sqrt{2pq}}{\sqrt{p^2 + q^2}} = \tan B,$$

$$\sec A = \frac{c}{b} = \frac{p + q}{\sqrt{2pq}} = \csc B,$$

$$\csc A = \frac{c}{a} = \frac{p + q}{\sqrt{p^2 + q^2}} = \sec B.$$

12. Compute the functions of A and B when

$$a = \sqrt{p^2 + pq},\ c = p + q.$$

$$\begin{aligned} b^2 &= c^2 - a^2 \\ &= q^2 + pq. \\ \therefore b &= \sqrt{q^2 + pq}. \end{aligned}$$

$$\sin A = \frac{a}{c} = \frac{\sqrt{p^2 + pq}}{p + q} = \cos B.$$

$$\cos A = \frac{b}{c} = \frac{\sqrt{q^2 + pq}}{p + q} = \sin B,$$

$$\tan A = \frac{a}{b} = \frac{\sqrt{p^2 + pq}}{\sqrt{q^2 + pq}} = \cot B,$$

$$\cot A = \frac{b}{a} = \frac{\sqrt{q^2 + pq}}{\sqrt{p^2 + pq}} = \tan B,$$

$$\sec A = \frac{c}{b} = \frac{p + q}{\sqrt{q^2 + pq}} = \csc B,$$

$$\csc A = \frac{c}{a} = \frac{p + q}{\sqrt{p^2 + pq}} = \sec B.$$

13. Compute the functions of A and B when

$$b = 2\sqrt{pq}, \quad c = p + q.$$

$$a^2 + b^2 = c^2,$$
$$a^2 + 4\,pq = p^2 + 2\,pq + q^2,$$
$$a^2 = p^2 - 2\,pq + q^2,$$
$$a = p - q.$$

$$\sin A = \frac{a}{c} = \frac{p - q}{p + q} = \cos B,$$

$$\cos A = \frac{b}{c} = \frac{2\sqrt{pq}}{p + q} = \sin B,$$

$$\tan A = \frac{a}{b} = \frac{p - q}{2\sqrt{pq}} = \cot B,$$

$$\cot A = \frac{b}{a} = \frac{2\sqrt{pq}}{p - q} = \tan B,$$

$$\sec A = \frac{c}{b} = \frac{p + q}{2\sqrt{pq}} = \csc B,$$

$$\csc A = \frac{c}{a} = \frac{p + q}{p - q} = \sec B.$$

14. Compute the functions of A when $a = 2\,b$.

$$a = 2\,b,$$
$$a^2 + b^2 = c^2,$$
$$4\,b^2 + b^2 = c^2,$$
$$5\,b^2 = c^2,$$
$$c = b\sqrt{5}.$$

$$\sin A = \frac{a}{c} = \frac{2\,b}{b\sqrt{5}} = \tfrac{2}{5}\sqrt{5} = 0.89443,$$

$$\cos A = \frac{b}{c} = \frac{b}{b\sqrt{5}} = \tfrac{1}{5}\sqrt{5},$$

$$\tan A = \frac{a}{b} = \frac{2\,b}{b} = 2,$$

$$\cot A = \frac{b}{a} = \frac{1}{2},$$

$$\sec A = \frac{c}{b} = \frac{b\sqrt{5}}{b} = \sqrt{5},$$

$$\csc A = \frac{c}{a} = \frac{b\sqrt{5}}{2\,b} = \tfrac{1}{2}\sqrt{5}.$$

15. Compute the functions of A when $a = \frac{2}{3}\,c$.

$$a = \tfrac{2}{3}\,c,$$
$$c = \tfrac{3}{2}\,a,$$
$$b^2 = c^2 - a^2,$$
$$b = \sqrt{c^2 - a^2},$$
$$= \sqrt{\tfrac{9}{4}\,a^2 - a^2},$$
$$= \frac{a}{2}\sqrt{5}.$$

$$\sin A = \frac{a}{c} = \frac{a}{\frac{3}{2}\,a} = \frac{2}{3},$$

$$\cos A = \frac{b}{c} = \frac{\frac{a}{2}\sqrt{5}}{\frac{3}{2}\,a} = \tfrac{1}{3}\sqrt{5},$$

$$\tan A = \frac{a}{b} = \frac{a}{\frac{a}{2}\sqrt{5}} = \tfrac{2}{5}\sqrt{5},$$

$$\cot A = \frac{b}{a} = \frac{\sqrt{5}}{2} = \tfrac{1}{2}\sqrt{5},$$

$$\sec A = \frac{c}{b} = \frac{\frac{3}{2}\,a}{\frac{a}{2}\sqrt{5}} = \tfrac{3}{5}\sqrt{5},$$

$$\csc A = \frac{c}{a} = \frac{3}{2}.$$

16. Compute the functions of A when $a + b = \frac{5}{4}c$.

$$a + b = \tfrac{5}{4}c,$$
$$a^2 + b^2 = c^2,$$
$$a^2 + b^2 + 2ab = \tfrac{25}{16}c^2,$$
$$2ab = \tfrac{9}{16}c^2,$$
$$a^2 - 2ab + b^2 = \tfrac{7}{16}c^2,$$
$$a - b = \frac{c}{4}\sqrt{7},$$
$$a + b = \tfrac{5}{4}c,$$
$$2b = \tfrac{5}{4}c - \frac{c}{4}\sqrt{7},$$
$$b = \tfrac{5}{8}c - \frac{c}{8}\sqrt{7},$$
$$2a = \tfrac{5}{4}c + \frac{c}{4}\sqrt{7},$$
$$a = \tfrac{5}{8}c + \frac{c}{8}\sqrt{7},$$
$$\frac{c}{8} = \frac{a}{5 + \sqrt{7}},$$
$$c = \frac{8a}{5 + \sqrt{7}}.$$

$$\sin A = \frac{a}{c} = \frac{a}{\frac{8a}{5 + \sqrt{7}}} = \frac{5 + \sqrt{7}}{8},$$

$$\cos A = \frac{b}{c} = \frac{\frac{5}{8}c - \frac{c}{8}\sqrt{7}}{c} = \frac{5 - \sqrt{7}}{8},$$

$$\tan A = \frac{a}{b} = \frac{5 + \sqrt{7}}{5 - \sqrt{7}},$$

$$\cot A = \frac{b}{a} = \frac{5 - \sqrt{7}}{5 + \sqrt{7}},$$

$$\sec A = \frac{c}{b} = \frac{8}{5 - \sqrt{7}},$$

$$\csc A = \frac{c}{a} = \frac{8}{5 + \sqrt{7}}.$$

17. Compute the functions of A when

$$a - b = \frac{c}{4}.$$

$$\begin{aligned} a^2 - 2ab + b^2 &= \frac{c^2}{16}. \\ a^2 \qquad + b^2 &= c^2 \\ \hline 2ab \qquad &= \frac{15c^2}{16} \\ a^2 \qquad + b^2 &= c^2 \\ \hline a^2 + 2ab + b^2 &= \frac{31c^2}{16} \end{aligned}$$

$$a + b = \frac{c}{4}\sqrt{31},$$
$$a - b = \frac{c}{4},$$
$$2a = \frac{c}{4}\sqrt{31} + \frac{c}{4}.$$
$$\therefore a = \frac{c}{8}(\sqrt{31} + 1),$$
$$2b = \frac{c}{4}\sqrt{31} - \frac{c}{4}.$$
$$\therefore b = \frac{c}{8}(\sqrt{31} - 1).$$

$$\sin A = \frac{a}{c} = \frac{\frac{c}{8}(\sqrt{31}+1)}{c} = \frac{\sqrt{31}+1}{8},$$

$$\cos A = \frac{b}{c} = \frac{\frac{c}{8}(\sqrt{31}-1)}{c} = \frac{\sqrt{31}-1}{8},$$

$$\tan A = \frac{a}{b} = \frac{\sqrt{31} + 1}{\sqrt{31} - 1},$$

$$\cot A = \frac{b}{a} = \frac{\sqrt{31} - 1}{\sqrt{31} + 1},$$

$$\sec A = \frac{c}{b} = \frac{8}{\sqrt{31} - 1},$$

$$\csc A = \frac{c}{a} = \frac{8}{\sqrt{31} + 1}.$$

18. Find a if $\sin A = \frac{3}{5}$ and $c = 20.5$.

$$\sin A = \frac{a}{c} = \frac{3}{5},$$
$$\frac{a}{20.5} = \frac{3}{5},$$
$$5a = 61.5,$$
$$a = 12.3.$$

19. Find b if $\cos A = 0.44$ and $c = 3.5$.

$$\frac{b}{c} = 0.44,$$
$$\frac{b}{3.5} = 0.44.$$
$$\therefore b = 1.54.$$

20. Find a if $\tan A = \frac{11}{3}$ and $b = 2\frac{5}{11}$.

$$\frac{a}{b} = \frac{a}{2\frac{5}{11}} = \frac{11}{3}.$$
$$\therefore \frac{11a}{27} = \frac{11}{3}.$$
$$\therefore a = 9.$$

21. Find b if $\cot A = 4$ and $a = 17$.

$$\frac{b}{a} = \frac{b}{17} = 4.$$
$$\therefore b = 68.$$

22. Find c if $\sec a = 2$ and $b = 20$.

$$\frac{c}{b} = \frac{c}{20} = 2.$$
$$\therefore c = 40.$$

23. Find c if $\csc A = 6.45$ and $a = 35.6$.

$$\csc A = \frac{c}{a} = \frac{c}{35.6} = 6.45.$$
$$\therefore c = 229.62.$$

24. Construct a right triangle; given $c = 6$, $\tan A = \frac{3}{2}$.

$$\tan A = \frac{a}{b}.$$
$$\therefore a = 3 \text{ and } b = 2.$$

Draw $AB = 2$, and $BC \perp$ to $AB = 3$; join C and A.

Prolong AC to D, making $AD = 6$.

Draw $DE \perp$ to AB produced.

Rt. $\triangle ADE$ will be similar to rt. $\triangle ACB$.

$\therefore ADE$ is the rt. $\triangle$ required.

25. Construct a right triangle; given $a = 3.5$, $\cos A = \frac{1}{2}$.

Construct $\triangle A'B'C'$ so that $b' = 1$, $c' = 2$. Then $\cos A = \frac{1}{2}$.

Construct $\triangle ABC$ similar to $A'B'C'$, and having $a = 3.5$.

26. Construct a right triangle; given $b = 2$, $\sin A = 0.6$.

Construct rt. $\triangle A'B'C'$, making $a' = 6$, $c' = 10$.

Then $\sin A' = \frac{6}{10}$.

Construct $\triangle ABC$ similar to $A'B'C'$, and having $B = 2$.

27. Construct a right triangle; given $b = 4$, $\csc A = 4$.

Construct rt. $\triangle A'B'C'$, having $c' = 4$, $a' = 1$.

Then construct $\triangle ABC$ similar to $A'B'C'$, and having $b = 4$.

28. In a right triangle, $c = 2.5$ miles, $\sin A = 0.6$, $\cos A = 0.8$; compute the legs.

$$\sin A = \frac{a}{c}. \qquad \cos A = \frac{b}{c}.$$
$$\therefore a = c \sin A. \qquad \therefore b = c \cos A.$$
$$\therefore a = 1.5. \qquad \therefore b = 2.$$

30. Find, by means of the table, the legs of a right triangle if $A = 20°$, $c = 1$; also, if $A = 20°$, $c = 4$.

$$A = 20°,\ c = 1.$$

$\sin A = \frac{a}{c}$. $\qquad \cos A = \frac{b}{c}$.

$\therefore a = c \sin A$. $\qquad \therefore b = c \cos A$.

$\therefore a = 0.342$. $\qquad \therefore b = 0.940$.

$$A = 20°,\ c = 4.$$

$\therefore a = 4 \times 0.342 = 1.368$. $\qquad \therefore b = 4 \times 0.940 = 3.760$.

31. In a right triangle, given $a = 3$ and $c = 5$; find the hypotenuse of a similar triangle in which $a = 240{,}000$ miles.

$$a : c :: 240{,}000 : x,$$
$$3 : 5 :: 240{,}000 : x.$$
$$\therefore x = 400{,}000.$$

32. By dividing the length of a vertical rod by the length of its horizontal shadow, the tangent of the angle of elevation of the sun at the time of observation was found to be 0.82. How high is a tower, if the length of its horizontal shadow at the same time is 174.3 yards?

$$\tan A = \frac{a}{b} = 0.82.$$
$$\therefore a = 0.82\, b.$$
$$b = 174.3 \text{ yards}.$$
$$\therefore a = 0.82 \text{ of } 174.3 \text{ yards} = 142.926.$$

Exercise III. Page 9.

1. Represent by lines the functions of a larger angle than that shown in Fig. 3.

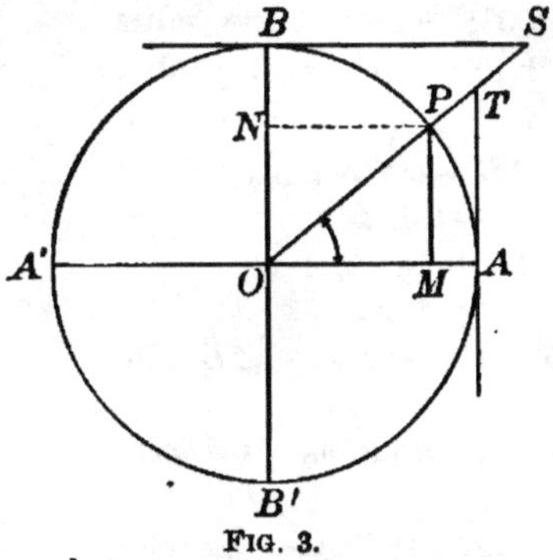

Fig. 3.

2. Show that if x is an acute angle, $\sin x$ is less than $\tan x$.

In Fig. 3, $OM : PM :: OA : AT$,

but $OM < OA$.

$\therefore PM < AT$.

3. Show that if x is an acute angle, $\sec x$ is greater than $\tan x$.

$OT = \sec$, $\quad AT = \tan$.

In rt. $\triangle OAT$, hyp. $OT >$ side AT.

$\therefore \sec > \tan$.

4. Show that if x is an acute angle, $\csc x$ is greater than $\cot x$.

$OS = \csc$, $\quad BS = \cot$.

In rt. $\triangle BOS$, hyp. $OS >$ side BS.

$\therefore \csc > \cot$.

5. Construct the angle x if $\tan x = 3$.

Let $\odot BAM$ be a unit circle, with centre O; then construct AT tangent to the circle at $A = 3\ OA$; then AOT is required angle.

6. Construct the angle x if $\csc x = 2$.

Let ⊙ ABM be a unit circle, with centre O; construct BT tangent to the circle at $B = 2\ OA$; connect OT; then AOT is required angle.

7. Construct the angle x if $\cos x = \frac{1}{4}$.

Take $OM = \frac{1}{4}$ radius OA. At M erect a ⊥ to meet the circumference at P. Draw OP.

Then is POM the angle required.

8. Construct the angle x if $\sin x = \cos x$.

Let $PM = \sin x$ and $OM = \cos x$.
But, by hypothesis, $PM = OM$.
∴ by Geometry, $x = 45°$.
Hence, construct an ∠ 45°.

9. Construct the angle x if $\sin x = 2 \cos x$.

Construct rt. ∠ PMO, making $PM = 2\ OM$. Draw OP.

Then POM is the angle required.

10. Construct the angle x if $4 \sin x = \tan x$.

Take $\frac{1}{4}$ of radius OA to M. At M erect a ⊥ to meet the circumference at P. Draw OP.

Then POM is the required angle.

11. Show that the sine of an angle is equal to one-half the chord of twice the angle.

Have given ∠ POA.

Construct $POB = 2\ POA$. Draw chord PB. Then it is ⊥ to OA; and PM, its half, is the sine of POA.

∴ $\sin x = \frac{1}{2}$ chord $2x$.

12. Find x if $\sin x$ is equal to one-half the side of a regular inscribed decagon.

Let AC be a side of a decagon.

Then $\frac{360°}{10} = 36°$ or AOC.

Draw OB bisecting AC. Then ∠AOC will be bisected, and ∠$AOB = 18°$.

But the sine of $AOB = \frac{1}{2} AC$.

∴ x or $AOB = 18°$.

13. Given x and y, $x + y$ being less than 90°; construct the value of $\sin(x + y) - \sin x$.

Let $AB = \sin(x + y)$ in a circle whose centre is O, and $CD = \sin x$.

Then, with a radius equal to CD, describe an arc from B, as centre, cutting AB at E.

Then EA will be the constructed value of $\sin(x + y) - \sin x$.

14. Given x and y, $x + y$ being less than 90°; construct the value of $\tan(x + y) - \sin(x + y) + \tan x - \sin x$.

Let $AB = \sin(x + y)$,
and $CD = \sin x$;
also $EF = \tan(x + y)$,
and $GF = \tan x$.

From F with a radius $= AB$ take FH.

From H with a radius $= GF$ take HI.

From I with a radius $= CD$ take IK.

Then EK will be the constructed value of $\tan(x + y) - \sin(x + y) + \tan x - \sin x$.

15. Given an angle x; construct an angle y such that $\sin y = 2 \sin x$.

Let AB be the sine of the $\angle x$ in a circle whose centre is O.

Draw AC perpendicular to the vertical diameter.

Then $CO = AB$.

Take CF on vertical diameter $= CO$. Draw FD perpendicular to vertical diameter, and meeting circumference at D.

Draw DE perpendicular to OB and draw OD.

$OF = 2\ CO$ by construction.

$ED = FO$; FO being the projection of the radius OD.

$\therefore DE = 2\ AB$, and DOB = angle required.

16. Given an angle x; construct an angle y such that $\cos y = \frac{1}{2} \cos x$.

Let $OB = \cos AOB$.

Erect a $\perp$ CD at C, the middle point of OB, and meeting the circumference at D. Draw DO.

Then DOB is the angle required.

17. Given an angle x; construct an angle y such that $\tan y = 3 \tan x$.

Let AB be the tangent of x.

Prolong AB to C, making $AC = 3\ AB$, and draw OC from O, the centre of the circle.

COA is the required angle.

18. Given an angle x; construct an angle y such that $\sec y = \csc x$.

Since $\sec y = \csc x$,

$$\frac{c}{b} = \frac{c}{a}.$$

$$\therefore a = b.$$

Hence, construct an isosceles right triangle.

The required angle will be 45°.

19. Show by construction that $2 \sin A > \sin 2A$.

Construct $\angle BOC$ and $\angle COA$ each equal to the given $\angle A$.

Then $AB = 2 \sin A$, and AD, the $\perp$ let fall from A to OB, $= \sin 2A$, But $AB > AD$.

Hence $2 \sin A > \sin 2A$.

20. Given two angles A and B, $A + B$ being less than 90°, show that $\sin (A + B) < \sin A + \sin B$.

Construct $HOK = \angle A$, and $COH = \angle B$.

Then $\sin (A + B) = CP$, $\sin A = HK$, $\sin B = CD$.

Now $\quad CP < CD + DE$,

and $\quad HK > DE$.

$$\therefore CP < CD + HK.$$

$\therefore \sin (A + B) < \sin A + \sin B$.

21. Given $\sin x$ in a unit circle; find the length of a line corresponding in position to $\sin x$ in a circle whose radius is r.

$1 : r :: \sin x$: the required line.

$\therefore$ length of line required $= r \sin x$.

22. In a right triangle, given the hypotenuse c, and also $\sin A = m$, $\cos A = n$; find the legs.

$$\sin A = \frac{a}{c} = m.$$

$$\therefore a = cm.$$

$$\cos A = \frac{b}{c} = n.$$

$$\therefore b = cn.$$

Exercise IV. Page 12.

1. Express the following functions as functions of the complementary angle:

sin 30°.	csc 18° 10′.
cos 45°.	cos 37° 24′.
tan 89°.	cot 82° 19′.
cot 15°.	csc 54° 46′.

$\sin 30° = \cos(90° - 30°) = \cos 60°.$
$\cos 45° = \sin(90° - 45°) = \sin 45°.$
$\tan 89° = \cot(90° - 89°) = \cot 1°.$
$\cot 15° = \tan(90° - 15°) = \tan 75°.$

$$\csc 18° 10' = \sec(90° - 18° 10') = \sec 71° 50'.$$
$$\cos 37° 24' = \sin(90° - 37° 24') = \sin 52° 36'.$$
$$\cot 82° 19' = \tan(90° - 82° 19') = \tan 7° 41'.$$
$$\csc 54° 46' = \sec(90° - 54° 46') = \sec 35° 14'.$$

2. Express the following functions as functions of an angle less than 45°:

sin 60°.	csc 69° 2′.
cos 75°.	cos 85° 39′.
tan 57°.	cot 89° 59′.
cot 84°.	csc 45° 1′.

$\sin 60° = \cos(90° - 60°) = \cos 30°.$
$\cos 75° = \sin(90° - 75°) = \sin 15°.$
$\tan 57° = \cot(90° - 57°) = \cot 33°.$
$\cot 84° = \tan(90° - 84°) = \tan 6°.$

$$\csc 69° 2' = \sec(90° - 69° 2') = \sec 20° 58'.$$
$$\cos 85° 39' = \sin(90° - 85° 39') = \sin 4° 21'.$$
$$\cot 89° 59' = \tan(90° - 89° 59') = \tan 0° 1'.$$
$$\csc 45° 1' = \sec(90° - 45° 1') = \sec 44° 59'.$$

3. Given $\tan 30° = \frac{1}{3}\sqrt{3}$; find cot 60°.

$$\tan 30° = \cot(90° - 30°) = \cot 60°.$$
$$\therefore \cot 60° = \tfrac{1}{3}\sqrt{3}.$$

4. Given $\tan A = \cot A$; find A.

$$\tan A = \cot(90° - A),$$
$$90° - A = A,$$
$$2A = 90°.$$
$$\therefore A = 45°.$$

5. Given $\cos A = \sin 2A$; find A.

$$\cos A = \sin(90° - A),$$
$$90° - A = 2A,$$
$$3A = 90°.$$
$$\therefore A = 30°.$$

6. Given $\sin A = \cos 2A$; find A.

$$\sin A = \cos(90° - A),$$
$$90° - A = 2A,$$
$$3A = 90°.$$
$$\therefore A = 30°.$$

7. Given $\cos A = \sin(45° - \frac{1}{2}A)$; find A.

$$\cos A = \sin(90° - A),$$
$$90° - A = 45° - \tfrac{1}{2}A,$$
$$180° - 2A = 90° - A.$$
$$\therefore A = 90°.$$

8. Given $\cot \frac{1}{2}A = \tan A$; find A.

$$\tan A = \cot(90° - A),$$
$$\tfrac{1}{2}A = 90° - A,$$
$$A = 180° - 2A,$$
$$3A = 180°.$$
$$\therefore A = 60°.$$

9. Given $\tan(45° + A) = \cot A$; find A.

$$\cot A = \tan(90° - A),$$
$$\tan(90° - A) = \tan(45° + A),$$
$$90° - A = 45° + A,$$
$$2A = 45°.$$
$$\therefore A = 22° 30'.$$

10. Find A if $\sin A = \cos 4A$.

$$\sin A = \cos(90° - A),$$
$$90° - A = 4A,$$
$$5A = 90°.$$
$$\therefore A = 18°.$$

11. Find A if $\cot A = \tan 8A$.

$$\cot A = \tan(90° - A),$$
$$8A = 90° - A,$$
$$9A = 90°.$$
$$\therefore A = 10°.$$

12. Find A if $\cot A = \tan nA$.

$$\cot A = \tan(90° - A),$$
$$90° - A = nA,$$
$$90° = A(n + 1).$$
$$\therefore A = \frac{90°}{n+1}.$$

Exercise V. Page 14.

1. Prove Formulas [1]–[3], using for the functions the line values in unit circle given in § 4.

[1]. $\sin^2 A + \cos^2 A = 1.$

[2]. $\tan A = \dfrac{\sin A}{\cos A}.$

[3]. $\sin A \times \csc A = 1,$
$\cos A \times \sec A = 1,$
$\tan A \times \cot A = 1.$

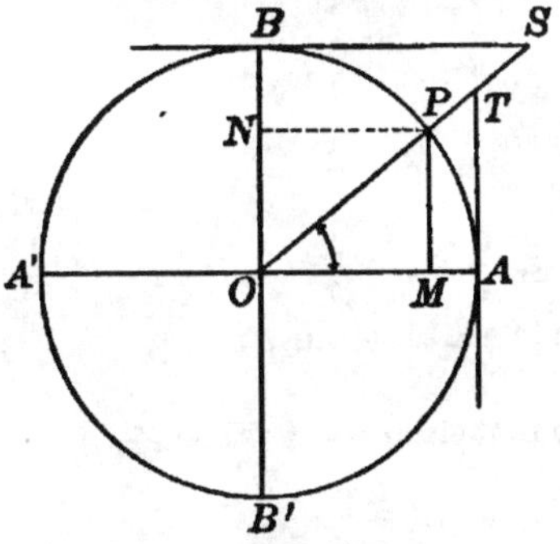

[1].
$$MP = \sin A,$$
$$OM = \cos A,$$
$$\overline{MP}^2 + \overline{OM}^2 = \overline{OP}^2;$$
but
$$\overline{OP}^2 = 1.$$
$$\therefore \overline{MP}^2 + \overline{OM}^2 = 1,$$
$$\therefore \sin^2 A + \cos^2 A = 1.$$

[2].
$$MP = \sin A,$$
$$OM = \cos A,$$
$$AT = \tan A.$$

△ OAT and OMP are similar.

$$\therefore TA : OA :: PM : OM.$$

Or,
$$\frac{TA}{OA} = \frac{PM}{OM};$$
but
$$OA = 1.$$
$$\therefore TA = \frac{PM}{OM}.$$
$$\therefore \tan A = \frac{\sin A}{\cos A}.$$

[3].
$$PM = \sin A,$$
$$OS = \csc A.$$

In similar △ OSB and POM,

$$OS : OB :: OP : PM.$$

or,
$$\frac{OS}{OB} = \frac{OP}{PM};$$
but
$$OB = 1,$$
$$OP = 1.$$
$$OS = \frac{1}{PM}.$$
$$\therefore OS \times PM = 1,$$
$$\csc A \times \sin A = 1.$$

Again,
$$\cos A = OM,$$
$$\sec A = OT.$$

In similar △ OTA and OPM,

$$OT : OA :: OP : OM.$$

or, $$\frac{OT}{OA} = \frac{OP}{OM};$$

but $$OA = 1,$$
$$OP = 1.$$

$$\therefore OT = \frac{1}{OM}.$$

$$\therefore OT \times OM = 1.$$

$$\therefore \sec A \times \cos A = 1.$$

Also, $$\tan A = AT,$$
$$\cot A = BS.$$

In similar △ SOB and TAO,

$$BS : BO :: AO : AT,$$

or, $$\frac{BS}{BO} = \frac{AO}{AT};$$

but $$BO = 1,$$
$$AO = 1.$$

$$\therefore BS = \frac{1}{AT},$$

$$\therefore BS \times AT = 1,$$

$$\therefore \cot A \times \tan A = 1.$$

2. Prove that $1 + \tan^2 A = \sec^2 A$.

$$\tan A = \frac{a}{b}, \qquad \sec A = \frac{b}{c},$$

$$a^2 + b^2 = c^2.$$

Dividing all the terms by b^2,

$$\frac{a^2}{b^2} + \frac{b^2}{b^2} = \frac{c^2}{b^2}.$$

Substituting for $\frac{a^2}{b^2}$ and $\frac{c^2}{b^2}$ their values $\tan^2 A$ and $\sec^2 A$, we have

$$\tan^2 A + 1 = \sec^2 A.$$

3. Prove that $1 + \cot^2 A = \csc^2 A$.

$$\cot A = \frac{b}{a},$$

$$\csc A = \frac{c}{a}.$$

$$a^2 + b^2 = c^2.$$

Dividing all the terms by a^2,

$$\frac{a^2}{a^2} + \frac{b^2}{a^2} = \frac{c^2}{a^2}.$$

Substituting for $\frac{b^2}{a^2}$ and $\frac{c^2}{a^2}$ their values $\cot^2 A$ and $\csc^2 A$, we have

$$1 + \cot^2 A = \csc^2 A.$$

4. Prove that $\cot A = \frac{\cos A}{\sin A}$.

$$\cot A = \frac{b}{a},$$

$$\sin A = \frac{a}{c},$$

$$\cos A = \frac{b}{c},$$

Substituting, $$\frac{b}{a} = \frac{b}{c} \div \frac{a}{c},$$

$$\therefore \cot A = \frac{\cos A}{\sin A}.$$

5. Prove that $\sin A \sec A = \tan A$.

$$\sin A = \frac{a}{c},$$

$$\sec A = \frac{c}{b},$$

$$\tan A = \frac{a}{b}.$$

Substituting, $$\frac{a}{c} \times \frac{c}{b} = \frac{a}{b}.$$

$$\therefore \sin A \sec A = \tan A.$$

6. Prove that $\sin A \cot A = \cos A$.

$$\sin A = \frac{a}{c},$$

$$\cot A = \frac{b}{a},$$

$$\cos A = \frac{b}{c}.$$

Substituting, $$\frac{a}{c} \times \frac{b}{a} = \frac{b}{c}.$$

$$\therefore \sin A \cot A = \cos A.$$

7. Prove that $\cos A \csc A = \cot A$.

$$\cos A = \frac{b}{c},$$

$$\csc A = \frac{c}{a},$$

$$\cot A = \frac{b}{a}.$$

Substituting, $\frac{b}{c} \times \frac{c}{a} = \frac{b}{a}.$

$\therefore \cos A \csc A = \cot A.$

8. Prove that $\tan A \cos A = \sin A$.

$$\tan A = \frac{a}{b},$$

$$\cos A = \frac{b}{c},$$

$$\sin A = \frac{a}{c}.$$

Substituting, $\frac{a}{b} \times \frac{b}{c} = \frac{a}{c}.$

$\therefore \tan A \cos A = \sin A.$

9. Prove that $\sin A \sec A \cot A = 1$.

$$\sin A = \frac{a}{c},$$

$$\sec A = \frac{c}{b},$$

$$\cot A = \frac{b}{a}.$$

Substituting,

$$\frac{a}{c} \times \frac{c}{b} \times \frac{b}{a} = 1.$$

$\therefore \sin A \sec A \cot A = 1.$

10. Prove that $\cos A \csc A \tan A = 1.$

$$\cos A = \frac{b}{c},$$

$$\csc A = \frac{c}{a},$$

$$\tan A = \frac{a}{b},$$

Substituting,

$$\frac{b}{c} \times \frac{c}{a} - \frac{a}{b} = 1.$$

$\therefore \cos A \csc A \tan A = 1.$

11. Prove that $(1 - \sin^2 A) \tan^2 A = \sin^2 A$.

From [1], § 6, $1 - \sin^2 A = \cos^2 A.$

$\therefore (1 - \sin^2 A) \tan^2 A = \cos^2 A \tan^2 A.$

But from Ex. 8, $\cos A \tan A = \sin A.$

$\therefore \cos^2 A \tan^2 A = \sin^2 A.$

12. Prove that $\sqrt{1 - \cos^2 A} \cot A = \cos A$.

From [1], § 6, $\sqrt{1 - \cos^2 A} = \sin A.$

$\therefore \sqrt{1 - \cos^2 A} \cot A = \sin A \cot A.$

But from Ex. 6, $\sin A \cot A = \cos A.$

$\therefore \sqrt{1 - \cos^2 A} \cot A = \cos A.$

13. Prove that $(1 + \tan^2 A) \sin^2 A = \tan^2 A$.

From Ex. 2, $1 + \tan^2 A = \sec^2 A.$

$\therefore (1 + \tan^2 A) \sin^2 A = \sec^2 A \sin^2 A.$

But from Ex. 5, $\sec A \sin A = \tan A.$

$\therefore (1 + \tan^2 A) \sin^2 A = \tan^2 A.$

14. Prove that $\csc^2 A\,(1 - \sin^2 A) = \cot^2 A$.

From [1], § 6, $1 - \sin^2 A = \cos^2 A$.

$\therefore \csc^2 A\,(1 - \sin^2 A) = \csc^2 A \cos^2 A$.

But from Ex. 7, $\csc A \cos A = \cot A$.

$\therefore \csc^2 A\,(1 - \sin^2 A) = \cot^2 A$.

15. Prove that $\tan^2 A \cos^2 A + \cos^2 A = 1$.

From Ex. 8, $\tan A \cos A = \sin A$.

$\therefore \tan^2 A \cos^2 A = \sin^2 A$.

And $\tan^2 A \cos^2 A + \cos^2 A = \sin^2 A + \cos^2 A$.

But from [1], § 6, $\sin^2 A + \cos^2 A = 1$.

$\therefore \tan^2 A \cos^2 A + \cos^2 A = 1$.

16. Prove that $(\sin^2 A - \cos^2 A)^2 = 1 - 4 \sin^2 A \cos^2 A$.

From [1], § 6, $\sin^2 A + \cos^2 A = 1$.

$\therefore (\sin^2 A + \cos^2 A)^2 = 1$.

But from Algebra, $(\sin^2 A - \cos^2 A)^2 = (\sin^2 A + \cos^2 A)^2 - 4 \sin^2 A \cos^2 A$.

$\therefore (\sin^2 A - \cos^2 A)^2 = 1 - 4 \sin^2 A \cos^2 A$.

17. Prove that $(1 - \tan^2 A)^2 = \sec^4 A - 4 \tan^2 A$.

From Ex. 2, $1 + \tan^2 A = \sec^2 A$.

$\therefore (1 + \tan^2 A)^2 = \sec^4 A$.

But from Algebra, $(1 - \tan^2 A)^2 = (1 + \tan^2 A)^2 - 4 \tan^2 A$.

$\therefore (1 - \tan^2 A)^2 = \sec^4 A - 4 \tan^2 A$.

18. Prove that $\dfrac{\sin A}{\cos A} + \dfrac{\cos A}{\sin A} = \sec A \csc A$.

$$\frac{\sin A}{\cos A} + \frac{\cos A}{\sin A} = \frac{\sin^2 A + \cos^2 A}{\cos A \sin A}.$$

But from [1], § 6, $\sin^2 A + \cos^2 A = 1$.

And from [3], § 6, $\dfrac{1}{\cos A} = \sec A$,

and $\dfrac{1}{\sin A} = \csc A$.

$$\therefore \frac{\sin A}{\cos A} + \frac{\cos A}{\sin A} = \sec A \csc A.$$

19. Prove that $\sin^4 A - \cos^4 A = \sin^2 A - \cos^2 A$.

$$\sin^4 A - \cos^4 A = (\sin^2 A + \cos^2 A)(\sin^2 A - \cos^2 A).$$

But from [1], § 6, $\sin^2 A + \cos^2 A = 1.$

$$\therefore \sin^4 A - \cos^4 A = \sin^2 A - \cos^2 A.$$

20. Prove that $\sec A - \cos A = \sin A \tan A$,

From [3], § 6, $\sec A = \dfrac{1}{\cos A}$,

$$\therefore \sec A - \cos A = \frac{1}{\cos A} - \cos A$$

$$= \frac{1 - \cos^2 A}{\cos A}$$

But from [1], § 6, $1 - \cos^2 A = \sin^2 A.$

$$\therefore \sec A - \cos A = \frac{\sin^2 A}{\cos A}.$$

Also, from [2], § 6, $\dfrac{\sin A}{\cos A} = \tan A.$

$$\therefore \sec A - \cos A = \sin A \tan A.$$

21. Prove that $\csc A - \sin A = \cos A \cot A$.

From [3], § 6, $\csc A = \dfrac{1}{\sin A}$,

$$\therefore \csc A - \sin A = \frac{1}{\sin A} - \sin A$$

$$= \frac{1 - \sin^2 A}{\sin A}.$$

But from [1], § 6, $1 - \sin^2 A = \cos^2 A.$

$$\therefore \csc A - \sin A = \frac{\cos^2 A}{\sin A}.$$

Also, from [2], § 6, $\dfrac{\cos A}{\sin A} = \cot A.$

$$\therefore \csc A - \sin A = \cos A \cot A.$$

22. Prove that $\dfrac{\cos x}{1 - \sin x} = \dfrac{1 + \sin x}{\cos x}$.

Clearing of fractions this becomes,

$$\cos^2 x = 1 - \sin^2 x,$$

which is correct ([1], § 6).

$$\therefore \frac{\cos x}{1 - \sin x} = \frac{1 + \sin x}{\cos x}.$$

Exercise VI. Page 16.

1. Find the values of the other functions when $\sin A = \frac{12}{13}$.

$$\sin^2 A + \cos^2 A = 1,$$

$$\cos^2 A = 1 - \left(\frac{12}{13}\right)^2,$$

$$\cos A = \sqrt{1 - \left(\frac{12}{13}\right)^2}$$

$$= \sqrt{\frac{25}{169}}.$$

$$\therefore \cos A = \frac{5}{13}.$$

$$\tan A = \frac{\sin A}{\cos A} = \frac{12}{5}.$$

$\cot A$ is reciprocal of $\tan A$.

$$\therefore \cot A = \frac{5}{12}.$$

$\sec A$ is reciprocal of $\cos A$.

$$\therefore \sec A = \frac{13}{5}.$$

$\csc A$ is reciprocal of $\sin A$.

$$\therefore \csc A = \frac{13}{12}.$$

2. Find the values of the other functions when $\sin A = 0.8$.

$$\sin^2 A + \cos^2 A = 1,$$

$$\cos^2 A = 1 - (0.8)^2,$$

$$\cos A = \sqrt{1 - 0.64}.$$

$$\therefore \cos A = 0.6.$$

$$\tan A = \frac{\sin A}{\cos A} = \frac{0.8}{0.6}.$$

$$\therefore \tan A = 1.3333.$$

$$\cot A = \frac{0.6}{0.8}.$$

$$\therefore \cot A = 0.75.$$

$$\sec A = \frac{1}{0.6}.$$

$$\therefore \sec A = 1.6667.$$

$$\csc A = \frac{1}{0.8}.$$

$$\therefore \csc A = 1.25.$$

3. Find the values of the other functions when $\cos A = \frac{60}{61}$.

$$\sin^2 A + \cos^2 A = 1,$$

$$\sin A = \sqrt{1 - \frac{3600}{3721}} = \sqrt{\frac{121}{3721}} = \frac{11}{61}.$$

$$\tan A = \frac{\sin A}{\cos A} = \frac{11}{60}.$$

$$\cot A = \frac{1}{\tan A} = \frac{60}{11}.$$

$$\sec A = \frac{1}{\cos A} = \frac{61}{60}.$$

$$\csc A = \frac{1}{\sin A} = \frac{61}{11}.$$

4. Find the values of the other functions when $\cos A = 0.28$.

$$\sin^2 A + \cos^2 A = 1.$$

$$\sin A = \sqrt{1 - (0.28)^2} = \sqrt{0.9216}.$$

$$= 0.96.$$

$$\tan A = \frac{\sin A}{\cos A} = \frac{0.96}{0.28} = 3.4285.$$

$$\cot A = \frac{1}{\tan A} = \frac{1}{3.4285} = 0.29167.$$

$$\sec A = \frac{1}{\cos A} = \frac{1}{0.28} = 3.5714.$$

$$\csc A = \frac{1}{\sin A} = \frac{1}{0.96} = 1.04167.$$

5. Find the values of the other functions when $\tan A = \frac{4}{3}$.

$$\tan A = \frac{4}{3}.$$
$$\therefore \cot A = \frac{3}{4}.$$
$$\tan A = \frac{\sin A}{\cos A},$$
$$\frac{4}{3} = \frac{\sin A}{\sqrt{1 - \sin^2 A}}.$$
$$3 \sin A = 4\sqrt{1 - \sin^2 A},$$
$$9 \sin^2 A = 16 - 16 \sin^2 A,$$
$$25 \sin^2 A = 16,$$
$$5 \sin A = 4.$$
$$\therefore \sin A = \frac{4}{5}.$$
$$\cos A = \frac{\sin A}{\tan A} = \frac{3}{5}.$$
$$\sec A = \frac{1}{\cos A} = \frac{5}{3}.$$
$$\csc A = \frac{1}{\sin A} = \frac{5}{4}.$$

6. Find the values of the other functions when $\cot A = 1$.

$$\cot A = 1,$$
$$\therefore \tan A = 1.$$
$$\tan A = \frac{\sin A}{\cos A},$$
$$1 = \frac{\sin A}{\sqrt{1 - \sin^2 A}},$$
$$\sin A = \sqrt{1 - \sin^2 A},$$
$$\sin^2 A = 1 - \sin^2 A,$$
$$2 \sin^2 A = 1,$$
$$\sin^2 A = \frac{1}{2}.$$
$$\therefore \sin A = \sqrt{\frac{1}{2}} = \tfrac{1}{2}\sqrt{2}.$$
$$\cos A = \frac{\sin A}{\tan A} = \tfrac{1}{2}\sqrt{2}.$$
$$\sec A = \frac{1}{\cos A} = \frac{1}{\frac{1}{2}\sqrt{2}} = \sqrt{2}.$$
$$\csc A = \frac{1}{\sin A} = \frac{1}{\frac{1}{2}\sqrt{2}} = \sqrt{2}.$$

7. Find the values of the other functions when $\cot A = 0.5$.

$$\tan A = \frac{1}{\cot A} = \frac{1}{0.5} = 2.$$
$$\tan A = \frac{\sin A}{\cos A} = 2.$$
$$2 \cos A = \sin A.$$
$$4 \cos^2 A - \sin^2 A = 0 \quad \text{(squaring)}$$
$$\underline{\cos^2 A + \sin^2 A = 1}$$
$$5 \cos^2 A \qquad = 1$$
$$\cos A = \sqrt{\frac{1}{5}} = 0.45.$$
$$4 \cos^2 A + 4 \sin^2 A = 4$$
$$\underline{4 \cos^2 A - \sin^2 A = 0}$$
$$5 \sin^2 A = 4$$
$$\sin A = \sqrt{\frac{4}{5}} = 0.90.$$
$$\sec A = \frac{1}{\cos A} = 2.22.$$
$$\csc A = \frac{1}{\sin A} = 1.11.$$

8. Find the values of the other functions when $\sec A = 2$.

$$\cos A = \frac{1}{\sec A} = \frac{1}{2}.$$
$$\sin A = \sqrt{1 - \cos^2 A}$$
$$= \sqrt{1 - \frac{1}{4}} = \sqrt{\frac{3}{4}}.$$
$$\therefore \sin A = \tfrac{1}{2}\sqrt{3}.$$
$$\tan A = \frac{\sin A}{\cos A} = \frac{\frac{1}{2}\sqrt{3}}{\frac{1}{2}} = \sqrt{3}.$$
$$\cot A = \frac{1}{\tan A} = \frac{1}{\sqrt{3}} = \tfrac{1}{3}\sqrt{3}.$$
$$\csc A = \frac{1}{\sin A} = \tfrac{2}{3}\sqrt{3}.$$

9. Find the values of the other functions when $\csc A = \sqrt{2}$.

$$\sin A = \frac{1}{\sqrt{2}} = \tfrac{1}{2}\sqrt{2},$$

$$\cos A = \sqrt{1-(\tfrac{1}{2}\sqrt{2})^2} = \sqrt{1-\tfrac{1}{2}}$$
$$= \sqrt{\tfrac{1}{2}} = \tfrac{1}{2}\sqrt{2},$$

$$\tan A = \frac{\tfrac{1}{2}\sqrt{2}}{\tfrac{1}{2}\sqrt{2}} = 1,$$

$$\cot A = \frac{1}{1} = 1,$$

$$\sec A = \frac{1}{\tfrac{1}{2}\sqrt{2}} = \sqrt{2}.$$

10. Find the values of the other functions when $\sin A = m$.

$$\cos A = \sqrt{1-\sin^2 A} = \sqrt{1-m^2},$$

$$\tan A = \frac{\sin A}{\cos A} = \frac{m}{\sqrt{1-m^2}}$$
$$= \frac{m\sqrt{1-m^2}}{1-m^2},$$

$$\cot A = \frac{1}{\tan A} = \frac{1}{m}\sqrt{1-m^2},$$

$$\sec A = \frac{1}{\cos A} = \frac{1}{\sqrt{1-m^2}},$$

$$\csc A = \frac{1}{\sin A} = \frac{1}{m}.$$

11. Find the values of the other functions when $\sin A = \dfrac{2m}{1+m^2}$.

$$\cos A = \sqrt{1-\sin^2 A}.$$

$$\therefore \cos A = \sqrt{1-\frac{4m^2}{1+2m^2+m^4}}$$
$$= \sqrt{\frac{1-2m^2+m^4}{1+2m^2+m^4}}$$
$$= \frac{1-m^2}{1+m^2}.$$

$$\tan A = \frac{\sin A}{\cos A} = \frac{2m}{1-m^2}.$$

$$\cot A = \frac{1}{\tan A} = \frac{1-m^2}{2m}.$$

$$\sec A = \frac{1}{\cos A} = \frac{1+m^2}{1-m^2}.$$

$$\csc A = \frac{1}{\sin A} = \frac{1+m^2}{2m}.$$

12. Find the values of the other functions when $\cos A = \dfrac{2mn}{m^2+n^2}$.

$$\sin A = \sqrt{1-\cos^2 A}.$$
$$= \sqrt{1-\frac{4m^2n^2}{m^4+2m^2n^2+n^4}}$$
$$= \sqrt{\frac{m^4-2m^2n^2+n^4}{m^4+2m^2n^2+n^4}}$$
$$= \frac{m^2-n^2}{m^2+n^2}.$$

$$\tan A = \frac{\sin A}{\cos A} = \frac{m^2-n^2}{2mn}.$$

$$\cot A = \frac{1}{\tan A} = \frac{2mn}{m^2-n^2}.$$

$$\sec A = \frac{1}{\cos A} = \frac{m^2+n^2}{2mn}.$$

$$\csc A = \frac{1}{\sin A} = \frac{m^2+n^2}{m^2-n^2}.$$

13. Given $\tan 45° = 1$; find the other functions of 45°.

$$\frac{\sin 45°}{\cos 45°} = \tan 45°.$$

$$\frac{\sin 45°}{\cos 45°} = 1. \qquad (1)$$

$$\sin^2 A + \cos^2 A = 1. \qquad (2)$$

By (1), $\sin 45° = \cos 45°$.

By (2), $\cos^2 45° + \cos^2 45° = 1.$

$$2\cos^2 45° = 1,$$

$$\cos^2 45° = \frac{1}{2},$$

$$\cos 45° = \sqrt{\frac{1}{2}} = \tfrac{1}{2}\sqrt{2}.$$

$$\sin 45° = \tfrac{1}{2}\sqrt{2}.$$

$$\cot 45° = \frac{1}{\tan 45°} = \frac{1}{1} = 1.$$

$$\sec 45° = \frac{1}{\frac{1}{2}\sqrt{2}} = \sqrt{2}.$$

$$\csc 45° = \frac{1}{\frac{1}{2}\sqrt{2}} = \sqrt{2}.$$

14. Given $\sin 30° = \frac{1}{2}$; find the other functions of 30°.

$$\sin^2 30° + \cos^2 30° = 1.$$

$$\cos 30° = \sqrt{1 - \frac{1}{4}}$$

$$= \sqrt{\frac{3}{4}} = \tfrac{1}{2}\sqrt{3}.$$

$$\tan 30° = \frac{\frac{1}{2}}{\frac{1}{2}\sqrt{3}} = \tfrac{1}{3}\sqrt{3}.$$

$$\cot 30° = \frac{1}{\frac{1}{3}\sqrt{3}} = \sqrt{3}.$$

$$\sec 30° = \frac{1}{\frac{1}{2}\sqrt{3}} = \tfrac{2}{3}\sqrt{3}.$$

$$\csc 30° = \frac{1}{\frac{1}{2}} = 2.$$

15. Given $\csc 60° = \frac{2}{3}\sqrt{3}$; find the other functions of 60°.

$$\sin 60° = \frac{1}{\csc 60°},$$

$$\sin 60° = \frac{1}{\frac{2}{3}\sqrt{3}} = \tfrac{1}{2}\sqrt{3}.$$

$$\cos 60° = \sqrt{1 - \sin^2 60°},$$

$$\cos 60° = \sqrt{1 - (\tfrac{1}{2}\sqrt{3})^2}$$

$$= \sqrt{1 - \frac{3}{4}} = \frac{1}{2}.$$

$$\tan 60° = \frac{\frac{1}{2}\sqrt{3}}{\frac{1}{2}} = \sqrt{3}.$$

$$\cot 60° = \frac{1}{\sqrt{3}} = \tfrac{1}{3}\sqrt{3}.$$

$$\sec 60° = \frac{1}{\frac{1}{2}} = 2.$$

16. Given $\tan 15° = 2 - \sqrt{3}$; find the other functions of 15°.

$$\frac{\sin 15°}{\cos 15°} = 2 - \sqrt{3}.$$

$$\sin^2 15° + \cos^2 15° = 1.$$

$$\sin 15° = \cos 15° (2 - \sqrt{3}).$$

$$[\cos (2 - \sqrt{3})]^2 + \cos^2 = 1,$$

$$\cos^2 (4 - 4\sqrt{3} + 3) + \cos^2 = 1,$$

$$\cos^2 (8 - 4\sqrt{3}) = 1.$$

$$\cos^2 15° = \frac{1}{4(2 - \sqrt{3})} = \frac{2 + \sqrt{3}}{4},$$

$$\cos 15° = \sqrt{\frac{2 + \sqrt{3}}{4}} = \tfrac{1}{2}\sqrt{2 + \sqrt{3}}.$$

$$\sin^2 15° = 1 - \cos^2 15°.$$

$$\sin^2 15° = 1 - \frac{2 + \sqrt{3}}{4} = \frac{2 - \sqrt{3}}{4},$$

$$\sin 15° = \tfrac{1}{2}\sqrt{2 - \sqrt{3}}.$$

$$\cot 15° = \frac{1}{\tan 15°} = \frac{1}{2 - \sqrt{3}}$$

$$= 2 + \sqrt{3}.$$

17. Given $\cot 22° 30' = \sqrt{2} + 1$; find the other functions of 22° 30'.

$$\tan 22\tfrac{1}{2}° = \frac{1}{\cot 22\frac{1}{2}°} = \frac{1}{\sqrt{2} + 1}$$

$$= \sqrt{2} - 1.$$

$$\frac{\sin 22\frac{1}{2}°}{\cos 22\frac{1}{2}°} = \tan 22\tfrac{1}{2}°, \quad (1)$$

$$\cos^2 22\tfrac{1}{2}° + \sin^2 22\tfrac{1}{2}° = 1. \quad (2)$$

From (1),

$$\cos 22\tfrac{1}{2}° \tan 22\tfrac{1}{2}° = \sin 22\tfrac{1}{2}°$$

Squaring,

$$\cos^2 22\tfrac{1}{2}° \tan^2 22\tfrac{1}{2}° = \sin^2 22\tfrac{1}{2}°$$

From (2),

$$\cos^2 22\tfrac{1}{2}° = -\sin^2 22\tfrac{1}{2}° + 1$$

Add,

$$\cos^2 22\tfrac{1}{2}° \tan^2 22\tfrac{1}{2}° + \cos^2 22\tfrac{1}{2}° = 1$$

$$\cos^2 22\tfrac{1}{2}^\circ(\tan^2 22\tfrac{1}{2}^\circ + 1) = 1,$$

$$\cos^2 22\tfrac{1}{2}^\circ(4 - 2\sqrt{2}) = 1,$$

$$\cos 22\tfrac{1}{2}^\circ \sqrt{4 - 2\sqrt{2}} = 1.$$

$$\therefore \cos 22\tfrac{1}{2}^\circ = \frac{1}{\sqrt{4 - 2\sqrt{2}}}$$

$$= \sqrt{\frac{4 + 2\sqrt{2}}{8}}$$

$$= \tfrac{1}{2}\sqrt{2 + \sqrt{2}}.$$

$$\sin 22\tfrac{1}{2}^\circ = \sqrt{1 - \frac{2 - \sqrt{2}}{4}}$$

$$= \sqrt{\frac{4 - 2 - \sqrt{2}}{4}}$$

$$= \sqrt{\frac{2 - \sqrt{2}}{4}}$$

$$= \tfrac{1}{2}\sqrt{2 - \sqrt{2}}.$$

18. Given $\sin 0^\circ = 0$; find the other functions of 0°.

$$\cos 0^\circ = \sqrt{1 - \sin^2 0^\circ}$$

$$= \sqrt{1 - 0}.$$

$$\therefore \cos 0^\circ = 1.$$

$$\tan 0^\circ = \frac{\sin 0^\circ}{\cos 0^\circ} = \frac{0}{1} = 0.$$

$$\cot 0^\circ = \frac{1}{\tan 0^\circ} = \frac{1}{0} = \infty.$$

$$\sec 0^\circ = \frac{1}{\cos 0^\circ} = \frac{1}{1} = 1.$$

$$\csc 0^\circ = \frac{1}{\sin 0^\circ} = \frac{1}{0} = \infty.$$

19. Given $\sin 90^\circ = 1$; find the other functions of 90°.

$$\sin 90^\circ = 1.$$

$$\cos 90^\circ = \sqrt{1 - \sin^2 90^\circ} = 0.$$

$$\tan 90^\circ = \frac{\sin 90^\circ}{\cos 90^\circ} = \frac{1}{0} = \infty.$$

$$\cot 90^\circ = \frac{1}{\tan 90^\circ} = \frac{1}{\infty} = 0.$$

$$\sec 90^\circ = \frac{1}{\cos 90^\circ} = \frac{1}{0} = \infty.$$

$$\csc 90^\circ = \frac{1}{\sin 90^\circ} = \frac{1}{1} = 1.$$

20. Given $\tan 90^\circ = \infty$; find the other functions of 90°.

$$\tan 90^\circ = \infty.$$

$$\cot 90^\circ = \frac{1}{\tan 90^\circ} = \frac{1}{\infty} = 0.$$

$$\frac{\cos 90^\circ}{\sin 90^\circ} = \cot 90^\circ = 0.$$

$$\therefore \cos 90^\circ = 0.$$

$$\sin^2 90^\circ + \cos^2 90^\circ = 1$$

$$\therefore \sin^2 90^\circ = 1,$$

$$\sin 90^\circ = 1.$$

$$\sec 90^\circ = \frac{1}{0} = \infty.$$

$$\csc 90^\circ = 1.$$

21. Express the values of all the other functions in terms of $\sin A$.

$$\cos A = \sqrt{1 - \sin^2 A},$$

$$\tan A = \frac{\sin A}{\sqrt{1 - \sin^2 A}},$$

$$\cot A = \frac{\sqrt{1 - \sin^2 A}}{\sin A},$$

$$\sec A = \frac{1}{\sqrt{1 - \sin^2 A}},$$

$$\csc A = \frac{1}{\sin A}.$$

22. Express the values of all the other functions in terms of cos A.

$$\sin A = \sqrt{1-\cos^2 A},$$

$$\tan A = \frac{\sqrt{1-\cos^2 A}}{\cos A},$$

$$\cot A = \frac{\cos A}{\sqrt{1-\cos^2 A}},$$

$$\sec A = \frac{1}{\cos A},$$

$$\csc A = \frac{1}{\sqrt{1-\cos^2 A}}.$$

23. Express the values of all the other functions in terms of tan A.

$$\cot A = \frac{1}{\tan A}.$$

$$\frac{a}{b} = \tan A.$$

$$a = b \tan A.$$

$$a^2 = b^2 \tan^2 A.$$

$$a^2 - b^2 \tan^2 A = 0$$

$$a^2 + b^2 = 1$$

$$b^2(1+\tan^2 A) = 1$$

$$b^2 = \frac{1}{1+\tan^2 A} = \cos^2 A.$$

$$\cos A = \sqrt{\frac{1}{1+\tan^2 A}}.$$

$$\sin A = \sqrt{1-\cos^2 A}$$

$$= \sqrt{1 - \frac{1}{1+\tan^2 A}}$$

$$= \frac{\tan A}{\sqrt{1+\tan^2 A}}.$$

$$\sec A = \frac{1}{\cos A} = \sqrt{1+\tan^2 A}.$$

$$\csc A = \frac{1}{\sin A} = \frac{\sqrt{1+\tan^2 A}}{\tan A}.$$

24. Express the values of all the other functions in terms of cot A.

$$\frac{1}{\cot A} = \tan A.$$

$$\frac{\sin A}{\cos A} = \tan A.$$

Let $x = \sin A$, $y = \cos A$.

$$\frac{x}{y} = \frac{1}{\cot A}.$$

$$x \cot A = y,$$

$$x^2 \cot^2 A = y^2.$$

$$x^2 \cot^2 A - y^2 = 0$$

$$x^2 + y^2 = 1$$

$$x^2(1+\cot^2 A) = 1$$

$$x^2 = \frac{1}{1+\cot^2 A}.$$

$$\sin A = \frac{1}{\sqrt{1+\cot^2 A}}.$$

$$\cos A = \sqrt{1-\sin^2 A}$$

$$= \sqrt{1 - \frac{1}{1+\cot^2 A}}$$

$$= \sqrt{\frac{1+\cot^2 A - 1}{1+\cot^2 A}}$$

$$= \frac{\cot A}{\sqrt{1+\cot^2 A}}.$$

$$\sec A = \frac{1}{\cos A} = \frac{\sqrt{1+\cot^2 A}}{\cot A}.$$

$$\csc A = \frac{1}{\sin A} = \sqrt{1+\cot^2 A}.$$

25. Given $2 \sin A = \cos A$; find $\sin A$ and $\cos A$.

$$\sin^2 A + \cos^2 A = 1.$$

$$\sin^2 A + 4 \sin^2 A = 1.$$

$$5 \sin^2 A = 1,$$

$$\sin^2 A = \frac{1}{5},$$

$$\sin A = \sqrt{\frac{1}{5}} = \tfrac{1}{5}\sqrt{5}.$$

$$\therefore \cos A = \tfrac{2}{5}\sqrt{5}.$$

26. Given $4 \sin A = \tan A$; find $\sin A$ and $\tan A$.

$$\tan A = \frac{\sin A}{\cos A}.$$

But $\tan A = 4 \sin A$.

$$\therefore 4 \sin A = \frac{\sin A}{\cos A},$$

$$4 \sin A \times \cos A = \sin A.$$

$$\therefore \cos A = \frac{\sin A}{4 \sin A} = \frac{1}{4}.$$

$$\sin^2 A + \cos^2 A = 1.$$

$$\therefore \sin A = \sqrt{1 - \frac{1}{16}} = \sqrt{\frac{15}{16}}$$

$$= \tfrac{1}{4}\sqrt{15}$$

$$\tan A = \frac{\sin A}{\cos A}$$

$$= \frac{\frac{1}{4}\sqrt{15}}{\frac{1}{4}} = \sqrt{15}.$$

27. If $\sin A : \cos A = 9 : 40$, find $\sin A$ and $\cos A$.

$$40 \sin A = 9 \cos A.$$

(sq.) $1600 \sin^2 A = 81 \cos^2 A$.

$$1600 \sin^2 A - 81 \cos^2 A = 0.$$

But $\sin^2 A + \cos^2 A = 1$.

Multiplying by 81 and adding,

$$1681 \sin^2 A = 81.$$

$$\therefore 41 \sin A = 9.$$

$$\sin A = \frac{9}{41}.$$

$$\sin^2 A + \cos^2 A = 1.$$

$$\cos A = \sqrt{1 - \sin^2 A}.$$

$$\therefore \cos A = \sqrt{1 - \left(\frac{9}{41}\right)^2} = \frac{40}{41}.$$

28. Transform the expression $\tan^2 A + \cot^2 A - \sin^2 A - \cos^2 A$ into a form containing only $\cos A$.

$$\tan^2 A = \frac{\sin^2 A}{\cos^2 A} = \frac{1 - \cos^2 A}{\cos^2 A}.$$

$$\cot^2 A = \frac{\cos^2 A}{\sin^2 A} = \frac{\cos^2 A}{1 - \cos^2 A}.$$

$$\frac{1 - \cos^2 A}{\cos^2 A} + \frac{\cos^2 A}{1 - \cos^2 A}$$

$$- 1 + \cos^2 A - \cos^2 A$$

$$= \frac{1 - 2\cos^2 A + 2\cos^4 A - \cos^2 A + \cos^4 A}{\cos^2 A - \cos^4 A}$$

$$= \frac{1 - 3\cos^2 A + 3\cos^4 A}{\cos^2 A - \cos^4 A}.$$

29. Prove that $\sin A + \cos A = (1 + \tan A) \cos A$.

$$\frac{\sin A}{\cos A} = \tan A.$$

$$\sin A = \tan A \cos A.$$

$$\sin A + \cos A = \tan A \cos A + \cos A$$

$$= (1 + \tan A) \cos A.$$

30. Prove that $\tan A + \cot A = \sec A \times \csc A$.

$$\tan A = \frac{\sin A}{\cos A}.$$

$$\cot A = \frac{\cos A}{\sin A}.$$

$$\tan A + \cot A = \frac{\sin A}{\cos A} + \frac{\cos A}{\sin A}$$

$$= \frac{\sin^2 A + \cos^2 A}{\cos A \sin A}.$$

But $\sin^2 A + \cos A = 1$.

$$\therefore \tan A + \cot A = \frac{1}{\cos A \sin A}$$

$$= \sec A \times \csc A.$$

EXERCISE VII. PAGE 18.

1. Solve the equation,

$$2\cos x = \sec x.$$

$$2\cos x = \frac{1}{\cos x}.$$

$$\therefore 2\cos^2 x = 1.$$

$$\cos x = \sqrt{\tfrac{1}{2}}.$$

$$\therefore x = 45°.$$

2. Solve the equation,

$$4\sin x = \csc x.$$

$$4\sin x = \frac{1}{\sin x}.$$

$$\therefore 4\sin^2 x = 1.$$

$$\sin x = \tfrac{1}{2}.$$

$$\therefore x = 30°.$$

3. Solve the equation,

$$\tan x = 2\sin x.$$

$$\frac{\sin x}{\cos x} = 2\sin x,$$

$$\sin x = 2\sin x\cos x.$$

$$\sin x(1 - 2\cos x) = 0.$$

$$\therefore \sin x = 0, \quad (1)$$

$$x = 0.$$

$$1 - 2\cos x = 0. \quad (2)$$

$$\cos x = \tfrac{1}{2}.$$

$$\therefore x = 60°.$$

$$\therefore x = 0° \text{ or } 60°.$$

4. Solve the equation,

$$\sec x = \sqrt{2}\tan x.$$

$$\frac{1}{\cos x} = \sqrt{2}\frac{\sin x}{\cos x},$$

$$1 = \sqrt{2}\sin x,$$

$$\sin x = \frac{1}{\sqrt{2}},$$

$$\therefore x = 45°.$$

5. Solve the equation,

$$\sin^2 x = 3\cos^2 x.$$

$$\frac{\sin^2 x}{\cos^2 x} = 3,$$

$$\tan^2 x = 3,$$

$$\tan x = \sqrt{3},$$

$$\therefore x = 60°.$$

6. Solve the equation,

$$2\sin^2 x + \cos^2 x = \tfrac{3}{2}.$$

$$\sin^2 x + (\sin^2 x + \cos^2 x) = \tfrac{3}{2},$$

$$\sin^2 x + 1 = \tfrac{3}{2},$$

$$\sin^2 x = \tfrac{1}{2},$$

$$\sin x = \frac{1}{\sqrt{2}}.$$

$$\therefore x = 45°.$$

7. Solve the equation,

$$3\tan^2 x - \sec^2 x = 1.$$

$$3\tan^2 x - (\tan^2 x + 1) = 1,$$

$$2\tan^2 x - 1 = 1,$$

$$2\tan^2 x = 2,$$

$$\tan x = 1.$$

$$\therefore x = 45°.$$

8. Solve the equation,

$$\tan x + \cot x = 2.$$

$$\tan x + \frac{1}{\tan x} = 2,$$

$$\tan^2 x - 2\tan x + 1 = 0,$$

$$(\tan x - 1)(\tan x - 1) = 0.$$

$$\therefore \tan x = 1.$$

$$\therefore x = 45°.$$

9. Solve the equation,

$$\sin^2 x - \cos x = \tfrac{1}{4}.$$

$$(1 - \cos^2 x) - \cos x = \tfrac{1}{4},$$

$$\cos^2 x + \cos x = \tfrac{3}{4},$$

$$\cos^2 x + \cos x + \tfrac{1}{4} = 1,$$

$$\cos x + \tfrac{1}{2} = 1,$$

$$\cos x = \tfrac{1}{2}.$$

$$\therefore x = 60°.$$

10. Solve the equation,

$$\tan^2 x - \sec x = 1.$$

$$(1 + \sec^2 x) - \sec x = 1,$$
$$\sec^2 x - \sec x = 0,$$
$$\sec x = 0 \text{ or } 1.$$

But $\sec x$ is never < 1.

$$\therefore \sec x = 1,$$
$$x = 0°.$$

11. Solve the equation,

$$\sin x + \sqrt{3} \cos x = 2,$$

$$\sqrt{3} \cos x = 2 - \sin x,$$
$$3 \cos^2 x = (2 - \sin x)^2,$$
$$3(1 - \sin^2 x) = 4 - 4 \sin x + \sin^2 x,$$
$$4 \sin^2 x - 4 \sin x + 1 = 0,$$
$$(2 \sin x - 1)^2 = 0,$$
$$\sin x = \tfrac{1}{2}.$$
$$\therefore x = 30°.$$

12. Solve the equation,

$$\tan^2 x + \csc^2 x = 3.$$

$$\tan^2 x + (1 + \cot^2 x) = 3,$$
$$\tan^2 x - 2 + \cot^2 x = 0,$$
$$\tan x - \cot x = 0,$$
$$\tan x - \frac{1}{\tan x} = 0,$$
$$\tan^2 x = 1,$$
$$\tan x = 1.$$
$$\therefore x = 45°.$$

13. Solve the equation,

$$2 \cos x + \sec x = 3.$$

$$2 \cos x + \frac{1}{\cos x} = 3,$$
$$2 \cos^2 x + 1 = 3 \cos x,$$
$$2 \cos^2 x - 3 \cos x + 1 = 0,$$
$$(2 \cos x - 1)(\cos x - 1) = 0,$$
$$\cos x = 1 \text{ or } \tfrac{1}{2}.$$
$$\therefore x = 0° \text{ or } 60°.$$

14. Solve the equation,

$$\cos^2 x - \sin^2 x = \sin x.$$

$$(1 - \sin^2 x) - \sin^2 x = \sin x,$$
$$2 \sin^2 x + \sin x - 1 = 0,$$
$$(2 \sin x - 1)(\sin x + 1) = 0,$$
$$\sin x = -1 \text{ or } \tfrac{1}{2}.$$
$$\therefore x = 30°.$$

15. Solve the equation,

$$2 \sin x + \cot x = 1 + 2 \cos x.$$

$$2 \sin x + \frac{\cos x}{\sin x} = 1 + 2 \cos x,$$
$$2 \sin^2 x + \cos x = \sin x + 2 \cos x \sin x,$$
$$2 \sin^2 x - \sin x = 2 \cos x \sin x - \cos x.$$
$$\sin x (2 \sin x - 1) = \cos x (2 \sin x - 1),$$
$$(\sin x - \cos x)(2 \sin x - 1) = 0.$$
$$\therefore \sin x = \cos x,$$
$$\tan x = 1.$$
$$\therefore x = 45°;$$
$$\sin x = \tfrac{1}{2}.$$
$$\therefore x = 30°.$$

Hence, $x = 30° \text{ or } 45°.$

16. Solve the equation,

$$\sin^2 x + \tan^2 x = 3 \cos^2 x.$$

$$\sin^2 x + \frac{\sin^2 x}{\cos^2 x} = 3 \cos^2 x,$$
$$1 - \cos^2 x + \frac{1 - \cos^2 x}{\cos^2 x} = 3 \cos^2 x,$$
$$-4 \cos^2 x + \frac{1}{\cos^2 x} = 0,$$
$$4 \cos^4 x = 1,$$
$$\cos x = \sqrt{\tfrac{1}{2}}.$$
$$\therefore x = 45°.$$

17. Solve the equation,

$$\tan x + 2 \cot x = \tfrac{5}{2} \csc x.$$

$$\frac{\sin x}{\cos x} + 2\frac{\cos x}{\sin x} = \frac{5}{2 \sin x},$$

$$\sin^2 x + 2 \cos^2 x = \tfrac{5}{2} \cos x,$$

$$1 - \cos^2 x + 2 \cos^2 x = \tfrac{5}{2} \cos x,$$

$$\cos^2 x - \tfrac{5}{2} \cos x + 1 = 0,$$

$$(\cos x - 2)(\cos x - \tfrac{1}{2}) = 0,$$

$$\cos x = 2 \text{ or } \tfrac{1}{2}.$$

$$\therefore x = 60°.$$

Exercise VIII. Page 23.

1. In Case II. give another way of finding c, after b has been found.

$$\cos A = \frac{b}{c},$$

$$b = c \cos A,$$

$$c = \frac{b}{\cos A}.$$

2. In Case III. give another way of finding c, after a has been found.

$$\sin A = \frac{a}{c},$$

$$c \sin A = a,$$

$$c = \frac{a}{\sin A}.$$

3. In Case IV. give another way of finding b, after the angles have been found.

$$\cos A = \frac{b}{c},$$

$$b = c \cos A.$$

4. In Case V. give another way of finding c, after the angles have been found.

$$\sin A = \frac{a}{c},$$

$$c \sin A = a,$$

$$c = \frac{a}{\sin A}.$$

5. Given B and c; find A, a, c.

$$A = 90° - B,$$

$$\cos B = \frac{a}{c},$$

$$a = c \cos B.$$

$$\sin B = \frac{b}{c},$$

$$b = c \sin B.$$

6. Given B and b; find A, a, c.

$$A = 90° - B,$$

$$\cot B = \frac{a}{b},$$

$$a = b \cot B.$$

$$\sin B = \frac{b}{c},$$

$$b = c \sin B,$$

$$c = \frac{b}{\sin B}.$$

7. Given B and a; find A, b, c.

$$A = 90° - B,$$

$$\cot B = \frac{a}{b},$$

$$b \cot B = a,$$

$$b = \frac{a}{\cot B}.$$

$$\cos B = \frac{a}{c},$$
$$c \cos B = a,$$
$$c = \frac{a}{\cos B}.$$

8. Given b and c; find A, B, a.

$$\cos A = \frac{b}{c},$$
$$B = 90° - A.$$
$$a = \sqrt{c^2 - b^2}$$
$$= \sqrt{(c+b)(c-b)}.$$

9. Given $a = 3$, $b = 4$; required $A = 36° 52'$, $B = 53° 8'$, $c = 5$.

$$\tan A = \frac{a}{b} = \tfrac{3}{4} = 0.7500.$$
$$\therefore A = 36° 52'.$$
$$B = 90° - A$$
$$= 53° 8',$$
$$c = \sqrt{a^2 + b^2}$$
$$= 5.$$

10. Given $a = 7$, $c = 13$; required $A = 32° 35'$, $B = 57° 25'$ $b = 10.954$.

$$\sin A = \frac{a}{c} = \tfrac{7}{13} = 0.5385.$$
$$\therefore A = 32° 35'.$$
$$B = 90° - A$$
$$= 57° 25'.$$
$$b = \sqrt{(c-a)(c+a)}$$
$$= \sqrt{120}$$
$$= 10.954.$$

11. Given $a = 5.3$, $A = 12° 17'$; required $B = 77° 43'$, $b = 24.342$, $c = 24.918$.

$$B = 90° - A$$
$$= 77° 43'.$$
$$\frac{b}{a} = \cot A,$$
$$b = a \cot A$$
$$= 5.3 \times 4.5928$$
$$= 24.342.$$
$$\frac{a}{c} = \sin A,$$
$$c = \frac{a}{\sin A}$$
$$= \frac{5.3}{0.2127}$$
$$= 24.918.$$

12. Given $a = 10.4$, $B = 43° 18'$; required $A = 46° 42'$, $b = 9.800$, $c = 14.290$.

$$A = 90° - B$$
$$= 46° 42',$$
$$\frac{b}{a} = \tan B,$$
$$b = a \tan B,$$
$$= 10.4 \times 0.9424$$
$$= 9.800.$$
$$\frac{a}{c} = \cos B,$$
$$c = \frac{a}{\cos B}$$
$$= \frac{10.4}{0.7278}$$
$$= 14.290.$$

13. Given $c = 26$, $A = 37° 42'$; required $B = 52° 18'$, $a = 15.900$, $b = 20.572$.

$$B = 90° - A$$
$$= 52° 18'.$$
$$\frac{a}{c} = \sin A,$$
$$a = c \sin A$$
$$= 26 \times 0.6115$$
$$= 15,900.$$
$$\frac{b}{c} = \cos A,$$
$$b = c \cos A$$
$$= 26 \times 0.7912$$
$$= 20.572.$$

14. Given $c = 140$, $B = 24° 12'$; required $A = 65° 48'$, $a = 127.694$, $b = 57.386$.

$$A = 90° - B,$$
$$= 65° 48'.$$
$$\frac{a}{c} = \cos B,$$
$$a = c \cos B$$
$$= 140 \times 0.9121$$
$$= 127.694.$$
$$\frac{b}{c} = \sin B,$$
$$b = c \sin B$$
$$= 140 \times 0.4099$$
$$= 57.386.$$

15. Given $b = 19$, $c = 23$; required A $34° 18'$, $B = 55° 42'$, $a = 12.961$.

$$\cos A = \frac{b}{c} = \frac{19}{23} = 0.8261,$$
$$A = 34° 18',$$
$$B = 90° - A$$
$$= 55° 42'.$$
$$a = \sqrt{(c - b)(c + b)}$$
$$= \sqrt{168}$$
$$= 12.961.$$

16. Given $b = 98$, $c = 135.2$; required $A = 43° 33'$, $B = 46° 27'$, $a = 93.139$.

$$\cos A = \frac{b}{c} = \frac{98}{135.2} = 0.7248,$$
$$A = 43° 33',$$
$$B = 90° - A$$
$$= 46° 27'.$$
$$a = \sqrt{(c - b)(c + b)}$$
$$= \sqrt{8675.04}$$
$$= 93.139.$$

17. Given $b = 42.4$, $A = 32° 14$; required $B = 57° 46'$, $a = 26.733$, $c = 50.124$.

$$B = 90° - A$$
$$= 57° 46'.$$
$$\frac{a}{b} = \tan A,$$
$$a = b \tan A$$
$$= 42.4 \times 0.6305$$
$$= 26.733,$$
$$\frac{b}{c} = \cos A,$$
$$c = \frac{b}{\cos A}$$
$$= \frac{42.4}{0.8459}$$
$$= 50.124.$$

18. Given $b = 200$, $B = 46° 11'$; required $A = 43° 49'$, $a = 191.900$, $c = 277.160$.

$$A = 90° - B$$
$$= 43° 49',$$
$$\frac{a}{b} = \cot B,$$
$$a = b \cot B$$
$$= 200 \times 0.9595$$
$$= 191.900,$$
$$\frac{b}{c} = \sin B,$$
$$c = \frac{b}{\sin B},$$
$$= \frac{200}{0.7216}$$
$$= 277.160.$$

19. Given $a = 95$, $b = 37$; required $A = 68° 43'$, $B = 21° 17'$, $c = 101.951$.

$$\tan A = \frac{a}{b} = \frac{95}{37} = 2.5676,$$
$$A = 68° 43',$$

$$B = 90^\circ - A,$$
$$= 21^\circ\ 17',$$
$$c = \sqrt{a^2 + b^2}$$
$$= \sqrt{10394}$$
$$= 101.951.$$

20. Given $a = 6$, $c = 103$; required $A = 3^\circ\ 21'$, $B = 86^\circ\ 39'$, $b = 102.825$.

$$\sin A = \frac{a}{c} = \frac{6}{103} = 0.0583,$$
$$A = 3^\circ\ 21',$$
$$B = 90^\circ - A$$
$$= 86^\circ\ 39',$$
$$b = \sqrt{c^2 - a^2}$$
$$= \sqrt{10573}$$
$$= 102.825.$$

21. Given $a = 3.12$, $B = 5^\circ\ 8'$; required $A = 84^\circ\ 52'$, $b = 0.280$, $c = 3.133$.

$$A = 90^\circ - A$$
$$= 84^\circ\ 52',$$
$$\frac{b}{a} = \tan B,$$
$$b = a \tan B$$
$$= 3.12 \times 0.0898$$
$$= 0.280,$$
$$\frac{a}{c} = \cos B,$$
$$c = \frac{a}{\cos B}$$
$$= \frac{3.12}{0.9960}$$
$$= 3.133.$$

22. Given $a = 17$, $c = 18$; required $A = 70^\circ\ 48'$, $B = 19^\circ\ 12'$, $b = 5.916$.

$$\tan \tfrac{1}{2} B = \sqrt{\frac{c - a}{c + a}}$$
$$= \sqrt{\frac{1}{35}}$$
$$= 0.1690.$$
$$\tfrac{1}{2} B = 9^\circ\ 36',$$
$$B = 19^\circ\ 12',$$
$$A = 90^\circ - B$$
$$= 70^\circ\ 48',$$
$$b = \sqrt{(c - a)(c + a)},$$
$$= \sqrt{35}$$
$$= 5.916.$$

23. Given $c = 57$, $A = 38^\circ\ 29'$; required $B = 51^\circ\ 31'$, $a = 35.471$, $b = 44.620$.

$$B = 90 - A$$
$$= 51^\circ\ 31',$$
$$\frac{a}{c} = \sin A,$$
$$a = c \sin A$$
$$= 57 \times 0.6223$$
$$= 35.471,$$
$$\frac{b}{c} = \cos A,$$
$$b = c \cos A$$
$$= 57 \times 0.7828$$
$$= 44.620.$$

24. Given $a + c = 18$, $b = 12$; required $A = 22^\circ\ 37'$, $B = 67^\circ\ 23'$, $a = 13$, $c = 5$.

$$c^2 - a^2 = b^2,$$
$$(c + a)(c - a) = b^2,$$
$$18(c - a) = 144,$$
$$c - a = 8,$$
$$c = 13,$$
$$a = 5.$$
$$\sin A = \frac{a}{c} = \tfrac{5}{13} = 0.3846,$$
$$A = 22^\circ\ 37',$$
$$B = 90^\circ - A$$
$$= 67^\circ\ 23'.$$

25. Given $a + b = 9$, $c = 8$; required $A = 82° \ 18'$, $B = 7° \ 42'$, $a = 7.928$, $b = 1.072$.

$$a^2 + b^2 = c^2 = 64,$$
$$a^2 + 2ab + b^2 = 9^2 = 81,$$
$$2ab = 17,$$
$$a^2 - 2ab + b^2 = 64 - 17 = 47,$$
$$a - b = \sqrt{47} = 6.856,$$
$$a + b = 9,$$
$$a = 7.928,$$
$$b = 1.072,$$
$$\tan \tfrac{1}{2} B = \sqrt{\frac{c-a}{c+a}}$$
$$= \sqrt{\frac{0.072}{15.928}}$$
$$= 0.0672,$$
$$\tfrac{1}{2} B = 3° \ 51',$$
$$B = 7° \ 42',$$
$$A = 90° - B$$
$$= 82° \ 18'.$$

Exercise IX. Page 27.

1. Given $a = 6$, $c = 12$; required $A = 30°$, $B = 60°$, $b = 10.392$.

$$\sin A = \frac{a}{c} = \frac{1}{2} = \sin 30°,$$
$$A = 30°.$$
$$B = (90° - A) = 60°.$$
$$\cos A = \frac{b}{c}.$$
$$\therefore b = c \cos A.$$
$$\log \cos A = 9.93753$$
$$\log 12 = 1.07918$$
$$\log b = 1.01671$$
$$b = 10.392.$$

2. Given $A = 60°$, $b = 4$; required $B = 30°$. $c = 8$, $a = 6.9282$.

Since $A = 60°$,
$$B = (90° - 60°) = 30°,$$
$$c^2 = a^2 + b^2.$$
$$\therefore c^2 - b^2 = a^2 = 48.$$
$$\log 48 = \log a^2 = 1.68124,$$
$$\log a = 0.84062,$$
$$a = 6.9282.$$

3. Given $A = 30°$, $a = 3$; required $B = 60°$, $c = 6$, $b = 5.1961$.

Since $A = 30°$,
$$B = (90° - 30°) = 60°,$$
$$c^2 = a^2 + b^2.$$
$$\therefore c^2 - a^2 = b^2 = 27.$$
$$\log 27 = \log b^2 = 1.43136,$$
$$\log b = 0.71568,$$
$$b = 5.1961.$$

4. Given $a = 4$, $b = 4$; required $A = B = 45°$, $c = 5.6568$.

Since a and b each $= 4$, the $\triangle$ is an isosceles $\triangle$, and the $\angle$s A and B are equal.

$$\therefore A = \tfrac{1}{2} \text{ of } 90° = 45°,$$
$$B = \tfrac{1}{2} \text{ of } 90° = 45°.$$
$$c^2 = a^2 + b^2 = 32.$$
$$\log 32 = \log c^2 = 1.50515,$$
$$\log c = 0.75257,$$
$$c = 5.6568.$$

5. Given $a = 2$, $c = 2.82843$; required $A = B = 45°$, $b = 2$.

$$b = \sqrt{c^2 - a^2}$$
$$= \sqrt{(c + a)(c - a)},$$
$$\log b^2 = \log (c + a) + \log (c - a).$$
$$\log (c + a) = 0.68381$$
$$\log (c - a) = 9.91826 - 10$$
$$\log b^2 = 0.60207$$
$$\log b = 0.30103,$$
$$b = 2.$$

$\therefore$ the $\triangle$ is an isosceles rt. $\triangle$.

$\therefore A = B = 45°$.

6. Given $c = 627$, $A = 23° 30'$; required $B = 66° 30'$, $a = 250.02$, $b = 575.0$.

$$B = (90° - A) = 66° 30'.$$
$$a = c \sin A.$$
$$\log a = \log c + \log \sin A.$$
$$\log c = 2.79727$$
$$\log \sin A = 9.60070$$
$$\log a = 2.39797$$
$$a = 250.02.$$
$$b = c \cos A.$$
$$\log b = \log c + \log \cos A.$$
$$\log c = 2.79727$$
$$\log \cos A = 9.96240$$
$$\log b = 2.75967$$
$$b = 575.$$

7. Given $c = 2280$, $A = 28° 5'$; required $B = 61° 55'$, $a = 1073.3$, $b = 2011.6$.

$$B = (90° - A) = 61° 55'.$$
$$a = c \sin A.$$
$$\log a = \log c + \log \sin A.$$
$$\log c = 3.35793$$
$$\log \sin A = 9.67280$$
$$\log a = 3.03073$$
$$a = 1073.3$$
$$b = c \cos A.$$
$$\log b = \log c + \log \cos A.$$
$$\log c = 3.35793$$
$$\log \cos A = 9.94560$$
$$\log b = 3.30353$$
$$b = 2011.6.$$

8. Given $c = 72.15$, $A = 39° 34'$; required $B = 50° 26'$, $a = 45.958$, $b = 55.620$.

$$B = (90° - A) = 50° 26'.$$
$$a = c \sin A.$$
$$\log a = \log c + \log \sin A.$$
$$\log c = 1.85824$$
$$\log \sin A = 9.80412$$
$$\log a = 1.66236$$
$$a = 45.958.$$
$$b = c \cos A.$$
$$\log b = \log c + \log \cos A.$$
$$\log c = 1.85824$$
$$\log \cos A = 9.88699$$
$$\log b = 1.74523$$
$$b = 55.620.$$

9. Given $c = 1$, $A = 36°$; required $B = 54°$, $a = 0.58779$, $b = 0.80902$.

$$B = (90° - A) = 54°.$$
$$\sin A = \frac{a}{c},$$
$$a = c \sin A.$$
$$\log a = \log c + \log \sin A.$$
$$\log c = 0.00000$$
$$\log \sin A = 9.76922$$
$$\log a = 9.76922$$
$$a = 0.58779.$$
$$\cos A = \frac{b}{c}.$$
$$b = c \cos A.$$
$$\log b = \log c + \log \cos A.$$
$$\log c = 0.00000$$
$$\log \cos A = 9.90796$$
$$\log b = 9.90796 - 10$$
$$b = 0.80902.$$

10. Given $c = 200$, $B = 21° 47'$; required $A = 68° 13'$, $a = 185.72$, $b = 74.219$.

$$A = (90° - B) = 68° 13'.$$

$$\sin A = \frac{a}{c},$$

$$a = c \sin A.$$

$$\log a = \log c + \log \sin A.$$

$$\log c = 2.30103$$
$$\log \sin A = 9.96783$$
$$\log a = 2.26886$$

$$a = 185.72.$$

$$\cos a = \frac{b}{c}.$$

$$b = c \cos A.$$

$$\log b = \log c + \log \cos A.$$

$$\log c = 2.30103$$
$$\log \cos A = 9.56949$$
$$\log b = 1.87052$$

$$b = 74.219.$$

11. Given $c = 93.4$, $B = 76° 25'$; required $A = 13° 35'$, $a = 21.936$, $b = 90.788$.

$$A = (90° - B) = 13° 35'.$$

$$a = c \sin A.$$

$$\log a = \log c + \log \sin A.$$

$$\log c = 1.97035$$
$$\log \sin A = 9.37081$$
$$\log a = 1.34116$$

$$a = 21.936.$$

$$b = a \cot A.$$

$$\log b = \log a + \log \cot A.$$

$$\log a = 1.34116$$
$$\log \cot A = 0.61687$$
$$\log b = 1.95803$$

$$b = 90.788.$$

12. Given $a = 637$, $A = 4° 35'$; required $B = 85° 25'$, $b = 7946$, $c = 7971.5$.

$$B = (90° - A) = 85° 25'.$$

$$b = a \cot A.$$

$$\log b = \log a + \log \cot A.$$

$$\log a = 2.80414$$
$$\log \cot A = 1.09601$$
$$\log b = 3.90015$$

$$b = 7946.$$

$$\log c = \log a + \text{colog} \sin A.$$

$$\log a = 2.80414$$
$$\text{colog} \sin A = 1.09740$$
$$\log c = 3.90154$$

$$c = 7971.5.$$

13. Given $a = 48.532$, $A = 36° 44'$; required $B = 53° 16'$, $b = 65.033$, $c = 81.144$.

$$B = 90° - A$$
$$= 90° - 36° 44'$$
$$= 53° 16'.$$

$$\sin A = \frac{a}{c}.$$

$$c = \frac{a}{\sin A}.$$

$$\log c = \log a + \text{colog} \sin A.$$

$$\log a = 1.68603$$
$$\text{colog} \sin A = 0.22323$$
$$\log c = 1.90926$$

$$c = 81.144.$$

$$\cos A = \frac{b}{c}.$$

$$b = c \cos A.$$

$$\log b = \log c + \log \cos A.$$

$$\log c = 1.90926$$
$$\log \cos A = 9.90386$$
$$\log b = 1.81312$$

$$b = 65.031.$$

14. Given $a = 0.0008$, $A = 86°$; required $B = 4°$, $b = 0.0000559$, $c = 0.000802$.

$$B = 90° - A$$
$$= 90° - 86°$$
$$= 4°.$$
$$\sin A = \frac{a}{c}.$$
$$c = \frac{a}{\sin A}.$$
$$\log c = \log a + \text{colog} \sin A.$$
$$\log a = 6.90309 - 10$$
$$\text{colog} \sin A = 0.00106$$
$$\log c = 6.90415 - 10$$
$$c = 0.000802.$$
$$\cos A = \frac{b}{c}.$$
$$b = c \cos A.$$
$$\log b = \log c + \log \cos A.$$
$$\log c = 6.90415 - 10$$
$$\log \cos A = 8.84358 - 10$$
$$\log b = 5.74773 - 10$$
$$b = 0.0000559.$$

15. Given $b = 50.937$, $B = 43° 48'$; required $A = 46° 12'$, $a = 53.116$, $c = 73.59$.

$$A = (90° - B) = 46° 12'.$$
$$\tan A = \frac{a}{b}.$$
$$a = b \tan A.$$
$$\log b = 1.70703$$
$$\log \tan A = 0.01820$$
$$\log a = 1.72523$$
$$a = 53.116.$$
$$\sin A = \frac{a}{c}.$$
$$c = \frac{a}{\sin A}.$$
$$\log a = 1.72523$$
$$\text{colog} \sin A = 0.14161$$
$$\log c = 1.86684$$
$$c = 73.593.$$

16. Given $b = 2$, $B = 3° 38'$; required $A = 86° 22'$, $a = 31.497$, $c = 31.560$.

$$A = (90° - B) = 86° 22'.$$
$$\tan A = \frac{a}{b}.$$
$$a = b \tan A.$$
$$\log b = 0.30103$$
$$\log \tan A = 1.19723$$
$$\log a = 1.49826$$
$$a = 31.496.$$
$$\sin A = \frac{a}{c}.$$
$$c = \frac{a}{\sin A}.$$
$$\log a = 1.49826$$
$$\text{colog} \sin A = 0.00087$$
$$\log c = 1.49913$$
$$c = 31.560.$$

17. Given $a = 992$, $B = 76° 19'$; $A = 13° 41'$, $b = 4074.45$, $c = 4193.55$.

$$A = 90° - 76° 19'$$
$$= 13° 41'.$$
$$\sin A = \frac{a}{c}.$$
$$\log c = \log a + \text{colog} \sin A.$$
$$\log a = 2.99651$$
$$\text{colog} \sin A = 0.62607$$
$$\log c = 3.62258$$
$$c = 4193.6.$$
$$\sin B = \frac{b}{c}.$$
$$\log b = \log c + \log \sin B.$$
$$\log c = 3.62258$$
$$\log \sin B = 9.98750$$
$$\log b = 3.61008$$
$$b = 4074.5.$$

18. Given $a = 73$, $B = 68° \ 52'$; required $A = 21° \ 8'$, $b = 188.86$, $c = 202.47$.

$$A = (90° - B) = 21° \ 8'.$$

$$\sin A = \frac{a}{c}.$$

$$\log c = \log a + \text{colog} \sin A.$$

$$\log a = 1.86332$$
$$\text{colog} \sin A = 0.44305$$
$$\log c = 2.30637$$
$$c = 202.47.$$

$$\sin B = \frac{b}{c}.$$

$$\log b = \log c + \log \sin B.$$

$$\log c = 2.30637$$
$$\log \sin B = 9.96976$$
$$\log b = 2.27613$$
$$b = 188.86.$$

19. Given $a = 2.189$, $B = 45° \ 25'$; required $A = 44° \ 35'$, $b = 2.2211$, $c = 3.1185$.

$$A = 90° - 45° \ 25'$$
$$= 44° \ 35'.$$

$$\sin A = \frac{a}{c}.$$

$$c = \frac{a}{\sin A}.$$

$$\log c = \log a + \text{colog} \sin A.$$

$$\log a = 0.34025$$
$$\text{colog} \sin A = 0.15370$$
$$\log c = 0.49395$$
$$c = 3.1185.$$

$$\cos A = \frac{b}{c}.$$

$$b = c \cos A.$$

$$\log b = \log c + \log \cos A.$$

$$\log c = 0.49395$$
$$\log \cos A = 9.85262$$
$$\log b = 0.34657$$
$$b = 2.2211.$$

20. Given $b = 4$, $A = 37° \ 56'$; required $B = 52° \ 4'$, $a = 3.1176$, $c = 5.0714$.

$$B = (90° - A) = 52° \ 4'.$$

$$\cos A = \frac{b}{c}.$$

$$b = c \cos A.$$

$$c = \frac{b}{\cos A}.$$

$$\log c = \log b + \text{colog} \cos A.$$

$$\log b = 0.60206$$
$$\text{colog} \cos A = 0.10307$$
$$\log c = 0.70513$$
$$c = 5.0714.$$

$$\tan A = \frac{a}{b}.$$

$$a = b \tan A.$$

$$\log a = \log b + \log \tan A.$$

$$\log b = 0.60206$$
$$\log \tan A = 9.89177$$
$$\log a = 0.49383$$
$$a = 3.1176.$$

21. Given $c = 8590$, $a = 4476$; required $A = 31° \ 24'$, $B = 58° \ 36'$, $b = 7332.8$.

$$\sin A = \frac{a}{c}.$$

$$\log \sin A = \log a + \text{colog} \, c.$$

$$\log a = 3.65089$$
$$\text{colog} \ c = 6.06601 - 10$$
$$\log \sin A = 9.71690 - 10$$
$$A = 31° \ 24'.$$
$$\therefore B = 58° \ 36'.$$

$$\cot A = \frac{b}{a}.$$

$$\log b = \log a + \log \cot A.$$

$$\log a = 3.65089$$
$$\log \cot A = 0.21438$$
$$\log b = 3.86527$$
$$b = 7332.8.$$

22. Given $c = 86.53$, $a = 71.78$; required $A = 56° \ 3'$, $B = 33° \ 57'$, $b = 48.324$.

$$\log \sin A = \log a + \text{colog}\, c.$$

$$\begin{aligned} \log a &= 1.85600 \\ \text{colog}\, c &= 8.06283 - 10 \\ \hline \log \sin A &= 9.91883 - 10 \end{aligned}$$

$$A = 56° \ 3'.$$
$$\therefore B = 33° \ 57'.$$
$$\log b = \log a + \log \cot A.$$

$$\begin{aligned} \log a &= 1.85600 \\ \log \cot A &= 9.82817 \\ \hline \log b &= 1.68417 \end{aligned}$$

$$b = 48.324.$$

23. Given $c = 9.35$, $a = 8.49$; required $A = 65° \ 14'$, $B = 24° \ 46'$, $b = 3.917$.

$$\sin A = \frac{a}{c}\cdot$$

$$\begin{aligned} \text{colog}\, c &= 9.02919 - 10 \\ \log a &= 0.92891 \\ \hline \log \sin A &= 9.95810 - 10 \end{aligned}$$

$$A = 65° \ 14'.$$
$$\therefore B = 24° \ 46'.$$
$$\cos A = \frac{b}{c}\cdot$$
$$b = c \cos A.$$

$$\begin{aligned} \log c &= 0.97081 \\ \log \cos A &= 9.62214 \\ \hline \log b &= 0.59295 \end{aligned}$$

$$b = 3.917.$$

24. Given $c = 2194$, $b = 1312.7$; required $A = 53° \ 15'$, $B = 36° \ 45'$, $a = 1758$.

$$\cos A = \frac{b}{c}\cdot$$

$$\begin{aligned} \log b &= 3.11816 \\ \text{colog}\, c &= 6.65876 - 10 \\ \hline \log \cos A &= 9.77692 - 10 \end{aligned}$$

$$A = 53° \ 15'.$$
$$B = (90° - A)$$
$$= 36° \ 45'.$$
$$\sin A = \frac{a}{c}\cdot$$
$$a = c \sin A.$$

$$\begin{aligned} \log c &= 3.34124 \\ \log \sin A &= 9.90377 \\ \hline \log a &= 3.24501 \end{aligned}$$

$$a = 1758.$$

25. Given $c = 30.69$, $b = 18.256$; required $A = 53° \ 30'$, $B = 36° \ 30'$, $a = 24.67$.

$$\cos A = \frac{b}{c}\cdot$$
$$\log \cos A = \log b + \text{colog}\, c.$$

$$\begin{aligned} \log b &= 1.26140 \\ \text{colog}\, c &= 8.51300 - 10 \\ \hline \log \cos A &= 9.77440 - 10 \end{aligned}$$

$$A = 53° \ 30'.$$
$$\therefore B = 36° \ 30'.$$
$$\tan A = \frac{a}{b}\cdot$$
$$\log a = \log \tan A + \log b.$$

$$\begin{aligned} \log \tan A &= 0.13079 \\ \log b &= 1.26140 \\ \hline \log a &= 1.39219 \end{aligned}$$

$$a = 24.671.$$

26. Given $a = 38.313$, $b = 19.522$; required $A = 63°$, $B = 27°$, $c = 43$.

$$\tan A = \frac{a}{b}.$$

$$\log \tan A = \log a + \text{colog } b.$$

$$\begin{aligned} \log a &= 1.58335 \\ \text{colog } b &= 8.70948 - 10 \\ \log \tan A &= 10.29283 - 10 \end{aligned}$$

$$A = 63°.$$

$$\therefore B = 27°.$$

$$\sin A = \frac{a}{c}.$$

$$\log c = \log a + \text{colog} \sin A.$$

$$\begin{aligned} \log a &= 1.58335 \\ \text{colog} \sin A &= 0.05012 \\ \log c &= 1.63347 \end{aligned}$$

$$c = 43.$$

27. Given $a = 1.2291$, $b = 14.950$; required $A = 4° 42'$, $B = 85° 18'$, $c = 15$.

$$\tan A = \frac{a}{b}.$$

$$\begin{aligned} \log a &= 0.08959 \\ \text{colog } b &= 8.82536 - 10 \\ \log \tan A &= 8.91495 - 10 \end{aligned}$$

$$A = 4° 42'.$$

$$\therefore B = 85° 18'.$$

$$\sin A = \frac{a}{c}.$$

$$a = c \sin A.$$

$$c = \frac{a}{\sin A}.$$

$$\begin{aligned} \log a &= 0.08959 \\ \text{colog} \sin A &= 1.08651 \\ \log c &= 1.17610 \end{aligned}$$

$$c = 15.$$

28. Given $a = 415.38$, $b = 62.080$; required $A = 81° 30'$, $B = 8° 30'$, $c = 420$.

$$\tan A = \frac{a}{b}.$$

$$\begin{aligned} \log a &= 2.61845 \\ \text{colog } b &= 8.20705 - 10 \\ \log \tan A &= 10.82550 - 10 \end{aligned}$$

$$A = 81° 30'.$$

$$\therefore B = 8° 30'.$$

$$\sin A = \frac{a}{c}.$$

$$a = c \sin A.$$

$$c = \frac{a}{\sin A}.$$

$$\begin{aligned} \log a &= 2.61845 \\ \text{colog} \sin A &= 0.00480 \\ \log c &= 2.62325 \end{aligned}$$

$$c = 420.$$

29. Given $a = 13.690$, $b = 16.926$; required $A = 38° 58'$, $B = 51° 2'$, $c = 21.77$.

$$\tan A = \frac{a}{b}.$$

$$\log \tan A = \log a + \text{colog } b.$$

$$\begin{aligned} \log a &= 1.13640 \\ \text{colog } b &= 8.77144 - 10 \\ \log \tan A &= 9.90784 - 10 \end{aligned}$$

$$A = 38° 58'.$$

$$\therefore B = 51° 2'.$$

$$\sin A = \frac{a}{c}.$$

$$c = \frac{a}{\sin A}.$$

$$\log c = \log a + \text{colog} \sin A.$$

$$\begin{aligned} \log a &= 1.13640 \\ \text{colog} \sin A &= 0.20144 \\ \log c &= 1.33784 \end{aligned}$$

$$c = 21.769.$$

30. Given $c = 91.92$, $a = 2.19$; required $A = 1° 22'$, $B = 88° 38'$, $b = 91.894$.

$$\sin A = \frac{a}{c}.$$

$$\log \sin A = \log a + \text{colog } c.$$

$$\begin{aligned} \log a &= 0.34044 \\ \text{colog } c &= 8.03659 - 10 \\ \log \sin A &= 8.37703 - 10 \end{aligned}$$

$$A = 1° 22'.$$
$$B = 88° 38'.$$

$$\cos A = \frac{b}{c}.$$

$$b = c \cos A.$$

$$\log b = \log c + \log \cos A.$$

$$\begin{aligned} \log c &= 1.96341 \\ \log \cos A &= 9.99988 \\ \log b &= 1.96329 \\ b &= 91.894. \end{aligned}$$

31. Compute the unknown parts and also the area, having given $a = 5$, $b = 6$.

$$\tan A = \frac{a}{b}.$$

$$\log \tan A = \log a + \text{colog } b.$$

$$\begin{aligned} \log a &= 0.69897 \\ \text{colog } b &= 9.22185 - 10 \\ \log \tan A &= 9.92082 - 10 \end{aligned}$$

$$A = 39° 48'.$$
$$B = 50° 12'.$$

$$\sin A = \frac{a}{c}.$$

$$c = \frac{a}{\sin A},$$

$$\log c = \log a + \text{colog} \sin A.$$

$$\begin{aligned} \log a &= 0.69897 \\ \text{colog } \sin A &= 0.19375 \\ \log c &= 0.89272 \end{aligned}$$

$$c = 7.8112$$

$$F = \frac{ab}{2} = \frac{30}{2} = 15.$$

32. Compute the unknown parts and also the area, having given $a = 0.615$, $c = 70$.

$$\sin A = \frac{a}{c}.$$

$$\log \sin A = \log a + \text{colog } c.$$

$$\begin{aligned} \log a &= 9.78888 - 10 \\ \text{colog } c &= 8.15490 - 10 \\ \log \sin A &= 7.94378 - 10 \end{aligned}$$

$$A = 30' 12''.$$
$$B = 89° 29' 48''.$$

$$b = \sqrt{(c + a)(c - a)}$$

$$\log b = \frac{\log (c + a) + \log (c - a)}{2}$$

$$\begin{aligned} \log (c + a) &= 1.84890 \\ \log (c - a) &= 1.84126 \\ \log b &= 1.84508 \\ b &= 69.997 \end{aligned}$$

$$F = \tfrac{1}{2} ab.$$

$$\begin{aligned} \log a &= 9.78888 - 10 \\ \log b &= 1.84508 \\ \text{colog } 2 &= 9.69897 - 10 \\ \log F &= 1.33293 \end{aligned}$$

$$F = 21.525.$$

33. Compute the unknown parts and also the area, having given $b = \sqrt[3]{2}$, $c = \sqrt{3}$.

$$\sqrt[3]{2} = 1.25991.$$
$$\sqrt{3} = 1.73205.$$

$$\cos A = \frac{b}{c}.$$

$$\log \cos A = \log b + \text{colog } c.$$

$$\begin{aligned} \log b &= 0.10034 \\ \text{colog } c &= 9.76144 - 10 \\ \log \cos A &= 9.86178 - 10 \end{aligned}$$

$$A = 43° 20'.$$
$$B = 46° 40'.$$

$$\sin A = \frac{a}{c}.$$

$$a = c \sin A.$$

$$\log a = \log c + \log \sin A.$$

$$\begin{aligned} c &= 0.23856 \\ \log \sin A &= 9.83648 \\ \log a &= 0.07504 \end{aligned}$$

$$a = 1.1886.$$

$$F = \tfrac{1}{2} ab.$$

$$\begin{aligned} \log a &= 0.07504 \\ \log b &= 0.10034 \\ \text{colog } 2 &= 9.69897 - 10 \\ \log F &= 9.87435 - 10 \end{aligned}$$

$$F = 0.74876.$$

34. Compute the unknown parts and also the area, having given $a = 7$, $A = 18° 14'$.

$$B = 71° 46'.$$

$$\sin A = \frac{a}{c}.$$

$$c = \frac{a}{\sin A}.$$

$$\log c = \log a + \text{colog} \sin A.$$

$$\begin{aligned} \log a &= 0.84510 \\ \text{colog} \sin A &= 0.50461 \\ \log c &= 1.34971 \end{aligned}$$

$$c = 22.372.$$

$$\tan A = \frac{a}{b}.$$

$$b = \frac{a}{\tan A}.$$

$$\log b = \log a + \text{colog} \tan A.$$

$$\begin{aligned} \log a &= 0.84510 \\ \text{colog} \tan A &= 0.48224 - 10 \\ \log b &= 1.32734 \end{aligned}$$

$$b = 21.249.$$

$$F = \tfrac{1}{2} ab.$$

$$\begin{aligned} \log a &= 0.84510 \\ \log b &= 1.32734 \\ \text{colog } 2 &= 9.69897 - 10 \\ \log F &= 1.87141 \end{aligned}$$

$$F = 74.371.$$

35. Compute the unknown parts and also the area, having given $b = 12$, $A = 29° 8'$.

$$A = 29° 8'.$$

$$\therefore B = 60° 52'.$$

$$\cos A = \frac{b}{c}.$$

$$c = \frac{b}{\cos A}.$$

$$\log c = \log b + \text{colog} \cos A.$$

$$\begin{aligned} \log b &= 1.07918 \\ \text{colog} \cos A &= 0.05874 \\ \log c &= 1.13792 \end{aligned}$$

$$c = 13.738.$$

$$\sin A = \frac{a}{c}.$$

$$a = c \sin A.$$

$$\log a = \log c + \log \sin A.$$

$$\begin{aligned} \log c &= 1.13792 \\ \log \sin A &= 9.68739 \\ \log a &= 0.82531 \end{aligned}$$

$$a = 6.6882.$$

$$F = \tfrac{1}{2} ab.$$

$$\log F = \log a + \log b + \text{colog } 2.$$

$$\begin{aligned} \log a &= 0.82531 \\ \log b &= 1.07918 \\ \text{colog } 2 &= 9.69897 - 10 \\ \log F &= 1.60346 \end{aligned}$$

$$F = 40.129.$$

36. Compute the unknown parts and also the area, having given $c = 68$, $A = 69° 54'$.

$$A = 69° 54'.$$

$$\therefore B = 20° 6'.$$

$\sin A = \dfrac{a}{c}$.

$a = c \sin A$.

$\log a = \log c + \log \sin A$.

$$\begin{aligned} \log c &= 1.83251 \\ \log \sin A &= 9.97271 \\ \log a &= 1.80522 \end{aligned}$$

$a = 63.859$.

$\cos A = \dfrac{b}{c}$.

$b = c \cos A$.

$\log b = \log c + \log \cos A$.

$$\begin{aligned} \log c &= 1.83251 \\ \log \cos A &= 9.53613 \\ \log b &= 1.36864 \end{aligned}$$

$b = 23.369$.

$F = \frac{1}{2} ab$.

$$\begin{aligned} \log a &= 1.80522 \\ \log b &= 1.36864 \\ \text{colog } 2 &= 9.69897 - 10 \\ \log F &= 2.87283 \end{aligned}$$

$F = 746.15$.

37. Compute the unknown parts and also the area, having given $c = 27$, $B = 44° 4'$.

$A = 45° 56'$.

$a = c \sin A$,

$\log a = \log c + \log \sin A$.

$$\begin{aligned} \log c &= 1.43136 \\ \log \sin A &= 9.85645 \\ \log a &= 1.28781 \end{aligned}$$

$a = 19.40$.

$b = c \cos A$.

$\log b = \log c + \log \cos A$.

$$\begin{aligned} \log c &= 1.43136 \\ \log \cos A &= 9.84229 \\ \log b &= 1.27365 \end{aligned}$$

$b = 18.778$.

$F = \frac{1}{2} ab$.

$$\begin{aligned} \log a &= 1.28781 \\ \log b &= 1.27365 \\ \text{colog } 2 &= 9.69897 - 10 \\ \log F &= 2.26043 \end{aligned}$$

$F = 182.15$.

38. Compute the unknown parts and also the area, having given $a = 47$, $B = 48° 49'$.

$A = 41° 11'$.

$b = a \cot A$.

$\log b = \log a + \log \cot A$.

$$\begin{aligned} \log a &= 1.67210 \\ \log \cot A &= 10.05803 \\ \log b &= 1.73013 \end{aligned}$$

$b = 53.719$.

$c = \dfrac{a}{\sin A}$.

$\log c = \log a + \text{colog} \sin A$.

$$\begin{aligned} \log a &= 1.67210 \\ \text{colog} \sin A &= 0.18146 \\ \log c &= 1.85356 \end{aligned}$$

$c = 71.377$.

$F = \frac{1}{2} ab$.

$$\begin{aligned} \log a &= 1.67210 \\ \log b &= 1.73013 \\ \text{colog } 2 &= 9.69897 - 10 \\ \log F &= 3.10120 \end{aligned}$$

$F = 1262.4$.

39. Compute the unknown parts and also the area, having given $b = 9$, $B = 34° 44'$.

$A = 55° 16'$.

$a = b \tan A$.

$\log a = \log b + \log \tan A$.

$$\begin{aligned} \log b &= 0.95424 \\ \log \tan A &= 10.15908 \\ \log a &= 1.11332 \end{aligned}$$

$a = 12.981$.

$$c = \frac{a}{\sin A}.$$

$$\log c = \log a + \text{colog} \sin A.$$

$$\log a = 1.11332$$
$$\text{colog} \sin A = 0.08523$$
$$\log c = 1.19855$$

$$c = 15.796$$
$$F = \tfrac{1}{2} ab.$$

$$\log a = 1.11332$$
$$\log b = 0.95424$$
$$\text{colog } 2 = 9.69897 - 10$$
$$\log F = 1.76653$$

$$F = 58.416.$$

40. Compute the unknown parts and also the area, having given $c = 8.462$, $B = 86° 4'$.

$$A = 3° 56'.$$
$$a = c \sin A.$$
$$\log a = \log c + \log \sin A.$$

$$\log c = 0.92747$$
$$\log \sin A = 8.83630 - 10$$
$$\log a = 9.76377 - 10$$

$$a = 0.58046.$$

$$b = c \cos A.$$
$$\log b = \log c + \log \cos A.$$

$$\log c = 0.92747$$
$$\log \cos A = 9.99898$$
$$\log b = 0.92645$$

$$b = 8.442.$$

$$F = \tfrac{1}{2} ab.$$

$$\log a = 9.76377 - 10$$
$$\log b = 0.92645$$
$$\text{colog } 2 = 9.69897 - 10$$
$$\log F = 0.38919$$

$$F = 2.4501.$$

41. Find the value of F in terms of c and A.

$$F = \tfrac{1}{2} ab.$$
$$\sin A = \frac{a}{c}.$$
$$a = c \sin A.$$
$$\cos A = \frac{b}{c}.$$
$$b = c \cos A.$$

Substitute,

$$F = \tfrac{1}{2} ab$$
$$= \tfrac{1}{2} (c^2 \sin A \cos A).$$

42. Find the value of F in terms of a and A.

$$F = \tfrac{1}{2} ab.$$
$$\cot A = \frac{b}{a}.$$
$$b = a \cot A.$$

Substitute,

$$F = \tfrac{1}{2} ab$$
$$= \tfrac{1}{2} (a^2 \cot A).$$

43. Find the value of F in terms of b and A.

$$F = \tfrac{1}{2} ab.$$
$$\tan A = \frac{a}{b}.$$
$$a = b \tan A.$$

Substitute,

$$F = \tfrac{1}{2} ab.$$
$$= \tfrac{1}{2} (b^2 \tan A).$$

44. Find the value of F in terms of a and c.

$$F = \tfrac{1}{2} ab$$
$$c^2 = a^2 + b^2.$$
$$b^2 = c^2 - a^2.$$
$$b = \sqrt{c^2 - a^2}.$$

Substitute,

$$F = \tfrac{1}{2} (a \sqrt{c^2 - a^2}).$$

45. Given $F = 58$, $a = 10$; solve the triangle.

$$F = \tfrac{1}{2}ab.$$
$$b = \frac{2F}{a}.$$
$$\log b = \log 2F + \text{colog } a$$

$$\begin{aligned} \log 2F &= 2.06446 \\ \text{colog } a &= 9.00000 - 10 \\ \log b &= 1.06446 \end{aligned}$$

$$b = 11.6.$$
$$\tan a = \frac{a}{b}.$$
$$\log \tan A = \log a + \text{colog } b.$$

$$\begin{aligned} \log a &= 1.00000 \\ \text{colog } b &= 8.93554 - 10 \\ \log \tan A &= 9.93554 - 10 \end{aligned}$$

$$A = 40^\circ\ 45'\ 48''.$$
$$B = 49^\circ\ 14'\ 12''.$$
$$c = \frac{a}{\sin A},$$
$$\log c = \log a + \text{colog} \sin A.$$

$$\begin{aligned} \log a &= 1.00000 \\ \text{colog} \sin A &= 0.18513 \\ \log c &= 1.18513 \end{aligned}$$

$$c = 15.315.$$

46. Given $F = 18$, $b = 5$; solve the triangle.

$$F = \tfrac{1}{2}ab.$$
$$a = \frac{2F}{b},$$
$$\log a = \log 2F + \text{colog } b.$$

$$\begin{aligned} \log 2F &= 1.55630 \\ \text{colog } b &= 9.30103 - 10 \\ \log a &= 0.85733 \end{aligned}$$

$$a = 7.2.$$

$$\tan A = \frac{a}{b}.$$
$$\log \tan A = \log a + \text{colog } b.$$

$$\begin{aligned} \log a &= 0.85733 \\ \text{colog } b &= 9.30103 - 10 \\ \log \tan A &= 10.15836 - 10 \end{aligned}$$

$$A = 55^\circ\ 13'\ 20''.$$
$$B = 34^\circ\ 46'\ 40''.$$
$$c = \frac{a}{\sin A},$$
$$\log c = \log a + \text{colog} \sin A.$$

$$\begin{aligned} \log a &= 0.85733 \\ \text{colog} \sin A &= 0.08546 \\ \log c &= 0.94279 \end{aligned}$$

$$c = 8.7658.$$

47. Given $F = 12$, $A = 29^\circ$; solve the triangle.

$$B = 61^\circ.$$
$$F = \tfrac{1}{2}ab = 12.$$
$$ab = 24.$$
$$a = \frac{24}{b}.$$
$$\tan A = \frac{a}{b}.$$
$$\tan 29^\circ = \frac{24}{b^2}.$$
$$b^2 = \frac{24}{\tan 29^\circ},$$
$$\log b = \tfrac{1}{2}(\log 24 + \text{colog} \tan 29^\circ).$$

$$\begin{aligned} \log 24 &= 1.38021 \\ \text{colog} \tan 29^\circ &= 0.25625 \\ 2)\ &1.63646 \\ \log b &= 0.81823 \end{aligned}$$

$$b = 6.58.$$

$$\tan 29° = \frac{a}{b}.$$

$$a = b \tan 29°.$$

$$\log a = \log b + \log \tan 29°.$$

$$\begin{aligned} \log b &= 0.81823 \\ \log \tan 29° &= \underline{9.74375} \\ \log a &= 0.56198 \end{aligned}$$

$$a = 3.6474.$$

$$\sin A = \frac{a}{c}.$$

$$c = \frac{a}{\sin 29°}.$$

$$\log c = \log a + \text{colog} \sin 29°.$$

$$\begin{aligned} \log a &= 0.56198 \\ \text{colog} \sin 29° &= \underline{0.31443} \\ \log c &= 0.87641 \end{aligned}$$

$$c = 7.5233.$$

48. Given $F = 100$, $c = 22$; solve the triangle.

$$F = \tfrac{1}{2} ab = 100.$$

$$ab = 200.$$

$$a = \frac{200}{b}.$$

$$a^2 = \frac{40000}{b^2}.$$

$$a^2 + b^2 = c^2 = 484.$$

Substitute,

$$\frac{40000}{b^2} + b^2 = 484.$$

$$40000 + b^4 = 484 b^2.$$

$$b^4 - 484 b^2 = -40000.$$

$$b^4 - (\;) + (242)^2 = 18564.$$

$$\log \sqrt{18564} = \tfrac{1}{2}(4.26867) = 2.13434;$$

but $2.13434 = \log 136.25$.

$$\therefore b^2 - 242 = 136.25.$$

$$b^2 = 378.25.$$

$$\log b = \tfrac{1}{2}(\log 378.25). = 1.28889.$$

$$b = 19.449.$$

$$\cos A = \frac{b}{c}.$$

$$\log \cos A = \log b + \text{colog}\, c.$$

$$\begin{aligned} \log b &= 1.28889 \\ \text{colog}\, c &= \underline{8.65758} \\ \log \cos A &= 9.94647 \end{aligned}$$

$$A = 27° \; 52'.$$

$$B = 62° \; 8'.$$

$$\sin A = \frac{a}{c}.$$

$$a = c \sin A.$$

$$\log a = \log c + \log \sin A.$$

$$\begin{aligned} \log c &= 1.34242 \\ \log \sin A &= \underline{9.66970} \\ \log a &= 1.01212 \end{aligned}$$

$$a = 10.283.$$

49. Find the angles of a right triangle if the hypotenuse is equal to three times one of the legs.

Let $c =$ hypotenuse,
and let $c =$ three times a, one of the legs.

$$\sin A = \frac{a}{c}.$$

$$\log \sin A = \log a + \text{colog}\, c.$$

$$\begin{aligned} \log a &= 0.00000 \\ \text{colog}\, c &= \underline{9.52288 - 10} \\ \log \sin A &= 9.52288 \end{aligned}$$

$$A = 19° \; 28' \; 17''.$$

$$B = 70° \; 31' \; 43''.$$

50. Find the legs of a right triangle if the hypotenuse $= 6$, and one angle is twice the other.

Let $c =$ hypotenuse $= 6$,
and let $B =$ twice A;

then $B = 60°,$
$A = 30°.$

$$\sin A = \frac{a}{c}.$$
$$a = c \sin A.$$
$$\log a = \log c + \log \sin A.$$

$$\log c = 0.77815$$
$$\log \sin A = 9.69897$$
$$\log a = 0.47712$$
$$a = 3.$$

$$\sin B = \frac{b}{c}.$$
$$b = c \sin B.$$
$$\log b = \log c + \log \sin B.$$

$$\log c = 0.77815$$
$$\log \sin B = 9.93753$$
$$\log b = 0.71568$$
$$b = 5.1961.$$

51. In a right triangle given c, and $A = nB$; find a and b.

$$B = 90° - A$$
$$= 90° - nB.$$
$$B(n+1) = 90°.$$
$$B = \frac{90°}{n+1}.$$
$$\cos B = \frac{a}{c}.$$
$$\cos \frac{90°}{n+1} = \frac{a}{c}.$$
$$a = c \cos \frac{90°}{n+1}.$$
$$\sin B = \frac{b}{c}.$$
$$\sin \frac{90°}{n+1} = \frac{b}{c}.$$
$$b = c \sin \frac{90°}{n+1}.$$

52. In a right triangle the difference between the hypotenuse and the greater leg is equal to the difference between the two legs; find the angles.

$$c - a = a - b.$$
$$2a - b = c. \quad (1)$$
$$a^2 + b^2 = c^2. \quad (2)$$

Squaring (1),

$$4a^2 - 4ab + b^2 = c^2$$
$$a^2 \qquad + b^2 = c^2$$
$$3a^2 - 4ab \qquad = 0$$

$$3a^2 = 4ab.$$
$$3a = 4b.$$
$$a = \frac{4b}{3}.$$
$$\tan A = \frac{a}{b} = \frac{4}{3}.$$
$$\log \tan A = \log 4 + \operatorname{colog} 3.$$

$$\log 4 = 0.60206$$
$$\operatorname{colog} 3 = 9.52288 - 10$$
$$\log \tan A = 10.12494$$

$$A = 53° \; 7' \; 48''.$$
$$B = 36° \; 52' \; 12''.$$

53. At a horizontal distance of 120 feet from the foot of a steeple, the angle of elevation of the top was found to be 60° 30′; find the height of the steeple.

$$\tan A = \frac{a}{b}.$$
$$a = b \tan A.$$
$$\log a = \log b + \log \tan A.$$

$$\log b = 2.07918$$
$$\log \tan A = 10.24736$$
$$\log a = 2.32654$$
$$a = 212.1.$$

54. From the top of a rock that rises vertically 326 feet out of the water, the angle of depression of a boat was found to be 24°; find the distance of the boat from the foot of the rock.

$$\cot A = \frac{b}{a}.$$

$$b = a + \cot A.$$

$$\log b = \log a + \log \cot A.$$

$$\begin{aligned} \log a &= 2.51322 \\ \log \cot A &= 10.35142 \\ \log b &= 2.86464 \end{aligned}$$

$$b = 732.22.$$

55. How far is a monument, in a level plain, from the eye, if the height of the monument is 200 feet and the angle of elevation of the top 3° 30′.

$$\cot A = \frac{b}{a}.$$

$$b = a \cot A.$$

$$\log b = \log a + \log \cot A.$$

$$\begin{aligned} \log a &= 2.30103 \\ \log \cot A &= 1.21351 \\ \log b &= 3.51454 \end{aligned}$$

$$b = 3270.$$

56. In order to find the breadth of a river a distance AB was measured along the bank, the point A being directly opposite a tree C on the other side. The angle ABC was also measured. If $AB = 96$ feet, and $ABC = 21° 14'$, find the breadth of the river.

If $ABC = 45°$, what would be the breadth of the river?

$$\tan B = AC \div AB.$$

$$AC = AB \times \tan B.$$

$$\log AC = \log AB + \log \tan B.$$

$$\begin{aligned} \log AB &= 1.98227 \\ \log \tan B &= 9.58944 \\ \log AC &= 1.57171 \end{aligned}$$

$$AC = 37.3 \text{ feet.}$$

$$\log AC = \log AB + \log \tan B.$$

$$\begin{aligned} \log AB &= 1.98227 \\ \log \tan B &= 10.00000 \\ \log AC &= 1.98227 \end{aligned}$$

$$AC = 96 \text{ feet.}$$

57. Find the angle of elevation of the sun when a tower a feet high casts a horizontal shadow b feet long. Find the angle when $a = 120$, $b = 70$.

$$\tan A = \frac{a}{b}.$$

$$\tan A = \frac{120}{70}.$$

$$\log \tan A = \log 120 + \text{colog } 70.$$

$$\begin{aligned} \log 120 &= 2.07918 \\ \text{colog } 70 &= 8.15490 - 10 \\ \log \tan A &= 10.23408 \end{aligned}$$

$$A = 59° 44' 35''.$$

58. How high is a tree that casts a horizontal shadow b feet in length when the angle of elevation of the sun is $A°$? Find the height of the tree when $b = 80°$, $A = 50°$.

$$\tan A = \frac{a}{b}.$$

$$a = b \tan A.$$

$$\log a = \log b + \log \tan A.$$

$$\begin{aligned} \log b &= 1.90309 \\ \log \tan A &= 10.07619 \\ \log a &= 1.97928 \end{aligned}$$

$$a = 95.34.$$

59. What is the angle of elevation of an inclined plane if it rises 1 foot in a horizontal distance of 40 feet?

$$\tan A = \frac{a}{b}.$$

$$\log \tan A = \log a + \text{colog}\, b.$$

$$\begin{aligned} \log a &= 0.00000 \\ \text{colog}\, b &= 8.39794 - 10 \\ \log \tan A &= 8.39794 \\ A &= 1° 25' 56''. \end{aligned}$$

60. A ship is sailing due northeast with a velocity of 10 miles an hour. Find the rate at which she is moving due north and also due east.

Let AB be the direction of the vessel, and equal one hour's progress $= 10$ miles.

$AC =$ distance due east passed over in one hour.

As the direction of the ship is northeast,

$$A = 45°.$$

$$b = c \cos A.$$

$$\log b = \log c + \log \cos A.$$

$$\begin{aligned} \log c &= 1.00000 \\ \log \cos A &= 9.84949 \\ \log b &= 0.84949 \end{aligned}$$

$b = 7.0712$ miles due east, and also due north, since

$$AP = AC.$$

61. In front of a window 20 feet high is a flower-bed 6 feet wide. How long must a ladder be to reach from the edge of the bed to the window?

$$\tan A = \frac{a}{b}.$$

$$\log \tan A = \log 20 + \text{colog}\, 6.$$

$$\begin{aligned} \log 20 &= 1.30103 \\ \text{colog}\, 6 &= 9.22185 - 10 \\ \log \tan A &= 10.52288 \\ A &= 73° 18'. \end{aligned}$$

$$c = \frac{a}{\sin A}.$$

$$\log c = \log 20 + \text{colog} \sin A.$$

$$\begin{aligned} \log a &= 1.30103 \\ \text{colog} \sin A &= 0.01871 \\ \log c &= 1.31974 \\ c &= 20.88. \end{aligned}$$

62. A ladder 40 feet long may be so placed that it will reach a window 33 feet high on one side of the street, and by turning it over without moving its foot it will reach a window 21 feet high on the other side. Find the breadth of the street.

$$\cos B = \frac{33}{40}.$$

$$\begin{aligned} \log 33 &= 1.51851 \\ \text{colog}\, 40 &= 8.39794 - 10 \\ \log \cos B &= 9.91645 - 10 \\ B &= 34° 24' 45''. \end{aligned}$$

$$\tan B = \frac{b}{33}.$$

$$b = 33 \tan B.$$

$$\begin{aligned} \log 33 &= 1.51851 \\ \log \tan B &= 9.83571 \\ \log b &= 1.35422 \\ b &= 22.605. \end{aligned}$$

$$\cos B' = \frac{21}{40}.$$

$$\begin{aligned} \log 21 &= 1.32222 \\ \text{colog}\, 40 &= 8.39794 - 10 \\ \log \cos B' &= 9.72016 \\ B' &= 58° 19' 54''. \end{aligned}$$

$$\tan B' = \frac{b}{21}.$$

$$b' = 21 \tan B'.$$

$$\begin{aligned} \log 21 &= 1.32222 \\ \log \tan B' &= \underline{0.20982} \\ \log b' &= 1.53204 \end{aligned}$$

$$\begin{aligned} b' &= 34.044 \\ b &= \underline{22.605} \\ b + b' &= 56.649 \end{aligned}$$

63. From the top of a hill the angles of depression of two successive milestones, on a straight level road leading to the hill, are observed to be 5° and 15°. Find the height of the hill.

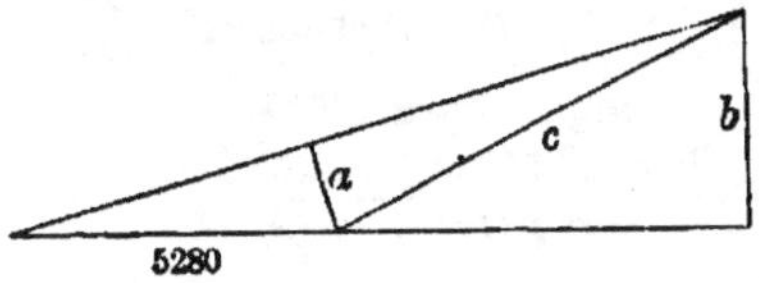

$$\sin 5° = \frac{a}{5280}.$$

$$a = 5280 \sin 5°.$$

$$\begin{aligned} \log 5280 &= 3.72263 \\ \log \sin 5° &= \underline{8.94030 - 10} \\ \log a &= 2.66293 \end{aligned}$$

$$\sin 10° = \frac{a}{c}.$$

$$a = c \sin 10°.$$

$$c = \frac{a}{\sin 10°}.$$

$$\begin{aligned} \log a &= 2.66293 \\ \text{colog} \sin 10° &= \underline{0.76033} \\ \log c &= 3.42326 \end{aligned}$$

$$\cos 75° = \frac{b}{c}.$$

$$b = c \cos 75°.$$

$$\begin{aligned} \log c &= 3.42326 \\ \log \cos 75° &= \underline{9.41300 - 10} \\ \log b &= 2.83626 \end{aligned}$$

$$\begin{aligned} b &= 685.9 \text{ feet} \\ &= 228.63 \text{ yards.} \end{aligned}$$

64. A fort stands on a horizontal plane. The angle of elevation at a certain point on the plane is 30°, and at a point 100 feet nearer the fort it has 45°. How high is the fort?

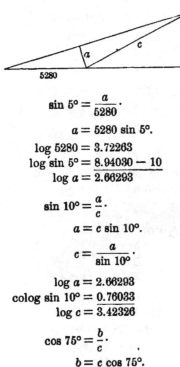

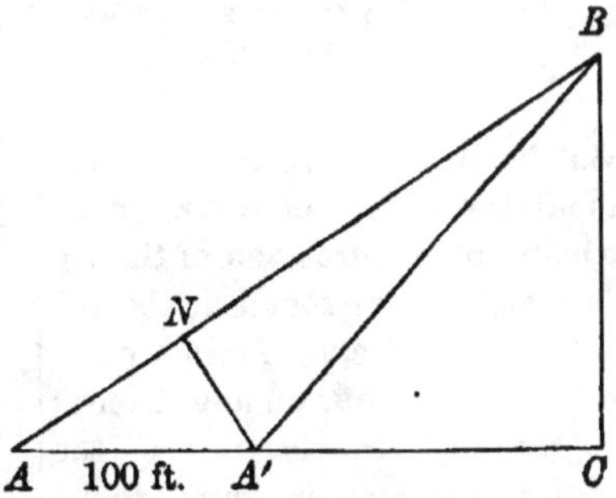

Let B represent the fort, AC the horizontal plane, BC a ⊥ from fort to plane.

BAC = angle made by line from eye of observer = 30°.

$BA'C$ = 45° = angle of elevation 100 feet nearer.

From A' draw $A'N$ ⊥ to AB.

In rt. △ $AA'N$,

$$\angle NAA' = 30°,$$

and $\angle NA'A = 60°$.

$$\therefore NA' = 50 \text{ feet.}$$

$$\begin{aligned} \therefore AN &= \sqrt{(100)^2 - (50)^2} \\ &= \sqrt{7500} = 50\sqrt{3} \\ &= 86.602. \end{aligned}$$

In rt. △ BNA',

$$\frac{BN}{NA'} = \cot NBA' = \cot 15°,$$

and $BN = NA' \cot 15°$.

$$\begin{aligned} \log NA' &= 1.69897 \\ \log \cot 15° &= 0.57195 \\ \log BN &= 2.27092 \end{aligned}$$

$$\begin{aligned} BN &= 186.60 \\ AN &= 86.60 \\ AB &= 273.20 \end{aligned}$$

In rt. $\triangle ABC$,

$\angle BAC = 30°$,

and $\angle ABC = 60°$.

$\therefore BC = \frac{1}{2} AB = \frac{1}{2} \times 273.20$
$= 136.60$ feet.

65. From a certain point on the ground the angles of elevation of the belfry of a church and of the top of a steeple were found to be 40° and 51° respectively. From a point 300 feet farther off, on a horizontal line, the angle of elevation of the top of the steeple is found to be 33° 45′. Find the distance from the belfry to the top of the steeple.

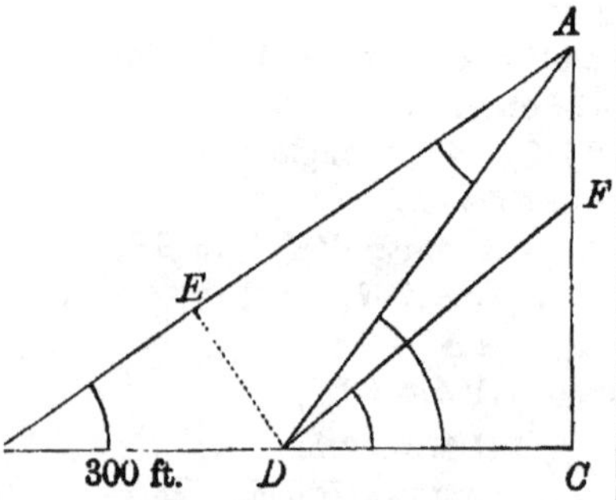

Draw $DE \perp$ to AB from D.

In $\triangle BED$,

$$\frac{ED}{BD} = \sin 33° 45'.$$

$$ED = 300 \times \sin 33° 45'.$$

$$\begin{aligned} \log 300 &= 2.47712 \\ \log \sin 33° 45' &= 9.74474 \\ \log ED &= 2.22186 \end{aligned}$$

$\angle EAD = 180° - 33° 45' - (180° - 51°) = 17° 15'$.

In $\triangle ADE$,

$$\frac{ED}{AD} = \sin 17° 15'.$$

$$AD = \frac{ED}{\sin 17° 15'}.$$

$$\begin{aligned} \log ED &= 2.22186 \\ \text{colog} \sin 17° 15' &= 0.52791 \\ \log AD &= 2.74977 \end{aligned}$$

In $\triangle ADC$,

$$\frac{DC}{AD} = \cos 51°.$$

$$DC = AD \cos 51°.$$

$$\begin{aligned} \log AD &= 2.74977 \\ \log \cos 51° &= 9.79887 \\ \log DC &= 2.54864 \end{aligned}$$

In $\triangle ADC$,

$$\frac{AC}{DC} = \tan 51°.$$

$$AC = DC \tan 51°.$$

$$\begin{aligned} \log DC &= 2.54864 \\ \log \tan 51° &= 10.09163 - 10 \\ \log AC &= 2.64027 \end{aligned}$$

$$AC = 436.79.$$

In $\triangle FDC$,

$$\frac{FC}{DC} = \tan 40°.$$

$$FC = DC \tan 40°.$$

$$\begin{aligned} \log DC &= 2.54864 \\ \log \tan 40° &= 9.92381 \\ \log FC &= 2.47245 \end{aligned}$$

$$FC = 296.79.$$

$$AC - FC = 140.$$

66. The angle of elevation of the top of an inaccessible fort C, observed from a point A, is 12°. At a point B, 219 feet from A and on a line AB perpendicular to AC, the angle ABC is 61° 45′. Find the height of the fort.

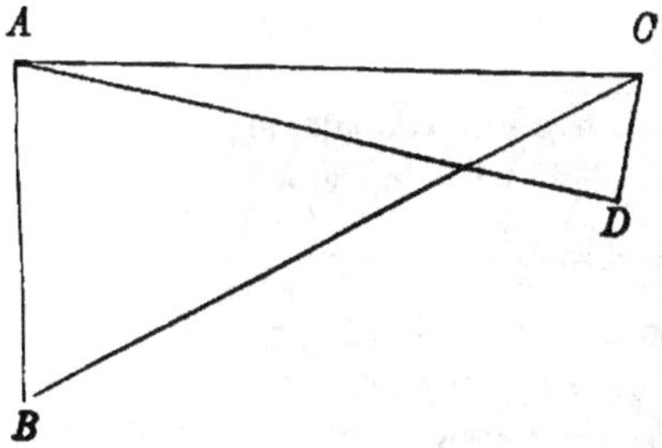

In rt. △ CAB,

$$\frac{AB}{AC} = \cot ABC.$$

$$\therefore AC = \frac{AB}{\cot ABC}.$$

$$\log AC = \log AB + \text{colog} \cot ABC.$$

$$\log AB = 2.34044$$
$$\text{colog} \cot ABC = 0.26977$$
$$\log AC = 2.61021$$

In rt. △ ADC,

$$\frac{CD}{AC} = \sin CAD.$$

$$CD = AC \sin CAD.$$

$$\log CD = \log AC + \log \sin CAD.$$

$$\log AC = 2.61021$$
$$\log \sin CAD = 9.31788 - 10$$
$$\log CD = 1.92809$$
$$CD = 84.74 \text{ feet.}$$

Exercise X. Page 32.

1. In an isosceles triangle, given a and A; find C, c, h.

$$C = 180° - 2A$$
$$= 2(90° - A).$$
$$\frac{\frac{1}{2}c}{a} = \cos A.$$
$$c = 2a \cos A.$$
$$\frac{h}{a} = \sin A.$$
$$h = a \sin A.$$

2. In an isosceles triangle, given a and C; find A, c, h.

$$C + 2A = 180°.$$
$$A = 90° - \tfrac{1}{2}C.$$
$$\frac{\frac{1}{2}c}{a} = \cos A.$$
$$c = 2a \cos A.$$
$$\frac{h}{a} = \sin A.$$
$$h = a \sin A.$$

3. In an isosceles triangle, given c and A; find C, a, h.

$$C = 180° - 2A$$
$$= 2(90° - A).$$
$$\frac{\frac{1}{2}c}{a} = \cos A.$$
$$2a = \frac{c}{\cos A}.$$
$$a = \frac{c}{2 \cos A}.$$
$$\frac{h}{a} = \sin A.$$
$$h = a \sin A.$$

4. In an isosceles triangle, given c and C; find A, a, h.

$$A = 90^\circ - \tfrac{1}{2} C.$$

$$\frac{\tfrac{1}{2} c}{a} = \cos A,$$

$$a = \frac{c}{2 \cos A}.$$

$$\frac{h}{a} = \sin A.$$

$$h = a \sin A.$$

5. In an isosceles triangle, given h and A; find C, a, c.

$$C = 2\,(90^\circ - A).$$

$$\sin A = \frac{h}{a}.$$

$$\therefore a = \frac{h}{\sin A}.$$

$$\cos A = \frac{\tfrac{1}{2} c}{a} = \frac{c}{2a}.$$

$$\therefore c = 2a \cos A.$$

6. In an isosceles triangle, given h and C; find A, a, c.

$$A = 90^\circ - \tfrac{1}{2} C.$$

$$\sin A = \frac{h}{a}.$$

$$\therefore a = \frac{h}{\sin A}.$$

$$\cos A = \frac{\tfrac{1}{2} c}{a} = \frac{c}{2a}.$$

$$\therefore c = 2a \cos A.$$

7. In an isosceles triangle, given a and h; find A, C, c.

$$\sin A = h \div a.$$

$$C = 2\,(90^\circ - A).$$

$$\cos A = \frac{\tfrac{1}{2} c}{a} = \frac{c}{2a}.$$

$$\therefore c = 2a \cos A.$$

8. In an isosceles triangle, given c and h; find A, C, a.

$$\tan A = \frac{h}{\tfrac{1}{2} c}.$$

$$C = 2\,(90^\circ - A).$$

$$\sin A = \frac{h}{a}.$$

$$a = \frac{h}{\sin A}.$$

9. In an isosceles triangle, given $a = 14.3$, $c = 11$; find A, C, h.

$$\cos A = \frac{\tfrac{1}{2} c}{a}.$$

$$\log \cos A = \log \tfrac{1}{2} c + \operatorname{colog} a,$$

$$\begin{aligned} \log \tfrac{1}{2} c &= 0.74036 \\ \operatorname{colog} a &= \underline{8.84466 - 10} \\ \log \cos A &= 9.58502 - 10 \end{aligned}$$

$$A = 67^\circ\ 22'\ 50''.$$

$$\begin{aligned} C &= 2\,(90^\circ - A) \\ &= 45^\circ\ 14'\ 20''. \end{aligned}$$

$$\sin A = \frac{h}{a}.$$

$$h = a \sin A.$$

$$\log h = \log a + \log \sin A.$$

$$\begin{aligned} \log a &= 1.15534 \\ \log \sin A &= \underline{9.96524} \\ \log h &= 1.12058 \end{aligned}$$

$$h = 13.2.$$

10. In an isosceles triangle, given $a = 0.295$, $A = 68^\circ\ 10'$; find c, h, F.

$$\sin A = \frac{h}{a}.$$

$$h = a \sin A.$$

$$\log h = \log a + \log \sin A.$$

$$\begin{aligned} \log a &= 9.46982 - 10 \\ \log \sin A &= \underline{9.96767 - 10} \\ \log h &= 9.43749 - 10 \end{aligned}$$

$$h = 0.27384.$$

$$\cos A = \frac{\frac{1}{2}c}{a}.$$

$$\tfrac{1}{2}c = a \cos A.$$

$$\log \tfrac{1}{2}c = \log a + \log \cos A.$$

$$\begin{aligned} \log a &= 9.46982 - 10 \\ \log \cos A &= \underline{9.57044 - 10} \\ \log \tfrac{1}{2}c &= 9.04026 - 10 \end{aligned}$$

$$\tfrac{1}{2}c = 0.109713.$$

$$c = 0.21943.$$

$$F = \tfrac{1}{2}ch.$$

$$2F = ch.$$

$$\log 2F = \log c + \log h.$$

$$\begin{aligned} \log c &= 9.34130 - 10 \\ \log h &= \underline{9.43749 - 10} \\ \log 2F &= 8.77879 - 10 \end{aligned}$$

$$2F = 0.060089.$$

$$F = 0.03004.$$

11. In an isosceles triangle, given $c = 2.352$, $C = 69° 49'$; find a, h, F.

$$\tfrac{1}{2}C = 34° 54' 30''.$$

$$\sin \tfrac{1}{2}C = \frac{\frac{1}{2}c}{a}.$$

$$a = \frac{\frac{1}{2}c}{\sin \frac{1}{2}C}.$$

$$\log a = \log \tfrac{1}{2}c + \text{colog} \sin \tfrac{1}{2}C.$$

$$\begin{aligned} \log \tfrac{1}{2}c &= 0.07041 \\ \text{colog} \sin \tfrac{1}{2}C &= \underline{0.24240} \\ \log a &= 0.31281 \end{aligned}$$

$$a = 2.0555.$$

$$\cos \tfrac{1}{2}C = \frac{h}{a}.$$

$$h = a \cos \tfrac{1}{2}C.$$

$$\log h = \log a + \log \cos \tfrac{1}{2}C.$$

$$\begin{aligned} \log a &= 0.31281 \\ \log \cos \tfrac{1}{2}C &= \underline{9.91385} \\ \log h &= 0.22666 \end{aligned}$$

$$h = 1.6852.$$

$$F = \tfrac{1}{2}ch.$$

$$2F = ch.$$

$$\log 2F = \log c + \log h.$$

$$\begin{aligned} \log c &= 0.37144 \\ \log h &= \underline{0.22666} \\ \log 2F &= 0.59810 \end{aligned}$$

$$2F = 3.9637.$$

$$F = 1.9819.$$

12. In an isosceles triangle, given $h = 7.4847$, $A = 76° 14'$; find a, c, F.

$$\sin A = \frac{h}{a}.$$

$$a = \frac{h}{\sin A}.$$

$$\log a = \log h + \text{colog} \sin A.$$

$$\begin{aligned} \log h &= 0.87417 \\ \text{colog} \sin A &= \underline{0.01266} \\ \log a &= 0.88683 \end{aligned}$$

$$a = 7.706.$$

$$\tan A = \frac{h}{\frac{1}{2}c}.$$

$$\tfrac{1}{2}c = \frac{h}{\tan A}.$$

$$\log \tfrac{1}{2}c = \log h + \text{colog} \tan A.$$

$$\begin{aligned} \log h &= 0.87417 \\ \text{colog} \tan A &= \underline{9.38918 - 10} \\ \log \tfrac{1}{2}c &= 0.26335 \end{aligned}$$

$$\tfrac{1}{2}c = 1.8338.$$

$$c = 3.6676.$$

$$F = \tfrac{1}{2}ch.$$

$$\log F = \log \tfrac{1}{2}c + \log h.$$

$$\begin{aligned} \log \tfrac{1}{2}c &= 0.26335 \\ \log h &= \underline{0.87417} \\ \log F &= 1.13752 \end{aligned}$$

$$F = 13.725.$$

13. In an isosceles triangle, given $a = 6.71$, $h = 6.60$; find A, C, c.

$$\sin A = \frac{h}{a}.$$

$$\log \sin A = \log h + \text{colog } a.$$

$$\log h = 0.81954$$
$$\text{colog } a = 9.17328 - 10$$
$$\log \sin A = 9.99282 - 10$$

$$A = 79^\circ 36' 30''.$$
$$C = 20^\circ 47'.$$

$$\cos A = \frac{\frac{1}{2}c}{a}.$$

$$\tfrac{1}{2}c = a \cos A.$$

$$\log \tfrac{1}{2}c = \log a + \log \cos A.$$

$$\log a = 0.82672$$
$$\log \cos A = 9.25617 - 10$$
$$\log \tfrac{1}{2}c = 0.08289$$

$$\tfrac{1}{2}c = 1.2103.$$
$$c = 2.4206.$$

14. In an isosceles triangle, given $c = 9$, $h = 20$; find A, c, a.

$$\tan \tfrac{1}{2}C = \frac{\frac{1}{2}c}{h}.$$

$$\log \tan \tfrac{1}{2}C = \log \tfrac{1}{2}c + \text{colog } h.$$

$$\log \tfrac{1}{2}c = 0.65321$$
$$\text{colog } h = 8.69897 - 10$$
$$\log \tan \tfrac{1}{2}C = 9.35218$$

$$\tfrac{1}{2}C = 12^\circ 40' 49''.$$
$$C = 25^\circ 21' 38''.$$
$$2A = 180^\circ - C.$$
$$A = 77^\circ 19' 11''.$$

$$\sin A = \frac{h}{a}.$$

$$a = \frac{h}{\sin A}.$$

$$\log a = \log h + \text{colog } \sin A.$$

$$\log h = 1.30103$$
$$\text{colog } \sin A = 0.01072$$
$$\log a = 1.31175$$

$$a = 20.5.$$

15. In an isosceles triangle, given $c = 147$, $F = 2572.5$; find A, C, a, h.

$$F = \tfrac{1}{2}ch.$$

$$h = \frac{2F}{c}.$$

$$\log h = \log 2F + \text{colog } c.$$

$$\log 2F = 3.71139$$
$$\text{colog } c = 7.83268 - 10$$
$$\log h = 1.54407$$

$$h = 35.$$

$$\tan A = \frac{h}{\frac{1}{2}c}.$$

$$\log \tan A = \log h + \text{colog } \tfrac{1}{2}c.$$

$$\log h = 1.54407$$
$$\text{colog } \tfrac{1}{2}c = 8.13371 - 10$$
$$\log \tan A = 9.67778 - 10$$

$$A = 25^\circ 28'.$$
$$C = 2(90^\circ - A)$$
$$= 129^\circ 4'.$$

$$a = \frac{h}{\sin A}.$$

$$\log a = \log h + \text{colog } \sin A.$$

$$\log h = 1.54407$$
$$\text{colog } \sin A = 0.36655$$
$$\log a = 1.91062$$

$$a = 81.40.$$

16. In an isosceles triangle, given $h = 16.8$, $F = 43.68$; find A, C, a, c.

$$F = \tfrac{1}{2}ch.$$

$$\tfrac{1}{2}c = \frac{F}{h},$$

$$\log \tfrac{1}{2}c = \log F + \text{colog } h.$$

$$\log F = 1.64028$$
$$\text{colog } h = 8.77469 - 10$$
$$\log \tfrac{1}{2}c = 0.41497$$

$$\tfrac{1}{2}c = 2.60.$$
$$c = 5.2.$$

$$\tan A = \frac{h}{\frac{1}{2}c}.$$

$$\log \tan A = \log h + \text{colog } \tfrac{1}{2}c.$$

$$\begin{aligned} \log h &= 1.22531 \\ \text{colog } \tfrac{1}{2}c &= 9.58503 - 10 \\ \log \tan A &= 10.81034 - 10 \end{aligned}$$

$$\begin{aligned} A &= 81° \, 12' \, 9''. \\ \tfrac{1}{2}C &= 8° \, 47' \, 51''. \\ C &= 17° \, 35' \, 42''. \end{aligned}$$

$$\cos A = \frac{\frac{1}{2}c}{a}.$$

$$\log a = \log \tfrac{1}{2}c + \text{colog} \cos A.$$

$$\begin{aligned} \log \tfrac{1}{2}c &= 0.41497 \\ \text{colog} \cos A &= 0.81547 \\ \log a &= 1.23044 \end{aligned}$$

$$a = 17.$$

17. In an isosceles triangle, find the value of F in terms of a and c.

$$\begin{aligned} F &= \tfrac{1}{2} ch. \\ h &= \sqrt{a^2 - \frac{c^2}{4}} \\ &= \sqrt{\frac{4a^2 - c^2}{4}} \\ &= \tfrac{1}{2}\sqrt{4a^2 - c^2}. \\ F &= \tfrac{1}{2}c \left(\tfrac{1}{2}\sqrt{4a^2 - c^2}\right) \\ &= \tfrac{1}{4}c\sqrt{4a^2 - c^2}. \end{aligned}$$

18. In an isosceles triangle, find the value of F in terms of a and C.

$$\begin{aligned} F &= \tfrac{1}{2}ch. \\ \tfrac{1}{2}c &= a \sin \tfrac{1}{2}C. \\ h &= a \cos \tfrac{1}{2}C. \\ F &= a \sin \tfrac{1}{2}C \times a \cos \tfrac{1}{2}C. \\ &= a^2 \sin \tfrac{1}{2}C \cos \tfrac{1}{2}C. \end{aligned}$$

19. In an isosceles triangle, find the value of F in terms of a and A.

$$\begin{aligned} F &= \tfrac{1}{2}ch. \\ \tfrac{1}{2}c &= a \cos A. \\ h &= a \sin A. \\ F &= a \cos A \times a \sin A \\ &= a^2 \sin A \cos A. \end{aligned}$$

20. In an isosceles triangle, find the value of F in terms of h and C.

$$\begin{aligned} F &= \tfrac{1}{2}ch. \\ \tfrac{1}{2}c &= h \tan \tfrac{1}{2}C. \\ F &= h\,(h \tan \tfrac{1}{2}C) \\ &= h^2 \tan \tfrac{1}{2}C. \end{aligned}$$

21. A barn is 40 × 80 feet, the pitch of the roof is 45°; find the length of the rafters and the area of both sides of the roof.

$$40 \div 2 = 20 = \tfrac{1}{2}c.$$

$$\cos A = \frac{\frac{1}{2}c}{a} = \frac{20}{a}.$$

$$20 = a \cos A.$$

$$a = \frac{20}{\cos A}.$$

$$\log a = \log 20 + \text{colog} \cos A.$$

$$\begin{aligned} \log 20 &= 1.30103 \\ \text{colog} \cos A &= 0.15051 \\ \log a &= 1.45154 \end{aligned}$$

$$a = 28.284.$$

$$28.284 \times 80 = 2262.72.$$

$$2262.72 \times 2 = 4525.44.$$

22. In a unit circle, what is the length of the chord corresponding to the angle 45° at the centre?

$$\sin \tfrac{1}{2}C = \frac{\frac{1}{2}c}{a}$$

$$\log \tfrac{1}{2}c = \log a + \log \sin \tfrac{1}{2}C.$$

$$\begin{aligned} \log a &= 0.00000 \\ \log \sin \tfrac{1}{2}C &= 9.58284 - 10 \\ \log \tfrac{1}{2}c &= 9.58284 - 10 \end{aligned}$$

$$\begin{aligned} \tfrac{1}{2}c &= 0.382683. \\ c &= 0.76537. \end{aligned}$$

23. If the radius of a circle = 30, and the length of a chord = 44, find the angle at the centre.

$$\sin \tfrac{1}{2} C = \frac{\tfrac{1}{2} c}{a}.$$

$$\log \sin \tfrac{1}{2} C = \log \tfrac{1}{2} c + \text{colog } a.$$

$$\log \tfrac{1}{2} c = 1.34242$$
$$\text{colog } a = 8.52288 - 10$$
$$\log \sin \tfrac{1}{2} C = 9.86530 - 10$$
$$\tfrac{1}{2} C = 47° 10'.$$
$$C = 94° 20'.$$

24. Find the radius of a circle if a chord whose length is 5 subtends at the centre an angle of 133°.

$$\sin \tfrac{1}{2} C = \frac{\tfrac{1}{2} c}{a}.$$

$$\log a = \log \tfrac{1}{2} c + \text{colog} \sin \tfrac{1}{2} C.$$

$$\log \tfrac{1}{2} c = 0.39794$$
$$\text{colog} \sin \tfrac{1}{2} C = 0.03760$$
$$\log a = 0.43554$$
$$a = 2.7261.$$

25. What is the angle at the centre of a circle if the corresponding chord is equal to $\tfrac{2}{3}$ of the radius?

Let $a = 3$, then $c = 2$, and $\tfrac{1}{2} c = 1$.

$$\sin \tfrac{1}{2} C = \frac{1}{3}.$$

$$\log \sin \tfrac{1}{2} C = \log 1 + \text{colog } 3.$$

$$\log 1 = 0.00000$$
$$\text{colog } 3 = 9.52288 - 10$$
$$\log \sin \tfrac{1}{2} C = 9.52288 - 10$$
$$\tfrac{1}{2} C = 19° 28' 17''.$$
$$C = 38° 56' 33''.$$

26. Find the area of a circular sector if the radius of the circle = 12 and the angle of the sector = 30°.

$$\text{Area } \odot = \pi R^2.$$

$$\text{Area sector} = \frac{30 \pi R^2}{360}.$$

log area sector = log 30 + colog 360 + log π + 2 log R.

$$\log 30 = 1.47712$$
$$\text{colog } 360 = 7.44370 - 10$$
$$\log \pi = 0.49715$$
$$2 \log R = 2.15836$$
$$\log \text{area} = 1.57633$$
$$\text{Area} = 37.699.$$

Exercise XI. Page 34.

1. In a regular polygon, given $n = 10$, $c = 1$; find r, h, F.

$$\tfrac{1}{2} C = \frac{180°}{10} = 18°.$$
$$\tfrac{1}{2} c = 0.5.$$
$$A = 72°.$$
$$h = \tfrac{1}{2} c \tan A.$$

$$\log h = \log \tfrac{1}{2} c + \log \tan A.$$

$$\log \tfrac{1}{2} c = 9.69897 - 10$$
$$\log \tan A = 10.48822 - 10$$
$$\log h = 0.18719$$
$$h = 1.5388.$$

$$\log r = \log \tfrac{1}{2} c + \text{colog} \cos A.$$

$$\log \tfrac{1}{2} c = 9.69897 - 10$$
$$\text{colog} \cos A = 0.51002$$
$$\log r = 0.20899$$

$$r = 1.618.$$
$$F = \tfrac{1}{2} hp.$$

$$\log h = 0.18719$$
$$\log p = 1.00000$$
$$\log 2F = 1.18719$$

$$2F = 15.388.$$
$$F = 7.694.$$

2. In a regular polygon, given $n = 12$, $p = 70$; find r, h, F.

$$\tfrac{1}{2}C = 15^\circ.$$
$$A = 75^\circ.$$
$$c = 70 \div 12 = 5.833.$$
$$\tfrac{1}{2}c = 2.917.$$
$$h = \tfrac{1}{2}c \tan A.$$

$$\begin{aligned} \log \tfrac{1}{2}c &= 0.46494 \\ \log \tan A &= \underline{10.57195} \\ \log h &= 1.03689 \end{aligned}$$

$$h = 10.886.$$

$$r = \tfrac{1}{2}c \cos A.$$

$$\begin{aligned} \log \tfrac{1}{2}c &= 0.46494 \\ \text{colog} \cos A &= \underline{0.58700} \\ \log r &= 1.05194 \end{aligned}$$

$$r = 11.269$$

$$F = \tfrac{1}{2}hp.$$

$$\begin{aligned} \log h &= 1.03689 \\ \log p &= \underline{1.84510} \\ \log 2F &= 2.88199 \end{aligned}$$

$$2F = 762.07.$$
$$F = 381.04.$$

3. In a regular polygon, given $n = 18$, $r = 1$; find h, p, F.

$$\tfrac{1}{2}C = 10^\circ.$$
$$A = 80^\circ.$$
$$h = r \sin A.$$

$$\begin{aligned} \log r &= 0.00000 \\ \log \sin A &= \underline{9.99335 - 10} \\ \log h &= 9.99335 - 10 \end{aligned}$$

$$h = 0.9848.$$

$$\tfrac{1}{2}c = r \cos A.$$

$$\begin{aligned} \log r &= 0.00000 \\ \log \cos A &= \underline{9.23967 - 10} \\ \log \tfrac{1}{2}c &= 9.23967 - 10 \end{aligned}$$

$$\tfrac{1}{2}c = 0.17365.$$
$$p = 6.2514.$$

$$F = \tfrac{1}{2}hp.$$

$$\begin{aligned} \log h &= 9.99335 - 10 \\ \log p &= \underline{0.79598} \\ \log 2F &= 0.78933 \end{aligned}$$

$$2F = 6.1564.$$
$$F = 3.0782.$$

4. In a regular polygon, given $n = 20$, $r = 20$; find h, c, F.

$$\tfrac{1}{2}C = 9^\circ.$$
$$A = 81^\circ.$$
$$h = r \sin A.$$

$$\begin{aligned} \log r &= 1.30103 \\ \log \sin A &= \underline{9.99462 - 10} \\ \log h &= 1.29565 \end{aligned}$$

$$h = 19.754.$$

$$\tfrac{1}{2}c = r \cos A.$$

$$\begin{aligned} \log r &= 1.30103 \\ \log \cos A &= \underline{9.19433 - 10} \\ \log \tfrac{1}{2}c &= 0.49536 \end{aligned}$$

$$\tfrac{1}{2}c = 3.1286.$$
$$c = 6.257.$$
$$p = 125.14.$$
$$F = \tfrac{1}{2}hp.$$

$$\begin{aligned} \log h &= 1.29565 \\ \log p &= \underline{2.09740} \\ \log 2F &= 3.39305 \end{aligned}$$

$$2F = 2472$$
$$F = 1236.$$

5. In a regular polygon, given $n = 8$, $h = 1$; find r, c, F.

$$\tfrac{1}{2}C = 22^\circ\, 30'.$$

$$\tan \tfrac{1}{2}C = \frac{\tfrac{1}{2}c}{h}.$$

$$\log \tfrac{1}{2}c = \log h + \log \tan \tfrac{1}{2}C.$$

$$\begin{aligned} \log h &= 0.00000 \\ \log \tan \tfrac{1}{2}C &= \underline{9.61722 - 10} \\ \log \tfrac{1}{2}c &= 9.61722 - 10 \end{aligned}$$

$$\tfrac{1}{2}c = 0.41421.$$
$$c = 0.82842.$$

$\cos\frac{1}{2}C = \frac{h}{r}.$

$\log r = \log h + \text{colog}\cos\frac{1}{2}C.$

$\log h = 0.00000$
$\text{colog}\cos\frac{1}{2}C = 0.03438$
$\log r = 0.03438$

$r = 1.0824.$

$F = \frac{1}{2}hp$
$= 3.3137.$

6. In a regular polygon, given $n = 11$, $F = 20$; find r, h, c.

$2F = ph.$

$40 = ph.$

$c = \frac{p}{11},$

$h = \frac{40}{p}.$

$\frac{1}{2}C = 16° 22'.$

$\tan\frac{1}{2}C = \frac{\frac{1}{2}c}{h}.$

Substituting values of h and c,

$\tan\frac{1}{2}C = \frac{p}{22} \div \frac{40}{p} = \frac{p^2}{880}.$

$\log p = \frac{1}{2}(\log 880 + \log\tan\frac{1}{2}C).$

$\log 880 = 2.94448$
$\log\tan\frac{1}{2}C = 9.46788 - 10$
$2)2.41236$
$\log p = 1.20618$

$p = 16.076.$
$c = 1.4615.$

$\sin\frac{1}{2}C = \frac{\frac{1}{2}c}{r}.$

$\log r = \log\frac{1}{2}c + \text{colog}\sin\frac{1}{2}C.$

$\log\frac{1}{2}c = 9.86376 - 10$
$\text{colog}\sin\frac{1}{2}C = 0.55008$
$\log r = 0.41384$

$r = 2.592.$

$\cos\frac{1}{2}C = \frac{h}{r}.$

$\log h = \log r + \log\cos\frac{1}{2}C.$

$\log r = 0.41384$
$\log\cos\frac{1}{2}C = 9.98204$
$\log h = 0.39588$

$h = 2.4882.$

7. In a regular polygon, given $n = 7$, $F = 7$; find r, h, p.

$14 = ph.$

$h = \frac{14}{p}.$

$c = \frac{p}{7}.$

$\frac{1}{2}C = 25° 43'.$

$\tan\frac{1}{2}C = \frac{\frac{1}{2}c}{h}.$

$\tan\frac{1}{2}C = \frac{p}{14} \div \frac{14}{p} = \frac{p^2}{196}.$

$\log p = \frac{1}{2}(\log 196 + \log\tan\frac{1}{2}C).$

$\log 196 = 2.29226$
$\log\tan\frac{1}{2}C = 9.68271 - 10$
$2)1.97497$
$\log p = 0.98749$

$p = 9.716$
$\frac{1}{2}c = 0.694.$

$\tan\frac{1}{2}C = \frac{\frac{1}{2}c}{h}.$

$\log h = \log\frac{1}{2}c + \text{colog}\tan\frac{1}{2}C.$

$\log\frac{1}{2}c = 9.84136 - 10$
$\text{colog}\tan\frac{1}{2}C = 0.31729$
$\log h = 0.15865$

$h = 1.441.$

$\sin\frac{1}{2}C = \frac{\frac{1}{2}c}{r}.$

$\log r = \log\frac{1}{2}c + \text{colog}\sin\frac{1}{2}C.$

$\log\frac{1}{2}c = 9.84136 - 10$
$\text{colog}\sin\frac{1}{2}C = 0.36259$
$\log r = 0.20395$

$r = 1.5994.$

8. Find the side of a regular decagon inscribed in a unit circle.

$$\tfrac{1}{2} C = 18°.$$

$$\sin \tfrac{1}{2} C = \frac{\tfrac{1}{2} c}{r}.$$

$$\log c = \log 2 + \log \sin \tfrac{1}{2} C.$$

$$\begin{aligned} \log 2 &= 0.30103 \\ \log \sin \tfrac{1}{2} C &= \underline{9.48998} \\ \log c &= 9.79101 - 10 \\ c &= 0.6181. \end{aligned}$$

9. Find the side of a regular decagon circumscribed about a unit circle.

$$\tfrac{1}{2} C = 18°.$$

$$\tan \tfrac{1}{2} C = \frac{\tfrac{1}{2} c}{h}.$$

$$\log \tfrac{1}{2} c = \log h + \log \tan \tfrac{1}{2} C.$$

$$\begin{aligned} \log h &= 0.00000 \\ \log \tan \tfrac{1}{2} C &= \underline{9.51178} \\ \log \tfrac{1}{2} c &= 9.51178 - 10 \\ \tfrac{1}{2} c &= 0.32492. \\ c &= 0.64984. \end{aligned}$$

10. If the side of an inscribed regular hexagon is equal to 1, find the side of an inscribed regular dodecagon.

Let O be the centre of the circle, BC a side of the hexagon, and BA a side of the dodecagon. Also let OD be $\perp$ to BA.

Then $OB = BC = 1$.

$\angle BOD = 15°$.

In rt. $\triangle ODB$,

$$\sin BOD = \tfrac{1}{2} AB \div OB$$

$$AB = 2\,OB \times \sin BOD.$$

$$\log AB = \log 2\,OB + \log \sin BOD.$$

$$\begin{aligned} \log 2\,OB &= 0.30103 \\ \log \sin 15° &= \underline{9.41300} \\ \log AB &= 9.71403 - 10 \\ AB &= 0.51764. \end{aligned}$$

11. Given n and c, and let b denote the side of the inscribed regular polygon having $2n$ sides; find b in terms of n and c.

Let O be the centre of the circle, BC the side of the polygon having n sides, BA the side of the polygon having $2n$ sides. Then OA is $\perp$ to BC at its middle point D.

$$\angle BOA = \frac{360°}{2n} = \frac{180°}{n}.$$

$$\angle OBC = 90° - \frac{180°}{n}.$$

The $\triangle BOA$ is isosceles.

$$\therefore \angle OBA = \frac{1}{2}\left(180° - \frac{180°}{n}\right)$$

$$= 90° - \frac{90°}{n}.$$

$$\angle ABC = \angle OBA - \angle OBC$$

$$= \left(90° - \frac{90°}{n}\right) - \left(90° - \frac{180°}{n}\right)$$

$$= \frac{90°}{n}.$$

$$\frac{\tfrac{1}{2} c}{b} = \cos \frac{90°}{n}.$$

$$\therefore \tfrac{1}{2} c = b \cos \frac{90°}{n}.$$

Whence,

$$b = \frac{\tfrac{1}{2} c}{\cos \frac{90°}{n}} = \frac{c}{2 \cos \frac{90°}{n}}.$$

12. Compute the difference between the areas of a regular octagon and a regular nonagon if the perimeter of each is 16.

$$\tfrac{1}{2} c = \frac{p}{2n} = \frac{16}{16} = 1.$$

$$A = \frac{180°}{n} = 22° \; 30'.$$

$$\log h = \log \tfrac{1}{2} c + \log \cot A.$$

$$\begin{aligned}
\log \tfrac{1}{2}c &= 0.00000 \\
\log \cot A &= 10.38278 - 10 \\
\log h &= 0.38278
\end{aligned}$$

$$\log F = \log h + \log \tfrac{1}{2}p.$$

$$\begin{aligned}
\log h &= 0.38278 \\
\log \tfrac{1}{2}p &= 0.90309 \\
\log F &= 1.28587
\end{aligned}$$

$$F = 19.3139.$$

$$\tfrac{1}{2}c' = \frac{p}{2n'} = \frac{16}{18} = 0.8889.$$

$$A' = \frac{180^\circ}{n'} = 20^\circ.$$

$$\log h' = \log \tfrac{1}{2}c' + \log \cot A'.$$

$$\begin{aligned}
\log \tfrac{1}{2}c' &= 9.94885 - 10 \\
\log \cot A' &= 10.43893 - 10 \\
\log h' &= 0.38778
\end{aligned}$$

$$\log F' = \log h' + \log \tfrac{1}{2}p.$$

$$\begin{aligned}
\log h' &= 0.38778 \\
\log \tfrac{1}{2}p &= 0.90309 \\
\log F' &= 1.29087
\end{aligned}$$

$$F' = 19.5377.$$

$$\begin{aligned}
F' - F &= 19.5377 - 19.3139 \\
&= 0.2238.
\end{aligned}$$

13. Compute the difference between the perimeters of a regular pentagon and a regular hexagon if the area of each is 12.

$$F = 12, \quad n = 5.$$

$$\tfrac{1}{2}C = \frac{180^\circ}{5} = 36^\circ.$$

$$F = \tfrac{1}{2}hp.$$

$$h = \frac{24}{p}.$$

$$\tfrac{1}{2}c = \frac{p}{2n} = \frac{p}{10}.$$

$$\tan \tfrac{1}{2}C = \frac{\tfrac{1}{2}c}{h} = \frac{\frac{p}{10}}{\frac{24}{p}} = \frac{p^2}{240}.$$

$$p^2 = 240 \tan \tfrac{1}{2}C.$$

$$\begin{aligned}
\log 240 &= 2.38021 \\
\log \tan \tfrac{1}{2}C &= 9.86126 - 10 \\
&\;\; 2)\,2.24147 \\
\log p &= 1.12074
\end{aligned}$$

$$p = 13.205.$$

$$n = 6, \; \tfrac{1}{2}C' = 30^\circ.$$

$$\tan \tfrac{1}{2}C' = \frac{\frac{p'}{12}}{\frac{24}{p'}} = \frac{p'^2}{288}.$$

$$p'^2 = 288 \tan \tfrac{1}{2}C'.$$

$$\begin{aligned}
\log 288 &= 2.45939 \\
\log \tan \tfrac{1}{2}C' &= 9.76144 - 10 \\
&\;\; 2)\,2.22083 \\
\log p' &= 1.11042
\end{aligned}$$

$$p' = 12.895.$$

$$p - p' = 0.310.$$

14. From a square whose side is equal to 1 the corners are cut away so that a regular octagon is left. Find the area of this octagon.

$$h = \tfrac{1}{2}.$$

$$\tfrac{1}{2}C = \frac{1}{2}\left(\frac{360^\circ}{8}\right) = 22^\circ\,30'.$$

$$\begin{aligned}
A &= 90^\circ - 22^\circ\,30' \\
&= 67^\circ\,30'.
\end{aligned}$$

$$\tan A = \frac{h}{\tfrac{1}{2}c}.$$

$$\tfrac{1}{2}c = \frac{h}{\tan A}.$$

$$\log \tfrac{1}{2}c = \log h + \operatorname{colog} \tan A.$$

$$\begin{aligned}
\log h &= 9.69897 - 10 \\
\operatorname{colog} \tan A &= 9.61722 - 10 \\
\log \tfrac{1}{2}c &= 9.31619 - 10
\end{aligned}$$

$$p = \tfrac{1}{2}c \times 2n = nc.$$

$$F = \tfrac{1}{2}ph = \tfrac{1}{2}c \times \tfrac{1}{2}n.$$

$$\log F = \log \tfrac{1}{2}c + \log \tfrac{1}{2}n.$$

$$\log \tfrac{1}{2}c = 9.31619 - 10$$
$$\log \tfrac{1}{2}n = 0.60206$$
$$\log F = 9.91825 - 10$$
$$F = 0.82842.$$

15. Find the area of a regular pentagon if its diagonals are each equal to 12.

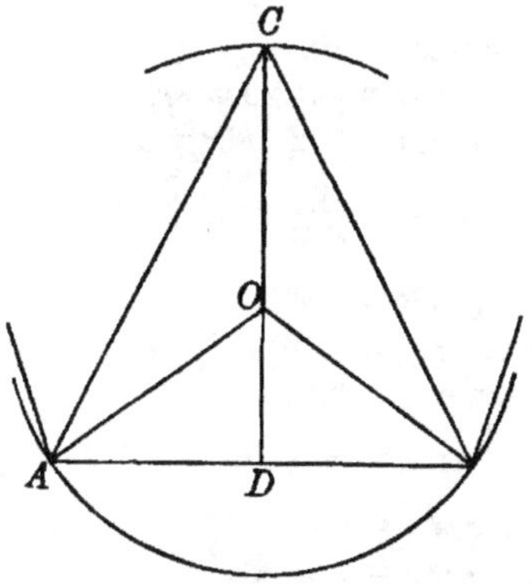

$$\angle AOD = \frac{180^\circ}{n} = 36^\circ.$$
$$\angle AOC = 180^\circ - 36^\circ = 144^\circ.$$
$$\angle ACD = \tfrac{1}{2}(180^\circ - 144^\circ)$$
$$= 18^\circ = \angle CAO.$$
$$\angle OAD = 90^\circ - \angle AOD = 54^\circ.$$
$$\angle DAC = 54^\circ + 18^\circ = 72^\circ.$$
$$\cos DAC = \frac{AD}{AC} = \frac{\tfrac{1}{2}c}{12}.$$
$$\log \tfrac{1}{2}c = \log 12 + \log \cos 72^\circ.$$

$$\log \cos 72^\circ = 9.48998$$
$$\log 12 = 1.07918$$
$$\log \tfrac{1}{2}c = 0.56916$$

$$\tan DAO = \frac{h}{\tfrac{1}{2}c}.$$
$$\log h = \log \tfrac{1}{2}c + \log \tan 54^\circ.$$

$$\log \tfrac{1}{2}c = 0.56916$$
$$\log \tan 54^\circ = 10.13874 - 10$$
$$\log h = 0.70790$$

$$p = \tfrac{1}{2}c \times 2n.$$
$$F = \tfrac{1}{2}ph = \tfrac{1}{2}c \times nh.$$
$$\log F = \log \tfrac{1}{2}c + \log n + \log h.$$
$$\log \tfrac{1}{2}c = 0.56916$$
$$\log n = 0.69897$$
$$\log h = 0.70790$$
$$\log F = 1.97603$$
$$F = 94.63.$$

16. The area of an inscribed regular pentagon is 331.8; find the area of a regular polygon of 11 sides inscribed in the same circle.

Let AB be a side of a regular inscribed pentagon, and AD the side of a regular inscribed polygon of 11 sides.

Let R be the radius of the circle whose centre is O, and h and h' the apothems of the 2 polygons, respectively.

Given F the area of pentagon $= 331.8$. Find F', the area of the 11-sided polygon.

Let p and p' and c and c' represent the perimeters and sides of the pentagon and the 11-sided polygon, respectively.

$$F = \tfrac{1}{2}ph.$$
$$331.8 = \tfrac{1}{2}ph.$$
$$ph = 663.6$$
$$h = \frac{663.6}{p}.$$
$$c = \frac{p}{5}.$$
$$\tfrac{1}{2}c = \frac{p}{10}.$$
$$\angle AOE = 36^\circ.$$
$$\tan 36^\circ = \frac{\tfrac{1}{2}c}{h} = \frac{p}{10} \times \frac{p}{663.6}$$
$$= \frac{p^2}{6636}.$$

$$\log p^2 = \log \tan 36° + \log 6636.$$

$$\begin{aligned} \log 6636 &= 3.82191 \\ \log \tan 36° &= \underline{9.86126 - 10} \\ \log p^2 &= 3.68317 \\ \log p &= 1.84159. \end{aligned}$$

$$\begin{aligned} \text{Since } \tfrac{1}{2}c &= \tfrac{1}{10} \text{ of } p, \\ \log \tfrac{1}{2}c &= 0.84159. \end{aligned}$$

$$\sin \angle AOE = \frac{\frac{1}{2}c}{R}.$$

$$\log R = \log \tfrac{1}{2}c + \text{colog} \sin 36°.$$

$$\begin{aligned} \log \tfrac{1}{2}c &= 0.84159 \\ \text{colog} \sin 36° &= \underline{0.23078} \\ \log R &= 1.07237 \end{aligned}$$

$$\angle AOC = \frac{360°}{22} = 16° \, 21' \, 49''.$$

$$\sin \angle AOC = \tfrac{1}{2}c' \div R.$$

$$\begin{aligned} \log R &= 1.07237 \\ \log \sin AOC &= \underline{9.44985 - 10} \\ \log \tfrac{1}{2}c' &= 0.52222 \end{aligned}$$

$$\tan AOC = \frac{\frac{1}{2}c'}{h'}.$$

$$\log h' = \log \tfrac{1}{2}c' + \text{colog} \tan AOC.$$

$$\begin{aligned} \log \tfrac{1}{2}c' &= 0.52222 \\ \text{colog} \tan AOC &= 0.53220 \\ \log h' &= 1.05442 \\ F &= \tfrac{1}{2}p'h' \\ &= \tfrac{1}{2}c' \times 11 \times h'. \end{aligned}$$

$$\begin{aligned} \log \tfrac{1}{2}c' &= 0.52222 \\ \log 11 &= 1.04139 \\ \log h' &= \underline{1.05442} \\ \log F &= 2.61803 \end{aligned}$$

$$F = 414.99.$$

17. The perimeter of an equilateral triangle is 20; find the area of inscribed circle.

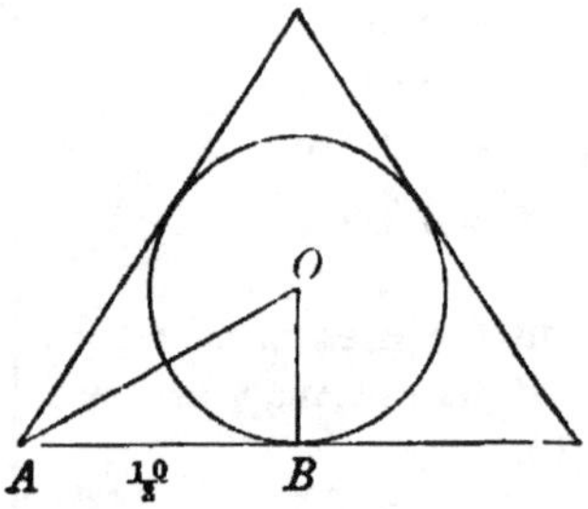

$$\begin{aligned} \text{Perimeter} &= 20. \\ AB &= \tfrac{1}{6} \times 20 = \tfrac{10}{3}. \\ \angle OAB &= \tfrac{1}{6} \text{ of } 180° = 30°. \\ \tan 30° &= \frac{r}{AB}. \end{aligned}$$

$$\begin{aligned} \log AB &= 0.52288 \\ \log \tan 30° &= \underline{9.76144 - 10} \\ \log r &= 0.28432 \end{aligned}$$

$$\text{Area} = \pi r^2.$$

$$\begin{aligned} \log \pi &= 0.49715 \\ \log r^2 &= \underline{0.56864} \\ \log \text{area} &= 1.06579 \end{aligned}$$

$$\text{Area} = 11.636.$$

18. The area of a regular polygon of 16 sides, inscribed in a circle, is 100; find the area of a regular polygon of 15 sides, inscribed in the same circle.

$$\tfrac{1}{2}C = \frac{360°}{32} = 11° \, 15'.$$

$$\tfrac{1}{2}C' = \frac{360°}{30} = 12°.$$

$$\begin{aligned} \text{Let} \quad AC &= h, \\ AB &= r, \\ BC &= \tfrac{1}{2}c. \\ F &= \tfrac{1}{2}hp. \\ 100 &= \tfrac{1}{2}hp. \\ h &= \frac{200}{p}. \end{aligned}$$

$$\tan \tfrac{1}{2} C = \frac{\frac{p}{32}}{\frac{200}{p}}.$$

$$p^2 = 6400 \tan \tfrac{1}{2} C.$$

$$\begin{aligned} \log 6400 &= 3.80618 \\ \log \tan \tfrac{1}{2} C &= \underline{9.29866 - 10} \\ 2)&\,\underline{3.10484} \\ \log p &= 1.55242 \end{aligned}$$

$$p = 35.68.$$

$$\tfrac{1}{2} c = 35.68 \div 32 = 1.115.$$

$$\sin \tfrac{1}{2} C = \frac{\frac{1}{2} c}{r} = \frac{1.115}{r}.$$

$$\begin{aligned} \log 1.115 &= 0.04727 \\ \text{colog} \sin \tfrac{1}{2} C &= \underline{0.70976} \\ \log r &= 0.75703 \end{aligned}$$

$$\frac{h'}{r} = \cos \tfrac{1}{2} C' \, (12^\circ).$$

$$h' = r \times \cos \tfrac{1}{2} C'.$$

$$\begin{aligned} \log r &= 0.75703 \\ \log \cos \tfrac{1}{2} C' &= \underline{9.99040 - 10} \\ \log h' &= 0.74743 \end{aligned}$$

$$\frac{\frac{1}{2} c'}{r} = \sin \tfrac{1}{2} C'.$$

$$\tfrac{1}{2} c' = r \times \sin \tfrac{1}{2} C'.$$

$$\begin{aligned} \log r &= 0.75703 \\ \log \sin \tfrac{1}{2} C' &= \underline{9.31788} \\ \log \tfrac{1}{2} c' &= 0.07491 \end{aligned}$$

$$F = \frac{1}{2}\left(\frac{c'}{2} \times 2\,nh'\right).$$

$$\log F = \log \tfrac{1}{2} c' + \log n + \log h'.$$

$$\begin{aligned} \log \tfrac{1}{2} c' &= 0.07491 \\ \log 15 &= 1.17609 \\ \log h' &= \underline{0.74743} \\ & 1.99843 \end{aligned}$$

$$F = 99.640.$$

19. A regular dodecagon is circumscribed about a circle, the circumference of which is equal to 1; find the perimeter of the dodecagon.

Given circumference of inscribed $\odot = 1$, $n = 12$; find p.

$$2\,\pi r = \text{circumference}.$$

$$r = \frac{\text{circ.}}{2\,\pi}.$$

$$\tfrac{1}{2} C = \frac{360^\circ}{24} = 15^\circ.$$

$$\tan 15^\circ = \frac{\frac{1}{2} c}{r} = \pi c.$$

$$c = \frac{\tan 15^\circ}{3.1416}.$$

$$\begin{aligned} \log \tan 15^\circ &= 9.42805 \\ \text{colog } 3.1416 &= \underline{9.50284 - 10} \\ \log c &= 8.93089 - 10 \\ \log 12 &= \underline{1.07918} \\ \log p &= 0.01007 \\ p &= 1.0235. \end{aligned}$$

20. The area of a regular polygon of 25 sides is equal to 40; find the area of the ring comprised between the circumferences of the inscribed and the circumscribed circles.

$$\tfrac{1}{2} ch = \frac{40}{25} = 1.6.$$

$$\tfrac{1}{2} C = 7^\circ\, 12'.$$

$$A = \frac{360^\circ}{2\,n} = 7^\circ\, 12'.$$

$$\frac{\frac{1}{2} c}{h} = \tan \tfrac{1}{2} C,$$

or

$$\frac{\frac{1}{2} ch}{h^2} = \tan \tfrac{1}{2} C.$$

$$h^2 = \frac{1.6}{\tan \tfrac{1}{2} C}.$$

$$\begin{aligned} \log 1.6 &= 0.20412 \\ \text{colog} \tan \tfrac{1}{2} C &= \underline{0.89850} \\ \log h^2 &= 1.10262 \\ \log h &= 0.55131 \end{aligned}$$

$$\frac{h}{r} = \cos \tfrac{1}{2} C.$$

$$r = \frac{h}{\cos \tfrac{1}{2} C}.$$

$$\begin{aligned} \log h &= 0.55131 \\ \text{colog} \cos \tfrac{1}{2} C &= \underline{0.00344} \\ \log r &= 0.55475 \\ \log r^2 &= 1.10950. \end{aligned}$$

πr^2 = area of circumscribed ⊙.

$$\begin{aligned} \log \pi &= 0.49715 \\ \log r^2 &= \underline{1.10950} \\ \log F &= 1.60665 \\ F &= 40.425 \end{aligned}$$

$$\begin{aligned} \log \pi &= 0.49715 \\ \log h^2 &= \underline{1.10262} \\ \log \pi h^2 &= 1.59977 \end{aligned}$$

Area = 39.790 (inscribed ⊙).

40.425 − 39.790 = 0.635.

Exercise XII. Page 44.

1. Construct the functions of an angle in Quadrant II. What are their signs?

Sines and tangents extending upwards from horizontal diameter are positive; downwards, negative. Cosines and cotangents extending from vertical diameter towards the right are positive; towards the left, negative. Signs of secant and cosecant are made to agree with cosine and sine, respectively. Hence,

sin and csc are +
cos and sec are −
tan and cot are −

2. Construct the functions of an angle in Quadrant III. What are their signs?

sin and csc are −
cos and sec are −
tan and cot are +

3. Construct the functions of an angle in Quadrant IV. What are their signs?

sin and csc are −
cos and sec are +
tan and cot are −

4. What are the signs of the functions of the following angles: 340°, 239°, 145°, 400°, 700°, 1200°, 3800°?

340° is in Quadrant IV.

sin = −	tan = −	sec = +
cos = +	cot = −	csc = −

239° is in Quadrant III.

sin = −	tan = +	sec = −
cos = −	cot = +	csc = −

145° is in Quadrant II.

sin = +	tan = −	sec = −
cos = −	cot = −	csc = +

$$400° = 360° + 40°.$$

Therefore,

400° is in Quadrant I.

sin = +	tan = +	sec = +
cos = +	cot = +	csc = +

$$700° = 360° + 340°.$$

Therefore,

700° is in Quadrant IV.

sin = −	tan = −	sec = +
cos = +	cot = −	csc = −

$$1200° = 3 \times 360° + 120°.$$

Therefore,

1200° is in Quadrant II.

sin = +	tan = −	sec = −
cos = −	cot = −	csc = +

$$3800° = 10 \times 360° + 200°.$$

Therefore,

3800° is in Quadrant III.

sin = −	tan = +	sec = −
cos = −	cot = +	csc = −

5. How many angles less than 360° have the value of the sine equal to $+\frac{5}{7}$, and in what quadrants do they lie?

Since the sine is +, by § 21, the angles can lie in but two quadrants, the first and second.

In the first quadrant, by § 4, the sine increases from 0 to 1, and in the second, decreases from 1 to 0. This is a continually increasing and decreasing quantity.

Therefore there can be but one angle whose sine is equal to $+\frac{5}{7}$ in each quadrant, the first and second.

6. How many values less than 720° can the angle x have if $\cos x = +\frac{2}{3}$, and in what quadrants do they lie?

720° is twice 360°; hence the moving radius will make exactly 2 complete revolutions.

The cosine has the + sign in the first and fourth quadrants, hence it will have four values: two in Quadrant I. and two in Quadrant IV.

7. If we take into account only angles less than 180°, how many values can x have if $\sin x = \frac{5}{7}$? if $\cos x = \frac{1}{5}$? if $\cos x = -\frac{1}{3}$? if $\tan x = \frac{3}{4}$? if $\cot x = -7$?

(i.) Sign being +, the angle can be in Quadrant I. or II.

∴ two values, one in Quadrant I. and one in Quadrant II.

(ii.) Sign being +, the angle is in Quadrant I. or IV.

∴ two values, one in Quadrant I. and one in Quadrant IV.

(iii.) Sign being −, the angle can be in Quadrant II. or III.

∴ two values, one in Quadrant II. and one in Quadrant III.

(iv.) Sign being +, the angle can be in Quadrant I. or III.

∴ two values, one in Quadrant I. and one in Quadrant III.

(v.) Sign being −, the angle can be in Quadrant II. or IV.

∴ two values, one in Quadrant II. and one in Quadrant IV.

8. Within what limits must the angle x lie if $\cos x = -\frac{2}{9}$? if $\cot x = 4$? if $\sec x = 80$? if $\csc x = -3$? (x to be less than 360°.)

If $\cos x = -\frac{2}{9}$, x must lie in the second or third quadrant, or between 90° and 270°.

If $\cot x = 4$, x is between 0° and 90° or 180° and 270°.

If $\sec x = 80$, x is between 0° and 90°, or 270° and 360°.

If $\csc x = -3$, x is between 180° and 360°.

9. In what quadrant does an angle lie if sine and cosine are both negative? if cosine and tangent are both negative? if the cotangent is positive and the sine negative?

(i.) Sine is negative in Quadrants II. and III.; cosine is negative in Quadrants III. and IV.

∴ angles having both sine and cosine negative are in Quadrant III.

(ii.) Cosine is negative in Quadrants II. and III.; tangent is negative in Quadrants II. and IV.

∴ angles having both cosine and tangent negative are in Quadrant II.

(iii.) Cotangent is positive in Quadrants I. and III.; sine is negative in quadrants III. and IV.

∴ angles having cotangent positive and sine negative are in Quadrant III.

10. Between 0° and 3600° how many angles are there whose sines have the absolute value $\frac{3}{5}$? Of these sines how many are positive and how many negative?

Between 0° and 3600° there are 10 revolutions, and in each there are 4 angles whose sines have the absolute value $\frac{3}{5}$. ∴ there are 40 angles. The sine is positive in Quadrants I. and II., and negative in Quadrants III. and IV. ∴ there are 20 angles with the sine positive, and 20 with the sine negative.

11. In finding $\cos x$ by means of the equation $\cos x = \pm\sqrt{1-\sin^2 x}$, when must we choose the positive sign and when the negative sign?

Since cosines only of angles in Quadrants I. or IV. are positive, we use the sign + only when angle x lies within these limits.

Also, since cosines of angles in Quadrants II. and III. are negative, we use the sign −, when x is known to lie in either of these.

12. Given $\cos x = -\sqrt{\frac{1}{2}}$; find the other functions when x is an angle in Quadrant II.

$$\sin^2 x + \cos^2 x = 1.$$

$$\sin x = \sqrt{1-\cos^2 x}$$

$$= \sqrt{1-(-\sqrt{\tfrac{1}{2}})^2} = \sqrt{\tfrac{1}{2}}.$$

$$\csc x = \frac{1}{\sin x} = \frac{1}{\sqrt{\frac{1}{2}}} = \sqrt{2}.$$

$$\sec x = \frac{1}{\cos x} = \frac{1}{-\sqrt{\frac{1}{2}}} = -\sqrt{2}.$$

$$\tan x = \frac{\sin x}{\cos x} = \frac{\sqrt{\frac{1}{2}}}{-\sqrt{\frac{1}{2}}} = -1.$$

$$\cot x = \frac{1}{\tan x} = \frac{1}{-1} = -1.$$

13. Given $\tan x = \sqrt{3}$; find the other functions when x is an angle in Quadrant III.

$$\tan x = \sqrt{3}.$$

$$\cot x = \frac{1}{\sqrt{3}} = \tfrac{1}{3}\sqrt{3}.$$

$$\tan x = \frac{\sin x}{\cos x}.$$

$$\tan x \times \cos x = \sin x.$$

$$\sqrt{3}\cos x = \sin x.$$

$$\begin{aligned} 3\cos^2 x - \sin^2 x &= 0 \\ \cos^2 x + \sin^2 x &= 1 \\ \hline 4\cos^2 x &= 1 \\ \cos^2 x &= \tfrac{1}{4} \\ \cos x &= \pm\tfrac{1}{2}. \end{aligned}$$

The angle being in Quadrant III. the cosine is negative.

$\therefore \cos x = -\frac{1}{2}.$

$$\sin x = \sqrt{1-(-\tfrac{1}{2})^2}$$

$$= \sqrt{\tfrac{3}{4}} = \pm\tfrac{1}{2}\sqrt{3}.$$

Sine is negative.

$\therefore \sin x = -\frac{1}{2}\sqrt{3}.$

$$\sec x = \frac{1}{-\frac{1}{2}} = -2.$$

$$\csc x = \frac{1}{-\frac{1}{2}\sqrt{3}} = -\tfrac{2}{3}\sqrt{3}.$$

14. Given $\sec x = +7$, and $\tan x$ negative; find the other functions of x.

x must be in Quadrant IV.

$\therefore$ sine, cosine, tangent, and cotangent will be negative, and cosine positive.

$$\cos x = \frac{1}{\sec x} = \frac{1}{7}.$$

$$\sin x = \pm\sqrt{1-\frac{1}{49}} = \pm\sqrt{\frac{48}{49}}$$

$$= -\tfrac{4}{7}\sqrt{3}.$$

$$\csc x = \frac{1}{\sin x} = \frac{1}{-\frac{4}{7}\sqrt{3}}$$

$$= -\tfrac{7}{12}\sqrt{3}.$$

$$\tan x = \frac{\sin x}{\cos x} = \frac{-\frac{4}{7}\sqrt{3}}{\frac{1}{7}}$$

$$= -4\sqrt{3}.$$

$$\cot x = \frac{1}{\tan x} = -\frac{1}{4\sqrt{3}}$$

$$= -\tfrac{1}{12}\sqrt{3}.$$

15. Given $\cot x = -3$; find all the possible values of the other functions.

By [3] $\tan x = -\frac{1}{3}$, and may be in Quadrant II. or IV.

By [1],

$$\sin^2 x = 1 - \cos^2 x.$$

$$\sin x = \sqrt{1-\cos^2 x}.$$

By [2],

$$\frac{1}{3} = \frac{\sqrt{1-\cos^2 x}}{\cos x}.$$

$$\frac{1}{9} = \frac{1-\cos^2 x}{\cos^2 x}.$$

$$\cos^2 x = 9 - 9\cos^2 x.$$

$$\cos^2 x = \frac{9}{10}.$$

$$\cos x = \frac{3}{\sqrt{10}} = \tfrac{3}{10}\sqrt{10},$$

and is − in Quadrant II., + in IV.

By [1],

$$\sin x = \sqrt{1-\frac{9}{10}} = \sqrt{\frac{1}{10}}.$$

$$= \tfrac{1}{10}\sqrt{10},$$

and is + in Quadrant II., − in IV.

$$\sec x = \frac{\sqrt{10}}{3} = \tfrac{1}{3}\sqrt{10}.$$

$$\csc x = \sqrt{10}.$$

16. What functions of an angle of a triangle may be negative? In what case are they negative?

When an angle of a triangle is acute, its functions are all positive. When an angle is obtuse, its functions are those of an angle in Quadrant II.

$\therefore$ sine and cosecant are positive, and cosine, tangent, cotangent, and secant are negative.

17. **What functions of an angle of a triangle determine the angle, and what functions fail to do so?**

The sine and cosecant being positive in the first and second quadrant, leave it doubtful whether the angle is obtuse or acute; but the other functions, if positive, determine an angle in the first quadrant, that is to say, an acute angle; if negative, an angle in the second quadrant, an obtuse angle.

18. **Why may cot 360° be considered equal either to $+\infty$ or to $-\infty$?**

The nearer an acute angle is to 0°, the greater the positive value of its cotangent; and the nearer an angle is to 360°, the greater the negative value of its cotangent. When the angle is 0° or 360°, the cotangent is parallel to the horizontal diameter and cannot meet it. But the cotangent of 360° may be regarded as extending either in the positive or in the negative direction; and hence either $+\infty$ or $-\infty$.

19. **Obtain by means of Formulas [1]–[3] the other functions of the angles given:**

(i.) $\tan 90° = \infty$.
(ii.) $\cos 180° = -1$.
(iii.) $\cot 270° = 0$.
(iv.) $\csc 360° = -\infty$.

(i.)

$$\tan 90° = \infty = \frac{1}{0}.$$

$$\cot 90° = \frac{1}{\infty} = 0.$$

$$\frac{\sin 90°}{\cos 90°} = \frac{1}{0}.$$

$$\cos 90° = 0 \sin 90° = 0.$$

$$\cos^2 90° + \sin^2 90° = 1.$$

$$\sin^2 90° = 1.$$

$$\sin 90° = 1.$$

(ii.)

$$\cos 180° = -1.$$

$$\sin^2 180° + \cos^2 180° = 1.$$

$$\sin^2 180° + 1 = 1.$$

$$\sin 180° = 0.$$

$$\tan 180° = \frac{\sin 180°}{\cos 180°} = \frac{0}{-1} = -0.$$

$$\cot 180° = \frac{\cos 180°}{\sin 180°} = \frac{-1}{0} = -\infty.$$

(iii.)

$$\cot 270° = 0.$$

$$\tan 270° = \frac{1}{0} = \infty.$$

$$\frac{\cos 270°}{\sin 270°} = 0.$$

$$\cos 270° = 0 \sin 270° = 0.$$

$$\sin^2 270° + \cos^2 270° = 1.$$

$$\sin^2 270° + 0 = 1.$$

$$\sin^2 270° = 1.$$

$$\sin 270° = -1.$$

(iv.)

$$\csc 360° = -\infty.$$

$$\sin 360° = \frac{1}{-\infty} = -0.$$

$$\sin^2 360° + \cos^2 360° = 1.$$

$$\cos^2 360° = 1.$$

$$\cos 360° = 1.$$

$$\tan 360° = \frac{-0}{1} = -0.$$

$$\cot 360° = \frac{1}{-0} = -\infty.$$

20. Find the values of sin 450°, tan 540°, cos 630°, cot 720°, sin 810°, csc 900°.

$$\sin 450^\circ = \sin(360^\circ + 90^\circ) = \sin 90^\circ = 1.$$

$$\tan 540^\circ = \tan(360^\circ + 180^\circ) = \tan 180^\circ = 0.$$

$$\cos 630^\circ = \cos(360^\circ + 270^\circ) = \cos 270^\circ = 0.$$

$$\cot 720^\circ = \cot(360^\circ + 360^\circ) = \cot 360^\circ = \infty.$$

$$\sin 810^\circ = \sin(2 \times 360^\circ + 90^\circ) = \sin 90^\circ = 1.$$

$$\csc 900^\circ = \csc(2 \times 360^\circ + 180^\circ) = \csc 180^\circ = \infty.$$

21. For what angle in each quadrant are the absolute values of the sine and cosine equal?

The sine and cosine of 45° are equal in absolute value. Corresponding to the angle of 45° in the first quadrant are the angles (90° + 45°), (180° + 45°), (270° + 45°) in the second, third, and fourth quadrants. Hence the sines and cosines of 45°, 135°, 225°, 315°, etc., are all equal in absolute value.

22. Compute the value of

$a \sin 0^\circ + b \cos 90^\circ - c \tan 180^\circ$.

$$\sin 0^\circ = 0.$$
$$\cos 90^\circ = 0.$$
$$\tan 180^\circ = 0.$$

Substituting,

$$a \times 0 + b \times 0 - c \times 0 = 0.$$

23. Compute the value of

$a \cos 90^\circ - b \tan 180^\circ + c \cot 90^\circ$.

$$\cos 90^\circ = 0.$$
$$\tan 180^\circ = 0.$$
$$\cot 90^\circ = 0.$$

Substituting,

$$a \times 0 - b \times 0 + c \times 0 = 0.$$

24. Compute the value of

$a \sin 90^\circ - b \cos 360^\circ + (a - b) \cos 180^\circ$.

$$\sin 90^\circ = 1.$$
$$\cos 360^\circ = 1.$$
$$\cos 180^\circ = -1.$$

Substituting,

$$a \times 1 - b \times 1 + (a - b) \times (-1) = 0.$$

25. Compute the value of

$(a^2 - b^2) \cos 360^\circ - 4ab \sin 270^\circ$.

$$\cos 360^\circ = 1.$$
$$\sin 270^\circ = -1.$$

Substituting,

$$(a^2 - b^2) \times 1 - 4ab \times (-1) = a^2 - b^2 + 4ab.$$

Exercise XIII. Page 49.

2. Express sin 172° in terms of the functions of angles less than 45°.

$$\sin 172° = \sin (180° - 8°)$$
$$= \sin 8°.$$

3. Express cos 100° in terms of the functions of angles less than 45°.

$$\cos 100° = \cos (90° + 10°)$$
$$= - \sin 10°.$$

4. Express tan 125° in terms of the functions of angles less than 45°.

$$\tan 125° = \tan (90° + 35°)$$
$$= - \cot 35°.$$

5. Express cot 91° in terms of the functions of angles less than 45°.

$$\cot 91° = \cot (90° + 1°)$$
$$= - \tan 1°.$$

6. Express sec 110° in terms of the functions of angles less than 45°.

$$\sec 110° = \sec (90° + 20°)$$
$$= - \csc 20°.$$

7. Express csc 157° in terms of the functions of angles less than 45°.

$$\csc 157° = \csc (180° - 23°)$$
$$= \csc 23°.$$

8. Express sin 204° in terms of the functions of angles less than 45°.

$$\sin 204° = \sin (180° + 24°)$$
$$= - \sin 24°.$$

9. Express cos 359° in terms of the functions of angles less than 45°.

$$\cos 359° = \cos (360° - 1°)$$
$$= \cos 1°.$$

10. Express tan 300° in terms of the functions of angles less than 45°.

$$\tan 300° = \tan (270° + 30°)$$
$$= - \cot 30°.$$

11. Express cot 264° in terms of the functions of angles less than 45°.

$$\cot 264° = \cot (270° - 6°)$$
$$= \tan 6°.$$

12. Express sec 244° in terms of the functions of angles less than 45°.

$$\sec 244° = \sec (270° - 26°)$$
$$= - \csc 26°.$$

13. Express csc 271° in terms of the functions of angles less than 45°.

$$\csc 271° = \csc (270° + 1°)$$
$$= - \sec 1°.$$

14. Express sin 163° 49′ in terms of the functions of angles less than 45°.

$$\sin 163° 49' = \sin (180° - 16° 11')$$
$$= \sin 16° 11'.$$

15. Express cos 195° 33′ in terms of the functions of angles less than 45°.

$$\cos 195° 33' = \cos (180° + 15° 33')$$
$$= - \cos 15° 33'.$$

16. Express tan 269° 15′ in terms of the functions of angles less than 45°.

$$\tan 269^\circ 15' = \tan(270^\circ - 45') = \cot 45'.$$

17. Express cot 139° 17′ in terms of the functions of angles less than 45°.

$$\cot 139^\circ 17' = \cot(180^\circ - 40^\circ 43') = -\cot 40^\circ 43'.$$

18. Express sec 299° 45′ in terms of the functions of angles less than 45°.

$$\sec 299^\circ 45' = \sec(270^\circ + 29^\circ 45') = \csc 29^\circ 45'.$$

19. Express csc 92° 25′ in terms of the functions of angles less than 45°.

$$\csc 92^\circ 25' = \csc(90^\circ + 2^\circ 25') = \sec 2^\circ 25'.$$

20. Express all the functions of − 75° in terms of those of positive angles less than 45°.

$$\sin(-75^\circ) = \sin(270^\circ + 15^\circ) = -\cos 15^\circ.$$
$$\cos(-75^\circ) = \cos(270^\circ + 15^\circ) = \sin 15^\circ.$$
$$\tan(-75^\circ) = \tan(270^\circ + 15^\circ) = -\cot 15^\circ.$$
$$\cot(-75^\circ) = \cot(270^\circ + 15^\circ) = -\tan 15^\circ.$$

21. Express all the functions of − 127° in terms of those of positive angles less than 45°.

$$\sin(-127^\circ) = \sin(270^\circ - 37^\circ) = -\cos 37^\circ.$$
$$\cos(-127^\circ) = \cos(270^\circ - 37^\circ) = -\sin 37^\circ.$$
$$\tan(-127^\circ) = \tan(270^\circ - 37^\circ = \cot 37^\circ.$$
$$\cot(-127^\circ) = \cot(270^\circ - 37^\circ) = \tan 37^\circ.$$

22. Express all the functions of − 200° in terms of those of positive angles less than 45°.

$$\sin(-200^\circ) = \sin(180^\circ - 20^\circ) = \sin 20^\circ.$$
$$\cos(-200^\circ) = \cos(180^\circ - 20^\circ) = -\cos 20^\circ.$$
$$\tan(-200^\circ) = \tan(180^\circ - 20^\circ) = -\tan 20^\circ.$$
$$\cot(-200^\circ) = \cot(180^\circ - 20^\circ) = -\cot 20^\circ.$$

23. Express all the functions of − 345° in terms of those of positive angles less than 45°.

$$\sin(-345^\circ) = \sin 15^\circ, \text{ etc.}$$

24. Express all the functions of − 52° 37′ in terms of those of positive angles less than 45°.

$$\sin(-52^\circ 37') = \sin(270^\circ + 37^\circ 23') = -\cos 37^\circ 23'.$$
$$\cos(-52^\circ 37') = \cos(270^\circ + 37^\circ 23') = \sin 37^\circ 23'.$$
$$\tan(-52^\circ 37') = \tan(270^\circ + 37^\circ 23') = -\cot 37^\circ 23'.$$
$$\cot(-52^\circ 37') = \cot(270^\circ + 37^\circ 23') = -\tan 37^\circ 23'.$$

25. Express all the functions of − 196° 54′ in terms of those of positive angles less than 45°.

$$\sin(-196^\circ 54') = \sin(180^\circ - 16^\circ 54') = \sin 16^\circ 54'.$$
$$\cos(-196^\circ 54') = \cos(180^\circ - 16^\circ 54') = -\cos 16^\circ 54'.$$

$\tan(-196^\circ 54') = \tan(180^\circ - 16^\circ 54')$
$= -\tan 16^\circ 54'.$
$\cot(-196^\circ 54') = \cot(180^\circ - 16^\circ 54')$
$= -\cot 16^\circ 54'.$

26. Find the functions of 120°.

$\sin 120^\circ = \sin(90^\circ + 30^\circ) = \cos 30^\circ$
$= \frac{1}{2}\sqrt{3}.$
$\cos 120^\circ = \cos(90^\circ + 30^\circ)$
$= -\sin 30^\circ = -\frac{1}{2}.$
$\tan 120^\circ = \tan(90^\circ + 30^\circ)$
$= -\tan 30^\circ = -\sqrt{3}.$
$\cot 120^\circ = \cot(90^\circ + 30^\circ)$
$= -\cot 30^\circ = -\frac{1}{3}\sqrt{3}.$
$\sec 120^\circ = -2.$
$\csc 120^\circ = \frac{2}{3}\sqrt{3}.$

27. Find the functions of 135°.

$\sin 135^\circ = \sin(90^\circ + 45^\circ)$
$= \cos 45^\circ = \frac{1}{2}\sqrt{2}.$
$\cos 135^\circ = \cos(90^\circ + 45^\circ)$
$= -\sin 45^\circ = -\frac{1}{2}\sqrt{2}.$

$$\tan 135^\circ = \frac{\sin 135^\circ}{\cos 135^\circ} = -1.$$

$$\cot 135^\circ = \frac{\cos 135^\circ}{\sin 135^\circ} = -1.$$

$$\sec 135^\circ = \frac{1}{\cos 135^\circ} = -\sqrt{2}.$$

$$\csc 135^\circ = \frac{1}{\sin 135^\circ} = \sqrt{2}.$$

28. Find the functions of 150°.

$\sin 150^\circ = \sin(180^\circ - 30^\circ)$
$= \sin 30^\circ = \frac{1}{2}.$
$\cos 150^\circ = \cos(180^\circ - 30^\circ)$
$= -\cos 30^\circ = -\frac{1}{2}\sqrt{3}.$
$\tan 150^\circ = \tan(180^\circ - 30^\circ)$

$$= \frac{\sin 30^\circ}{-\cos 30^\circ}$$

$= -\frac{1}{3}\sqrt{3}.$
$\cot 150^\circ = \cot(180^\circ - 30^\circ)$

$$= \frac{-\cos 30^\circ}{\sin 30^\circ}$$

$= -\sqrt{3}.$
$\sec 150^\circ = \sec(180^\circ - 30^\circ)$

$$= \frac{1}{-\cos 30^\circ}$$

$= -\frac{2}{3}\sqrt{3}.$
$\csc 150^\circ = \csc(180^\circ - 30^\circ)$

$$= \frac{1}{\sin 30^\circ}$$

$= 2.$

29. Find the functions of 210°.

$\sin 210^\circ = \sin(180^\circ + 30^\circ)$
$= -\sin 30^\circ = -\frac{1}{2}.$
$\cos 210^\circ = \cos(180^\circ + 30^\circ)$
$= -\cos 30^\circ = -\frac{1}{2}\sqrt{3}.$
$\tan 210^\circ = \tan(180^\circ + 30^\circ)$
$= \tan 30^\circ = \frac{1}{3}\sqrt{3}.$
$\cot 210^\circ = \cot(180^\circ + 30^\circ)$
$= \cot 30^\circ = \sqrt{3}.$

30. Find the functions of 225°.

$\sin 225^\circ = \sin(180^\circ + 45^\circ)$
$= -\sin 45^\circ = -\frac{1}{2}\sqrt{2}.$
$\cos 225^\circ = \cos(180^\circ + 45^\circ)$
$= -\cos 45^\circ = -\frac{1}{2}\sqrt{2}.$
$\tan 225^\circ = \tan(180^\circ + 45^\circ)$
$= \tan 45^\circ = 1.$
$\cot 225^\circ = \cot(180^\circ + 45^\circ)$
$= \cot 45^\circ = 1.$

31. Find the functions of 240°.

$\sin 240^\circ = \sin(270^\circ - 30^\circ)$
$= -\cos 30^\circ = -\frac{1}{2}\sqrt{3}.$
$\cos 240^\circ = \cos(270^\circ - 30^\circ)$
$= -\sin 30^\circ = -\frac{1}{2}.$
$\tan 240^\circ = \tan(270^\circ - 30^\circ)$
$= \cot 30^\circ = \sqrt{3}.$
$\cot 240^\circ = \cot(270^\circ - 30^\circ)$
$= \tan 30^\circ = \frac{1}{3}\sqrt{3}.$

32. Find the functions of 300°.

$$\sin 300° = \sin (270° + 30°)$$
$$= -\cos 30° = -\tfrac{1}{2}\sqrt{3}.$$
$$\cos 300° = \cos (270° + 30°)$$
$$= \sin 30° = \tfrac{1}{2}.$$
$$\tan 300° = \tan (270° + 30°)$$
$$= -\cot 30° = -\sqrt{3}.$$
$$\cot 300° = \cot (270° + 30°)$$
$$= -\tan 30° = -\tfrac{1}{3}\sqrt{3}.$$

33. Find the functions of $-30°$.

$$\sin -30° = -\sin 30° = -\tfrac{1}{2}.$$
$$\cos -30° = \cos 30° = \tfrac{1}{2}\sqrt{3}.$$
$$\tan -30° = -\tan 30° = -\tfrac{1}{3}\sqrt{3}.$$
$$\cot -30° = -\cot 30° = -\sqrt{3}.$$
$$\sec -30° = \sec 30° = \tfrac{2}{3}\sqrt{3}.$$
$$\csc -30° = -\csc 30° = -2.$$

34. Find the functions of $-225°$.

$$-225° = 90° + 45°.$$
$$\sin -225° = \sin (90° + 45°)$$
$$= \cos 45° = \tfrac{1}{2}\sqrt{2}.$$
$$\cos -225° = \cos (90° + 45°)$$
$$= -\sin 45° = -\tfrac{1}{2}\sqrt{2}.$$
$$\tan -225° = \tan (90° + 45°)$$
$$= -\cot 45° = -1.$$
$$\cot -225° = \cot (90° + 45°)$$
$$= -\tan 45° = -1.$$
$$\sec -225° = \frac{1}{\cos (90° + 45°)}$$
$$= -\sqrt{2}.$$
$$\csc -225° = \frac{1}{\sin (90° + 45°)} = \sqrt{2}.$$

35. Given $\sin x = -\sqrt{\tfrac{1}{2}}$, and $\cos x$ negative; find the other functions of x, and the value of x.

Since $\sin 45° = \sqrt{\tfrac{1}{2}}$, and the signs of both the sine and cosine are negative, the angle must be in Quadrant III., and must be, therefore,

$$180° + 45° = 225°.$$

Then $\cos 45° = \sqrt{\tfrac{1}{2}}$.

Hence $\cos (180° + 45°) = -\sqrt{\tfrac{1}{2}}$.

$$\tan (180° + 45°) = \frac{\sin 225°}{\cos 225°}$$
$$= \frac{-\sqrt{\tfrac{1}{2}}}{-\sqrt{\tfrac{1}{2}}} = 1.$$
$$\cot (180° + 45°) = \frac{1}{\tan 225°} = 1.$$
$$\sec 225° = \frac{1}{\cos 225°} = \frac{1}{-\sqrt{\tfrac{1}{2}}}$$
$$= -\sqrt{2}.$$
$$\csc 225° = \frac{1}{\sin 225°} = \frac{1}{-\sqrt{\tfrac{1}{2}}}$$
$$= -\sqrt{2}.$$

36. Given $\cot x = -\sqrt{3}$, and x in Quadrant II.; find the other functions of x, and the value of x.

Since $\cot 30° = \sqrt{3}$, and the sign is negative, the angle is $180° - 30° = 150°$.

$$\tan x = \frac{1}{\cot x} = \frac{1}{-\sqrt{3}} = -\tfrac{1}{3}\sqrt{3}.$$
$$\frac{\sin x}{\cos x} = -\tfrac{1}{3}\sqrt{3}.$$
$$\sin x = -\tfrac{1}{3}\sqrt{3}\cos x.$$
$$\sin^2 x = \tfrac{1}{3}\cos^2 x.$$

But $\sin^2 x + \cos^2 x = 1$.

$\therefore \tfrac{1}{3}\cos^2 x + \cos^2 x = 1;$

and $$\cos^2 x = \frac{3}{4}.$$
$$\therefore \cos x = -\tfrac{1}{2}\sqrt{3};$$

and $$\sin^2 x = \frac{1}{4}.$$
$$\therefore \sin x = \frac{1}{2}.$$
$$\sec x = \frac{1}{\cos x} = -\tfrac{2}{3}\sqrt{3}.$$
$$\csc x = \frac{1}{\sin x} = 2.$$

37. Find the functions of 3540°.

$$3540° = 9 \times 360° + 300°.$$

$$\sin 300° = \sin (360° - 60°) = -\sin 60° = -\tfrac{1}{2}\sqrt{3}.$$

$$\cos 300° = \cos (360° - 60°) = \cos 60° = \frac{1}{2}.$$

$$\tan 300° = \frac{\sin 300°}{\cos 300°} = \frac{-\frac{1}{2}\sqrt{3}}{\frac{1}{2}} = -\sqrt{3}.$$

$$\cot 300° = \frac{1}{\tan 300°} = \frac{1}{-\sqrt{3}} = -\tfrac{1}{3}\sqrt{3}.$$

$$\sec 300° = \frac{1}{\cos 300°} = \frac{1}{\frac{1}{2}} = 2.$$

$$\csc 300° = \frac{1}{\sin 300°} = \frac{1}{-\frac{1}{2}\sqrt{3}} = -\tfrac{2}{3}\sqrt{3}.$$

38. What angles less than 360° have a sine equal to $-\frac{1}{2}$? a tangent equal to $-\sqrt{3}$?

(i.) Since $\sin 30° = \frac{1}{2}$ and the sign is negative, the angle must be in Quadrant III. or IV., and must be therefore $180° + 30° = 210°$, or $360° - 30° = 330°$.

(ii.) Since $\tan 60° = \sqrt{3}$ and the sign is negative, the angle must be in Quadrant II. or IV., and must be therefore $180° - 60° = 120°$, or $360° - 60° = 300°$.

39. Which of the angles mentioned in Examples 27–34 have a cosine equal to $-\sqrt{\frac{1}{2}}$? a cotangent equal to $-\sqrt{3}$?

(i.) Since $\cos 45° = \sqrt{\frac{1}{2}}$ and the sign is negative, the angle must be in Quadrant II. or III., and must be therefore $180° - 45° = 135°$, or $180° + 45° = 225°$. Also, the functions of $-225°$ are the same as the functions of $360° - 225° = 135°$. Hence the angles are 135°, 225°, or $-225°$.

(ii.) Since $\cot 30° = \sqrt{3}$ and the sign is negative, the angle must be in Quadrant II. or IV., and must be therefore $180° - 30° = 150°$, or $360° - 30° = 330°$, or $-30°$. Hence the angles are 150° or $-30°$.

40. What values of x between 0° and 720° will satisfy the equation $\sin x = +\frac{1}{2}$?

Since $\sin 30° = \frac{1}{2}$ and the sign is positive, the angle must be in Quadrant I. or II., and must be therefore 30° or $180° - 30° = 150°$, the first revolution. In the second revolution these angles must be increased by 360°. Hence the angles are 30°, 150°, 390°, and 510°.

41. In each of the following cases find the other angle between 0° and 360° for which the corresponding function (sign included) has the same value: sin 12°, cos 26°, tan 45°, cot 72°; sin 191°, cos 120°, tan 244°, cot 357°.

In order that the sign shall be the same,

sin 12° must be in Quadrant II.
$$= \sin (180° - 12°) = \sin 168°.$$
cos 26° must be in Quadrant IV,
$$= \cos (360° - 26°) = \cos 334°.$$
tan 45° must be in Quadrant III.
$$= \tan (180° + 45°) = \tan 225°.$$
cot 72° must be in Quadrant III.
$$= \cot (180° + 72°) = \cot 252°.$$

sin 191° must be in Quadrant IV.

$= \sin(360° - 11°) = \sin 349°.$

cos 120° must be in Quadrant III.

$= \cos(180° + 60°) = \cos 240°.$

tan 244° must be in Quadrant I.

$= \tan(244° - 180°) = \tan 64°.$

cot 357° must be in Quadrant II.

$= \cot(357° - 180°) = \cot 177°.$

42. Given $\tan 238° = 1.6$; find $\sin 122°$.

$$\tan 238° = (\tan 180° + 58°)$$
$$= \tan 58°.$$
$$\sin 122° = \sin(180° - 58°)$$
$$= \sin 58°.$$

But $\tan 238° = 1.6.$

$\therefore \tan 58° = 1.6.$

$$\tan 58° = \frac{\sin 58°}{\cos 58°}.$$

$$1.6 = \frac{\sin 58°}{\sqrt{1 - \sin^2 58°}}.$$

$$2.56 - 2.56 \sin^2 58° = \sin^2 58°.$$
$$3.56 \sin^2 58° = 2.56.$$
$$\sin 58° = \sqrt{\frac{2.56}{3.56}}$$
$$= 0.848.$$

43. Given $\cos 333° = 0.89$; find $\tan 117°$.

$$\cos 333° = 0.89.$$
$$= \cos(270° + 63°)$$
$$= \sin 63°.$$
$$\therefore \tan 117° = \tan(180° - 63°)$$
$$= -\tan 63°.$$
$$\sin^2 63° + \cos^2 63° = 1.$$
$$(0.89)^2 + \cos^2 63° = 1.$$
$$\cos^2 63° = 0.2079.$$
$$\cos 63° = 0.456.$$
$$-\tan 63° = -\frac{\sin 63°}{\cos 63°}$$
$$= -\frac{0.89}{0.456} = -1.952.$$

44. Simplify the expression

$$a\cos(90° - x) + b\cos(90° + x)$$
$$= a\sin x + b(-\sin x)$$
$$= (a - b)\sin x.$$

45. Simplify the expression

$$m\cos(90° - x)\sin(90° - x).$$
$$\cos(90° - x) = \sin x.$$
$$\sin(90° - x) = \cos x.$$

$\therefore$ the expression $= m \sin x \cos x.$

46. Simplify the expression

$$(a - b)\tan(90° - x) + (a + b)\cot(90° + x).$$
$$\tan(90° - x) = \cot x.$$
$$\cot(90° + x) = -\tan x.$$

$\therefore$ the expression equals

$$(a - b)\cot x - (a + b)\tan x.$$

47. Simplify the expression

$$a^2 + b^2 - 2ab\cos(180° - x)$$
$$= a^2 + b^2 - 2ab(-\cos x)$$
$$= a^2 + b^2 + 2ab\cos x.$$

48. Simplify the expression

$$\sin(90° + x)\sin(180° + x) + \cos(90° + x)\cos(180° - x)$$
$$= (\cos x)(-\sin x) + (-\sin x)(-\cos x)$$
$$= -\sin\cos x + \sin\cos x$$
$$= 0.$$

49. Simplify the expression

$$\cos(180° + x)\cos(270° - y) - \sin(180° + x)\sin(270° - y).$$
$$\cos(180° + x) = -\cos x.$$
$$\cos(270° - y) = -\sin y.$$
$$\sin(180° + x) = -\sin x.$$
$$\sin(270° - y) = -\cos y.$$

Hence the expression

$$= \cos x \sin y - \sin x \cos y.$$

50. Simplify the expression

$\tan x + \tan(-y) - \tan(180° - y)$.

$\tan(-y) = -\tan y$.

$-\tan(180° - y) = \tan y$.

Hence the expression $= \tan x$.

51. For what values of x is the expression $\sin x + \cos x$ positive, and for what values negative? Represent the result by a drawing in which the sectors corresponding to the negative values are shaded.

If x be any angle in Quadrant I., $\sin x + \cos x$ must be positive since both the sine and cosine are positive. In Quadrant II. the sine is positive and cosine negative; hence, so long as the sine is greater than, or equal to, the cosine, the expression $\sin x + \cos x$ is positive; but after passing the middle of Quadrant II., viz., 135°, the cosine of x is greater than sine, and the expression is negative. In Quadrant III. both sine and cosine are negative, and hence their sum must be negative. In Quadrant IV. the sine is negative and cosine positive. The sine and cosine are equal at 315°, after which the cosine is greater than sine. Hence the expression $\sin x + \cos x$ is negative from 135° to 315°, and positive between 0° and 135°, and 315° and 360°.

52. Answer the question of last example for $\sin x - \cos x$.

As x increases from 0° to 45°, the sine increases in value, and cosine decreases, until at 45° sin = cosine. Hence up to this point $\sin x - \cos x$ is negative. For the remainder of Quadrant I. sine is greater than cosine, and consequently the expression $\sin x - \cos x$ is positive. In Quadrant II. sine is positive and cosine negative, so the expression $\sin x - \cos x$ is uniformly positive. In Quadrant III. sine is negative and cosine negative; hence, so long as sine is less than cosine, the expression is positive, viz., to 225°; after that point, sine is greater than cosine, and $\sin x - \cos x$ is negative. In Quadrant IV. sine is negative and cosine positive: therefore $\sin x - \cos x$ is uniformly negative. The expression is, then, negative between 0° and 45°, and 225° and 360°; positive between 45° and 225°.

53. Find the functions of $(x - 90°)$ in terms of the functions of x.

$$x - 90° = 360° - (90° - x)$$
$$= 270° + x.$$

$$\sin(x - 90°) = \sin(270° + x)$$
$$= -\cos x.$$

$$\cos(x - 90°) = \cos(270° + x)$$
$$= \sin x.$$

$$\tan(x - 90°) = \tan(270° + x)$$
$$= -\cot x.$$

$$\cot(x - 90°) = \cot(270° + x)$$
$$= -\tan x.$$

54. Find the functions of $(x - 180°)$ in terms of the functions of x.

$$x - 180° = 360° - (180° - x)$$
$$= 180° + x.$$
$$\sin(x - 180°) = \sin(180° + x)$$
$$= -\sin x.$$
$$\cos(x - 180°) = \cos(180° + x)$$
$$= -\cos x.$$
$$\tan(x - 180°) = \tan(180° + x)$$
$$= \tan x.$$
$$\cot(x - 180°) = \cot(180° + x)$$
$$= \cot x.$$

EXERCISE XIV. PAGE 56.

1. Find the value of $\sin(x + y)$ and $\cos(x + y)$ when $\sin x = \frac{3}{5}$, $\cos x = \frac{4}{5}$, $\sin y = \frac{5}{13}$, $\cos y = \frac{12}{13}$.

$$\sin(x + y) = \sin x \cos y + \cos x \sin y$$
$$= \left(\frac{3}{5} \times \frac{12}{13}\right) + \left(\frac{4}{5} \times \frac{5}{13}\right)$$
$$= \frac{36}{65} + \frac{20}{65} = \frac{56}{65}.$$
$$\cos(x + y) = \cos x \cos y - \sin x \sin y$$
$$= \left(\frac{4}{5} \times \frac{12}{13}\right) - \left(\frac{3}{5} \times \frac{5}{13}\right)$$
$$= \frac{48}{65} - \frac{15}{65} = \frac{33}{65}.$$

2. Find $\sin(90° - y)$ and $\cos(90° - y)$ by making $x = 90°$ in Formulas [8] and [9].

$$(\sin 90° - y)$$
$$= \sin 90° \cos y - \cos 90° \sin y.$$
$$\sin 90° = 1. \quad \cos 90° = 0.$$
$$\therefore \sin(90° - y)$$
$$= (1 \times \cos y) - (0 \times \sin y)$$
$$= \cos y.$$
$$\cos(90° - y)$$
$$= \cos 90° \cos y + \sin 90° \sin y$$
$$= (0 \times \cos y) + (1 \times \sin y)$$
$$= \sin y.$$

3. Find, by Formulas [4]-[11], the first four functions of $90° + y$.

$$\sin(90° + y)$$
$$= \sin 90° \cos y + \cos 90° \sin y$$
$$= (1 \times \cos y) + (0 \times \sin y)$$
$$= \cos y.$$
$$\cos(90° + y)$$
$$= \cos 90° \cos y - \sin 90° \sin y$$
$$= (0 \times \cos y) - (1 \times \sin y)$$
$$= -\sin y.$$
$$(\tan 90° + y)$$
$$= -\frac{\cos y}{\sin y} = -\cot y.$$
$$\cot(90° + y)$$
$$= -\frac{\sin y}{\cos y} = -\tan y.$$

4. Find, by Formulas [4]-[11], the first four functions of $180° - y$.

$$\sin(180° - y)$$
$$= \sin 180° \cos y - \cos 180° \sin y$$
$$= (0 \times \cos y) - (-1 \times \sin y)$$
$$= \sin y.$$
$$\cos(180° - y)$$
$$= \cos 180° \cos y + \sin 180° \sin y$$
$$= (-1 \times \cos y) + (0 \times \sin y)$$
$$= -\cos y.$$

$\tan(180° - y)$

$$= -\frac{\sin y}{\cos y} = -\tan y.$$

$\cot(180° - y)$

$$= -\frac{\cos y}{\sin y} = -\cot y.$$

5. Find, by Formulas [4]–[11], the first four functions of $180° + y$.

$\sin(180° + y)$

$$= \sin 180° \cos y + \cos 180° \sin y$$
$$= (0 \times \cos y) + (-1 \times \sin y)$$
$$= -\sin y.$$

$\cos(180° + y)$

$$= \cos 180° \cos y - \sin 180° \sin y$$
$$= (-1 \times \cos y) - (0 \times \sin y)$$
$$= -\cos y$$

$\tan(180° + y)$

$$= \frac{-\sin y}{-\cos y} = \tan y.$$

$\cot(180° + y)$

$$= \frac{-\cos y}{-\sin y} = \cot y.$$

6. Find, by Formulas [4]–[11], the first four functions of $270° - y$.

$\sin(270° - y)$

$$= \sin 270° \cos y - \cos 270° \sin y$$
$$= (-1 \times \cos y) - (0 \times \sin y)$$
$$= -\cos y.$$

$\cos(270° - y)$

$$= \cos 270° \cos y + \sin 270° \sin y$$
$$= (0 \times \cos y) + (-1 \times \sin y)$$
$$= -\sin y.$$

$\tan(270° - y)$

$$= \frac{-\cos y}{-\sin y} = \cot y.$$

$\cot(270° - y)$

$$= \frac{-\sin y}{-\cos y} = \tan y.$$

7. Find, by Formulas [4]–[11], the first four functions of $270° + y$.

$\sin(270° + y)$

$$= \sin 270° \cos y + \cos 270° \sin y$$
$$= (-1 \times \cos y) + (0 \times \sin y)$$
$$= -\cos y.$$

$\cos(270° + y)$

$$= \cos 270° \cos y - \sin 270° \sin y$$
$$= (0 \times \cos y) - (-1 \times \sin y)$$
$$= \sin y.$$

$\tan(270° + y)$

$$= \frac{-\cos y}{\sin y} = -\cot y.$$

$\cot(270° + y)$

$$= \frac{\sin y}{-\cos y} = -\tan y.$$

8. Find, by Formulas [4]–[11], the first four functions of $360° - y$.

$\sin(360° - y)$

$$= \sin 360° \cos y - \cos 360° \sin y$$
$$= (0 \times \cos y) - (1 \times \sin y)$$
$$= -\sin y.$$

$\cos(360° - y)$

$$= \cos 360° \cos y + \sin 360° \sin y$$
$$= (1 \times \cos y) + (0 \times \sin y)$$
$$= \cos y.$$

$\tan(360° - y)$

$$= \frac{-\sin y}{\cos y} = -\tan y.$$

$\cot(360° - y)$

$$= \frac{\cos y}{-\sin y} = -\cot y.$$

9. Find, by Formulas [4]–[11], the first four functions of $360° + y$.

$\sin(360° + y)$

$$= \sin 360° \cos y + \cos 360° \sin y$$
$$= (0 \times \cos y) + (1 \times \sin y)$$
$$= \sin y.$$

$\cos(360° + y)$
$= \cos 360° \cos y - \sin 360° \sin y$
$= (1 \times \cos y) - (0 \times \sin y)$
$= \cos y.$

$\tan(360° + y)$
$= \frac{\sin y}{\cos y} = \tan y.$

$\cot(360° + y)$
$= \frac{\cos y}{\sin y} = \cot y.$

10. Find, by Formulas [4]–[11], the first four functions of $x - 90°$.

$\sin(x - 90°)$
$= \sin x \cos 90° - \cos x \sin 90°$
$= (0 \times \sin x) - (1 \times \cos x)$
$= -\cos x.$

$\cos(x - 90°)$
$= \cos x \cos 90° + \sin x \sin 90°$
$= (0 \times \cos x) + (1 \times \sin x)$
$= \sin x.$

$\tan(x - 90°)$
$= \frac{-\cos x}{\sin x} = -\cot x.$

$\cot(x - 90°)$
$= \frac{\sin x}{-\cos x} = -\tan x.$

11. Find, by Formulas [4]–[11], the first four functions of $x - 180°$.

$\sin(x - 180°)$
$= \sin x \cos 180° - \cos x \sin 180°$
$= \sin x(-1) - \cos x \times 0$
$= -\sin x.$

$\cos(x - 180°)$
$= \cos x \cos 180° + \sin x \sin 180°$
$= \cos x(-1) + \sin x \times 0$
$= -\cos x.$

$\tan(x - 180°)$
$= \frac{-\sin x}{-\cos x} = \tan x.$

$\cot(x - 180°)$
$= \frac{-\cos x}{-\sin x} = \cot x.$

12. Find, by Formulas [4]–[11], the first four functions of $x - 270°$.

$\sin(x - 270°)$
$= \sin x \cos 270° - \cos x \sin 270°$
$= \sin x \times 0 - \cos x \times (-1)$
$= \cos x.$

$\cos(x - 270°)$
$= \cos x \cos 270° + \sin x \sin 270°$
$= \cos x \times 0 + \sin x(-1)$
$= -\sin x.$

$\tan(x - 270°)$
$= \frac{\cos x}{-\sin x} = -\cot x.$

$\cot(x - 270°)$
$= \frac{-\sin x}{\cos x} = -\tan x.$

13. Find, by Formulas [4]–[11], the first four functions of $-y$.

$\sin(0° - y)$
$= \sin 0° \cos y - 0° \sin y$
$= (0 \times \cos y) - (1 \times \sin y)$
$= -\sin y.$

$\cos(0° - y)$
$= \cos 0° \cos y + \sin 0° \sin y$
$= (1 \times \cos y) + (0 + \sin y)$
$= \cos y.$

$\tan(0° - y)$
$= \frac{-\sin y}{\cos y} = -\tan y.$

$\cot(0° - y)$
$= \frac{\cos y}{-\sin y} = -\cot y.$

14. Find, by Formulas [4]–[11], the first four functions of $45° - y$.

$\sin(45° - y)$
$= \sin 45° \cos y - \cos 45° \sin y$
$= \frac{1}{2}\sqrt{2} \cos y - \frac{1}{2}\sqrt{2} \sin y$
$= \frac{1}{2}\sqrt{2}(\cos y - \sin y).$

$\cos(45^\circ - y)$
$= \cos 45^\circ \cos y + \sin 45^\circ \sin y$
$= \frac{1}{2}\sqrt{2}\cos y + \frac{1}{2}\sqrt{2}\sin y$
$= \frac{1}{2}\sqrt{2}(\cos y + \sin y)$.

$\tan(45^\circ - y)$
$= \frac{\cos y - \sin y}{\cos y + \sin y} = \frac{1 - \tan y}{1 + \tan y}$.

$\cot(45^\circ - y)$
$= \frac{\cos y + \sin y}{\cos y - \sin y} = \frac{\cot y + 1}{\cot y - 1}$.

15. Find, by Formulas [4]–[11], the first four functions of $45^\circ + y$.

$\sin(45^\circ + y)$
$= \sin 45^\circ \cos y + \cos 45^\circ \sin y$
$= \frac{1}{2}\sqrt{2}\cos y + \frac{1}{2}\sqrt{2}\sin y$
$= \frac{1}{2}\sqrt{2}(\cos y + \sin y)$.

$\cos(45^\circ + y)$
$= \cos 45^\circ \cos y - \sin 45^\circ \sin y$
$= \frac{1}{2}\sqrt{2}\cos y - \frac{1}{2}\sqrt{2}\sin y$
$= \frac{1}{2}\sqrt{2}(\cos y - \sin y)$.

$\tan(45^\circ + y)$
$= \frac{\cos y + \sin y}{\cos y - \sin y} = \frac{1 + \tan y}{1 - \tan y}$.

$\cot(45^\circ + y)$
$= \frac{\cos y - \sin y}{\cos y + \sin y} = \frac{\cot y - 1}{\cot y + 1}$.

16. Find, by Formulas [4]–[11], the first four functions of $30^\circ + y$.

$\sin(30^\circ + y)$
$= \sin 30^\circ \cos y + \cos 30^\circ \sin y$
$= \frac{1}{2}(\cos y + \sqrt{3}\sin y)$.

$\cos(30^\circ + y)$
$= \cos 30^\circ \cos y - \sin 30^\circ \sin y$
$= \frac{1}{2}(\sqrt{3}\cos y - \sin y)$.

$\tan(30^\circ + y)$
$= \frac{\cos y + \sqrt{3}\sin y}{\sqrt{3}\cos y - \sin y}$;

divide each term by $\sqrt{3}\cos y$,

$= \frac{\frac{1}{3}\sqrt{3} + \tan y}{1 - \frac{1}{3}\sqrt{3}\tan y}$.

$\cot(30^\circ + y)$
$= \frac{\sqrt{3}\cos y - \sin y}{\cos y + \sqrt{3}\sin y}$;

divide each term by $\sin y$,

$= \frac{\sqrt{3}\cot y - 1}{\cot y + \sqrt{3}}$.

17. Find, by Formulas [4]–[11], the first four functions of $60^\circ - y$.

$\sin(60^\circ - y)$
$= \sin 60^\circ \cos y - \cos 60^\circ \sin y$
$= \frac{1}{2}(\sqrt{3}\cos y - \sin y)$.

$\cos(60^\circ - y)$
$= \cos 60^\circ \cos y + \sin 60^\circ \sin y$
$= \frac{1}{2}(\cos y + \sqrt{3}\sin y)$.

$\tan(60^\circ - y)$
$= \frac{\sqrt{3}\cos y - \sin y}{\cos y + \sqrt{3}\sin y}$
$= \frac{\sqrt{3} - \tan y}{1 + \sqrt{3}\tan y}$.

$\cot(60^\circ - y)$
$= \frac{\cos y + \sqrt{3}\sin y}{\sqrt{3}\cos y - \sin y}$
$= \frac{\frac{1}{3}\sqrt{3}\cot y + 1}{\cot y - \frac{1}{3}\sqrt{3}}$.

18. Find $\sin 3x$ in terms of $\sin x$.

$\sin 3x = \sin(2x + x)$
$= \sin 2x \cos x + \cos 2x \sin x$.

$\sin 2x = 2\sin x \cos x$.

$\cos 2x = \cos^2 x - \sin^2 x$.

Substituting,

$\sin 3x = 2\sin x \cos^2 x + \sin x \cos^2 x - \sin^3 x$
$= 3\sin x \cos^2 x - \sin^3 x$.

But $\cos^2 x = 1 - \sin^2 x$.

Substituting,

$\sin 3x = 3\sin x - 3\sin^3 x - \sin^3 x$
$= 3\sin x - 4\sin^3 x$.

19. Find $\cos 3x$ in terms of $\cos x$.

$$\cos 3x = \cos(2x + x)$$
$$= \cos 2x \cos x - \sin 2x \sin x.$$
$$\sin 2x = 2 \sin x \cos x.$$
$$\cos 2x = \cos^2 x - \sin^2 x.$$

Substituting,

$$\cos 3x = \cos^3 x - \sin^2 x \cos x - 2 \sin^2 x \cos x$$
$$= \cos^3 x - 3 \sin^2 x \cos x.$$

But $\sin^2 x = 1 - \cos^2 x$.

Substituting,

$$\cos 3x = \cos^3 x - 3\cos x + 3\cos^3 x$$
$$= 4\cos^3 x - 3\cos x.$$

20. Given $\tan \frac{1}{2}x = 1$; find $\cos x$.

$$\tan \tfrac{1}{2}x = \sqrt{\frac{1 - \cos x}{1 + \cos x}}.$$
$$1 = \sqrt{\frac{1 - \cos x}{1 + \cos x}}.$$
$$1 = \frac{1 - \cos x}{1 + \cos x}.$$
$$1 + \cos x = 1 - \cos x.$$
$$2\cos x = 0.$$
$$\cos x = 0.$$

21. Given $\cot \frac{1}{2}x = \sqrt{3}$; find $\sin x$.

$$\cot \tfrac{1}{2}x = \sqrt{\frac{1 + \cos x}{1 - \cos x}}.$$
$$\sqrt{3} = \sqrt{\frac{1 + \cos x}{1 - \cos x}}.$$
$$3 = \frac{1 + \cos x}{1 - \cos x}.$$
$$3 - 3\cos x = 1 + \cos x.$$
$$-4\cos x = -2.$$
$$\cos x = \frac{1}{2}.$$
$$\sin^2 x = 1 - \cos^2 x$$
$$= 1 - \frac{1}{4} = \frac{3}{4}.$$
$$\sin x = \sqrt{\frac{3}{4}} = \tfrac{1}{2}\sqrt{3}.$$

22. Given $\sin x = 0.2$; find $\sin \frac{1}{2}x$ and $\cos \frac{1}{2}x$.

$$\sin x = 0.2.$$
$$\cos^2 x = 1 - \sin^2 x$$
$$= 1 - 0.04.$$
$$\cos x = \sqrt{0.96}.$$
$$\sin \tfrac{1}{2}x = \sqrt{\frac{1 - \cos x}{2}}$$
$$= \sqrt{\frac{1 - \sqrt{0.96}}{2}}$$
$$= \sqrt{\frac{1 - 0.4\sqrt{6}}{2}}$$
$$= 0.10051.$$
$$\cos \tfrac{1}{2}x = \sqrt{\frac{1 + \cos x}{2}}$$
$$= \sqrt{\frac{1 + 0.4\sqrt{6}}{2}}$$
$$= 0.99494.$$

23. Given $\cos x = 0.5$; find $\cos 2x$ and $\tan 2x$.

$$\cos 2x = \cos^2 x - \sin^2 x.$$
$$\sin x = \sqrt{1 - \left(\frac{1}{2}\right)^2} = \tfrac{1}{2}\sqrt{3}.$$
$$\therefore \cos 2x = 0.25 - 0.75.$$
$$= -0.50 = -\frac{1}{2}.$$
$$\tan x = \frac{\sin x}{\cos x} = \frac{\frac{1}{2}\sqrt{3}}{\frac{1}{2}} = \sqrt{3}.$$
$$\tan 2x = \frac{2\tan x}{1 - \tan^2 x} = \frac{2\sqrt{3}}{1-3}$$
$$= -\sqrt{3}.$$

24. Given $\tan 45^\circ = 1$; find the functions of $22^\circ\ 30'$.

Let $x = 45^\circ$.

$$\tan x = \frac{\sin x}{\cos x} = 1.$$

$$\therefore \sin x = \cos x.$$

$$\sin^2 x + \cos^2 x = 1.$$

$$2\sin^2 x = 1.$$

$$\sin^2 x = \tfrac{1}{2}.$$

$$\sin x = \tfrac{1}{2}\sqrt{2} = \cos x.$$

$\sin \frac{1}{2}x$ or $\sin 22^\circ\ 30'$

$$= \sqrt{\frac{1 - \frac{1}{2}\sqrt{2}}{2}}$$

$$= \tfrac{1}{2}\sqrt{2 - \sqrt{2}}$$

$$= 0.3827.$$

$\cos \frac{1}{2}x$ or $\cos 22^\circ\ 30'$

$$= \sqrt{\frac{1 + \frac{1}{2}\sqrt{2}}{2}}$$

$$= \tfrac{1}{2}\sqrt{2 + \sqrt{2}}$$

$$= 0.9239.$$

$$\tan \tfrac{1}{2}x = \frac{\sin \frac{1}{2}x}{\cos \frac{1}{2}x}$$

$$= \sqrt{\frac{2 - \sqrt{2}}{2 + \sqrt{2}}};$$

multiply by $\dfrac{2 - \sqrt{2}}{2 - \sqrt{2}}$,

$$= \sqrt{\frac{(2 - \sqrt{2})^2}{4 - 2}}$$

$$= \tfrac{1}{2}\sqrt{(2 - \sqrt{2})^2} \times \sqrt{2}$$

$$= (1 - \tfrac{1}{2}\sqrt{2}) \times \sqrt{2}$$

$$= \sqrt{2} - 1 = 0.4142.$$

$$\cot \tfrac{1}{2}x = \frac{\cos \frac{1}{2}x}{\sin \frac{1}{2}x}$$

$$= \frac{\frac{1}{2}\sqrt{2 + \sqrt{2}}}{\frac{1}{2}\sqrt{2 - \sqrt{2}}}$$

$$= \sqrt{\frac{2 + \sqrt{2}}{2 - \sqrt{2}}}$$

$$= \sqrt{2} + 1 = 2.4142.$$

25. Given $\sin 30^\circ = 0.5$; find the functions of 15°.

$$\sin 30^\circ = 0.5 = \frac{1}{2}.$$

$$\therefore \cos 30^\circ = \sqrt{1 - \frac{1}{4}} = \sqrt{\frac{3}{4}}$$

$$= \tfrac{1}{2}\sqrt{3}.$$

$$\sin \tfrac{1}{2}x = \sqrt{\frac{1 - \cos x}{2}}.$$

$$\therefore \sin 15^\circ = \sqrt{\frac{1 - \frac{1}{2}\sqrt{3}}{2}}$$

$$= \tfrac{1}{2}\sqrt{2 - \sqrt{3}} = 0.2588.$$

$$\cos 15^\circ = \sqrt{\frac{1 + \frac{1}{2}\sqrt{3}}{2}}$$

$$= \tfrac{1}{2}\sqrt{2 + \sqrt{3}} = 0.96592.$$

$$\tan 15^\circ = \sqrt{\frac{1 - \frac{1}{2}\sqrt{3}}{1 + \frac{1}{2}\sqrt{3}}}$$

$$= \sqrt{\frac{2 - \sqrt{3}}{2 + \sqrt{3}}}$$

$$= \sqrt{\frac{2 - \sqrt{3}}{2 + \sqrt{3}} \times \frac{2 - \sqrt{3}}{2 - \sqrt{3}}}$$

$$= \sqrt{\frac{(2 - \sqrt{3})^2}{4 - 3}}$$

$$= 2 - \sqrt{3} = 0.2679.$$

$$\cot 15^\circ = \sqrt{\frac{1 + \frac{1}{2}\sqrt{3}}{1 - \frac{1}{2}\sqrt{3}}}$$

$$= 2 + \sqrt{3} = 3.7321.$$

26. Prove that

$$\tan 18^\circ = \frac{\sin 33^\circ + \sin 3^\circ}{\cos 33^\circ + \cos 3^\circ}.$$

Let $x = 18^\circ$,

$y = 15^\circ$.

Then

(1) $2 \sin x \cos y$

$$= \sin(x+y) + \sin(x-y).$$

(2) $2 \cos x \cos y$

$$= \cos(x+y) + \cos(x-y).$$

Divide (1) by (2),

$$\tan x = \frac{\sin(x+y)+\sin(x-y)}{\cos(x+y)+\cos(x-y)}.$$

Substitute values of x and y,

$$\tan 18^\circ = \frac{\sin 33^\circ + \sin 3^\circ}{\cos 33^\circ + \cos 3^\circ}.$$

27. Prove the formula

$$\sin 2x = \frac{2 \tan x}{1 + \tan^2 x}.$$

$$\sin 2x = 2 \sin x \cos x.$$

$$2 \tan x = \frac{2 \sin x}{\cos x}.$$

$$1 + \tan^2 x = 1 + \frac{\sin^2 x}{\cos^2 x}$$

$$= \frac{\cos^2 x + \sin^2 x}{\cos^2 x}.$$

But $\cos^2 x + \sin^2 x = 1$.

$$\therefore 1 + \tan^2 x = \frac{1}{\cos^2 x}.$$

$$2 \sin x \cos x = \frac{2 \sin x}{\cos x} \times \frac{\cos^2 x}{1}.$$

$$\therefore 2 \sin x \cos x = 2 \sin x \cos x.$$

28. Prove the formula

$$\cos 2x = \frac{1 - \tan^2 x}{1 + \tan^2 x}.$$

$$\cos 2x = \frac{1 - \frac{\sin^2 x}{\cos^2 x}}{1 + \frac{\sin^2 x}{\cos^2 x}}.$$

$$\cos 2x = \cos^2 x - \sin^2 x.$$

$$\cos^2 x - \sin^2 x = \frac{1 - \frac{\sin^2 x}{\cos^2 x}}{1 + \frac{\sin^2 x}{\cos^2 x}}.$$

$$= \frac{\cos^2 x - \sin^2 x}{\cos^2 x + \sin^2 x}$$

$$= \frac{\cos^2 x - \sin^2 x}{1}$$

$$= \cos^2 x - \sin^2 x.$$

29. Prove the formula

$$\tan \tfrac{1}{2} x = \frac{\sin x}{1 + \cos x}.$$

$$\tan \tfrac{1}{2} x = \frac{\sqrt{1 - \cos x}}{\sqrt{1 + \cos x}}.$$

$$\frac{\sin x}{1 + \cos x} = \frac{\sqrt{1 - \cos^2 x}}{1 + \cos x}.$$

$$\therefore \frac{\sqrt{1 - \cos x}}{\sqrt{1 + \cos x}} = \frac{\sqrt{1 - \cos^2 x}}{1 + \cos x}.$$

$$\frac{1 - \cos x}{1 + \cos x} = \frac{1 - \cos^2 x}{(1 + \cos x)^2}$$

$$= \frac{1 - \cos x}{1 + \cos x}.$$

30. Prove the formula

$$\cot \tfrac{1}{2} x = \frac{\sin x}{1 - \cos x}.$$

$$\sin x = \sqrt{1 - \cos^2 x}.$$

$$\cot \tfrac{1}{2} x = \sqrt{\frac{1 + \cos x}{1 - \cos x}}.$$

By substituting,

$$\sqrt{\frac{1 + \cos x}{1 - \cos x}} = \frac{\sqrt{1 - \cos^2 x}}{1 - \cos x}.$$

$$\frac{1 + \cos x}{1 - \cos x} = \frac{1 - \cos^2 x}{(1 - \cos x)^2}$$

$$= \frac{1 + \cos x}{1 - \cos x}.$$

31. Prove the formula

$$\sin \tfrac{1}{2}x \pm \cos \tfrac{1}{2}x = \sqrt{1 \pm \sin x}.$$

By squaring,

$$\sin^2 \tfrac{1}{2}x \pm 2 \sin \tfrac{1}{2}x \cos \tfrac{1}{2}x + \cos^2 \tfrac{1}{2}x = 1 \pm \sin x.$$

But $\sin \tfrac{1}{2}x = \sqrt{\dfrac{1 - \cos x}{2}}$,

and $\cos \tfrac{1}{2}x = \sqrt{\dfrac{1 + \cos x}{2}}$.

Substitute values of $\sin \tfrac{1}{2}x$ and $\cos \tfrac{1}{2}x$,

$$\frac{1 - \cos x}{2} \pm 2 \sin \tfrac{1}{2}x \cos \tfrac{1}{2}x + \frac{1 + \cos x}{2} = 1 \pm \sin x.$$

$$1 \pm 2 \sin \tfrac{1}{2}x \cos \tfrac{1}{2}x = 1 \pm \sin x.$$

$$\pm 2 \sin \tfrac{1}{2}x \cos \tfrac{1}{2}x = \pm \sin x.$$

$$2 \sin \tfrac{1}{2}x \cos \tfrac{1}{2}x = \pm 2\sqrt{\frac{1 - \cos^2 x}{4}}.$$

$$\pm \sin x = \pm \sqrt{\frac{1 - \cos 2x}{2}}.$$

$$\therefore \pm 2\sqrt{\frac{1 - \cos^2 x}{4}} = \pm \sqrt{\frac{1 - \cos 2x}{2}}.$$

$$1 - \cos^2 x = \frac{1 - \cos 2x}{2}.$$

$$\sqrt{1 - \cos^2 x} = \sqrt{\frac{1 - \cos 2x}{2}}.$$

$$\therefore \sin x = \sin x.$$

32. Prove the formula

$$\frac{\tan x \pm \tan y}{\cot x \pm \cot y} = \pm \tan x \tan y.$$

$$\tan x \pm \tan y = \pm \tan x \cot x \tan y + \cot x \tan y \tan x.$$

But $\tan x \cot x = 1$,

and $\tan y \cot y = 1$.

$$\therefore \tan x \pm \tan y = \tan x \pm \tan y.$$

33. Prove the formula

$$\tan (45° - x) = \frac{1 - \tan x}{1 + \tan x}.$$

$$\tan (45° - x) = \frac{\sin (45° - x)}{\cos (45° - x)}.$$

$$\sin (45° - x) = \sin 45° \cos x - \cos 45° \sin x = \tfrac{1}{2}\sqrt{2} \cos x - \tfrac{1}{2}\sqrt{2} \sin x = \tfrac{1}{2}\sqrt{2} (\cos x - \sin x).$$

$$\cos (45° - x) = \cos 45° \cos x + \sin 45° \sin x = \tfrac{1}{2}\sqrt{2} \cos x + \tfrac{1}{2}\sqrt{2} \sin x. = \tfrac{1}{2}\sqrt{2} (\cos x + \sin x).$$

$$\tan (45° - x) = \frac{\cos x - \sin x}{\cos x + \sin x}.$$

Dividing numerator and denominator by $\cos x$,

$$\tan (45° - x) = \frac{1 - \tan x}{1 + \tan x}.$$

34. If A, B, C are the angles of a triangle, prove that

$$\sin A + \sin B + \sin C = 4 \cos \tfrac{1}{2} A \cos \tfrac{1}{2} B \cos \tfrac{1}{2} C.$$

$$\sin A + \sin B + \sin C = \sin A + \sin B + \sin[180° - (A + B)] = \sin A + \sin B + \sin (A + B).$$

By [20] and [12],

$$= 2 \sin \tfrac{1}{2}(A + B) \cos \tfrac{1}{2}(A - B) + 2 \sin \tfrac{1}{2}(A + B) \cos \tfrac{1}{2}(A + B)$$
$$= 2 \sin \tfrac{1}{2}(A + B) [\cos \tfrac{1}{2}(A - B) + \cos \tfrac{1}{2}(A + B)].$$

By [22],

$$= 2 \sin \tfrac{1}{2}(A + B)\, 2 \cos \tfrac{1}{2} A \cos \tfrac{1}{2} B.$$

$$\therefore = 4 \sin \tfrac{1}{2}(A + B) \cos \tfrac{1}{2} A \cos \tfrac{1}{2} B.$$

But $\cos \tfrac{1}{2} C = \cos [90° - \tfrac{1}{2}(A + B)] = \sin \tfrac{1}{2}(A + B)$.

$$\therefore \sin A + \sin B + \sin C = 4 \cos \tfrac{1}{2} A \cos \tfrac{1}{2} B \cos \tfrac{1}{2} C.$$

35. If A, B, C are the angles of a triangle, prove that

$\cos A + \cos B + \cos C$
$= 1 + 4 \sin \frac{1}{2} A \sin \frac{1}{2} B \sin \frac{1}{2} C.$

$\cos C = \cos [180° - (A + B)]$
$= - \cos (A + B).$

$\therefore \cos A + \cos B + \cos C$
$= \cos A + \cos B - \cos (A + B).$

By [22],
$= 2 \cos \frac{1}{2} (A + B) \cos \frac{1}{2} (A - B)$
$- \cos (A + B).$

By [17],
$= 2 \cos \frac{1}{2} (A + B) \cos \frac{1}{2} (A - B)$
$- 2 \cos^2 \frac{1}{2} (A + B) + 1$
$= [2 \cos \frac{1}{2} (A + B)]$
$\times [\cos \frac{1}{2} (A - B) - \cos \frac{1}{2} (A + B)] + 1.$

By [23],
$= [2 \cos \frac{1}{2} (A + B)]$
$\times [2 \sin \frac{1}{2} A \sin \frac{1}{2} B] + 1$
$= (2 \sin \frac{1}{2} C) (2 \sin \frac{1}{2} A \sin \frac{1}{2} B) + 1$
$= 1 + 4 \sin \frac{1}{2} A \sin \frac{1}{2} B \sin \frac{1}{2} C.$

36. If A, B, C are the angles of a triangle, prove that

$\tan A + \tan B + \tan C$
$= \tan A \times \tan B \times \tan C.$

Since $A + B + C = 180°$,
$C = 180° - (A + B).$

$\therefore \tan C = \tan [180° - (A + B)]$
$= - \tan (A + B).$

Again,
$\tan A + \tan B$
$= \tan (A + B) (1 - \tan A \tan B)$
$= \tan (A + B)$
$- \tan (A + B) \tan A \tan B.$

$\therefore \tan A + \tan B + \tan C$
$= \tan (A + B) - \tan (A + B)$
$- \tan (A + B) \tan A \tan B$
$= - \tan (A + B) \tan A \tan B$
$= \tan A \tan B \tan C.$

37. If A, B, C are the angles of a triangle, prove that

$\cot \frac{1}{2} A + \cot \frac{1}{2} B + \cot \frac{1}{2} C$
$= \cot \frac{1}{2} A \times \cot \frac{1}{2} B \times \cot \frac{1}{2} C.$

Since $\frac{1}{2} A + \frac{1}{2} B + \frac{1}{2} C = 90°$,
$\frac{1}{2} C = 90° - \frac{1}{2} (A + B).$

$\therefore \cot \frac{1}{2} C = \tan \frac{1}{2} (A + B),$
and $\cot \frac{1}{2} B = \tan \frac{1}{2} (A + C),$
and $\cot \frac{1}{2} A = \tan \frac{1}{2} (B + C).$

$\therefore \cot \frac{1}{2} A + \cot \frac{1}{2} B + \cot \frac{1}{2} C$
$= \tan \frac{1}{2} (A + B) + \tan \frac{1}{2} (A + C)$
$+ \tan \frac{1}{2} (B + C)$
$= \tan \frac{1}{2} (A + B) \times \tan \frac{1}{2} (A + C)$
$\times \tan \frac{1}{2} (B + C).$

By substitution,
$\cot \frac{1}{2} A + \cot \frac{1}{2} B + \cot \frac{1}{2} C$
$= \cot \frac{1}{2} A \times \cot \frac{1}{2} B \times \cot \frac{1}{2} C.$

38. Change to a form more convenient for logarithmic computation $\cot x + \tan x$.

$$\cot x + \tan x$$
$$= \frac{\cos x}{\sin x} + \frac{\sin x}{\cos x}$$
$$= \frac{\cos^2 x + \sin^2 x}{\sin x \cos x}$$
$$= \frac{2 (\cos^2 x + \sin^2 x)}{2 \sin x \cos x}$$
$$= \frac{2}{\sin 2x}. \qquad [12]$$

39. Change to a form more convenient for logarithmic computation $\cot x - \tan x$.

$$\cot x - \tan x$$
$$= \frac{\cos x}{\sin x} - \frac{\sin x}{\cos x}$$
$$= \frac{\cos^2 x - \sin^2 x}{\sin x \cos x}$$
$$= \frac{\cos 2x}{\sin x \cos x} \qquad [13]$$

$$= \frac{2\cos 2x}{2\sin x\cos x}$$

$$= \frac{2\cos 2x}{\sin 2x} \qquad [12]$$

$$= 2\cot 2x.$$

40. Change to a form more convenient for logarithmic computation $\cot x + \tan y$.

$$\cot x = \frac{\cos x}{\sin x}, \ \tan y = \frac{\sin y}{\cos y}. \qquad [2]$$

$$\therefore \cot x + \tan y = \frac{\cos x}{\sin x} + \frac{\sin y}{\cos y}$$

$$= \frac{\cos x\cos y + \sin x\sin y}{\sin x\cos y}.$$

$$= \frac{\cos(x-y)}{\sin x\cos y}. \qquad [9]$$

41. Change to a form more convenient for logarithmic computation $\cot x - \tan y$.

$$\tan y = \frac{\sin y}{\cos y}.$$

$$\cot x = \frac{\cos x}{\sin x}.$$

$$\cot x - \tan y$$

$$= \frac{\cos x}{\sin x} - \frac{\sin y}{\cos y}$$

$$= \frac{\cos x\cos y - \sin x\sin y}{\sin x\cos y}$$

$$= \frac{\cos(x+y)}{\sin x\cos y}.$$

42. Change to a form more convenient for logarithmic computation $\frac{1-\cos 2x}{1+\cos 2x}$.

$$\frac{1-\cos 2x}{1+\cos 2x} = \frac{\frac{1-\cos 2x}{2}}{\frac{1+\cos 2x}{2}}.$$

$$= \frac{\sin^2 x}{\cos^2 x} \qquad [16], [17]$$

$$= \tan^2 x.$$

43. Change to a form more convenient for logarithmic computation $1 + \tan x\tan y$.

$$1 + \tan x\tan y$$

$$= 1 + \frac{\sin x}{\cos x} \times \frac{\sin y}{\cos y}$$

$$= \frac{\cos x\cos y + \sin x\sin y}{\cos x\cos y}$$

$$= \frac{\cos(x-y)}{\cos x\cos y}.$$

44. Change to a form more convenient for logarithmic computation $1 - \tan x\tan y$.

$$1 - \tan x\tan y$$

$$= 1 - \frac{\sin x\sin y}{\cos x\cos y}$$

$$= \frac{\cos x\cos y - \sin x\sin y}{\cos x\cos y}$$

$$= \frac{\cos(x+y)}{\cos x\cos y}.$$

45. Change to a form more convenient for logarithmic computation $\cot x\cot y + 1$.

$$\cot x\cot y + 1$$

$$= \frac{\cos x}{\sin x} \times \frac{\cos y}{\sin y} + 1$$

$$= \frac{\cos x\cos y + \sin x\sin y}{\sin x\sin y}.$$

By [9] $= \frac{\cos(x-y)}{\sin x\sin y}.$

46. Change to a form more convenient for logarithmic computation $\cot x \cot y - 1$.

$$\cot x \cot y - 1 = \frac{\cos x \cos y}{\sin x \sin y} - 1 = \frac{\cos x \cos y - \sin x \sin y}{\sin x \sin y} = \frac{\cos(x+y)}{\sin x \sin y}.$$

47. Change to a form more convenient for logarithmic computation $\frac{\tan x + \tan y}{\cot x + \cot y}$.

$$\frac{\tan x + \tan y}{\cot x + \cot y} = \frac{\frac{\sin x}{\cos x} + \frac{\sin y}{\cos y}}{\frac{\cos x}{\sin x} + \frac{\cos y}{\sin y}} = \frac{\frac{\sin x \cos y + \cos x \sin y}{\cos x \cos y}}{\frac{\sin x \cos y + \cos x \sin y}{\sin x \sin y}} = \frac{\sin x \sin y}{\cos x \cos y} = \tan x \tan y.$$

Exercise XV. Page 59.

1. Find all the values of the following functions: $\sin^{-1} \frac{1}{2}\sqrt{3}$, $\tan^{-1}\frac{1}{\sqrt{3}}$, $\text{vers}^{-1}\frac{1}{2}$, $\cos^{-1}\left(-\frac{1}{\sqrt{2}}\right)$, $\csc^{-1}\sqrt{2}$, $\tan^{-1}\infty$, $\sec^{-1}2$, $\cos^{-1}(-\frac{1}{2}\sqrt{3})$.

$$\sin^{-1}\tfrac{1}{2}\sqrt{3} = 60° + 2n\pi \text{ or } 120° + 2n\pi.$$
$$\tan^{-1}\frac{1}{\sqrt{3}} = 30° + 2n\pi \text{ or } 210° + 2n\pi.$$
$$\text{vers}^{-1}\tfrac{1}{2} = 60° + 2n\pi \text{ or } -60° + 2n\pi.$$
$$\cos^{-1}\left(-\frac{1}{\sqrt{2}}\right) = 135° + 2n\pi \text{ or } 225° + 2n\pi.$$
$$\csc^{-1}\sqrt{2} = 45° + 2n\pi \text{ or } 135° + 2n\pi.$$
$$\tan^{-1}\infty = 90° + 2n\pi \text{ or } 270° + 2n\pi.$$
$$\sec^{-1}2 = 60° + 2n\pi \text{ or } -60° + 2n\pi.$$
$$\cos^{-1}(-\tfrac{1}{2}\sqrt{3}) = 150° + 2n\pi \text{ or } 210° + 2n\pi.$$

2. Prove that $\sin^{-1}(-x) = -\sin^{-1}x$; $\cos^{-1}(-x) = \pi - \cos^{-1}x$.

$\sin^{-1}(-x) =$ the angle whose sine is $-x$

$= -$ the angle whose sine is x

$= -\sin^{-1}x$.

$\cos^{-1}(-x) =$ the angle whose cosine is $-x$

$= \pi -$ the angle whose cosine is x

$= \pi - \cos^{-1}x$.

3. If $\sin^{-1}x + \sin^{-1}y = \pi$, prove that $x = y$.

If the sum of two angles is 180°, the sine of the one is equal to that of the other.

Hence $\sin(\sin^{-1}x) = \sin(\sin^{-1}y)$.

$$x = y.$$

4. If $y = \sin^{-1}\frac{1}{3}$, find $\tan y$.

$$y = \sin^{-1}\tfrac{1}{3}.$$

$$\therefore \sin y = \tfrac{1}{3},$$

$$\cos y = \sqrt{1 - \sin^2 y} = \sqrt{\tfrac{8}{9}} = \tfrac{2}{3}\sqrt{2}.$$

$$\tan y = \frac{\sin y}{\cos y} = \frac{\frac{1}{3}}{\frac{2}{3}\sqrt{2}} = \frac{1}{2\sqrt{2}}.$$

5. Prove that

$$\cos(\sin^{-1}x) = \sqrt{1 - x^2}.$$

Let $\sin^{-1}x = y$,

then $x = \sin y$,

$$\sqrt{1 - x^2} = \cos y$$

$$= \cos(\sin^{-1}x).$$

6. Prove that

$$\cos(2\sin^{-1}x) = 1 - 2x^2.$$

Let $\sin^{-1}x = y$,

then $x = \sin y$

$$\cos(2\sin^{-1}x) = \cos 2y$$

$$= 1 - 2\sin^2 y$$

$$= 1 - x^2.$$

7. Prove that

$$\tan(\tan^{-1}x + \tan^{-1}y) = \frac{x + y}{1 - xy}.$$

Let $\tan^{-1}x = u$,

and $\tan^{-1}y = v$.

Then $x = \tan u$,

$y = \tan v$,

$$\tan(\tan^{-1}x + \tan^{-1}y)$$

$$= \tan(u + v)$$

$$= \frac{\tan u + \tan v}{1 - \tan u \tan v}$$

$$= \frac{x + y}{1 - xy}.$$

8. If $x = \sqrt{\frac{1}{2}}$, find all the values of $\sin^{-1}x + \cos^{-1}x$.

$$\sin^{-1}\sqrt{\tfrac{1}{2}} = 45° + 2n\pi \text{ or } 135° + 2n\pi$$

$$\cos^{-1}\sqrt{\tfrac{1}{2}} = 45° + 2n\pi \text{ or } -45° + 2n\pi$$

$$\therefore \sin^{-1}\sqrt{\tfrac{1}{2}} + \cos^{-1}\sqrt{\tfrac{1}{2}}$$

$$= 90° + 2n\pi,\ 2n\pi,$$

$$180° + 2n\pi, \text{ or } 90° + 2n\pi$$

$$= 0°,\ 90°, \text{ or } 180°.$$

9. Prove that

$$\tan^{-1}\frac{x}{\sqrt{1 - x^2}} = \sin^{-1}x.$$

Let

$$\tan^{-1}\frac{x}{\sqrt{1 - x^2}} = y,$$

then

$$\frac{x}{\sqrt{1 - x^2}} = \tan y,$$

$$\sec^2 y = 1 + \tan^2 y$$

$$= 1 + \left(\frac{x}{\sqrt{1 - x^2}}\right)^2$$

$$= \frac{1}{1 - x^2},$$

$$\cos^2 y = \frac{1}{\sec^2 y} = 1 - x^2,$$

$$\sin^2 y = 1 - \cos^2 y = x^2,$$

$$\sin y = x,$$

$$y = \sin^{-1}x.$$

$$\therefore \tan^{-1}\frac{x}{\sqrt{1 - x^2}} = \sin^{-1}x.$$

10. Find the value of
$\sin(\tan^{-1}\frac{5}{12})$.

Let $\tan^{-1}\frac{5}{12} = x$,
then $\frac{5}{12} = \tan x$,

$$\sec^2 x = 1 + \tan^2 x = 1 + (\tfrac{5}{12})^2 = \tfrac{169}{144},$$
$$\cos^2 x = \tfrac{144}{169},$$
$$\sin^2 x = \tfrac{25}{169}.$$
$$\therefore \sin x = \pm\tfrac{5}{13}.$$

11. Find the value of
$\cot(2\sin^{-1}\frac{3}{5})$.

Let $\sin^{-1}\frac{3}{5} = x$,
then $\sin x = \frac{3}{5}$,

$$\cos x = \pm\tfrac{4}{5},$$
$$\cot x = \pm\tfrac{4}{3},$$
$$\cot(2\sin^{-1}\tfrac{3}{5}) = \cot 2x.$$
$$= \frac{\cot^2 x - 1}{2\cot x}$$
$$= \frac{\frac{16}{9} - 1}{\pm\frac{8}{3}}$$
$$= \pm\frac{7}{24}.$$

12. Find the value of
$\sin(\tan^{-1}\frac{1}{2} + \tan^{-1}\frac{1}{3})$.

Let $\tan^{-1}\frac{1}{2} = x$,
$\tan^{-1}\frac{1}{3} = y$,
then $\tan x = \frac{1}{2}$,

$$\sec^2 x = 1 + \tan^2 x = \tfrac{5}{4},$$
$$\cos^2 x = \tfrac{4}{5},$$
$$\cos x = \frac{\pm 2}{\sqrt{5}},$$
$$\sin x = \frac{\pm 1}{\sqrt{5}},$$
$$\cos y = \frac{\pm 3}{\sqrt{10}},$$
$$\sin y = \frac{\pm 1}{\sqrt{10}}.$$

$$\sin(\tan^{-1}\tfrac{1}{2} + \tan^{-1}\tfrac{1}{3})$$
$$= \sin(x + y)$$
$$= \sin x \cos y + \cos x \sin y$$
$$= \frac{1}{\sqrt{5}} \times \frac{3}{\sqrt{10}} + \frac{2}{\sqrt{5}} \times \frac{1}{\sqrt{10}}$$
$$= \frac{\pm 5}{\sqrt{50}}$$
$$= \pm\tfrac{1}{2}\sqrt{2}.$$

13. If $\sin^{-1}x = 2\cos^{-1}x$, find x.

$$\sin^{-1}x = 90° - \cos^{-1}x.$$
$$\therefore 90° - \cos^{-1}x = 2\cos^{-1}x,$$
$$3\cos^{-1}x = 90°,$$
$$\cos^{-1}x = 30°, 150°, \text{ or } 270°,$$
$$x = \pm\tfrac{1}{2}\sqrt{3}, \text{ or } 0.$$

14. Prove that
$$\tan(2\tan^{-1}x) = \frac{2x}{1 + x^2}.$$

Let $\tan^{-1}x = y$,
then $x = \tan y$,

$$\tan(2\tan^{-1}x) = \tan 2y$$
$$= \frac{2\tan y}{1 - \tan^2 y}$$
$$= \frac{2x}{1 - x^2}.$$

15. Prove that
$$\sin(2\tan^{-1}x) = \frac{2x}{1 + x^2}.$$

Let $\tan^{-1}\frac{1}{2}x = y$,
then $x = \tan y$,

$$\sin(2\tan^{-1}x) = \sin 2y$$
$$= 2\sin y\cos y$$
$$= 2\frac{\sin y}{\cos y}\cos^2 y$$
$$= \frac{2\tan y}{\sec^2 y}$$
$$= \frac{2x}{1 + x^2}.$$

Exercise XVI. Page 63.

1. What do the formulas of § 33 become when one of the angles is a right angle?

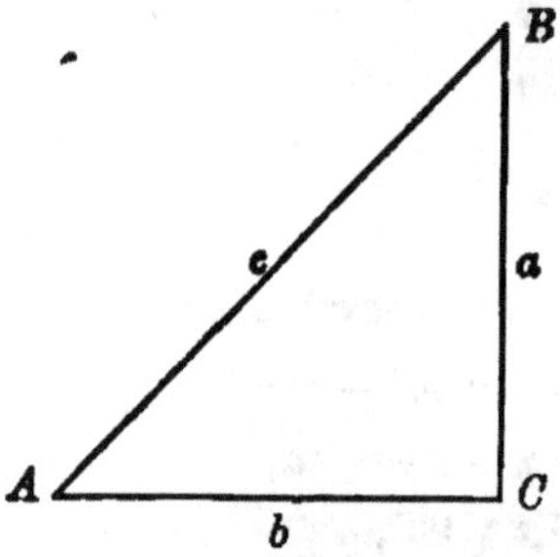

If angle C is a right angle,

$$\frac{a}{c}=\frac{\sin A}{\sin C}=\sin A;$$

$$\frac{c}{b}=\frac{\sin C}{\sin B}=\frac{1}{\sin B};$$

$$\frac{a}{b}=\frac{\sin A}{\sin B}=\tan A;$$

$$\frac{a}{\sin A}=\frac{c}{\sin C}=c;$$

$$\frac{b}{\sin B}=\frac{c}{\sin C}=c.$$

2. Prove by means of the Law of Sines that the bisector of an angle of a triangle divides the opposite side into parts proportional to the adjacent sides.

Let CD bisect angle C.

Then $$\frac{AD}{CD}=\frac{\sin \frac{1}{2} C}{\sin A},$$

and $$\frac{DB}{CD}=\frac{\text{sih} \frac{1}{2} C}{\sin B}.$$

By division,

$$\frac{AD}{DB}=\frac{\sin B}{\sin A}.$$

But $$\frac{\sin B}{\sin A}=\frac{b}{a}.$$

$$\therefore \frac{AD}{DB}=\frac{b}{a}.$$

3. What does Formula [26] become when $A=90°$? when $A=0°$? when $A=180°$? What does the triangle become in each of these cases?

Formula [26] is

$$a^2=b^2+c^2-2bc\cos A.$$

When $A=90°$, $\cos A=0°$.

$\therefore a^2=b^2+c^2.$

When $A=0°$, $\cos A=1$.

$\therefore a^2=b^2+c^2-2bc.$

When $A=180°$, $\cos A=-1$.

$\therefore a^2=b^2+c^2+2bc.$

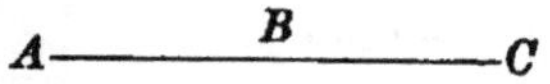

$a=BC.$ $\quad c=AB.$

$b=AC.$ $\quad a=b-c.$

$a=BC.$ $\quad c=BA.$

$b=AC.$ $\quad a=b+c.$

4. Prove that whether the angle B is acute or obtuse $c=a\cos B+b\cos A$. What are the two symmetrical formulas obtained by changing the letters? What does the formula become when $B=90°$?

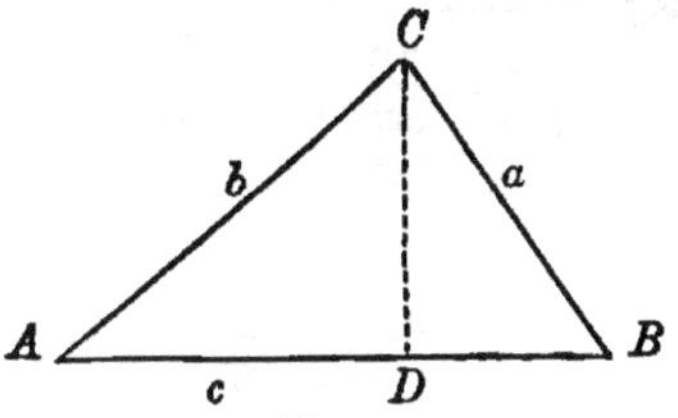

Fig. 1.

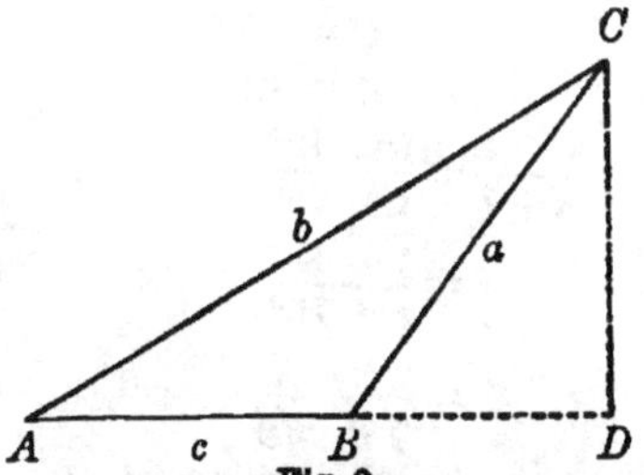

Fig. 2.

CASE I. When angle B is acute (Fig. 1).

(1) $\cos B = \frac{DB}{a}$.

$\cos A = \frac{AD}{b}$.

$\therefore DB = a \cos B$,

and $AD = b \cos A$.

Add, $DB + AD = a \cos B + b \cos A$.

But $DB + AD = c$.

$\therefore c = a \cos B + b \cos A$.

CASE II. When angle B is obtuse (Fig. 2).

(2) $\frac{AD}{b} = \cos A$.

$\frac{BD}{a} = \cos (180° - B)$

$= -\cos B$.

$\therefore AD = b \cos A$,

and $BD = -a \cos B$.

Subtract, observing that the sign of $\cos B$ is minus.

$AD - BD = b \cos A + a \cos B$.

But $AD - BD = c$.

$\therefore c = a \cos B + b \cos A$.

The symmetrical formulas are

$b = a \cos C + c \cos A$,

$a = b \cos C + c \cos B$.

When $B = 90°$.

(3) $\cos A = \frac{c}{b}$.

$\therefore c = b \cos A$.

5. From the three following equations (found in the last exercise) prove the theorem of § 34:

$c = a \cos B + b \cos A$,

$b = a \cos C + c \cos A$,

$a = b \cos C + c \cos B$.

$c^2 = ac \cos B + bc \cos A$. (1)

$b^2 = ab \cos C + bc \cos A$. (2)

$a^2 = ab \cos C + ac \cos B$. (3)

Add (2) and (3),

$a^2 + b^2 = 2\,ab \cos C + bc \cos A + ac \cos B$. (4)

Subtract (4) from (1),

$c^2 - a^2 - b^2 = -2\,ab \cos C$.

$\therefore c^2 = a^2 + b^2 - 2\,ab \cos C$. § 34

6. In Formula [27] what is the maximum value of $\frac{1}{2}(A - B)$? of $\frac{1}{2}(A + B)$?

$$\frac{a-b}{a+b} = \frac{\tan \frac{1}{2}(A - B)}{\tan \frac{1}{2}(A + B)}.$$

The limit of $A - B$ is 180°.

$\therefore$ the limit of the maximum value of $\frac{1}{2}(A - B)$

$= \frac{180°}{2} = 90°$.

The limit of $A + B$ is 180°.

$\therefore$ the limit of the maximum value of $\frac{1}{2}(A + B)$

$= \frac{180°}{2} = 90°$.

7. Find the form to which Formula [27] reduces, and describe the nature of the triangle when

(i.) $C = 90°$;

(ii.) $A - B = 90°$, and $B = C$.

$$\frac{a-b}{a+b} = \frac{\tan \frac{1}{2}(A-B)}{\tan \frac{1}{2}(A+B)}.$$

(i.) When $C = 90°$.

$$A + B = 90°.$$
$$B = 90° - A.$$
$$\frac{a-b}{a+b} = \frac{\tan \frac{1}{2}[A-(90°-A)]}{\tan 45°}$$
$$= \frac{\tan (A - 45°)}{1}$$
$$= \tan (A - 45°).$$

Since C is a right angle, the triangle is a right triangle.

(ii.) When $A - B = 90°$, and $B = C$.

$$\frac{a-b}{a+b} = \frac{\tan \frac{1}{2}(A-B)}{\tan \frac{1}{2}(A+B)}.$$

$$A + B + C = 180°,$$

or

$$A + 2B = 180°$$
$$A - B = 90°$$
$$\therefore 3B = 90°,$$
$$B = 30°,$$
$$C = 30°,$$

and

$$A = 120°.$$
$$\frac{a-b}{a+b} = \frac{\tan 45°}{\tan 75°}$$
$$= \frac{\tan 45°}{\cot 15°}$$
$$= \frac{1}{2+\sqrt{3}}.$$
$$\therefore a + b = (a - b)(2 + \sqrt{3}).$$

Since $A = B$, the triangle is isosceles.

Exercise XVII. Page 65.

1. Given $a = 500$, $A = 10° 12'$, $B = 46° 36'$; Find $C = 123° 12'$, $b = 2051.48$, $c = 2362.61$.

$$a = 500.$$
$$A = 10° 12'$$
$$B = 46° 36'$$
$$A + B = 56° 48'$$
$$\therefore C = 123° 12'.$$

$\log a = 2.69897$
colog $\sin A = 0.75182$
$\log \sin B = 9.86128$
$\log b = 3.31207$
$b = 2051.48.$

$\log a = 2.69897$
colog $\sin A = 0.75182$
$\log \sin C = 9.92260$
$\log c = 3.37339$
$c = 2362.61$

2. Given $a = 795$, $A = 79° 59'$, $B = 44° 41'$; Find $C = 55° 20'$, $b = 567.688$, $c = 663.986$.

$$a = 795.$$
$$A = 79° 59'$$
$$B = 44° 41'$$
$$A + B = 124° 40'$$
$$\therefore C = 55° 20'.$$

$$\begin{aligned}\log a &= 2.90037\\ \text{colog}\sin A &= 0.00667\\ \log\sin B &= 9.84707\\ \log b &= 2.75411\\ b &= 567.688\end{aligned}$$

$$\begin{aligned}\log a &= 2.90037\\ \text{colog}\sin A &= 0.00667\\ \log\sin C &= 9.91512\\ \log c &= 2.82216\\ c &= 663.986.\end{aligned}$$

3. Given Find
$a = 804,$ $C = 35° 4',$
$A = 99° 55',$ $b = 577.31,$
$B = 45° 1';$ $c = 468.93.$

$$a = 804.$$

$$\begin{aligned}A &= 99°\ 55'\\ B &= 45°\ 1'\\ A + B &= 144°\ 56'\\ \therefore C &= 35°\ 4'.\end{aligned}$$

$$\begin{aligned}\log a &= 2.90526\\ \text{colog}\sin A &= 0.00654\\ \log\sin B &= 9.84961\\ \log b &= 2.76141\\ b &= 577.31.\end{aligned}$$

$$\begin{aligned}\log a &= 2.90526\\ \text{colog}\sin A &= 0.00654\\ \log\sin C &= 9.75931\\ \log c &= 2.67111\\ c &= 468.93.\end{aligned}$$

4. Given Find
$a = 820,$ $C = 25° 12',$
$A = 12° 49',$ $b = 2276.63,$
$B = 141° 59';$ $c = 1573.89.$

$$a = 820.$$

$$\begin{aligned}A &= 12°\ 49'\\ B &= 141°\ 59'\\ A + B &= 154°\ 48'\\ \therefore C &= 25°\ 12'.\end{aligned}$$

$$\begin{aligned}\log a &= 2.91381\\ \text{colog}\sin A &= 0.65398\\ \log\sin B &= 9.78950\\ \log b &= 3.35729\\ b &= 2276.63.\end{aligned}$$

$$\begin{aligned}\log a &= 2.91381\\ \text{colog}\sin A &= 0.65398\\ \log\sin C &= 9.62918\\ \log c &= 3.19697\\ c &= 1573.89.\end{aligned}$$

5. Given Find
$c = 1005,$ $C = 47° 14',$
$A = 78° 19',$ $a = 1340.6,$
$B = 54° 27';$ $b = 1113.8.$

$$c = 1005.$$

$$\begin{aligned}A &= 78°\ 19'\\ B &= 54°\ 27'\\ A + B &= 132°\ 46'\\ \therefore C &= 47°\ 14'.\end{aligned}$$

$$\begin{aligned}\log c &= 3.00217\\ \text{colog}\sin C &= 0.13423\\ \log\sin A &= 9.99091\\ \log a &= 3.12731\\ a &= 1340.6.\end{aligned}$$

$$\begin{aligned}\log c &= 3.00217\\ \text{colog}\sin C &= 0.13423\\ \log\sin B &= 9.91042\\ \log b &= 3.04682\\ b &= 1113.8.\end{aligned}$$

6. Given Find
$b = 13.57,$ $A = 108° 50',$
$B = 13° 57',$ $a = 53.276,$
$C = 57° 13';$ $c = 47.324.$

$$b = 13.57.$$

$$\begin{aligned}B &= 13°\ 57'\\ C &= 57°\ 13'\\ B + C &= 71°\ 10'\\ \therefore A &= 108°\ 50'.\end{aligned}$$

$$\begin{aligned} \log b &= 1.13258 \\ \text{colog} \sin B &= 0.61785 \\ \log \sin A &= \underline{9.97610} \\ \log a &= 1.72653 \\ a &= 53.276. \end{aligned}$$

$$\begin{aligned} \log a &= 1.72653 \\ \text{colog} \sin A &= 0.02390 \\ \log \sin C &= \underline{9.92465} \\ \log c &= 1.67508 \\ c &= 47.324. \end{aligned}$$

7. Given $a = 6412$, $A = 70° 55'$, $C = 52° 9'$; Find $B = 56° 56'$, $b = 5685.9$, $c = 5357.5$.

$$a = 6412.$$

$$\begin{aligned} A &= 70° 55' \\ C &= \underline{52° \ 9'} \\ A + C &= 123° \ 4' \\ \therefore B &= 56° 56'. \end{aligned}$$

$$\begin{aligned} \log a &= 3.80699 \\ \log \sin B &= 9.92326 \\ \text{colog} \sin A &= \underline{0.02455} \\ \log b &= 3.75480 \\ b &= 5685.9. \end{aligned}$$

$$\begin{aligned} \log a &= 3.80699 \\ \log \sin C &= 9.89742 \\ \text{colog} \sin A &= \underline{0.02455} \\ \log c &= 3.72896 \\ c &= 5357.5. \end{aligned}$$

8. Given $b = 999$, $A = 37° 58'$, $C = 65° 2'$; Find $B = 77°$, $a = 630.77$, $c = 929.48$.

$$b = 999.$$

$$\begin{aligned} A &= 37° 58' \\ C &= \underline{65° \ 2'} \\ A + C &= 103° \\ \therefore B &= 77°. \end{aligned}$$

$$\begin{aligned} \log b &= 2.99957 \\ \text{colog} \sin B &= 0.01128 \\ \log \sin A &= \underline{9.78902} \\ \log a &= 2.79987 \\ a &= 630.77. \end{aligned}$$

$$\begin{aligned} \log b &= 2.99957 \\ \text{colog} \sin B &= 0.01128 \\ \log \sin C &= \underline{9.95739} \\ \log c &= 2.96824 \\ c &= 929.48. \end{aligned}$$

9. In order to determine the distance of a hostile fort A from a place B, a line BC and the angles ABC and BCA were measured, and found to be 322.55 yards, 60° 34′, and 56° 10′, respectively. Find the distance AB.

$$a = 322.55.$$

$$\begin{aligned} B &= 60° 34' \\ C &= \underline{56° 10'} \\ B + C &= 116° 44' \\ \therefore A &= 63° 16'. \end{aligned}$$

$$\begin{aligned} \log a &= 2.50860 \\ \text{colog} \sin A &= 0.04910 \\ \log \sin C &= \underline{9.91942} \\ \log c &= 2.47712 \\ c &= 300. \end{aligned}$$

10. In making a survey by triangulation, the angles B and C of a triangle ABC were found to be 50° 30′ and 122° 9′, respectively, and the length BC is known to be 9 miles. Find AB and AC.

$$\begin{aligned} C &= 122° \ 9' \\ B &= \underline{50° 30'} \\ B + C &= 172° 39' \\ \therefore A &= 7° 21'. \end{aligned}$$

$$\begin{aligned} \log BC &= 0.95424 \\ \text{colog}\sin A &= 0.89303 \\ \log\sin B &= 9.88741 \\ \log b &= 1.73468 \end{aligned}$$

$$b = AC = 54.285.$$

$$\begin{aligned} \log BC &= 0.95424 \\ \text{colog}\sin A &= 0.89303 \\ \log\sin C &= 9.92771 \\ \log c &= 1.77498 \end{aligned}$$

$$c = AB = 59.564.$$

11. Two observers 5 miles apart on a plain, and facing each other, find that the angles of elevation of a balloon in the same vertical plane with themselves are 55° and 58°, respectively. Find the distance from the balloon to each observer, and also the height of the balloon above the plain.

$$\begin{aligned} B &= 58° \\ A &= 55° \\ A + B &= 113° \end{aligned}$$

$$\therefore C = 67°.$$

$$\begin{aligned} \log c &= 0.69897 \\ \text{colog}\sin C &= 0.03597 \\ \log\sin A &= 9.91336 \\ \log a &= 0.64830 \end{aligned}$$

$$a = BC = 4.4494.$$

$$\begin{aligned} \log c &= 0.69897 \\ \text{colog}\sin C &= 0.03597 \\ \log\sin B &= 9.92842 \\ \log b &= 0.66336 \end{aligned}$$

$$b = AC = 4.6064.$$

To find h.

$$\frac{h}{a} = \sin B.$$

$$\therefore h = a \sin B.$$

$$\begin{aligned} \log a &= 0.64830 \\ \log\sin B &= 9.92842 \\ \log h &= 0.57672 \end{aligned}$$

$$h = 3.7733.$$

12. In a parallelogram, given a diagonal d and the angles x and y which this diagonal makes with the sides. Find the sides. Compute the results when $d = 11.237$, $x = 19° 1'$, and $y = 42° 54'$.

$$d = 11.237.$$

$$\begin{aligned} x &= 19° \ 1' \\ y &= 42° \ 54' \\ x + y &= 61° \ 55' \end{aligned}$$

$$\therefore z = 118° \ 5'$$

$$\begin{aligned} \log d &= 1.05065 \\ \text{colog}\sin z &= 0.05440 \\ \log\sin x &= 9.51301 \\ \log a &= 0.61806 \end{aligned}$$

$$a = 4.1501.$$

$$\begin{aligned} \log d &= 1.05065 \\ \text{colog}\sin z &= 0.05440 \\ \log\sin y &= 9.84297 \\ \log c &= 0.93802 \end{aligned}$$

$$c = 8.67.$$

13. A lighthouse was observed from a ship to bear N. 34° E.; after sailing due south 3 miles, it bore N. 23° E. Find the distance from the lighthouse to the ship in both positions.

$$c = 3.$$

$$\begin{aligned} A &= 23° \\ B = (180° - 34°) &= 146° \\ A + B &= 169° \end{aligned}$$

$$\therefore C = 11°.$$

$$\log c = 0.47712$$
$$\text{colog} \sin C = 0.71940$$
$$\log \sin A = 9.59188$$
$$\log a = 0.78840$$
$$a = 6.1433.$$
$$\log c = 0.47712$$
$$\text{colog} \sin C = 0.71940$$
$$\log \sin B = 9.74756$$
$$\log b = 0.94408$$
$$b = 8.7918.$$

14. In a trapezoid, given the parallel sides a and b, and the angles x and y at the ends of one of the parallel sides. Find the non-parallel sides. Compute the results when $a=15$, $b=7$, $x=70°$, $y=40°$.

Given parallel sides,

$$AB = 7 \quad \text{and} \quad DC = 15;$$

also, $ADC = 40°$ and $BCD = 70°$;

required AD and BC.

Draw $AE \parallel BC$;

then $AB = EC$ (||s comp. bet. ||s),

and $DE = DC - AB$

$= 15 - 7 = 8.$

Also $AED = BCD = 70°$ (ext. int. ∠s).

Now

$$DAE = 180° - (40° + 70°)$$
$$= 70°.$$

But since

$$AED = DAE = 70°,$$

the △ is isosceles, and side

$$DA = DE = 8.$$

Now $AE = BC$, and we are to find BC.

$$\frac{AE}{DE} = \frac{\sin ADE}{\sin DAE}.$$

$$\log DE = 0.90309$$
$$\log \sin ADE = 9.80807$$
$$\text{colog} \sin DAE = 0.02701$$
$$\log AE = 0.73817$$
$$AE = BC = 5.4723.$$

15. Given $b = 7.07107$, $A = 30°$, $C = 105°$; find a and c without using logarithms.

Let p and q denote the segments of c made by the ⊥ dropped from C.

$$A = 30°,$$
$$C = 105°,$$
$$B = 45°.$$

$$\therefore \frac{a}{b} = \frac{\sin A}{\sin B} = \frac{\frac{1}{2}}{\frac{1}{2}\sqrt{2}}.$$

$$a = \frac{b}{\sqrt{2}}$$

$$= \frac{7.07107}{1.41421} = 5.$$

$$\frac{p}{b} = \cos A = \tfrac{1}{2}\sqrt{3} = 0.86602.$$

$$p = b \times 0.86602$$
$$= 7.07107 \times 0.86602$$
$$= 6.12369.$$

$$\frac{q}{a} = \cos B = \tfrac{1}{2}\sqrt{2} = 0.70711.$$

$$q = a \times 0.70711$$
$$= 5 \times 0.70711 = 3.53555.$$

$$c = p + q$$
$$= 6.12369 + 3.53555$$
$$= 9.6592.$$

16. Given $c = 9.562$, $A = 45°$, $B = 60°$; find a and b without using logarithms.

$$C = 75°.$$

$$a = \frac{c \sin A}{\sin C}.$$

$$\sin C = \sin(45° + 30°)$$
$$= \sin 45° \cos 30° + \cos 45° \sin 30°.$$
$$= \tfrac{1}{2}\sqrt{2} \times \tfrac{1}{2}\sqrt{3} + \tfrac{1}{2}\sqrt{2} \times \tfrac{1}{2}$$
$$= \tfrac{1}{4}(\sqrt{6} + \sqrt{2}).$$

$$\therefore a = \frac{9.562 \times \frac{1}{2}\sqrt{2}}{\frac{1}{4}(\sqrt{6} + \sqrt{2})}$$
$$= \frac{19.124 \times \sqrt{2}}{\sqrt{6} + \sqrt{2}}$$
$$= \frac{(19.124 \times \sqrt{2})(\sqrt{6} - \sqrt{2})}{6 - 2}$$
$$= 9.562(\sqrt{3} - 1)$$
$$= 6.999 = 7.$$

$$b = \frac{a \sin B}{\sin A} = \frac{7 \times \frac{1}{2}\sqrt{3}}{\frac{1}{2}\sqrt{2}}$$
$$= \frac{7\sqrt{3}}{\sqrt{2}} = \frac{7\sqrt{6}}{2}$$
$$= 3.5\sqrt{6} = 8.573.$$

17. The base of a triangle is 600 feet, and the angles at the base are 30° and 120°; find the other sides and the altitude without using logarithms.

$$AB = 600$$
$$A = 30°.$$
$$B = 120°.$$
$$\therefore C = 30°.$$
$$A = C.$$
$$a = c = 600 \text{ feet.}$$

$$b = \frac{a \sin B}{\sin A}$$
$$= \frac{600 \times \sin(180° - 60°)}{\sin 30°}$$
$$= \frac{600 \times \frac{1}{2}\sqrt{3}}{\frac{1}{2}}$$
$$= 600 \times 1.732051$$
$$= 1039.2.$$
$$h = a \sin B = 600 \times \tfrac{1}{2}\sqrt{3}$$
$$= 519.6 \text{ feet.}$$

18. Two angles of a triangle are, the one 20°, the other 40°; find the ratio of the opposite sides without using logarithms.

Let $x = 20°.$
$y = 40°,$

and a and b be opposite sides.

Then $\dfrac{\sin x}{\sin y} = \dfrac{a}{b}.$

nat $\sin x = 0.3420.$
nat $\sin y = 0.6428.$

$$\therefore a : b :: 3420 : 6428.$$
$$:: 855 : 1607.$$

19. The angles of a triangle are as 5 : 10 : 21, and the side opposite the smallest angle is equal to 3; find the other sides without using logarithms.

Since the angles A, B, C, are as 5 : 10 : 21,

$$A = \tfrac{5}{36} \text{ of } 180° = 25°.$$
$$B = \tfrac{10}{36} \text{ of } 180° = 50°.$$
$$C = \tfrac{21}{36} \text{ of } 180° = 105°.$$

$$b = \frac{a \sin B}{\sin A} = \frac{3 \times 0.766}{0.4226}$$
$$= 5.438.$$
$$c = \frac{a \sin C}{\sin A} = \frac{3 \times 0.9659}{0.4226}$$
$$= 6.857.$$

20. Given one side of a triangle equal to 27, the adjacent angles equal each to 30°; find the radius of the circumscribed circle without using logarithms.

$$2R = \frac{a}{\sin A}.$$

$$\sin A = \sin 120°$$
$$= \sin(180° - 60°)$$
$$= \sin 60°.$$

$$\sin 60° = \tfrac{1}{2}\sqrt{3}.$$

$$\therefore 2R = \frac{27}{\frac{1}{2}\sqrt{3}} = \frac{54}{\sqrt{3}} = \frac{54 \times \sqrt{3}}{3}$$
$$= 18\sqrt{3}.$$

$$\therefore R = 9\sqrt{3} = 15.588.$$

Exercise XVIII. Page 69.

1. Determine the number of solutions in each of the following cases:

(i.) $a = 80$, $b = 100$, $A = 30°$.

$\therefore a < b$,

but $a > b \sin A = 100 \times \frac{1}{2}$,

and $A < 90°$.

$\therefore$ two solutions.

(ii.) $a = 50$, $b = 100$, $A = 30°$.

$\therefore a = b \sin A = 100 \times \frac{1}{2}$.

$\therefore$ one solution.

(iii.) $a = 40$, $b = 100$, $A = 30°$.

$\therefore a < b \sin A = 100 \times \frac{1}{2}$,

and $A < 90°$.

$\therefore$ no solution.

(iv.) $a = 13.4$ $b = 11.46$, $A = 77° 20'$.

$\therefore a > b$.

$\therefore$ one solution.

(v.) $a = 70$, $b = 75$, $A = 60°$.

$\therefore a < b$,

but $a > b \sin A = 75 \times \frac{1}{2}\sqrt{3}$,

and $A < 90°$.

$\therefore$ two solutions.

(vi.) $a = 134.16$, $b = 84.54$, $B = 52° 9' 11''$.

$b < a$,

$B < 90°$,

nat $\sin B = 0.7897$.

$84.54 < 134.16 \times 0.7897$.

$\therefore b < a \sin B$.

$\therefore$ no solution.

(vii.) $a = 200$, $b = 100$, $A = 30°$.

$a > b$.

$\therefore$ one solution.

2. Given $a = 840$, $b = 485$, $A = 21° 31'$; Find $B = 12° 13' 34''$, $C = 146° 15' 26''$, $c = 1272.15$.

Here $a > b$. $\therefore$ one solution.

$$\begin{aligned}
\text{colog } a &= 7.07572 - 10 \\
\log b &= 2.68574 \\
\log \sin A &= 9.56440 \\
\hline
\log \sin B &= 9.32586 \\
B &= 12° 13' 34''. \\
\therefore C &= 146° 15' 26''.
\end{aligned}$$

$$\begin{aligned}
\log a &= 2.92428 \\
\log \sin C &= 9.74466 \\
\text{colog } \sin A &= 0.43560 \\
\hline
\log c &= 3.10454 \\
c &= 1272.15.
\end{aligned}$$

3. Given $a = 9.399$, $b = 9.197$, $A = 120° 35'$; Find $B = 57° 23' 40''$, $C = 2° 1' 20''$, $c = 0.38525$.

$$\begin{aligned}
\text{colog } a &= 9.02692 - 10 \\
\log b &= 0.96365 \\
\log \sin A &= 9.93495 \\
\hline
\log \sin B &= 9.92552 \\
B &= 57° 23' 40''. \\
\therefore C &= 2° 1' 20''.
\end{aligned}$$

$$\begin{aligned} \log a &= 0.97308 \\ \log \sin C &= 8.54761 \\ \text{colog} \sin A &= \underline{0.06505} \\ \log c &= 9.58574 - 10 \end{aligned}$$

$c = 0.38525.$

4. Given $a = 91.06$, $b = 77.04$, $A = 51° 9' 6''$;

Find $B = 41° 13'$, $C = 87° 37' 54''$, $c = 116.82$.

$$\begin{aligned} \text{colog } a &= 8.04067 - 10 \\ \log b &= 1.88672 \\ \log \sin A &= \underline{9.89143} \\ \log \sin B &= 9.81882 \end{aligned}$$

$B = 41° 13'.$

$\therefore C = 87° 37' 54''.$

$$\begin{aligned} \log a &= 1.95933 \\ \log \sin C &= 9.99963 \\ \text{colog} \sin A &= \underline{0.10857} \\ \log c &= 2.06753 \end{aligned}$$

$c = 116.82.$

5. Given $a = 55.55$, $b = 66.66$, $B = 77° 44' 40''$;

Find $A = 54° 31' 13''$, $C = 47° 44' \ 7''$, $c = 54.481$.

Here $b > a$. $\therefore$ one solution.

$$\begin{aligned} \log a &= 1.74468 \\ \log \sin B &= 9.98999 \\ \text{colog } b &= \underline{8.17613 - 10} \\ \log \sin A &= 9.91080 \end{aligned}$$

$A = 54° 31' 13''.$

$\therefore C = 47° 44' \ 7''.$

$$\begin{aligned} \log a &= 1.74468 \\ \log \sin C &= 9.86925 \\ \text{colog} \sin A &= \underline{0.08920} \\ \log c &= 1.70313 \end{aligned}$$

$c = 50.481.$

6. Given

$a = 309$, $b = 360$, $A = 21° 14' 25''$;

find $B = 24° 57' 54''$, $B' = 155° \ 2' \ 6''$, $C = 133° 47' 41''$, $C' = 3° 43' 29''$, $c = 615.67$, $c' = 55.41$.

There are two solutions,

for $a < b$,

but $a > b \sin A$,

and $A < 90°$.

$$\begin{aligned} \log b &= 2.55630 \\ \log \sin A &= 9.55904 \\ \text{colog } a &= \underline{7.51004 - 10} \\ \log \sin B &= 9.62538 \end{aligned}$$

$B = 24° 57' 54''.$

$\therefore C = 133° 47' 41''.$

$$\begin{aligned} \log a &= 2.48996 \\ \log \sin C &= 9.85843 \\ \text{colog} \sin A &= \underline{0.44096} \\ \log c &= 2.78935 \end{aligned}$$

$c = 615.67.$

Second Solution.

$B' = 180° - B = 155° 2' 6''.$ $C' = B - A = 3° 43' 29''.$

$$\begin{aligned} \log a &= 2.48996 \\ \log \sin C' &= 8.81267 \\ \text{colog} \sin A &= \underline{0.44096} \\ \log c' &= 1.74359 \end{aligned}$$

$c' = 55.41.$

7. Given $a = 8.716$, $b = 9.787$, $A = 38° 14' 12''$;

find $B = 44° \ 1' 28''$, $B' = 135° 58' 32''$, $C = 97° 44' 20''$, $C' = 5° 47' 16''$, $c = 13.954$, $c' = 1.4203$.

There are two solutions,

for $a < b$, and $\log \sin B < 0$.

colog $a = 9.05968 - 10$
log $b = 0.99065$
log sin $A = 9.79163$

log sin $B = 9.84196$

$B = \ \ 44° \ \ 1' \ 28''$.
$B' = 135° \ 58' \ 32''$.
$C = \ \ 97° \ 44' \ 20''$.
$C' = \ \ \ 5° \ 47' \ 16''$.

log $a = 0.94032$
log sin $C = 9.99602$
colog sin $A = 0.20837$

log $c = 1.14471$

$c = 13.954$.

log $a = 0.94032$
log sin $C' = 9.00365$
colog sin $A = 0.20837$

log $c' = 0.15234$

$c' = 1.4203$.

8. Given   Find
$a = 4.4$, $B = 90°$.
$b = 5.21$, $C = 32° \ 22' \ 43''$,
$A = 57° \ 37' \ 17''$; $c = 2.79$.

log sin $A = \ \ 9.92661$
log $b = \ \ 0.71684$
colog $a = \ \ 9.35655 - 10$

log sin $B = 10.00000$

$B = 90°$.
$\therefore C = 32° \ 22' \ 43''$.

log $b = 0.71684$
log cos $A = 9.72877$

log $c = 0.44561$

$c = 2.791$.

9. Given $a = 34$,
                    $b = 22$,
                    $B = 30° \ 20'$;

find $A = \ \ 51° \ 18' \ 27''$,
              $A' = 128° \ 41' \ 33''$,
              $C = \ \ 98° \ 21' \ 33''$,
              $C' = \ \ 20° \ 58' \ 27''$,
              $c = 43.098$,
              $c' = 15.593$.

Here $b < a$, but $> a$ sin B, and $B < 90°$.

$\therefore$ two solutions.

log $a = 1.53148$
log sin $B = 9.70332$
colog $b = 8.65758 - 10$

log sin $A = 9.89238$

$A = \ \ 51° \ 18' \ 27''$.
$A' = 128° \ 41' \ 33''$.

$\therefore C = \ \ 98° \ 21' \ 33''$.

$\therefore C' = \ \ 20° \ 58' \ 27''$.

log $a = 1.53148$
log sin $C = 9.99536$
colog sin $A = 0.10762$

log $c = 1.63446$

$c = 43.098$.

log $a = 1.53148$
log sin $C' = 9.55382$
colog sin $A = 0.10762$

log $c' = 1.19292$

$c' = 15.593$.

10. Given $b = 19,$
$c = 18,$
$C = 15° 49';$

find $B = 16° 43' 13'',$
$B' = 163° 16' 47'',$
$A = 147° 27' 47'',$
$A' = 0° 54' 13'',$
$a = 35.519,$
$a' = 1.0415.$

There are two solutions,

for $c < b,$
but $c > b \sin C,$
and $C < 90°.$

$$\begin{aligned} \log b &= 1.27875 \\ \log \sin C &= 9.43546 \\ \text{colog } c &= \underline{8.74473 - 10} \\ \log \sin B &= 9.45894 \end{aligned}$$

$$\begin{aligned} B &= 16° 43' 13''. \\ B' &= 163° 16' 47''. \\ A &= 147° 27' 47''. \\ A' &= 0° 54' 13''. \end{aligned}$$

$$\begin{aligned} \log b &= 1.27875 \\ \text{colog} \sin B &= 0.54106 \\ \log \sin A &= \underline{9.73065} \\ \log a &= 1.55046 \end{aligned}$$

$$a = 35.519.$$

$$\begin{aligned} \log b &= 1.27875 \\ \text{colog} \sin B' &= 0.54106 \\ \log \sin A' &= \underline{8.19784} \\ \log a' &= 0.01765 \end{aligned}$$

$$a' = 1.0415.$$

11. Given $a = 75$, $b = 29$, $B = 16° 15' 36''$; find the difference between the areas of the two corresponding triangles, without computing their areas separately.

The triangle which is the difference of the two triangles has for its altitude $a \sin B$, and two of its sides are of length 29.

$$\begin{aligned} \log a &= 1.87506 \\ \log \sin B &= \underline{9.44715} \\ \log (a \sin B) &= 1.32221 \end{aligned}$$

$$a \sin B = 21.$$

$$\begin{aligned} 29^2 - 21^2 &= (29-21)(29+21) \\ &= 8 \times 50 \\ &= 400. \end{aligned}$$

$$\therefore \sqrt{29^2 - 21^2} = 20.$$

Hence the base of the triangle is $2 \times 20 = 40$, and its altitude 21. Its area is therefore $\frac{1}{2} \times 40 \times 21 = 420$.

12. Given in a parallelogram the side a, a diagonal d, and the angle A made by the two diagonals; find the other diagonal.

Special case: $a = 35$, $d = 63$, $A = 21° 36' 30''$.

$$\begin{aligned} a &= 35. \\ \tfrac{1}{2} d &= 31.5. \\ A &= 21° 36' 30''. \end{aligned}$$

$$\begin{aligned} \text{colog } a &= 8.45593 - 10 \\ \log \tfrac{1}{2} d &= 1.49831 \\ \log \sin A &= \underline{9.56615} \\ \log \sin B &= 9.52039 \end{aligned}$$

$$\begin{aligned} B &= 19° 21' 20''. \\ C &= 139° 2' 10''. \end{aligned}$$

$$\begin{aligned} \log a &= 1.54407 \\ \log \sin C &= 9.81663 \\ \text{colog} \sin A &= \underline{0.43385} \\ \log \tfrac{1}{2} d' &= 1.79455 \end{aligned}$$

$$\begin{aligned} \tfrac{1}{2} d' &= 62.3085. \\ d' &= 124.617. \end{aligned}$$

EXERCISE XIX. PAGE 73.

1. Given $a = 77.99$, $b = 83.39$, $C = 72° 15'$; Find $A = 51° 15'$, $B = 56° 30'$, $c = 95.24$.

$$b + a = 161.38.$$
$$b - a = 5.4.$$
$$B + A = 107° 45'.$$
$$\tfrac{1}{2}(B + A) = 53° 52' 30''$$

$$\log(b - a) = 0.73239$$
$$\text{colog}(b + a) = 7.79215 - 10$$
$$\log\tan\tfrac{1}{2}(B + A) = 0.13675$$
$$\log\tan\tfrac{1}{2}(B - A) = 8.66129$$

$$\tfrac{1}{2}(B - A) = 2° 37' 30''.$$
$$A = 51° 15'.$$
$$B = 56° 30'.$$

$$\log b = 1.92111$$
$$\log\sin C = 9.97882$$
$$\text{colog}\sin B = 0.07889$$
$$\log c = 1.97882$$

$$c = 95.24.$$

2. Given $b = 872.5$, $c = 632.7$, $A = 80°$; Find $B = 60° 45'$, $C = 39° 15'$, $a = 984.83$.

$$b - c = 239.8.$$
$$b + c = 1505.2.$$
$$B + C = 100°.$$
$$\tfrac{1}{2}(B + C) = 50°.$$

$$\log(b - c) = 2.37985$$
$$\log\tan\tfrac{1}{2}(B + C) = 0.07619$$
$$\text{colog}(b + c) = 6.82240 - 10$$
$$\log\tan\tfrac{1}{2}(B - C) = 9.27844$$

$$\tfrac{1}{2}(B - C) = 10° 45'.$$
$$B = 60° 45'.$$
$$C = 39° 15'.$$

$$\log b = 2.94077$$
$$\log\sin A = 9.99335$$
$$\text{colog}\sin B = 0.05924$$
$$\log a = 2.99336$$

$$a = 984.83.$$

3. Given $a = 17$, $b = 12$, $C = 59° 17'$; Find $A = 77° 12' 53''$, $B = 43° 30' 7''$, $c = 14.987$.

$$a + b = 29.$$
$$a - b = 5.$$
$$A + B = 120° 43'.$$
$$\tfrac{1}{2}(A + B) = 60° 21' 30''.$$

$$\log(a - b) = 0.69897$$
$$\text{colog}(a + b) = 8.53760 - 10$$
$$\log\tan\tfrac{1}{2}(A + B) = 10.24486$$
$$\log\tan\tfrac{1}{2}(A - B) = 9.48143$$

$$\tfrac{1}{2}(A - B) = 16° 51' 23''.$$
$$A = 77° 12' 53''.$$
$$B = 43° 30' 7''.$$

$$\log b = 1.07918$$
$$\log\sin C = 9.93435$$
$$\text{colog}\sin B = 0.16218$$
$$\log c = 1.17571$$

$$c = 14.987.$$

4. Given | Find

$b = \sqrt{5}$, $B = 93° 28' 36''$,
$c = \sqrt{3}$, $C = 50° 38' 24''$,
$A = 35° 53'$; $a = 1.313$.

$\sqrt{5} = 2.2361.$
$\sqrt{3} = 1.7321.$
$b + c = 3.9681.$
$b - c = 0.5040.$
$B + C = 144° 7'.$
$\frac{1}{2}(B + C) = 72° 3' 30''.$

$\log(b - c) = 9.70243 - 10$
$\text{colog}(b + c) = 9.40142 - 10$
$\log\tan\frac{1}{2}(B + C) = 10.48973$
$\log\tan\frac{1}{2}(B - C) = 9.59358$
$\frac{1}{2}(B - C) = 21° 25' 6''.$
$B = 93° 28' 36''.$
$C = 50° 38' 24''.$

$\log c = 0.23856$
$\log\sin A = 9.76800$
$\text{colog}\sin C = 0.11172$
$\log a = 0.11828$
$a = 1.313.$

5. Given | Find

$a = 0.917$, $A = 132° 18' 27''$,
$b = 0.312$, $B = 14° 34' 24''$,
$C = 33° 7' 9''$; $c = 0.67748$.

$a + b = 1.229.$
$a - b = 0.605.$
$A + B = 146° 52' 51''.$
$\frac{1}{2}(A + B) = 73° 26' 25''.$
$A = 132° 18' 27''.$
$B = 14° 34' 24''.$

$\log(a - b) = 9.78176 - 10$
$\log\tan\frac{1}{2}(A + B) = 10.52674$
$\text{colog}(a + b) = 9.91045 - 10$
$\log\tan\frac{1}{2}(A - B) = 10.21895$
$\frac{1}{2}(A - B) = 58° 52' 1''.$

$\log b = 9.49415 - 10$
$\log\sin C = 9.73750$
$\text{colog}\sin B = 0.59925$
$\log c = 9.83090 - 10$
$c = 0.67748.$

6. Given | Find

$a = 13.715$, $A = 118° 55' 49''$,
$c = 11.214$, $C = 45° 41' 35''$,
$B = 15° 22' 36''$; $b = 4.1554$.

$a - c = 2.501.$
$a + c = 24.929.$
$A + C = 164° 37' 24''.$
$\frac{1}{2}(A + C) = 82° 18' 42''.$

$\log(a - c) = 0.39811$
$\log\tan\frac{1}{2}(A + C) = 10.86968$
$\text{colog}(a + c) = 8.60330 - 10$
$\log\tan\frac{1}{2}(A - C) = 9.87109$

$\frac{1}{2}(A - C) = 36° 37' 7''.$
$A = 118° 55' 49''.$
$C = 45° 41' 35''.$

$\log\sin B = 9.42352$
$\log a = 1.13720$
$\text{colog}\sin A = 0.05789$
$\log b = 0.61861$
$b = 4.1554.$

7. Given | Find

$b = 3000.9$, $B = 65° 13' 51''$,
$c = 1587.2$, $C = 28° 42' 5''$,
$A = 86° 4' 4''$; $a = 3297.2$.

$b + c = 4588.1.$
$b - c = 1413.7.$
$B + C = 93° 55' 56''.$
$\frac{1}{2}(B + C) = 46° 57' 58''.$

$\log(b - c) = 3.15036$
$\text{colog}(b + c) = 6.33837 - 10$
$\log\tan\frac{1}{2}(B + C) = 10.02983$
$\log\tan\frac{1}{2}(B - C) = 9.51856$

$\frac{1}{2}(B-C) = 18° 15' 53''.$
$C = 28° 42' 5''.$
$B = 65° 13' 51''.$

$\log b = 3.47726$
$\log \sin A = 9.99898$
$\text{colog} \sin B = 0.04191$

$\log a = 3.51815$

$a = 3297.2.$

8. Given / Find

$a = 4527,$ $A = 68° 29' 15'',$
$b = 3465,$ $B = 45° 24' 18'',$
$C = 66° 6' 27'';$ $c = 4449.$

$a + b = 7992.$
$a - b = 1062.$
$A + B = 113° 53' 33''.$
$\frac{1}{2}(A+B) = 56° 56' 47''.$

$\log (a-b) = 3.02612$
$\text{colog} (a+b) = 6.09734-10$
$\log \tan \frac{1}{2}(A+B) = 10.18659$

$\log \tan \frac{1}{2}(A-B) = 9.31005$

$\frac{1}{2}(A-B) = 11° 32' 28''.$
$A = 68° 29' 15''.$
$B = 45° 24' 18''.$

$\log \sin C = 9.96109$
$\text{colog} \sin A = 0.03136$
$\log a = 3.65581$

$\log c = 3.64826$

$c = 4449.$

9. Given / Find

$a = 55.14,$ $A = 117° 24' 33'',$
$b = 33.09,$ $B = 32° 11' 27'',$
$C = 30° 24';$ $c = 31.431.$

$a + b = 88.23.$
$a - b = 22.05.$
$A + B = 149° 36'.$
$\frac{1}{2}(A+B) = 74° 48'.$

$\log (a-b) = 1.34341$
$\text{colog} (a+b) = 8.05438-10$
$\log \tan \frac{1}{2}(A+B) = 10.56592$

$\log \tan \frac{1}{2}(A-B) = 9.96371$

$\frac{1}{2}(A-B) = 42° 36' 33''.$
$A = 117° 24' 33''.$
$B = 32° 11' 27''.$

$\log b = 1.51970$
$\log \sin C = 9.70418$
$\text{colog} \sin B = 0.27348$

$\log c = 1.49736$

$c = 31.431.$

10. Given / Find

$a = 47.99,$ $A = 2° 46' 8'',$
$b = 33.14,$ $B = 1° 54' 42'',$
$C = 175° 19' 10'';$ $c = 81.066.$

$a + b = 81.13.$
$a - b = 14.85.$
$A + B = 4° 40' 50''.$
$\frac{1}{2}(A+B) = 2° 20' 25''.$

$\log (a-b) = 1.17173$
$\text{colog} (a+b) = 8.09082-10$
$\log \tan \frac{1}{2}(A+B) = 8.61138$

$\log \tan \frac{1}{2}(A-B) = 7.87393$

$\frac{1}{2}(A-B) = 0° 25' 43''.$
$A = 2° 46' 8''.$
$B = 1° 54' 42''.$

$\log b = 1.52035$
$\log \sin C = 8.91169$
$\text{colog} \sin B = 1.47680$

$\log c = 1.90884$

$c = 81.066.$

11. If two sides of a triangle are each equal to 6, and the included angle is 60°, find the third side.

Since $a = b$,

$$A = B,$$

$$A + B = 120°.$$

$$\therefore A = B = C = 60°.$$

$$\therefore a = b = c = 6.$$

12. If two sides of a triangle are each equal to 6, and the included angle is 120°, find the third side.

$$A + B = 60°.$$

$$\therefore A = B = 30°,$$

$$a = 6 = b.$$

$$\log a = 0.77815$$

$$\log \sin C = 9.93753$$

$$\text{colog} \sin A = 0.30103$$

$$\log c = 1.01671$$

$$c = 10.392.$$

13. Apply Solution I. to the case in which $a = b$, that is, the case in which the triangle is isosceles.

If $a = b$, the formula

$$\tan \tfrac{1}{2}(A-B) = \frac{a-b}{a+b} \times \tan \tfrac{1}{2}(A+B)$$

will become

$$\tan \tfrac{1}{2}(A - B) = 0.$$

$$\therefore A - B = 0,$$

$$A = B$$

$$= \tfrac{1}{2}(180° - C)$$

$$= 90° - \tfrac{1}{2}C.$$

$$c = \frac{a \sin C}{\sin A}.$$

14. If two sides of a triangle are 10 and 11, and the included angle is 50°, find the third side.

$$a + b = 21.$$

$$a - b = 1.$$

$$A + B = 130°.$$

$$\tfrac{1}{2}(A + B) = 65°.$$

$$\log (a - b) = 0.00000$$

$$\text{colog} (a + b) = 8.67778 - 10$$

$$\log \tan \tfrac{1}{2}(A + B) = 10.33133$$

$$\log \tan \tfrac{1}{2}(A - B) = 9.00911$$

$$\tfrac{1}{2}(A - B) = 5° 49' 51''.$$

$$A = 70° 49' 51''.$$

$$B = 59° 10' 9''.$$

$$\log b = 1.00000$$

$$\log \sin C = 9.88425$$

$$\text{colog} \sin B = 0.06617$$

$$\log c = 0.95042$$

$$c = 8.9212.$$

15. If two sides of a triangle are 43.301 and 25, and the included angle is 30°, find the third side.

$$a + b = 68.301.$$

$$a - b = 18.301.$$

$$A + B = 150°.$$

$$\tfrac{1}{2}(A + B) = 75°.$$

$$\log (a - b) = 1.26247$$

$$\text{colog} (a + b) = 8.16557 - 10$$

$$\log \tan \tfrac{1}{2}(A + B) = 10.57195$$

$$\log \tan \tfrac{1}{2}(A - B) = 9.99999$$

$$\tfrac{1}{2}(A - B) = 45°.$$

$$A = 120°.$$

$$B = 30°.$$

$\therefore$ in isosceles triangle ABC

$$c = b = 25.$$

16. In order to find the distance between two objects A and B separated by a swamp, a station C was chosen, and the distances $CA = 3825$ yards, $CB = 3475.6$ yards, together with the angle $ACB = 62° \; 31'$, were measured. Find the distance from A to B.

$$b + a = 7300.6.$$
$$b - a = 349.4.$$
$$B + A = 117° \; 29'.$$
$$\tfrac{1}{2}(B + A) = 58° \; 44' \; 30''.$$

$$\begin{aligned} \log(b - a) &= 2.54332 \\ \text{colog}(b + a) &= 6.13664 - 10 \\ \log\tan\tfrac{1}{2}(B + A) &= 10.21680 \\ \hline \log\tan\tfrac{1}{2}(B - A) &= 8.89676 \end{aligned}$$

$$\tfrac{1}{2}(B - A) = 4° \; 30' \; 30''.$$
$$B = 63° \; 15'.$$
$$A = 54° \; 14'.$$

$$\begin{aligned} \log b &= 3.58263 \\ \log\sin C &= 9.94799 \\ \text{colog}\sin B &= 0.04916 \\ \hline \log c &= 3.57978 \\ c &= 3800. \end{aligned}$$

17. Two inaccessible objects A and B are each viewed from two stations C and D 562 yards apart. The angle ACB is $62° \; 12'$, BCD $41° \; 8'$, ADB $60° \; 49'$, and ADC $34° \; 51$; required the distance AB.

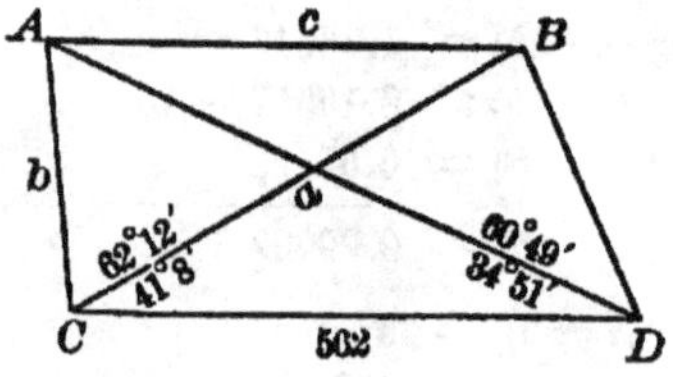

In triangle ACD

$$A = 180° - (C + D)$$
$$= 41° \; 49'.$$

$$\frac{b}{562} = \frac{\sin 34° \; 51'}{\sin 41° \; 49'}.$$

$$\therefore b = \frac{562 \sin 34° \; 51'}{\sin 41° \; 49'}.$$

$$\begin{aligned} \log 562 &= 2.74974 \\ \log\sin 34° \; 51' &= 9.75696 \\ \text{colog}\sin 41° \; 49' &= 0.17604 \\ \hline \log b &= 2.68274 \\ b &= 481.65. \end{aligned}$$

In triangle CBD

$$B = 180° - (C + D)$$
$$= 43° \; 12'.$$

$$\frac{a}{562} = \frac{\sin 95° \; 40'}{\sin 43° \; 12'}.$$

$$\therefore a = \frac{562 \cos 5° \; 40'}{\sin 43° \; 12'}.$$

$$\begin{aligned} \log 562 &= 2.74974 \\ \log\cos 5° \; 40' &= 9.99787 \\ \text{colog}\sin 43° \; 12' &= 0.16460 \\ \hline \log a &= 2.91221 \\ a &= 816.98. \end{aligned}$$

In triangle ACB

$$\tan\tfrac{1}{2}(A - B) = \frac{a - b}{a + b} \times \tan\tfrac{1}{2}(A + B)$$

$$\tfrac{1}{2}(A + B) = \tfrac{1}{2}(180° - C)$$
$$= 58° \; 54'.$$
$$a - b = 816.98 - 481.65$$
$$= 335.33.$$
$$a + b = 816.98 + 481.65$$
$$= 1298.63.$$

$$\begin{aligned} \log(a - b) &= 2.52547 \\ \text{colog}(a + b) &= 6.88651 - 10 \\ \log\tan\tfrac{1}{2}(A + B) &= 10.21951 \\ \hline \log\tan\tfrac{1}{2}(A - B) &= 9.63149 \end{aligned}$$

$$\tfrac{1}{2}(A - B) = 23° \; 10' \; 26''.$$
$$A = 82° \; 4' \; 26''.$$

$$\begin{aligned} \log a &= 2.91221 \\ \log\sin C &= 9.94674 \\ \text{colog}\sin A &= 0.00418 \\ \hline \log c &= 2.86313 \\ c &= 729.68. \end{aligned}$$

18. Two trains start at the same time from the same station, and move along straight tracks that form an angle of 30°, one train at the rate of 30 miles an hour, the other at the rate of 40 miles an hour. How far apart are the trains at the end of half an hour?

$$a + b = 35.$$
$$a - b = 5.$$
$$A + B = 150°.$$
$$\tfrac{1}{2}(A + B) = 75'.$$

$$\log(a - b) = 0.69897$$
$$\text{colog}(a + b) = 8.45593 - 10$$
$$\log\tan\tfrac{1}{2}(A + B) = 10.57197$$
$$\log\tan\tfrac{1}{2}(A - B) = 9.72687$$
$$\tfrac{1}{2}(A - B) = 28°\ 4'.$$
$$B = 46°\ 56'.$$
$$A = 103°\ 4'.$$

$$\log b = 1.17609$$
$$\log\sin C = 9.69897$$
$$\text{colog}\sin B = 0.13634$$
$$\log c = 1.01140$$
$$c = 10.266.$$

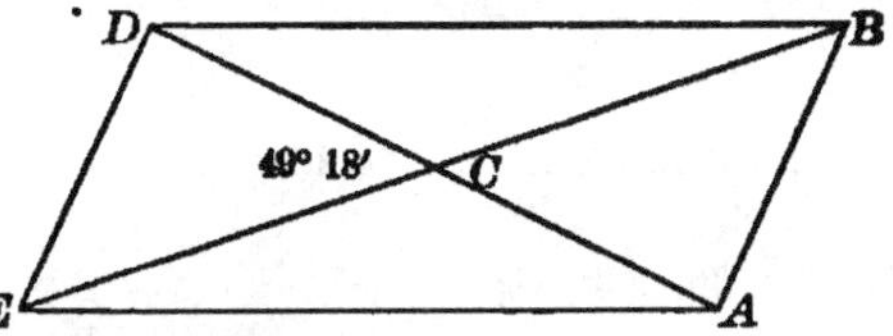

19. In a parallelogram given the two diagonals 5 and 6, and the angle that they form 49° 18'. Find the sides.

In the parallelogram $ABDE$

let $EB = 6$, and $AD = 5$,

and $\angle BCA = 49°\ 18'$.

In triangle ACB

let $BC = a = 3$.

$AC = b = 2.5$.

Find $AB = c$.

$$a - b = 0.5.$$
$$a + b = 5.5.$$
$$A + B = 130°\ 42'.$$
$$\tfrac{1}{2}(A + B) = 65°\ 21'.$$

$$\log(a - b) = 9.69897 - 10$$
$$\text{colog}(a + b) = 9.25964 - 10$$
$$\log\tan\tfrac{1}{2}(A + B) = 10.33829$$
$$\log\tan\tfrac{1}{2}(A - B) = 9.29690$$
$$\tfrac{1}{2}(A - B) = 11°\ 12'\ 20''.$$
$$A = 76°\ 33'\ 20''.$$
$$B = 54°\ 8'\ 40''.$$

$$\log a = 0.47712$$
$$\text{colog}\sin A = 0.01207$$
$$\log\sin C = 9.87975$$
$$\log c = 0.36894$$

$$c = AB = 2.3385.$$

In triangle AEC

$$EC = a = 3,$$
$$AC = b = 2.5,$$
$$\angle ACE = 130°\ 42'.$$
$$A + E = 49°\ 18'$$
$$\tfrac{1}{2}(A + E) = 24°\ 39'$$
$$\tfrac{1}{2}(A - E) = 2°\ 23'\ 20''$$
$$A = 27°\ 2'\ 20''.$$

$$\log(a - b) = 9.69897 - 10$$
$$\text{colog}(a + b) = 9.25964 - 10$$
$$\log\tan\tfrac{1}{2}(A + E) = 9.66171$$
$$\log\tan\tfrac{1}{2}(A - E) = 8.62032$$

$$\tfrac{1}{2}(A - E) = 2°\ 23'\ 20''$$
$$A = 27°\ 2'\ 20''$$

$\log a = 0.47712$
$\text{colog} \sin A = 0.34238$
$\log \sin c = 9.87975 - 10$
$\log c = 0.69925$

$c = EA = 5.0032.$

20. In a triangle one angle equals 139° 54′, and the sides forming the angle have the ratio 5 : 9. Find the other two angles.

$a = 9.$
$b = 5.$
$a + b = 14.$
$a - b = 4.$
$A + B = 40° 6′.$

$\log (a - b) = 0.60206$
$\text{colog} (a + b) = 8.85387 - 10$
$\log \tan \frac{1}{2}(A + B) = 9.56224$
$\log \tan \frac{1}{2}(A - B) = 9.01817$

$\frac{1}{2}(A - B) = 5° 57′ 10″.$
$A = 26° 0′ 10″.$
$B = 14° 5′ 50″.$

Exercise XX. Page 77.

1. Given $a = 51$, $b = 65$, $c = 20$; find the angles.

$a = 51$
$b = 65$
$c = 20$
$2s = 136$

$s = 68.$
$s - a = 17.$
$s - b = 3.$
$s - c = 48.$

$\text{colog}\ s = 8.16749 - 10$
$\text{colog} (s - a) = 8.76955 - 10$
$\log (s - b) = 0.47712$
$\log (s - c) = 1.68124$
$2)\ 19.09540 - 20$
$\log \tan \frac{1}{2}A = 9.54770$

$\frac{1}{2}A = 19° 26′ 24″.$
$A = 38° 52′ 48″.$

$\text{colog}\ s = 8.16749 - 10$
$\text{colog} (s - b) = 9.52288 - 10$
$\log (s - a) = 1.23045$
$\log (s - c) = 1.68124$
$2)\ 20.60206 - 20$
$\log \tan \frac{1}{2}B = 10.30103$

$\frac{1}{2}B = 63° 26′ 6″.$
$B = 126° 52′ 12″.$
$A + B = 165° 45′.$

$\therefore C = 14° 15′.$

2. Given $a = 78$, $b = 101$, $c = 29$; find the angles.

$a = 78$
$b = 101$
$c = 29$

$2s = 208$

$s = 104.$
$s - a = 26.$
$s - b = 3.$
$s - c = 75.$

$\text{colog}\ s = 7.98297 - 10$
$\text{colog} (s - a) = 8.58503 - 10$
$\log (s - b) = 0.47712$
$\log (s - c) = 1.87506$
$2)\ 18.92018 - 20$
$\log \tan \frac{1}{2}A = 9.46009$

$\frac{1}{2}A = 16° 5′ 27″.$
$A = 32° 10′ 54″.$

$$\begin{aligned}
\text{colog } s &= 7.98297 - 10\\
\text{colog } (s-b) &= 9.52288 - 10\\
\log (s-a) &= 1.41497\\
\log (s-c) &= 1.87506\\
2)\ &\overline{20.79588 - 20}\\
\log \tan \tfrac{1}{2} B &= 10.39794
\end{aligned}$$

$$\begin{aligned}
\tfrac{1}{2} B &= 68^\circ\ 11'\ 55''.\\
B &= 136^\circ\ 23'\ 50''.\\
A + B &= 168^\circ\ 34'\ 44''.\\
\therefore C &= 11^\circ\ 25'\ 16''.
\end{aligned}$$

3. Given $a = 111$, $b = 145$, $c = 40$; find the angles.

$$\begin{aligned}
a &= 111\\
b &= 145\\
c &= 40\\
2s &= 296
\end{aligned}$$

$$\begin{aligned}
s &= 148.\\
s - a &= 37.\\
s - b &= 3.\\
s - c &= 108.
\end{aligned}$$

$$\begin{aligned}
\text{colog } s &= 7.82974 - 10\\
\text{colog } (s-a) &= 8.43180 - 10\\
\log (s-b) &= 0.47712\\
\log (s-c) &= 2.03342\\
2)\ &\overline{18.77208 - 20}\\
\log \tan \tfrac{1}{2} A &= 9.38604
\end{aligned}$$

$$\begin{aligned}
\tfrac{1}{2} A &= 13^\circ\ 40'\ 16''.\\
A &= 27^\circ\ 20'\ 32''.
\end{aligned}$$

$$\begin{aligned}
\text{colog } s &= 7.82974 - 10\\
\log (s-a) &= 1.56820\\
\text{colog } (s-b) &= 9.52288 - 10\\
\log (s-c) &= 2.03342\\
2)\ &\overline{20.95424 - 20}\\
\log \tan \tfrac{1}{2} B &= 10.47712
\end{aligned}$$

$$\begin{aligned}
\tfrac{1}{2} B &= 71^\circ\ 33'\ 54''.\\
B &= 143^\circ\ 7'\ 48''.\\
B + A &= 170^\circ\ 28'\ 20''.\\
\therefore C &= 9^\circ\ 31'\ 40''.
\end{aligned}$$

4. Given $a = 21$, $b = 26$, $c = 31$; find the angles.

$$\begin{aligned}
a &= 21\\
b &= 26\\
c &= 31\\
2s &= 78
\end{aligned}$$

$$\begin{aligned}
s &= 39.\\
s - a &= 18.\\
s - b &= 13.\\
s - c &= 8.
\end{aligned}$$

$$\begin{aligned}
\text{colog } s &= 8.40894 - 10\\
\text{colog } (s-a) &= 8.74473 - 10\\
\log (s-b) &= 1.11394\\
\log (s-c) &= 0.90309\\
2)\ &\overline{19.17070 - 20}\\
\log \tan \tfrac{1}{2} A &= 9.58535
\end{aligned}$$

$$\begin{aligned}
\tfrac{1}{2} A &= 21^\circ\ 3'\ 6.3''.\\
\therefore A &= 42^\circ\ 6'\ 13''.
\end{aligned}$$

$$\begin{aligned}
\text{colog } s &= 8.40894 - 10\\
\log (s-a) &= 1.25527\\
\text{colog } (s-b) &= 8.88606 - 10\\
\log (s-c) &= 0.90309\\
2)\ &\overline{19.45336 - 20}\\
\log \tan \tfrac{1}{2} B &= 9.72668
\end{aligned}$$

$$\begin{aligned}
\tfrac{1}{2} B &= 28^\circ\ 3'\ 18''.\\
\therefore B &= 56^\circ\ 6'\ 36''.\\
A + B &= 98^\circ\ 12'\ 49''.\\
\therefore C &= 81^\circ\ 47'\ 11''.
\end{aligned}$$

5. Given $a = 19$, $b = 34$, $c = 49$; find the angles.

$$\begin{aligned}
a &= 19\\
b &= 34\\
c &= 49\\
2s &= 102
\end{aligned}$$

$$\begin{aligned}
s &= 51.\\
s - a &= 32.\\
s - b &= 17.\\
s - c &= 2.
\end{aligned}$$

$\operatorname{colog} s = 8.29243 - 10$
$\operatorname{colog} (s - a) = 8.49485 - 10$
$\log (s - b) = 1.23045$
$\log (s - c) = 0.30103$
$2) \; 18.31876 - 20$
$\log \tan \frac{1}{2} A = 9.15938$
$\frac{1}{2} A = 8° \, 12' \, 48''.$
$A = 16° \, 25' \, 36''.$

$\operatorname{colog} s = 8.29243 - 10$
$\operatorname{colog} (s - b) = 8.76955 - 10$
$\log (s - c) = 0.30103$
$\log (s - a) = 1.50515$
$2) \; 18.86816 - 20$
$\log \tan \frac{1}{2} B = 9.43408$
$\frac{1}{2} B = 15° \, 12'.$
$B = 30° \, 24'.$
$\therefore C = 133° \, 10' \, 24''.$

6. Given $a = 43$, $b = 50$, $c = 57$; find the angles.

$a = 43$
$b = 50$
$c = 57$
$2s = 150$
$s = 75.$
$s - a = 32.$
$s - b = 25.$
$s - c = 18.$

$\operatorname{colog} s = 8.12494 - 10$
$\operatorname{colog} (s - a) = 8.49485 - 10$
$\log (s - b) = 1.39794$
$\log (s - c) = 1.25527$
$2) \; 19.27300 - 20$
$\log \tan \frac{1}{2} A = 9.63650$
$\frac{1}{2} A = 23° \, 24' \, 47''.$
$A = 46° \, 49' \, 35''.$

$\operatorname{colog} s = 8.12494 - 10$
$\log (s - a) = 1.50515$
$\operatorname{colog} (s - b) = 8.60206 - 10$
$\log (s - c) = 1.25527$
$2) \; 19.48742 - 20$
$\log \tan \frac{1}{2} B = 9.74371$
$\frac{1}{2} B = 28° \, 59' \, 52''.$
$B = 57° \, 59' \, 44''.$
$\therefore C = 75° \, 10' \, 41''.$

7. Given $a = 37$, $b = 58$, $c = 79$; find the angles.

$a = 37$
$b = 58$
$c = 79$
$2s = 174$
$s = 87.$
$s - a = 50.$
$s - b = 29.$
$s - c = 8.$

$\operatorname{colog} s = 8.06048 - 10$
$\operatorname{colog} (s - a) = 8.30103 - 10$
$\log (s - b) = 1.46240$
$\log (s - c) = 0.90309$
$2) \; 18.72700 - 20$
$\log \tan \frac{1}{2} A = 9.36350$
$\frac{1}{2} A = 13° \, 0' \, 14''.$
$A = 26° \, 0' \, 29''.$

$\operatorname{colog} s = 8.06048 - 10$
$\log (s - a) = 1.69897$
$\operatorname{colog} (s - b) = 8.53760 - 10$
$\log (s - c) = 0.90309$
$2) \; 19.20014 - 20$
$\log \tan \frac{1}{2} B = 9.60007$
$\frac{1}{2} B = 21° \, 42' \, 40''.$
$B = 43° \, 25' \, 20''.$
$\therefore C = 110° \, 34' \, 11''.$

8. Given $a = 73$, $b = 82$, $c = 91$; find the angles.

$a = 73$
$b = 82$
$c = 91$
$2s = 246$
$s = 123.$
$s - a = 50.$
$s - b = 41.$
$s - c = 32.$

$$\begin{aligned} \text{colog}\, s &= 7.91009 - 10 \\ \text{colog}\,(s-a) &= 8.30103 - 10 \\ \log\,(s-b) &= 1.61278 \\ \log\,(s-c) &= 1.50515 \\ 2)\ &\underline{19.32905 - 20} \\ \log \tan \tfrac{1}{2}A &= 9.66453 \end{aligned}$$

$$\begin{aligned} \tfrac{1}{2}A &= 24^\circ\, 47'\, 29''. \\ A &= 49^\circ\, 34'\, 58''. \end{aligned}$$

$$\begin{aligned} \text{colog}\, s &= 7.91009 - 10 \\ \log\,(s-a) &= 1.69897 \\ \text{colog}\,(s-b) &= 8.38722 - 10 \\ \log\,(s-c) &= 1.50515 \\ 2)\ &\underline{19.50143 - 20} \\ \log \tan \tfrac{1}{2}B &= 9.75072 \end{aligned}$$

$$\begin{aligned} \tfrac{1}{2}B &= 29^\circ\, 23'\, 29''. \\ B &= 58^\circ\, 46'\, 58''. \\ \therefore C &= 71^\circ\, 38'\, 4''. \end{aligned}$$

9. Given $a = 14.493$, $b = 55.4363$, $c = 66.9129$; find the angles.

$$\begin{aligned} a &= 14.493 \\ b &= 55.4363 \\ c &= 66.9129 \\ 2s &= 136.8422 \\ s &= 68.4211. \\ s-a &= 53.9281. \\ s-b &= 12.9848. \\ s-c &= 1.5082. \end{aligned}$$

$$\begin{aligned} \text{colog}\, s &= 8.16481 - 10 \\ \text{colog}\,(s-a) &= 8.26819 - 10 \\ \log\,(s-b) &= 1.11344 \\ \log\,(s-c) &= 0.17846 \\ 2)\ &\underline{17.72490 - 20} \\ \log \tan \tfrac{1}{2}A &= 8.8624 \end{aligned}$$

$$\begin{aligned} \tfrac{1}{2}A &= 4^\circ\, 10'. \\ A &= 8^\circ\, 20'. \end{aligned}$$

$$\begin{aligned} \text{colog}\, s &= 8.16481 - 10 \\ \log\,(s-a) &= 1.73181 \\ \text{colog}\,(s-b) &= 8.88656 - 10 \\ \log\,(s-c) &= 0.17846 \\ 2)\ &\underline{18.96164 - 20} \\ \log \tan \tfrac{1}{2}B &= 9.48082 \end{aligned}$$

$$\begin{aligned} \tfrac{1}{2}B &= 16^\circ\, 50'. \\ B &= 33^\circ\, 40'. \\ \therefore C &= 138^\circ. \end{aligned}$$

10. Given $a=\sqrt{5}$, $b=\sqrt{6}$, $c=\sqrt{7}$; find the angles.

$$\begin{aligned} a &= \sqrt{5} = 2.2361 \\ b &= \sqrt{6} = 2.4495 \\ c &= \sqrt{7} = 2.6458 \\ 2s &= 7.3314 \\ s &= 3.6657 \\ s-a &= 1.4296. \\ s-b &= 1.2162. \\ s-c &= 1.0199. \end{aligned}$$

$$\begin{aligned} \log\,(s-b) &= 0.08500 \\ \log\,(s-c) &= 0.00856 \\ \text{colog}\, s &= 9.43585 - 10 \\ \text{colog}\,(s-a) &= 9.84478 - 10 \\ 2)\ &\underline{19.37419 - 20} \\ \log \tan \tfrac{1}{2}A &= 9.68709 \end{aligned}$$

$$\begin{aligned} \tfrac{1}{2}A &= 25^\circ\, 56'\, 36''. \\ A &= 51^\circ\, 53'\, 12''. \end{aligned}$$

$$\begin{aligned} \text{colog}\,(s-b) &= 9.91500 - 10 \\ \log\,(s-c) &= 0.00856 \\ \text{colog}\, s &= 9.43585 - 10 \\ \log\,(s-a) &= 0.15522 \\ 2)\ &\underline{19.51463 - 20} \\ \log \tan \tfrac{1}{2}B &= 9.75732 \end{aligned}$$

$$\begin{aligned} \tfrac{1}{2}B &= 29^\circ\, 45'\, 54''. \\ B &= 59^\circ\, 31'\, 48''. \\ \therefore C &= 68^\circ\, 35'. \end{aligned}$$

11. Given $a=6$, $b=8$, $c=10$; find the angles.

$$\begin{aligned}
a &= 6.\\
b &= 8.\\
c &= 10.\\
s &= 12.\\
s-a &= 6.\\
s-b &= 4.\\
s-c &= 2.\\
\text{colog } s &= 8.92082-10\\
\text{colog }(s-a) &= 9.22185-10\\
\log(s-b) &= 0.60206\\
\log(s-c) &= \underline{0.30103}\\
2)\ & \underline{19.04576-20}\\
\log\tan\tfrac{1}{2}A &= 9.52288\\
\tfrac{1}{2}A &= 18^\circ\ 26'\ 6''.\\
A &= 36^\circ\ 52'\ 12''.
\end{aligned}$$

Since this is a right triangle,

$$\begin{aligned}
C &= 90^\circ.\\
B &= 90^\circ - A\\
&= 53^\circ\ 7'\ 48''.
\end{aligned}$$

12. Given $a=6$, $b=6$, $c=10$; find the angles.

$$\begin{aligned}
a &= 6\\
b &= 6\\
c &= \underline{10}\\
2s &= 22\\
s &= 11.\\
s-a &= 5.\\
s-b &= 5.\\
s-c &= 1.\\
\text{colog } s &= 8.95861-10\\
\text{colog }(s-c) &= 0.00000\\
\log(s-b) &= 0.69897\\
\log(s-a) &= \underline{0.69897}\\
2)\ & \underline{20.35655-20}\\
\log\tan\tfrac{1}{2}C &= 10.17828\\
\tfrac{1}{2}C &= 56^\circ\ 26'\ 33''.\\
C &= 112^\circ\ 53'\ 6''.
\end{aligned}$$

Since this is an isosceles triangle,

$$\begin{aligned}
A = B &= \tfrac{1}{2}(180^\circ - C)\\
&= 33^\circ\ 33'\ 27''.
\end{aligned}$$

13. Given $a=6$, $b=6$, $c=6$; find the angles.

The triangle is equilateral and also equiangular.

$\therefore A = B = C = \frac{1}{3}$ of $180^\circ = 60^\circ$.

14. Given $a=6$, $b=5$, $c=12$; find the angles.

The sum of the two sides a and b is less than the side c.

$\therefore$ the triangle is impossible.

15. Given $a=2$, $b=\sqrt{6}$, $c=\sqrt{3}-1$; find the angles.

$$\begin{aligned}
a &= 2\\
b &= \sqrt{6} = 2.4495\\
c &= \sqrt{3}-1 = \underline{0.7320}\\
2s &= 5.1815\\
s &= 2.5908.\\
s-a &= 0.5908.\\
s-b &= 0.1413.\\
s-c &= 1.8588.\\
\log(s-a) &= 9.77144-10\\
\log(s-b) &= 9.15014-10\\
\log(s-c) &= 0.26923\\
\text{colog } s &= \underline{9.58656-10}\\
\log r^2 &= 18.77737-20\\
\log r &= 9.38869-10\\
\log\tan\tfrac{1}{2}A &= 9.61725.\\
\log\tan\tfrac{1}{2}B &= 10.23855.\\
\log\tan\tfrac{1}{2}C &= 9.11946.\\
\tfrac{1}{2}A &= 22^\circ\ 30'.\\
\tfrac{1}{2}B &= 60^\circ.\\
\tfrac{1}{2}C &= 7^\circ\ 30'.\\
A &= 45^\circ.\\
B &= 120^\circ.\\
C &= 15^\circ.
\end{aligned}$$

16. Given $a = 2$, $b = \sqrt{6}$, $c = \sqrt{3} + 1$; find the angles.

$$a = 2$$
$$b = \sqrt{6} = 2.4495$$
$$c = \sqrt{3} + 1 = 2.7320$$
$$2s = 7.1815$$
$$s = 3.5908$$
$$s - a = 1.5908$$
$$s - b = 1.1413$$
$$s - c = 0.8588$$

$$\log(s-a) = 0.20162$$
$$\log(s-b) = 0.05740$$
$$\log(s-c) = 9.93385 - 10$$
$$\text{colog}\, s = 9.44481 - 10$$
$$\log r^2 = 19.63768 - 20$$
$$\log r = 9.81884 - 10$$

$$\log\tan\tfrac{1}{2}A = 9.61721.$$
$$\log\tan\tfrac{1}{2}B = 9.76146.$$
$$\log\tan\tfrac{1}{2}C = 9.88494.$$
$$\tfrac{1}{2}A = 22°\ 30'.$$
$$\tfrac{1}{2}B = 30°.$$
$$\tfrac{1}{2}C = 37°\ 30'.$$
$$A = 45°.$$
$$B = 60°.$$
$$C = 75°.$$

17. The distances between three cities A, B, and C are as follows: $AB = 165$ miles, $AC = 72$ miles, and $BC = 185$ miles. B is due east from A. In what direction is C from A? What two answers are admissible?

$$a = 185$$
$$b = 72$$
$$c = 165$$
$$2s = 422$$
$$s = 211.$$
$$(s-a) = 26.$$
$$(s-b) = 139.$$
$$(s-c) = 46.$$

$$\text{colog}\, s = 7.67572 - 10$$
$$\text{colog}(s-a) = 8.58503 - 10$$
$$\log(s-b) = 2.14301$$
$$\log(s-c) = 1.66276$$
$$2)\underline{20.06652 - 20}$$
$$\log\tan\tfrac{1}{2}A = 10.03326$$
$$\tfrac{1}{2}A = 47°\ 11'\ 30''.$$
$$A = 94°\ 23'.$$

Angle $BAC = 94°\ 23'$. Subtract $90°$ of the quadrant E to N, and we obtain $4°\ 23'$ W. of N.

But C may be to the southward of A. Hence two answers are admissible: W. of N. or W. of S.

18. Under what visual angle is an object 7 feet long seen by an observer whose eye is 5 feet from one end of the object and 8 feet from the other end?

$$a = 5$$
$$b = 8$$
$$c = 7$$
$$2s = 20$$
$$s = 10.$$
$$s - a = 5.$$
$$s - b = 2.$$
$$s - c = 3.$$
$$\text{colog}\, s = 9.00000 - 10$$
$$\log(s-a) = 0.69897$$
$$\log(s-b) = 0.30103$$
$$\text{colog}(s-c) = 9.52288 - 10$$
$$2)\underline{19.52288 - 20}$$
$$\log\tan\tfrac{1}{2}C = 9.76144$$
$$\tfrac{1}{2}C = 30°.$$
$$C = 60°.$$

19. When Formula [28] is used for finding the value of an angle, why does the ambiguity that occurs in Case II. not exist?

When Formula [28] is used for finding the value of an angle, the ambiguity that occurs in Case II. does not exist because the sides are all known and the angle can have but one value; while in Case II. the side opposite the angle is not known, and may have two values; therefore the angle also may have two values.

20. If the sides of a triangle are 3, 4, and 6, find the sine of the largest angle.

$$a = 3$$
$$b = 4$$
$$c = 6$$
$$2s = 13$$

$$s = 6.5.$$
$$s - a = 3.5.$$
$$s - b = 2.5.$$
$$s(s - c) = 3.25.$$

$$\log (s - a) = 0.54407$$
$$\log (s - b) = 0.39794$$
$$\text{colog}\, s(s - c) = 9.48812 - 10$$
$$2) \; 20.43013 - 20$$
$$\log \tan \tfrac{1}{2} C = 10.21507$$

$$\tfrac{1}{2} C = 58° \, 38' \, 25''.$$
$$C = 117° \, 16' \, 50''.$$
$$\log \sin C = 9.94879.$$
$$\sin C = 0.88877.$$

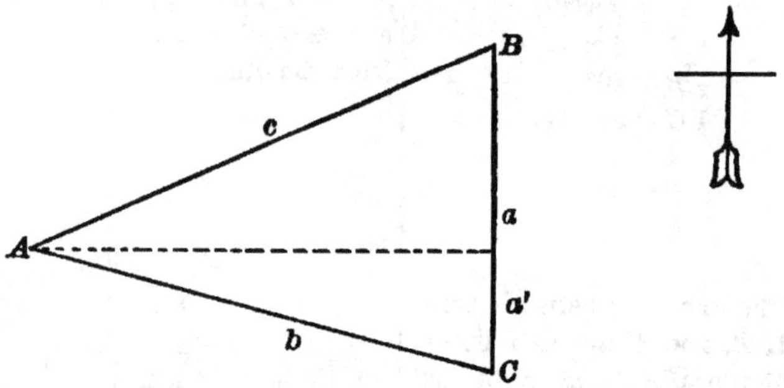

21. Of three towns A, B, and C, A is 200 miles from B and 184 miles from C, B is 150 miles due north from C; how far is A north of C?

$$a = 150$$
$$b = 184$$
$$c = 200$$
$$2s = 534$$

$$s = 267.$$
$$s - a = 117.$$
$$s - b = 83.$$
$$s - c = 67.$$

$$\text{colog}\, s = 7.57349 - 10$$
$$\text{colog}\,(s - c) = 8.17393 - 10$$
$$\log (s - a) = 2.06819$$
$$\log (s - b) = 1.91908$$
$$2) \; 19.73469 - 20$$
$$\log \tan \tfrac{1}{2} A = 9.86735$$

$$\tfrac{1}{2}A = 36° 22' 58''.$$
$$A = 72° 45' 56''.$$

Draw ⊥ from A to BC. To find a' (part cut off by ⊥ on BC from c).

$$a' = b \cos C.$$

$$\log b = 2.26482$$
$$\log \cos C = 9.47171$$
$$\log a' = 1.73653$$
$$a' = 54.516.$$

Exercise XXI. Page 80.

1. Given $a = 4474.5$, $b = 2164.5$, $C = 116° 30' 20''$; find the area.

$$F = \tfrac{1}{2} ab \sin C.$$

$$\log a = 3.65075$$
$$\log b = 3.33536$$
$$\text{colog } 2 = 9.69897 - 10$$
$$\log \sin C = 9.95177$$
$$\log F = 6.63685$$
$$F = 4333600.$$

2. Given $b = 21.66$, $c = 36.94$, $A = 66° 4' 19''$; find the area.

$$F = \tfrac{1}{2} bc \sin A.$$

$$\log b = 1.33566$$
$$\log c = 1.56750$$
$$\log \sin A = 9.96097$$
$$\log 2F = 2.86413$$
$$2F = 731.36.$$
$$F = 365.68.$$

3. Given $a = 510$, $c = 173$, $B = 162° 30' 28''$; find the area.

$$\log a = 2.70757$$
$$\log c = 2.23805$$
$$\log \sin B = 9.47795$$
$$\text{colog } 2 = 9.69897 - 10$$
$$\log F = 4.12254$$
$$F = 13260.$$

4. Given $a = 408$, $b = 41$, $c = 401$; find the area.

$$a = 408$$
$$b = 41$$
$$c = 401$$
$$2s = 850$$
$$s = 425.$$
$$s - a = 17.$$
$$s - b = 384.$$
$$s - c = 24.$$

$$\log s = 2.62839$$
$$\log (s - a) = 1.23045$$
$$\log (s - b) = 2.58433$$
$$\log (s - c) = 1.38021$$
$$2)\ 7.82338$$
$$\log F = 3.91169$$
$$F = 8160.$$

5. Given $a = 40$, $b = 13$, $c = 37$; find the area.

$$a = 40$$
$$b = 13$$
$$c = 37$$
$$2s = 90$$
$$s = 45.$$
$$s - a = 5.$$
$$s - b = 32.$$
$$s - c = 8.$$

$\log s = 1.65321$
$\log (s - a) = 0.69897$
$\log (s - b) = 1.50515$
$\log (s - c) = 0.90309$
$2)\underline{4.76042}$
$\log F = 2.38021$

$F = 240.$

6. Given $a=624$, $b=205$, $c=445$; find the area.

$a = 624$
$b = 205$
$c = 445$
$2s = 1274$

$s = 637.$
$s - a = 13.$
$s - b = 432.$
$s - c = 192.$

$\log s = 2.80414$
$\log (s - a) = 1.11394$
$\log (s - b) = 2.63548$
$\log (s - c) = 2.28330$
$2 \log F = 8.83686$

$\log F = 4.41843$
$F = 26208.$

7. Given $b = 149$, $A = 70° 42' 30''$, $B = 39° 18' 28''$; find the area.

$A = 70° 42' 30''.$
$B = 39° 18' 28''.$
$\therefore C = 69° 59' 2''.$

$\log b = 2.17319$
$\text{colog} \sin B = 0.19827$
$\log \sin A = 9.97490$
$\log a = 2.34636$

$\text{colog } 2 = 9.69897 - 10$
$\log a = 2.34636$
$\log b = 2.17319$
$\log \sin C = 9.97294$
$\log F = 4.19146$

$F = 15540.$

8. Given $a = 215.9$, $c = 307.7$, $A = 25° 9' 31''$; find the area.

$a < c$ and $> c \sin A$.
$A < 90°$. $\therefore$ two solutions.

$\log c = 2.48813$
$\log \sin A = 9.62852$
$\text{colog } a = 7.66575 - 10$
$\log \sin C = 9.78240$

$C = 37° 17' 38''.$
$\therefore B = 117° 32' 51''.$
Or, $C' = 142° 42' 22''.$
$\therefore B' = 12° 8' 7''.$

$\text{colog } 2 = 9.69897 - 10$
$\log a = 2.33425$
$\log c = 2.48813$
$\log \sin B = 9.94774$
$\log F = 4.46909$
$F = 29450.$

$\text{colog } 2 = 9.69897 - 10$
$\log a = 2.33425$
$\log c = 2.48813$
$\log \sin B' = 9.32269$
$\log F' = 3.84404$

$F' = 6983.$

9. Given $b = 8$, $c = 5$, $A = 60°$; find the area.

$F = \frac{1}{2} bc \sin A$
$= \frac{1}{2}(8 \times 5)(0.86602)$
$= 20 \times 0.86602$
$= 17.3204.$

10. Given $a = 7$, $c = 3$, $A = 60°$; find the area.

$$\text{colog } a = 9.15490 - 10$$
$$\log c = 0.47712$$
$$\log \sin A = 9.93753$$
$$\log \sin C = 9.56955$$

$$C = 21° \, 47' \, 12''.$$
$$\therefore B = 98° \, 12' \, 48''.$$
$$F = \tfrac{1}{2} ac \sin B$$
$$= \tfrac{1}{2} \times 21 \times 0.9897$$
$$= 10.3923.$$

11. Given $a = 60$, $B = 40° \, 35' \, 12''$, area $= 12$; find the radius of the inscribed circle.

$$\tfrac{1}{2} ac \sin B = 12.$$
$$c = \frac{24}{a \sin B}.$$

$$\log 24 = 1.38021$$
$$\text{colog } a = 8.22185 - 10$$
$$\text{colog } \sin B = 0.18665$$
$$\log c = 9.78875 - 10$$

$$c = 0.61483.$$

$$\tan \tfrac{1}{2}(A - C)$$
$$= \frac{a - c}{a + c} \times \tan \tfrac{1}{2}(A + C)$$
$$= \frac{59.38517}{60.61483} \times \tan (69° \, 42' \, 24'').$$

$$\log (a - c) = 1.77368$$
$$\text{colog } (a + c) = 8.21742 - 10$$
$$\log \tan \tfrac{1}{2}(A + C) = 0.43206$$
$$\log \tan \tfrac{1}{2}(A - C) = 0.42316$$

$$\tfrac{1}{2}(A - C) = 69° \, 19' \, 19''$$
$$\tfrac{1}{2}(A + C) = 69° \, 42' \, 24''$$
$$\therefore A = 139° \, 1' \, 43''$$

$$\frac{b}{a} = \frac{\sin B}{\sin A}.$$
$$\therefore b = \frac{a \sin B}{\sin A}.$$

$$\log a = 1.77815$$
$$\log \sin B = 9.81331$$
$$\text{colog } \sin A = 0.18331$$
$$\log b = 1.77477$$

$$b = 59.534.$$

$$a = 60$$
$$b = 59.534$$
$$c = 0.61483$$
$$2s = 120.14883$$

$$s = 60.07442.$$
$$F = rs.$$
$$\therefore r = \frac{F}{s}$$
$$= \frac{12}{60.07442}$$
$$= 0.19952.$$

12. Obtain a formula for the area of a parallelogram in terms of two adjacent sides and the included angle.

By Geometry, area of parallelogram = base × height.

In this case, area $= bh$.

But $h = a \sin A$.

$\therefore$ area of $\square = ab \sin A$.

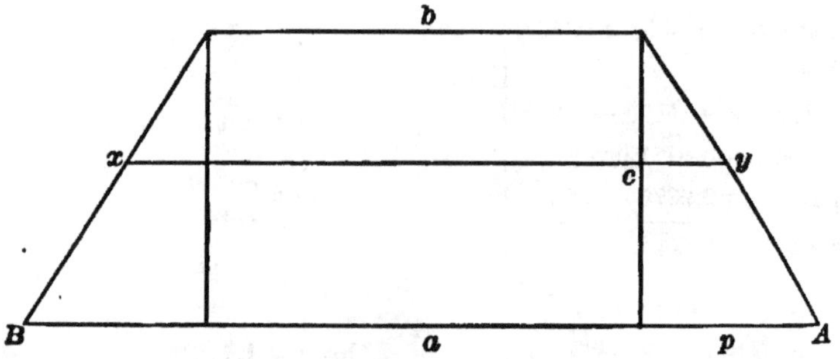

13. Obtain a formula for the area of an isosceles trapezoid in terms of the two parallel sides and an acute angle.

Let $AB = a.$

$$F = \tfrac{1}{2}(a+b)\,c.$$

$$\frac{c}{p} = \tan A.$$

$$c = p \tan A.$$

$$p = \tfrac{1}{2}(a-b).$$

$$\therefore F = \tfrac{1}{2}(a+b) \times \tfrac{1}{2}(a-b)\tan A$$
$$= \tfrac{1}{4}(a^2 - b^2)\tan A.$$

14. Two sides and included angle of a triangle are 2416, 1712, and 30°; and two sides and included angle of another triangle are 1948, 2848, and 150°; find the sum of their areas.

Let $a = 2416,\ c = 1712,\ B = 30°.$
$F = \tfrac{1}{2}ac \sin B.$

$$\begin{aligned} \log a &= 3.38310 \\ \log c &= 3.23350 \\ \text{colog } 2 &= 9.69897 - 10 \\ \log \sin B &= 9.69897 \\ \hline \log F &= 6.01454 \end{aligned}$$

$$F = 1034000.$$

Let $a' = 1948,\ c' = 2848,\ B' = 150°.$
$F' = \tfrac{1}{2}a'c' \sin B'.$

$$\begin{aligned} \log a' &= 3.28959 \\ \log c' &= 3.45454 \\ \text{colog } 2 &= 9.69897 - 10 \\ \log \sin B' &= 9.69897 \\ \hline \log F' &= 6.14207 \end{aligned}$$

$$F' = 1387000.$$

$$F + F' = 2421000.$$

15. The base of an isosceles triangle is 20, and its area is $100 \div \sqrt{3}$; find its angles.

$$a = b.$$

$$c = 20.$$

$$F = 100 \div \sqrt{3}.$$

$$\tfrac{1}{2}ch = \frac{100}{\sqrt{3}}.$$

$$10\,h = \frac{100}{\sqrt{3}}.$$

$$h = \frac{10}{\sqrt{3}}.$$

$$\frac{h}{\tfrac{1}{2}c} = \tan A.$$

$$\begin{aligned} \log h &= 0.76144 \\ \text{colog } \tfrac{1}{2}c &= 9.00000 - 10 \\ \hline \log \tan A &= 9.76144 \end{aligned}$$

$$A = 30°.$$

$$B = 30°.$$

$$C = 120°.$$

16. Show that the area of a quadrilateral is equal to one-half the product of its diagonals into the sine of their included angle.

The diagonals divide the quadrilateral into four equal triangles. If the lengths of the diagonals are a and b, and their included angle C, the area of each of the four triangles is

$$\tfrac{1}{2} \times \tfrac{1}{2}a \times \tfrac{1}{2}b \times \sin C$$
$$= \tfrac{1}{8}ab \sin C.$$

Hence the area of the four triangles is

$$4 \times \tfrac{1}{8}ab \sin C = \tfrac{1}{2}ab \sin C.$$

Exercise XXII. Page 80.

1. From a ship sailing down the English Channel the Eddystone was observed to bear N. 33° 45′ W.; and after the ship had sailed 18 miles S. 67° 30′ W. it bore N. 11° 15′ E. Find its distance from each position of the ship.

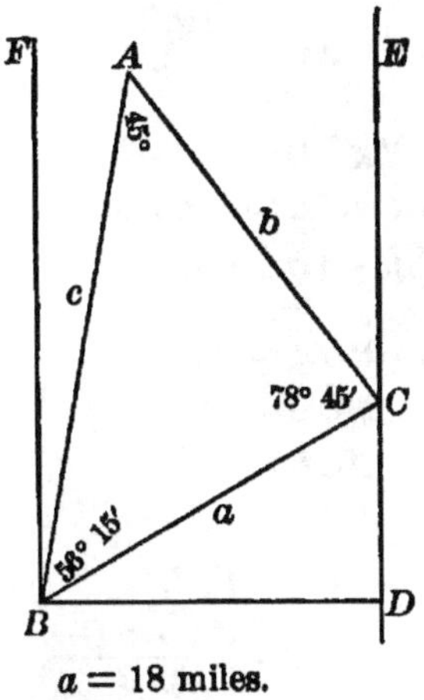

$a = 18$ miles.
$ACE = 33° 45′.$
$DCB = 67° 30′.$
$ABF = 11° 15′.$
$ACB = 180° - (ACE + DCB)$
$= 78° 45′.$
$CBD = 90° - DCB$
$= 22° 30′.$
$ABC = 90° - (CBD + ABF)$
$= 56° 15′.$
$\therefore BAC = 45°.$

$$\frac{b}{a} = \frac{\sin B}{\sin A}.$$

$$\frac{c}{a} = \frac{\sin C}{\sin A}.$$

$\log a = 1.25527$
$\log \sin B = 9.91985$
$\text{colog} \sin A = 0.15051$
$\log b = 1.32563$
$b = 21.166.$

$\log a = 1.25527$
$\log \sin C = 9.99157$
$\text{colog} \sin A = 0.15051$
$\log c = 1.39735$
$c = 24.966.$

2. Two objects, A and B, were observed from a ship to be at the same instant in a line bearing N. 15° E. The ship then sailed northwest 5 miles, when it was found that A bore due east and B bore northeast. Find the distance from A to B.

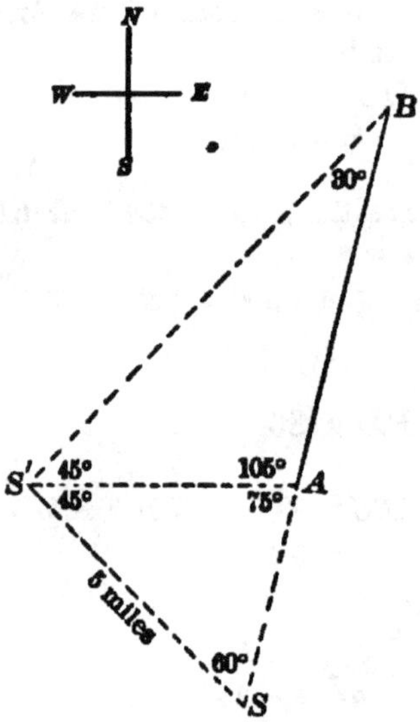

$$\frac{S'A}{SS'} = \frac{\sin ASS'}{\sin S'AS}.$$

$\log SS' = 0.69897$

$\text{colog} \sin SAS' = 0.01506$

$\log \sin ASS' = 9.93753$

$\log S'A = 0.65156$

$$\frac{AB}{S'A} = \frac{\sin BS'A}{\sin S'BA}.$$

$\log S'A = 0.65156$

$\text{colog} \sin S'BA = 0.30103$

$\log \sin BS'A = 9.84949$

$\log AB = 0.80208$

$AB = 6.3399.$

3. A castle and a monument stand on the same horizontal plane. The angles of depression of the top and the bottom of the monument viewed from the top of the castle are 40° and 80°; the height of the castle is 140 feet. Find the height of the monument.

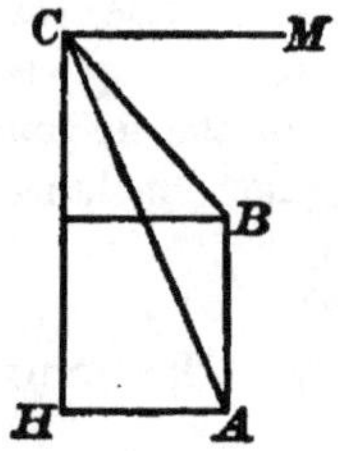

$HC =$ height of castle.

$AB =$ height of monument.

$MCB = 40°.$

$HCA = 10°.$

$HAC = 80°.$

$HC = 140$ ft.

$$AC = \frac{140}{\sin A}.$$

$\log 140 = 2.14613$

$\text{colog} \sin A = 0.00665$

$\log AC = 2.15278$

$HCA = 10°,$

$MCB = 40°.$

$\therefore ACB = 40°,$

$CAB = 10°.$

$\therefore ABC = 130°.$

$$AB = \frac{AC \sin C}{\sin B}.$$

$\log AC = 2.15278$

$\log \sin C = 9.80807$

$\text{colog} \sin B = 0.11575$

$\log AB = 2.07660$

$AB = 119.29.$

4. If the sun's altitude is 60°, what angle must a stick make with the horizon in order that its shadow in a horizontal plane may be the longest possible?

The shadow of the stick will be the longest when the stick is perpendicular to the rays of the sun.

Let BC represent the stick, and AC the horizontal plane.

$$B = 90°.$$
$$A = 60°.$$
$$\therefore C = 30°.$$

5. If the sun's altitude is 30°, find the length of the longest shadow cast on a horizontal plane by a stick 10 feet in length.

Let a be a stick $\perp$ to rays of sun, and c be the longest shadow.

$$\frac{a}{c} = \sin A = \tfrac{1}{2}.$$
$$c = 2a = 20.$$

6. In a circle with the radius 3 find the area of the part comprised between parallel chords whose lengths are 4 and 5. (Two solutions.)

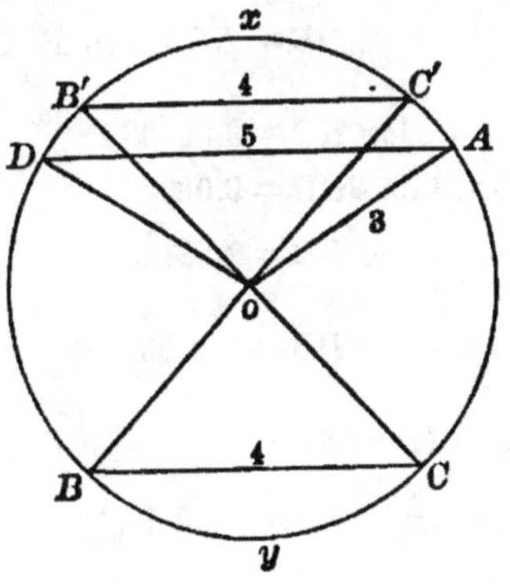

In triangle BOC,

$$h = \sqrt{3^2 - 2^2}$$
$$= \sqrt{5}.$$
$$F = \tfrac{1}{2} \times \sqrt{5} \times 4$$
$$= 2\sqrt{5}.$$
$$\sin \tfrac{1}{2} BOC = \tfrac{2}{3}.$$

$$\log 2 = 0.30103$$
$$\text{colog } 3 = 9.52288 - 10$$
$$\log \sin \tfrac{1}{2} BOC = 9.82391$$

$$\tfrac{1}{2} BOC = 41° \, 48' \, 38''.$$
$$BOC = 83° \, 37' \, 16''.$$

By Table VI.,

$$R = 3.$$
$$\therefore \text{area } \odot = 28.274.$$

Area of sector BOC

$$= \frac{83° \, 37' \, 16''}{360°} \times 28.274$$
$$= \frac{301036}{1296000} \times 28.274$$
$$= \frac{75259}{324000} \times 28.274.$$

$$\log 75259 = 4.87656$$
$$\log 28.274 = 1.45139$$
$$\text{colog } 324000 = 4.48945 - 10$$
$$\log \text{area} = 0.81740$$

$$\text{Area} = 6.5675.$$

Area of segment ByC

$$= 6.5675 - 2\sqrt{5}$$
$$= 6.5675 - 4.4722$$
$$= 2.0953.$$

In triangle DOA,

$$h = \sqrt{3^2 - 2.5^2}$$
$$= 1.6583.$$
$$F = \tfrac{1}{2} \times 1.6583 \times 5$$
$$= 4.1458.$$
$$\sin \tfrac{1}{2} DOA = \tfrac{5}{6}.$$

$$\log 5 = 0.69897$$
$$\log 6 = 9.22185 - 10$$
$$\log \sin \tfrac{1}{2} DOA = 9.92082$$

$$\tfrac{1}{2} DOA = 56° 26' 35''.$$
$$DOA = 112° 53' 10''.$$

Area of sector DOA

$$= \frac{406391}{1296000} \times 28.274.$$

$$\log 406391 = 5.60894$$
$$\log 28.274 = 1.45139$$
$$\text{colog } 1296000 = 3.88739 - 10$$
$$\log \text{area} = 0.94772$$

Area sector = 8.8658.

Area segment DxA

$$= 4.72.$$

Area segment $DACB$

$$= \text{area} \odot - [ByC + DxA]$$
$$= 21.4587.$$

Area segment $DAC'B'$

$$= DxA - B'xC'$$
$$= 2.6247.$$

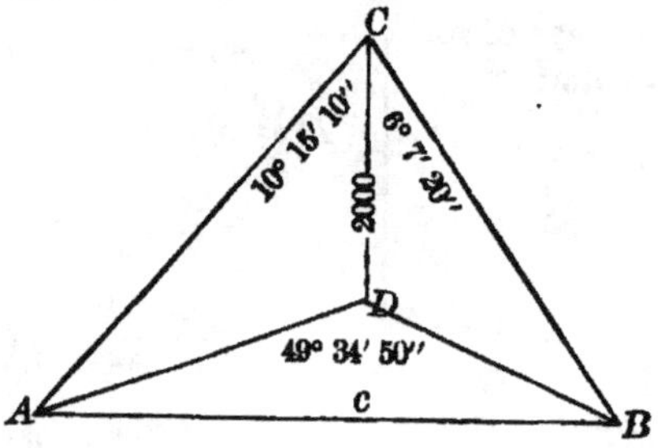

7. A and B, two inaccessible objects in the same horizontal plane, are observed from a balloon at C and from a point D directly under the balloon, and in the same horizontal plane with A and B. If $CD = 2000$ yards, $\angle ACD = 10° 15' 10''$, $\angle BCD = 6° 7' 20''$, $\angle ADB = 49° 34' 50''$, find AB.

$$AD = DC \times \tan ACD.$$

$$\log \tan ACD = 9.25739$$
$$\log DC = 3.30103$$
$$\log AD = 2.55842$$

$$AD = 361.76.$$

$$DB = DC \times \tan BCD.$$

$$\log DC = 3.30103$$
$$\log \tan BCD = 9.03045$$
$$\log DB = 2.33148$$

$$DB = 214.53.$$

$$\tan \tfrac{1}{2}(B - A) = \frac{b - a}{b + a} \times \tan \tfrac{1}{2}(B + A).$$

$\frac{1}{2}(B+A) = 65° 12' 35''.$

$\log(b-a) = 2.16800$
$\text{colog}(b+a) = 7.23936 - 10$
$\log\tan\frac{1}{2}(B+A) = 0.33549$
$\log\tan\frac{1}{2}(B-A) = 9.74285$

$\frac{1}{2}(B-A) = 28° 56' 58''$

$B = 94° 9' 33''.$

$\log AD = 2.55842$
$\text{colog}\sin B = 0.00115$
$\log\sin C = 9.88156$
$\log c = 2.44113$

$c = AB = 276.14.$

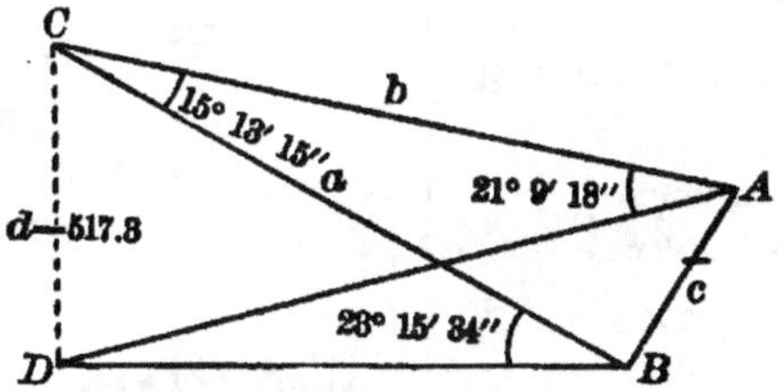

8. *A* and *B* are two objects whose distance, on account of intervening obstacles cannot be directly measured. At the summit *C* of a hill, whose height above the common horizontal plane of the objects is known to be 517.3 yards, ∠ *ACB* is found to be 15° 13′ 15″. The angles of elevation of *C* viewed from *A* and *B* are 21° 9′ 18″ and 23° 15′ 34″ respectively. Find the distance from *A* to *B*.

In triangle *DCA*, being a rt. △,

$$\frac{d}{b} = \sin A.$$

$$b = \frac{d}{\sin A}.$$

$\log d = 2.71374$
$\text{colog}\sin A = 0.44262$

$\log b = 3.15636$

$b = 1433.4.$

In right triangle *CDB*,

$$\frac{d}{a} = \sin B.$$

$$a = \frac{d}{\sin B}.$$

$\log d = 2.71374$
$\text{colog}\sin B = 0.40352$
$\log a = 3.11726$

$a = 1310.$

$$\tan\frac{1}{2}(B-A) = \frac{b-a}{b+a} \times \tan\frac{1}{2}(B+A).$$

$\frac{1}{2}(B+A) = 82° 23' 22.5''.$

$\log(b-a) = 2.09132$
$\text{colog}(b+a) = 6.56171 - 10$
$\log\tan\frac{1}{2}(B+A) = 10.87415$
$\log\tan\frac{1}{2}(B-A) = 9.52718$

$\frac{1}{2}(B-A) = 18° 36' 21''.$
$B = 100° 59' 43.5''.$
$A = 63° 47' \ 1.5''.$

$$c = \frac{a\sin C}{\sin A}.$$

$\log a = 3.11726$
$\log\sin C = 9.41920$
$\text{colog}\sin A = 0.04714$
$\log c = 2.58360$

$c = 383.35.$

MISCELLANEOUS EXAMPLES. PAGE 82.

2. The angle of elevation of a tower is 48° 19′ 14″, and the distance of the base from the point of observation is 95 ft. Find the height of the tower, and the distance of the top from the point of observation.

Given $A = 48° 19' 14''$, $b = 95$ ft.; required a and c.

$$a = b \tan A.$$
$$c = b \sec A.$$

$$\begin{aligned} \log b &= 1.97772 \\ \log \tan A &= \underline{10.05045} \\ \log a &= 2.02817 \end{aligned}$$

$$a = 106.70.$$

$$\begin{aligned} \log b &= 1.97772 \\ \log \sec A &= \underline{0.17720} \\ \log c &= 2.15492 \end{aligned}$$

$$c = 142.86.$$

Height of tower, 106.70 ft.; distance of top from point of observation, 142.86 ft.

3. From a mountain 1000 ft. high, the angle of depression of a ship is 77° 35′ 11″. Find the distance of the ship from the summit of the mountain.

Given $B = 12° 24' 49''$, $a = 1000$ ft.; required c.

$$c = a \sec B.$$

$$\begin{aligned} \log a &= 3.00000 \\ \log \sec B &= \underline{0.01027} \\ \log c &= 3.01027 \end{aligned}$$

$$c = 1023.9.$$

Reouired distance, 1023.9 ft.

4. A flag-staff 90 ft. high, on a horizontal plane, casts a shadow of 117 ft. Find the altitude of the sun.

Given $a = 90$ ft., $b = 117$ ft.; required A.

$$\tan A = \frac{a}{b}.$$

$$\begin{aligned} \log a &= 1.95424 \\ \text{colog } b &= \underline{7.93181 - 10} \\ \log \tan A &= 9.88605 \end{aligned}$$

$$A = 37° 34' 5''.$$

Altitude of sun, 37° 34′ 5″.

5. When the moon is setting at any place, the angle at the moon subtended by the earth's radius passing through that place is 57′ 3″. If the earth's radius is 3956.2 miles, what is the moon's distance from the earth's centre?

Let C represent the place, A the moon, and B the earth's centre. Then in the right triangle ABC. given $A = 57' 3''$, $a = 3956.2$ miles; required c.

$$c = a \csc A.$$

$$\begin{aligned} \log a &= 3.59728 \\ \log \csc A &= \underline{1.78004} \\ \log c &= 5.37732 \end{aligned}$$

$$c = 238400.$$

Moon's distance, 238400 miles.

6. The angle at the earth's centre subtended by the sun's radius is 16′ 2″, and the sun's distance is 92,400,000 miles. Find the sun's diameter in miles.

Let A represent the centre of the earth, B that of the sun, and C a point on the edge of the sun's disk. Then in the right triangle ABC, given $A = 16' \ 2''$, $C = 92{,}400{,}000$ miles; required $2a$.

$$a = c \sin A.$$

$$\log c = 7.96567$$
$$\log \sin A = 7.66875$$
$$\log a = 5.63442$$
$$a = 430940.$$

Sun's diameter, 861880 miles.

7. The latitude of Cambridge, Mass., is 42° 22′ 49″. What is the length of the radius of that parallel of latitude?

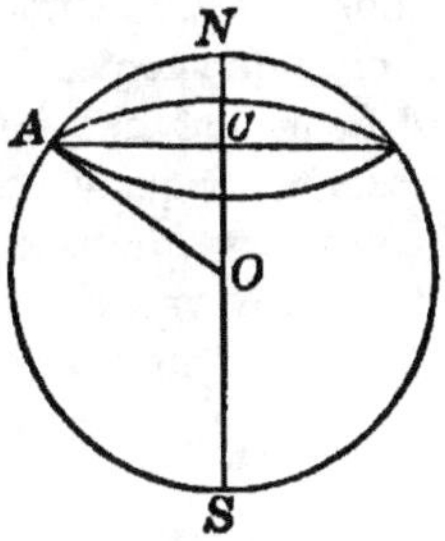

Let O be the centre of the earth, NS the axis, NAS the meridian of Cambridge, A the position of Cambridge, and C the centre of its parallel of latitude. Then, in the right triangle OAC, given $O = 90° - 42° \ 22' \ 49'' = 47° \ 37' \ 11''$, $OA = 3956.2$ miles; required AC.

$$AC = AO \sin O.$$

$$\log AO = 3.59728$$
$$\log \sin O = 9.86846$$
$$\log AC = 3.46574$$
$$AC = 2922.4.$$

Radius of parallel of latitude, 2922.4 miles.

8. At what latitude is the circumference of the parallel of latitude half of that of the equator?

The radius of the parallel will be half of the radius of the earth.

In the figure of Ex. 7, given $AC = \frac{1}{2} AO$; required $90° -$ angle O, *i.e.* angle A.

$$\cos A = \frac{AC}{AO} = \frac{1}{2}.$$

$$\therefore A = 60°.$$

The required latitude is 60°.

9. In a circle with a radius of 6.7 is inscribed a regular polygon of thirteen sides. Find the length of one of the sides.

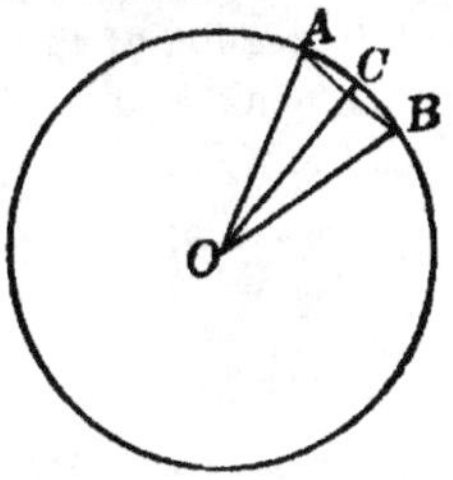

Let O be the centre of the circle, AB a side of the polygon, and C the middle point. Then in the right triangle OCB, given $O = \frac{360°}{26} = 13° \ 50' \ 46''$, $OB = 6.7$; required $AB = 2\,CB$.

$$CB = OB \sin BOC.$$

$$\log OB = 0.82607$$
$$\log \sin BOC = 9.37897$$
$$\log CB = 0.20504$$
$$CB = 1.6034.$$
$$AB = 3.2068.$$

Length of a side of the polygon, 3.2068.

10. A regular heptagon one side of which is 5.73 is inscribed in a circle. Find the radius of the circle.

In the figure of Ex. 9, given $BC = \frac{1}{2} \times 5.73 = 2.865$ and angle $BOC = \frac{360^\circ}{14} = 25^\circ\,42'\,51''$; required OB.

$$OB = BC \csc BOC.$$

$$\begin{aligned} \log BC &= 0.45712 \\ \log \csc BOC &= 0.36263 \\ \log OB &= 0.81975 \end{aligned}$$

$$OB = 6.6031.$$

Radius of circle, 6.6031.

11. A tower 93.97 ft. high is situated on the bank of a river. The angle of depression of an object on the opposite bank is $25^\circ\,12'\,54''$. Find the breadth of the river.

Given $A = 90^\circ - 25^\circ\,12'\,54'' = 64^\circ\,47'\,6''$, $b = 93.97$; required a.

$$a = b \tan A.$$

$$\begin{aligned} \log b &= 1.97299 \\ \log \tan A &= 10.32708 \\ \log a &= 2.30007 \end{aligned}$$

$$a = 199.56.$$

Breadth of river, 199.56 ft.

12. From a tower 58 ft. high the angles of depression of two objects situated in the same horizontal line with the base of the tower, and on the same side, are $30^\circ\,13'\,18''$ and $45^\circ\,46'\,14''$. Find the distance between the objects.

(i.) Given $A = 90^\circ - 30^\circ\,13'\,18'' = 59^\circ\,46'\,42''$, $b = 58$; required a.

$$a = b \tan A.$$

$$\begin{aligned} \log b &= 1.76343 \\ \log \tan A &= 10.23469 \\ \log a &= 1.99812 \end{aligned}$$

$$a = 99.568.$$

(ii.) Given $A' = 90^\circ - 45^\circ\,46'\,14'' = 44^\circ\,13'\,46''$, $b = 58$, required a'.

$$a' = b \tan A'.$$

$$\begin{aligned} \log b &= 1.76343 \\ \log \tan A' &= 9.98832 \\ \log a' &= 1.75175 \end{aligned}$$

$$a' = 56.461.$$

$$a - a' = 43.107.$$

Distance between the objects, 43.107 ft.

13. Standing directly in front of one corner of a flat-roofed house which is 150 ft. in length, I observe that the horizontal angle which the length subtends has for its cosine $\sqrt{\frac{1}{5}}$, and that the vertical angle subtended by its height has for its sine $\frac{3}{\sqrt{34}}$. What is the height of the house?

Let a = distance of observer from house,
b' = height of house,
B = horizontal angle subtended by length of house,
B' = vertical angle subtended by height of house.

Then,
$$\begin{aligned} a &= 150 \operatorname{ctn} B. \\ b &= a \tan B' \\ &= 150 \operatorname{ctn} B \tan B'. \end{aligned}$$

But $\cos B = \sqrt{\frac{1}{5}}$,

hence
$$\sin B = \sqrt{1 - \tfrac{1}{5}} = \frac{2}{\sqrt{5}}.$$

$$\text{ctn } B = \frac{\cos B}{\sin B}$$
$$= \tfrac{1}{2}.$$

Also, $\sin B' = \dfrac{3}{\sqrt{34}}$.

$$\therefore \cos B' = \frac{5}{\sqrt{34}},$$
$$\tan B' = \tfrac{3}{5}.$$

Hence, $b = 150 \times \frac{1}{2} \times \frac{3}{5}$
$$= 45.$$

Height of house, 45 ft.

14. A regular pyramid with a square base has an edge 150 feet in length, and the length of a side of its base is 200 ft. Find the inclination of the face of the pyramid to the base.

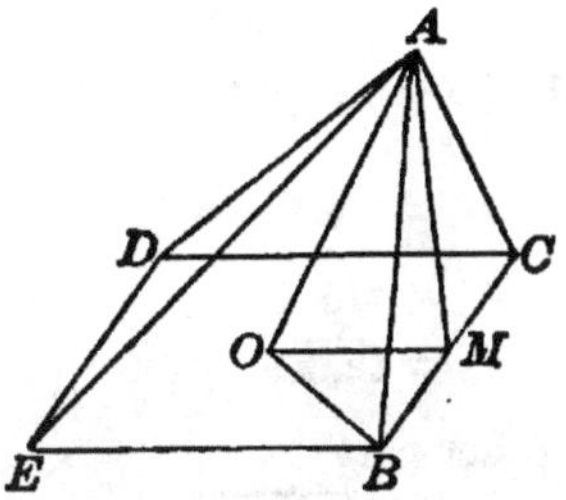

Let A be the vertex of the pyramid, $BCDE$ its base, O the centre of the base, and M the middle point of the side BC. Required the angle AMO.

In the right triangle AOB,

$$AB = 150,$$
$$OB = \tfrac{1}{2} BD$$
$$= 100\sqrt{2}.$$
$$\therefore AO = \sqrt{\overline{AB}^2 - \overline{OB}^2}$$
$$= 50.$$

In the right triangle AOM,

$$\tan OMA = \frac{AM}{OM}$$
$$= \tfrac{50}{100}$$
$$= 0.5.$$
$$OMA = 26° \ 34'.$$

Inclination of face of pyramid to base, 26° 34'.

15. From one edge of a ditch 36 ft. wide the angle of elevation of a wall on the opposite edge is 62° 39' 10". Find the length of a ladder which will reach from the point of observation to the top of the wall.

Given $b = 36$, $A = 62° \ 39' \ 10''$; required c.

$$c = b \sec A.$$

$$\log b = 1.55630$$
$$\log \sec A = 0.33783$$
$$\log c = 1.89413$$

$$c = 78.367.$$

Length of ladder, 78.367 ft.

16. The top of a flag-staff has been broken off, and touches the ground at a distance of 15 ft. from the foot of the staff. The length of the broken part being 39 ft., find the whole length of the staff.

Given $c = 39$, $b = 15$; required $c + a$.

$$a = \sqrt{(c+b)(c-b)}.$$
$$= \sqrt{1296}$$
$$= 36.$$
$$c + a = 75.$$

Whole length of flag-staff, 75 ft.

17. From a balloon, which is directly above one town, is observed the angle of depression of another town, 10° 14′ 9″. The towns being 8 miles apart, find the height of the balloon.

Given $A = 90° - 10°\ 14'\ 9'' = 79°\ 45'\ 51''$, $a = 8$; required b.

$$b = a \cot A.$$

$$\log a = 0.90309$$
$$\log \cot A = 9.25666$$
$$\log b = 0.15975$$
$$b = 1.4446.$$

Height of balloon, 1.4446 miles.

18. From the top of a mountain 3 miles high the angle of depression of the most distant object which is visible on the earth's surface is found to be 2° 13′ 50.″ Find the diameter of the earth.

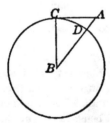

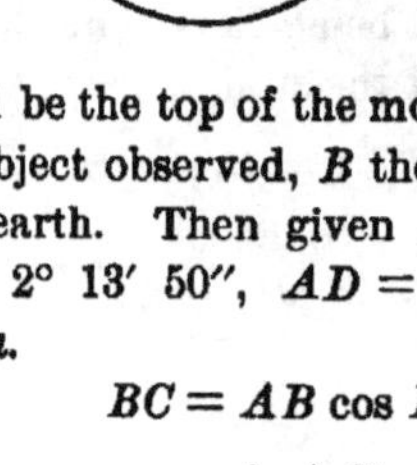

Let A be the top of the mountain, C the object observed, B the centre of the earth. Then given $B = 90° - A = 2°\ 13'\ 50''$, $AD = 3$; required a.

$$BC = AB \cos B,$$

$$a = (a + 3) \cos B.$$

$$\therefore a(1 - \cos B) = 3 \cos B.$$

$$a = \frac{3 \cos B}{1 - \cos B}$$
$$= \frac{3 \cos B}{2 \sin^2 \frac{B}{2}}.$$

$$\log \tfrac{3}{2} = 0.17609$$
$$\log \cos B = 9.99967$$
$$\text{colog} \sin^2 \frac{B}{2} = 3.42152$$
$$\log a = 4.59728$$
$$a = 3956.2.$$

Radius of earth, 3956.2 miles.

19. A ladder 40 ft. long reaches a window 33 ft. high on one side of a street. Being turned over upon its foot, it reaches another window 21 ft. high, on the opposite side of the street. Find the width of the street.

Width of the one part of the street

$$= \sqrt{40^2 - 33^2}$$
$$= \sqrt{511}$$
$$= 22.605.$$

Width of other part

$$= \sqrt{40^2 - 21^2}$$
$$= \sqrt{1159}$$
$$= 34.044.$$

Total width of the street, 56.649 ft.

20. The height of a house subtends a right angle at a window on the other side of the street, and the elevation of the top of the house from the same point is 60°. The street is 30 ft. wide. How high is the house?

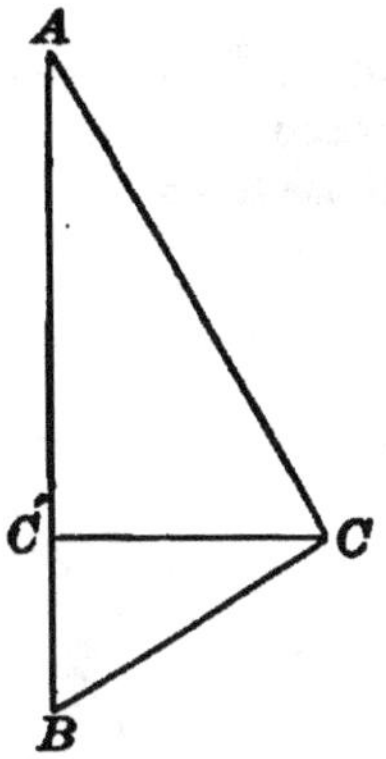

Given $CC' = 30$, $ACC' = 60°$, $BCC' = 30°$; required AB.

$$AC' = CC' \tan ACC'$$
$$= 30\sqrt{3}.$$
$$BC' = CC' \tan BCC'$$
$$= 30 \times \frac{1}{\sqrt{3}}$$
$$= 10\sqrt{3}.$$
$$\therefore AB = 40\sqrt{3}.$$
$$= 69.282.$$

Height of house, 69.282 ft.

21. A lighthouse 54 feet high is situated on a rock. The elevation of the top of the lighthouse, as observed from a ship, is 4° 52′, and the elevation of the top of the rock is 4° 2′. Find the height of the rock and its distance from the ship.

Let h = height of rock.
a = distance of ship.

Then $$\frac{h+54}{h} = \frac{\tan 4° 52'}{\tan 4° 2'}.$$

$$1 + \frac{54}{h} = \frac{\tan 4° 52'}{\tan 4° 2'}.$$

$$\frac{54}{h} = \frac{\tan 4° 52' - \tan 4° 2'}{\tan 4° 2'}.$$

$$h = 54 \frac{\tan 4° 2'}{\tan 4° 52' - \tan 4° 2'}$$

$$= 54 \frac{\cos 4° 52' \sin 4° 2'}{\sin(4° 52' - 4° 2')}$$

$$= 54 \frac{\cos 4° 52' \sin 4° 2'}{\sin 50'}.$$

$$\begin{aligned} \log 54 &= 1.73239 \\ \log \cos 4° 52' &= 9.99843 \\ \log \sin 4° 2' &= 8.84718 \\ \text{colog} \sin 50' &= 1.83732 \\ \log h &= 2.41532 \\ h &= 260.20. \end{aligned}$$

Also $a = h \text{ ctn } 4° 2'.$

$$\begin{aligned} \log h &= 2.41532 \\ \log \text{ctn } 4° 2' &= 11.15174 \\ \log a &= 3.56706 \\ a &= 3690.3. \end{aligned}$$

Height of rock, 260.20 ft.; distance of ship, 3690.3 ft.

22. A man in a balloon observes the angle of depression of an object on the ground, bearing south, to be 35° 30′; the balloon drifts $2\frac{1}{2}$ miles east at the same height, when the angle of depression of the same object is 23° 14′. Find the height of the balloon.

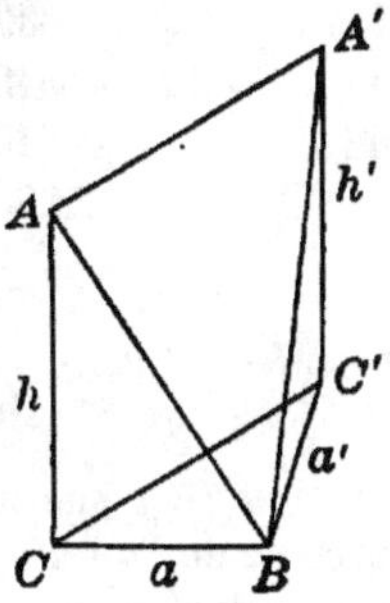

Let A and A' be the first and second positions of the balloon, respectively, C and C' the points on the ground directly under A and A', and B the object observed.

Then $A = 54° 30'$,

$$A' = 66° 46',$$

$$CC' = AA'$$

$$= 2\tfrac{1}{2}.$$

$$a = h \tan A,$$

$$a' = h \tan A'.$$

$$a'^2 - a^2 = (2\tfrac{1}{2})^2.$$

$$h^2 \tan^2 A' - h^2 \tan^2 A = (2\tfrac{1}{2})^2.$$

$$h^2 = \frac{(2\tfrac{1}{2})^2}{\tan^2 A' - \tan^2 A}.$$

$$h = \frac{2\tfrac{1}{2}}{\sqrt{\tan^2 A' - \tan^2 A}}.$$

But $\tan^2 A' - \tan^2 A$

$$= (\tan A' + \tan A)(\tan A' - \tan A)$$

$$= \frac{\sin(A'+A)}{\cos A' \cos A} \times \frac{\sin(A'-A)}{\cos A' \cos A}$$

$$= \frac{\sin(A'+A)\sin(A'-A)}{\cos^2 A' \cos^2 A}.$$

Hence

$$h = \frac{2\tfrac{1}{2}\cos A' \cos A}{\sqrt{\sin(A'+A)\sin(A'-A)}}.$$

$$\log 2\tfrac{1}{2} = 0.39794$$
$$\log \cos A' = 9.59602$$
$$\log \cos A = 9.76395$$
$$\text{colog}\sqrt{\sin(A'+A)} = 0.03408$$
$$\text{colog}\sqrt{\sin(A'-A)} = 0.33636$$
$$\log h = 0.12835$$

$$h = 1.3438.$$

Height of balloon, 1.3438 miles.

23. A man standing south of a tower, on the same horizontal plane, observes its elevation to be 54° 16′; he goes east 100 yds., and then finds its elevation is 50° 8′. Find the height of the tower.

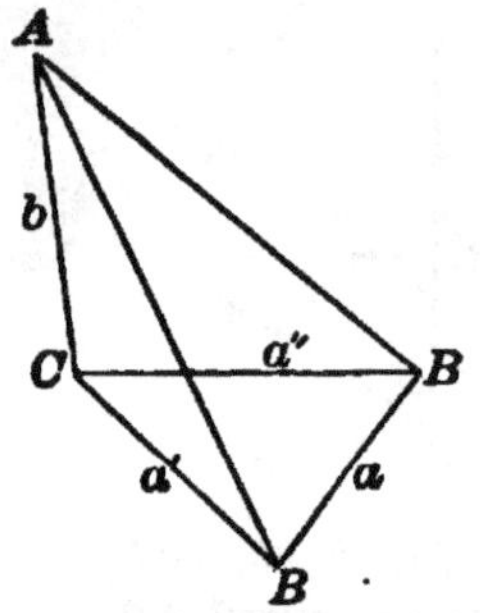

Let AC be the tower, B and B' the first and second positions of the observer.

Then, $BB' = 100.$

$$a' = b \cot ABC.$$

$$a'' = b \cot AB'C.$$

$$a''^2 - a'^2 = a^2.$$

$$b^2(\cot^2 AB'C - \cot^2 ABC) = 100^2.$$

$$b = \frac{100}{\sqrt{\cot^2 50° 8' - \cot^2 54° 16'}}$$

$$= \frac{100 \sin 54° 16' \sin 50° 8'}{\sqrt{\sin 104° 24' \sin 4° 8'}}.$$

$$\log 100 = 2.00000$$
$$\log \sin 54° 16' = 9.90942$$
$$\log \sin 50° 8' = 9.88510$$
$$\text{colog}\sqrt{\sin 104° 24'} = 0.00693$$
$$\text{colog}\sqrt{\sin 4° 8'} = 0.57110$$
$$\log b = 2.37255$$

$$b = 235.80.$$

Height of tower, 235.80 yds.

24. The elevation of a tower at a place A south of it is 30°; and at a place B, west of A, and at a distance a from it, the elevation is 18°. Show that the height of the tower is $\frac{a}{\sqrt{2+2\sqrt{5}}}$, the tangent of 18° being $\frac{\sqrt{5}-1}{\sqrt{10+2\sqrt{5}}}$.

With the figure and notation of the last example,

$$b = \frac{a}{\sqrt{\cot^2 18° - \cot^2 30°}}.$$

$$\text{But } \cot^2 18° = \frac{10+2\sqrt{5}}{6-2\sqrt{5}}$$

$$= \frac{(10+2\sqrt{5})(6+2\sqrt{5})}{6^2-(2\sqrt{5})^2}$$

$$= 5+2\sqrt{5},$$

and $\cot^2 30° = 3$.

$$\text{Hence} \quad b = \frac{a}{\sqrt{2+2\sqrt{5}}}.$$

25. A pole is fixed on the top of a mound, and the angles of elevation of the top and bottom of the pole are 60° and 30°. Prove that the length of the pole is twice the height of the mound.

Let l = length of pole.
h = height of mound.
a = horizontal distance of observer.

Then $h = a \tan 30°$.

$$h + l = a \tan 60°.$$

$$\frac{h+l}{h} = \frac{\tan 60°}{\tan 30°}$$

$$= \frac{\sqrt{3}}{\frac{1}{\sqrt{3}}}$$

$$= 3.$$

$$h + l = 3h.$$

$$\therefore l = 2h.$$

26. At a distance (a) from the foot of a tower, the angle of elevation (A) of the tower is the complement of the angle of elevation of a flag-staff on top of it. Show that the length of the staff is $2a \cot 2A$.

Let h = height of tower.
l = length of staff.

Then $h = a \tan A$.

$$h + l = a \cot A.$$

$$l = a(\cot A - \tan A)$$

$$= a\frac{\cot^2 A - 1}{\cot A}$$

$$= 2a \cot 2A.$$

27. A line of true level is a line every point of which is equally distant from the centre of the earth. A line drawn tangent to a line of true level at any point is a line of apparent level. If at any point both these lines are drawn and extended one mile, find the distance they are then apart.

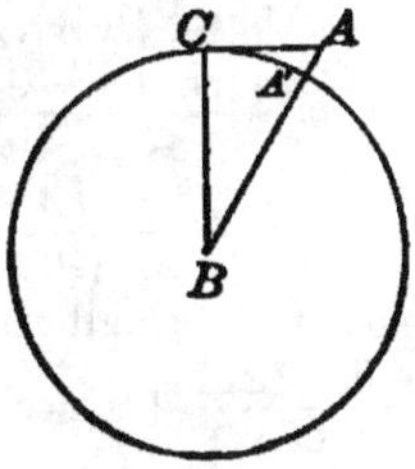

Given $CA = 1$ mile, $BC =$ radius of the earth $= 3956.2$ miles; required $AA' = AB - CB$.

The required distance is much too small to be obtained by the usual process of solution. It is most easily found as follows:

$$\overline{AC}^2 = \overline{AB}^2 - \overline{BC}^2$$
$$= (AB - BC)(AB + BC).$$
$$\therefore AB - BC = \frac{\overline{AC}^2}{AB + BC}.$$

Now, as AB differs very little from BC, and both are very large in comparison with $\overline{AC}^2$, we may assume as a close approximation that $AB = BC$. Then

$$AA' = AB - BC$$
$$= \frac{\overline{AC}^2}{2\,BC}$$
$$= \frac{1}{7912.4} \text{ miles}$$
$$= \frac{5280 \times 12}{7912.4} \text{ inches.}$$

$$\log 5280 = 3.72263$$
$$\log \quad 12 = 1.07918$$
$$\text{colog } 7912.4 = 6.10169 - 10$$
$$\log AA' = 0.90350$$
$$AA' = 8.0076 \text{ inches.}$$

The required distance is 8 inches.

28. In problem 2, determine the effect upon the computed height of the tower, of an error in either the angle of elevation or the measured distance.

With the notation of Ex. 2, suppose that the error in the angle is e_1 and that in the measured distance is e_2. Then the formulas

$$a = b \tan A, \qquad c = b \sec A$$

become $$a = (b + e_2) \tan (A + e_1), \qquad c = (b + e_2) \sec (A + e_1),$$

and the error in the computed value of a is

$$(b + e_2) \tan (A + e_1) - b \tan A$$
$$= b\,\{\tan (A + e_1) - \tan A\} + e_2 \tan (A + e_1)$$
$$= \frac{b \sin e_1}{\cos (A + e_1) \cos A} + e_2 \tan (A + e_1),$$

or, approximately, for small errors,

$$\frac{be_1}{\cos^2 A} + e_2 \tan A,$$

where e_1 is measured in radians.

The error in c is

$$b\,\{\sec (A + e_1) - \sec A\} + e_2 \sec (A + e_1)$$
$$= \frac{b\,\{\cos (A + e_1) - \cos A\}}{\cos (A + e_1) \cos A} + e_2 \sec (A + e_1).$$

[23] $$= \frac{-2\,b \sin (A + \tfrac{1}{2} e_1) \sin (\tfrac{1}{2} e_1)}{\cos (A + e_1) \cos A} + e_2 \sec (A + e_1),$$

or, approximately, for small errors,

$$\frac{-be_1 \sin A}{\cos^2 A} + e_2 \sec A = (-be_1 \tan A + e_2) \sec A.$$

29. To determine the height of an inaccessible object, situated on a horizonal plane, by observing its angles of elevation at two points in the same line with its base, and measuring the distance of these two points.

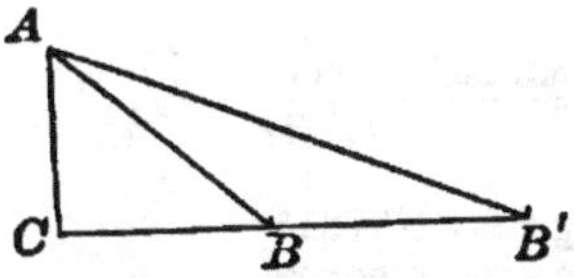

Let AC be the object, B and B' the two points of observation. Then given the angles B' and ABC, and the side BB'; required AC.

$$AB = BB' \frac{\sin B'}{\sin BAB'}$$

$$= BB' \frac{\sin B'}{\sin (ABC - B')}.$$

$$AC = AB \sin ABC$$

$$= BB' \frac{\sin B' \sin ABC}{\sin (ABC - B')}.$$

30. The angle of elevation of an inaccessible tower, situated on a horizontal plane, is 63° 26′; at a point 500 ft. farther from the base of the tower the elevation of its top is 32° 14′. Find the height of the tower.

From the solution of Ex. 29,

$$AC = 500 \frac{\sin 32° 14' \sin 63° 26'}{\sin (63° 26' - 32° 14')}$$

$$= 500 \frac{\sin 32° 14' \sin 63° 26'}{\sin 31° 12'}.$$

$$\begin{aligned} \log 500 &= 2.69897 \\ \log \sin 32° 14' &= 9.72703 \\ \log \sin 63° 26' &= 9.95154 \\ \text{colog} \sin 31° 12' &= 0.28565 \\ \log AC &= 2.66319 \\ AC &= 460.46. \end{aligned}$$

Height of the tower, 460.46 ft.

31. A tower is situated on the bank of a river. From the opposite bank the angle of elevation of the tower is 60° 13′, and from a point 40 ft. more distant the elevation is 50° 19′. Find the breadth of the river.

In the figure for the solution of Ex. 29,

$$CB = AB \cos ABC$$

$$= BB' \frac{\sin B' \cos ABC}{\sin (ABC - B')}.$$

Hence,

$$CB = 40 \frac{\sin 50° 19' \cos 60° 13'}{\sin 9° 54'}.$$

$$\begin{aligned} \log 40 &= 1.60206 \\ \log \sin 50° 19' &= 9.88626 \\ \log \cos 60° 13' &= 9.69611 \\ \text{colog} \sin 9° 54' &= 0.76465 \\ \log CB &= 1.94908 \\ CB &= 88.936. \end{aligned}$$

Breadth of river, 88.936 ft.

32. A ship sailing north sees two lighthouses 8 miles apart, in a line due west; after an hour's sailing, one lighthouse bears S. W., the other S.S.W. Find the ship's rate.

In the figure for the solution of Ex. 29, let B and B' be the lighthouses, C the original position of its ship, and A its final position.

Then $CBA = 22° 30'$ and $CAB' = 45°$; hence $ABC = 67° 30'$ and $B' = 45°$.

$$AC = BB' \frac{\sin 45° \sin 67° 30'}{\sin 22° 30'}$$

$$= BB' \sin 45° \cot 22° 30'.$$

$$\begin{aligned} \log 8 &= 0.90309 \\ \log \sin 45° &= 9.84949 \\ \log \cot 22° 30' &= 10.38278 \\ \hline \log AC &= 1.13536 \end{aligned}$$

$$AC = 13.657.$$

Ship's rate, 13.657 miles per hour.

33. To determine the height of an accessible object situated on an inclined plane.

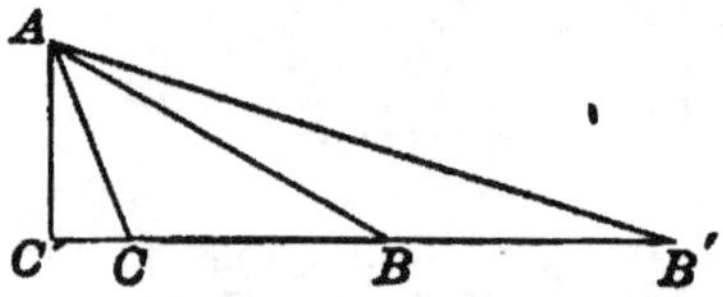

Let CBB' be the inclined plane, AC the object, B and B' two points of observation, AC' the perpendicular from A on CBB'. Then, given CB, BB', and the angles ABC, B', required AC.

From the solution of Ex. 29,

$$AC' = BB' \frac{\sin B' \sin ABC}{\sin (ABC - B')},$$

and $$C'B = BB' \frac{\sin B' \cos ABC}{\sin (ABC - B')}.$$

Then $C'C = C'B - CB$,

and $$AC = \sqrt{\overline{AC'}^2 + \overline{C'C}^2}.$$

34. At a distance of 40 ft. from the foot of a tower on an inclined plane, the tower subtends an angle of 41° 19'; at a point 60 ft. farther away, the angle subtended by the tower is 23° 45'. Find the height of the tower.

From the solution of Ex. 33,

$$AC' = 60 \frac{\sin 23° 45' \sin 41° 19'}{\sin 17° 34'}.$$

$$C'B = 60 \frac{\sin 23° 45' \cos 41° 19'}{\sin 17° 34'}.$$

$$\begin{aligned} \log 60 &= 1.77815 \\ \log \sin 23° 45' &= 9.60503 \\ \log \sin 41° 19' &= 9.81969 \\ \text{colog} \sin 17° 34' &= 0.52026 \\ \hline \log AC' &= 1.72313 \end{aligned}$$

$$AC' = 52.860.$$

$$\begin{aligned} \log 60 &= 1.77815 \\ \log \sin 23° 45' &= 9.60503 \\ \log \cos 41° 19' &= 9.87568 \\ \text{colog} \sin 17° 34' &= 0.52026 \\ \hline \log C'B &= 1.77912 \end{aligned}$$

$$C'B = 60.134.$$
$$C'C = 20.134.$$

$$\tan ACC' = \frac{AC'}{C'C}.$$

$$AC = AC' \csc ACC'.$$

$$\begin{aligned} \log AC' &= 1.72313 \\ \log C'C &= 1.30393 \\ \hline \log \tan ACC' &= 0.41920 \end{aligned}$$

$$ACC' = 69° 8' 55''.$$

$$\begin{aligned} \log AC' &= 1.72313 \\ \log \csc ACC' &= 0.02941 \\ \hline \log AC &= 1.75254 \end{aligned}$$

$$AC = 56.564.$$

Height of tower, 56.564 ft.

35. A tower makes an angle of 113° 12′ with the inclined plane on which it stands; and at a distance of 89 ft. from its base, measured down the plane, the angle subtended by the tower is 23° 27′. Find the height of the tower.

In the triangle ACB, given $CB =$ 89 ft., $C = 113° \ 12'$, $B = 23° \ 27'$; required AC.

$$A = 180° - (B + C)$$
$$= 43° \ 21'.$$

$$AC = CB \frac{\sin B}{\sin A}.$$

$$\log 89 = 1.94939$$
$$\log \sin 23° \ 27' = 9.59983$$
$$\text{colog} \sin 43° \ 21' = 0.16339$$
$$\log AC = 1.71261$$

$$AC = 51.595.$$

Height of tower, 51.595 ft.

36. From the top of a house 42 ft. high, the angle of elevation of the top of a pole is 14° 13′; at the bottom of the house it is 23° 19′. Find the height of the pole.

Let A be the top of the pole, B and B' the top and bottom of the house, and C the foot of the perpendicular from A on BB'; required $B'C$.

From the solution of Exs. 29 and 31,

$$CB = BB' \frac{\sin AB'C \cos ABC}{\sin (ABC - AB'C)}$$

$$= 42 \frac{\sin 66° \ 41' \cos 75° \ 47'}{\sin 9° \ 6'}.$$

$$\log 42 = 1.62325$$
$$\log \sin 66° \ 41' = 9.96300$$
$$\log \cos 75° \ 47' = 9.39021$$
$$\text{colog} \sin 9° \ 6' = 0.80091$$
$$\log CB = 1.77737$$

$$CB = 59.892.$$
$$B'C = CB + BB'$$
$$= 59.892 + 42$$
$$= 101.892.$$

Height of pole, 101.892 ft.

37. The sides of a triangle are 17, 21, 28; prove that the length of a line bisecting the greatest side and drawn to the opposite angle is 13.

Let

$$a = 28, \ b = 21, \ c = 17,$$

then

$$17^2 = 28^2 + 21^2 - 2 \times 28 \times 21 \cos C;$$

to prove that

$$13^2 = 14^2 + 21^2 - 2 \times 14 \times 21 \cos C.$$

Subtract the first equation from twice the second,

$$2 \times 13^2 - 17^2 = 2 \times 14^2 - 28^2 + 21^2$$
$$= 21^2 - 2 \times 14^2,$$
$$2 \times 169 - 289 = 441 - 2 \times 196,$$
$$49 = 49.$$

38. A privateer, 10 miles S.W. of a harbor, sees a ship sail from it in a direction S. 80° E. at a rate of 9 miles an hour. In what direction, and at what rate, must the privateer sail in order to come up with the ship in $1\frac{1}{2}$ hours?

Let A be the harbor, B the original position of the privateer, and C the point where the vessels are to meet. Then $A=125°$, $b=13\frac{1}{2}$, $C=10$; required B and $\frac{a}{1\frac{1}{2}}$.

$$\tan\tfrac{1}{2}(B-C)=\frac{b-c}{b+c}\tan\tfrac{1}{2}(B+C)$$
$$=\frac{3.5}{23.5}\tan 27°\,30'.$$

$$\begin{aligned}
\log 3.5 &= 0.54407\\
\text{colog } 23.5 &= 8.62803-10\\
\log\tan 27°\,30' &= 9.71648\\
\hline
\log\tan\tfrac{1}{2}(B-C) &= 8.88948\\
\tfrac{1}{2}(B-C) &= 4°\,26'\\
B-C &= 8°\,52'\\
B+C &= 55°\\
\hline
B &= 31°\,56'
\end{aligned}$$

$$a=b\frac{\sin A}{\sin B}$$
$$=13.5\frac{\sin 125°}{\sin 31°56'}.$$

$$\begin{aligned}
\log 13.5 &= 1.13033\\
\log\sin 125° &= 9.91336\\
\text{colog}\sin 31°\,56' &= 0.27660\\
\hline
\log a &= 1.32029\\
a &= 20.907.\\
\frac{a}{1\frac{1}{2}} &= 13.938.
\end{aligned}$$

Privateer's course, 31° 56′ E. of N.E., or N. 76° 56′ E.; rate 13.938 miles per hour.

39. A person goes 70 yards up a slope of 1 in $3\frac{1}{2}$ from the edge of a river, and observes the angle of depression of an object on the opposite shore to be $2\frac{1}{4}°$. Find the breadth of the river.

Let A and B be the original and final positions of the observer, and C the object observed. Then, given $c=70$, $C=2\frac{1}{4}°$, $A=180°-\tan^{-1}\frac{1}{3\frac{1}{2}}$; required b.

$$\begin{aligned}
A &= 180°-\tan^{-1}\tfrac{2}{7}\\
&= 180°-\tan^{-1}0.2857\\
&= 180°-15°\,56'\,40''\\
&= 164°\,3'\,20''.\\
B &= 180°-(A+C)\\
&= 13°\,41'\,40''.\\
b &= c\frac{\sin B}{\sin C}.
\end{aligned}$$

$$\begin{aligned}
\log 70 &= 1.84510\\
\log\sin 13°\,41'\,40'' &= 9.37428\\
\text{colog}\sin 2°\,15' &= 1.40605\\
\hline
\log b &= 2.62543\\
b &= 422.11.
\end{aligned}$$

Breadth of river, 422.11 yds.

40. The length of a lake subtends, at a certain point, an angle of 46° 24″, and the distances from this point to the two extremities of the lake are 346 and 290 feet. Find the length of the lake.

Given $A=46.24'$, $b=346$, $c=290$; required a.

$$\tan\tfrac{1}{2}(B-C)=\frac{b-c}{b+c}\tan\tfrac{1}{2}(B+C)$$
$$=\frac{56}{636}\tan 66°\,48'.$$

$$\begin{aligned}
\log 56 &= 1.74819\\
\text{colog } 636 &= 7.19654-10\\
\log\tan 66°\,48' &= 10.36795\\
\hline
\log\tan\tfrac{1}{2}(B-C) &= 9.31268\\
\tfrac{1}{2}(B-C) &= 11°\,36'\,33''\\
B-C &= 23°\,13'\,6''\\
B+C &= 133°\,36'\\
\hline
B &= 78°\,24'\,33''
\end{aligned}$$

$$a = b\,\frac{\sin A}{\sin B}$$

$$= 346\,\frac{\sin 46^\circ\, 24''}{\sin 78^\circ\, 24'\, 33''}.$$

$$\log 346 = 2.53908$$
$$\log \sin 46^\circ\, 24' = 9.85984$$
$$\text{colog} \sin 78^\circ\, 24'\, 33'' = 0.00895$$
$$\log a = 2.40787$$
$$a = 255.78.$$

Length of lake, 255.78 ft.

41. Two ships are a mile apart. The angular distance of the first ship from a fort on shore, as observed from the second ship, is 35° 14′ 10″; the angular distance of the second ship from the fort, observed from the first ship, is 42° 11′ 53″. Find the distance in feet from each ship to the fort.

Given $B = 35^\circ\, 14'\, 10''$, $C = 42^\circ\, 11'\, 53''$, $a = 5280$; required b and c.

$$A = 180 - (B + C)$$
$$= 102^\circ\, 33'\, 57''.$$

$$b = a\,\frac{\sin B}{\sin A}.$$

$$c = a\,\frac{\sin C}{\sin A}.$$

$$\log 5280 = 3.72263$$
$$\log \sin 35^\circ\, 14'\, 10'' = 9.76116$$
$$\text{colog} \sin 102^\circ\, 33'\, 57'' = 0.01053$$
$$\log b = 3.49432$$
$$b = 3121.2.$$

$$\log 5280 = 3.72263$$
$$\log \sin 42^\circ\, 11'\, 53'' = 9.82717$$
$$\text{colog} \sin 102^\circ\, 33'\, 57'' = 0.01053$$
$$\log c = 3.56033$$
$$c = 3633.5.$$

Distance of first ship from fort, 3121.2 ft.; of second ship from fort, 3633.5 ft.

42. Along the bank of a river is drawn a base line of 500 ft. The angular distance of one end of this line from an object on the opposite side of the river, as observed from the other end of the line, is 53°; that of the second extremity from the same object, observed at the first, is 79° 12′. Find the perpendicular breadth of the river.

Given $B = 53^\circ$, $C = 79^\circ\, 12'$, $a = 500$; required p, the perpendicular from A on a.

$$b = a\,\frac{\sin B}{\sin A}.$$

$$p = b \sin C$$

$$= a\,\frac{\sin B \sin C}{\sin A}$$

$$= 500\,\frac{\sin 53^\circ \sin 79^\circ\, 12'}{\sin 47^\circ\, 48'}.$$

$$\log 500 = 2.69897$$
$$\log \sin 53^\circ = 9.90235$$
$$\log \sin 79^\circ\, 12' = 9.99224$$
$$\text{colog} \sin 47^\circ\, 48' = 0.13030$$
$$\log p = 2.72386$$
$$p = 529.49.$$

Perpendicular breadth of river, 529.49 ft.

43. A vertical tower stands on a declivity inclined 15° to the horizon. A man ascends the declivity 80 ft. from the base of the tower, and finds the angle then subtended by the tower to be 30°. Find the height of the tower.

Let A and B be the top and bottom of the tower, and C the position of observation. Then, given $a = 80$, $B = 75^\circ$, $C = 30^\circ$; required c.

$$A = 180° - (B + C)$$
$$= 75°.$$

$\therefore$ the triangle is isosceles, and

$$C = 2a \cos B$$
$$= 160 \cos 75°.$$

$$\log 160 = 2.20412$$
$$\log \cos 75° = 9.41300$$
$$\log C = 1.61712$$
$$C = 41.411.$$

Height of tower, 41.411 ft.

44. The angle subtended by a tower on an inclined plane is, at a certain point, 42° 17′; 325 ft. farther down, it is 21° 47′. The inclination of the plane is 8° 53′. Find the height of the tower.

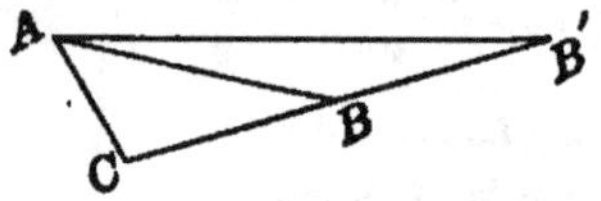

$$AB = BB' \frac{\sin B'}{\sin BAB'}$$
$$= BB' \frac{\sin B'}{\sin (B - B')}.$$
$$AC = \frac{AB \sin B}{\sin C}$$
$$= BB' \frac{\sin B \sin B'}{\sin C \sin (B - B')}$$
$$= 325 \frac{\sin 42° 17' \sin 21° 47'}{\sin 81° 7' \sin 20° 30'}.$$

$$\log 325 = 2.51188$$
$$\log \sin 42° 17' = 9.82788$$
$$\log \sin 21° 47' = 9.56949$$
$$\text{colog} \sin 81° 7' = 0.00524$$
$$\text{colog} \sin 20° 30' = 0.45567$$
$$\log AC = 2.37016$$
$$AC = 234.51.$$

Height of tower, 234.51 ft.

45. A cape bears north by east, as seen from a ship. The ship sails northwest 30 miles, and then the cape bears east. How far is it from the second point of observation?

Let A be the cape, B and C the first and second positions of the ship. Then, given $B = 56° 15'$, $C = 45°$, $a = 30$; required b.

$$A = 180° - (B + C)$$
$$= 78° 45'.$$
$$b = \frac{a \sin B}{\sin A}$$
$$= \frac{30 \sin 56° 15'}{\sin 78° 45'}.$$

$$\log 30 = 1.47712$$
$$\log \sin 56° 15' = 9.91985$$
$$\text{colog} \sin 78° 45' = 0.00843$$
$$\log b = 1.40540$$
$$b = 25.433.$$

Distance of cape from second point of observation, 25.433 miles.

46. Two observers, stationed on *opposite* sides of a cloud, observe its angle of elevation to be 44° 56′ and 36° 4′. Their distance from each other is 700 ft. What is the linear height of the cloud?

Given $A = 44° 56'$, $B = 36° 4'$, $c = 700$; required the perpendicular p from C on c.

$$C = 180° - (A + B)$$
$$= 99°.$$
$$p = b \sin A$$
$$= c \frac{\sin B \sin A}{\sin C}$$
$$= 700 \frac{\sin 36° 4' \sin 44° 56'}{\sin 99°}.$$

$$\begin{aligned} \log 700 &= 2.84510 \\ \log \sin 36° 4' &= 9.76991 \\ \log \sin 44° 56' &= 9.84898 \\ \text{colog} \sin 99° &= 0.00538 \\ \hline \log p &= 2.46937 \\ p &= 294.69. \end{aligned}$$

Linear height of cloud, 294.69 ft.

47. From a point B at the foot of a mountain, the elevation of the top A is 60°. After ascending the mountain one mile, at an inclination of 30° to the horizon, and reaching a point C, the angle ACB is found to be 135°. Find the height of the mountain in feet.

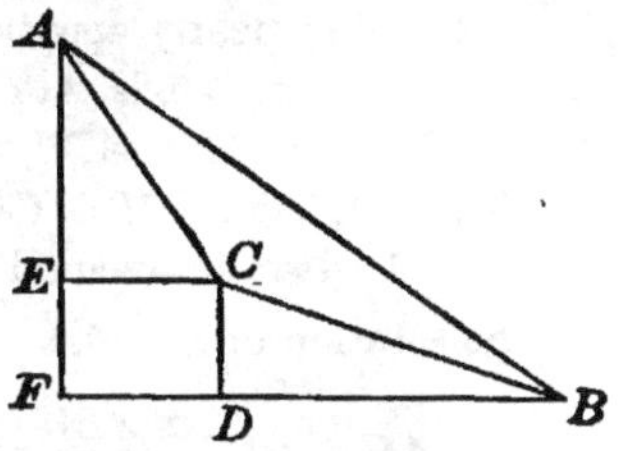

$$\begin{aligned} CD &= CB \sin CBD \\ &= 5280 \times \tfrac{1}{2} \\ &= 2640. \\ AE &= AC \sin ECA \\ &= \frac{CB \sin CBA \sin ECA}{\sin CAB} \\ &= \frac{5280 \sin 30° \sin 75°}{\sin 15°} \\ &= \frac{5280 \times \frac{1}{2} \cos 15°}{\sin 15°} \\ &= 2640 \cot 15°. \end{aligned}$$

$$\begin{aligned} \log 2640 &= 3.42160 \\ \log \cot 15° &= 10.57195 \\ \hline \log AE &= 3.99355 \\ AE &= 9852.6. \\ AF &= AE + CD \\ &= 12492.6. \end{aligned}$$

Height of the mountain, 12492.6 ft.

48. From a ship two rocks are seen in the same right line with the ship, bearing N. 15° E. After the ship has sailed northwest 5 miles, the first rock bears east, and the second northeast. Find the distance between the rocks.

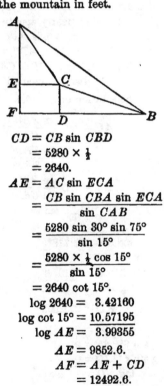

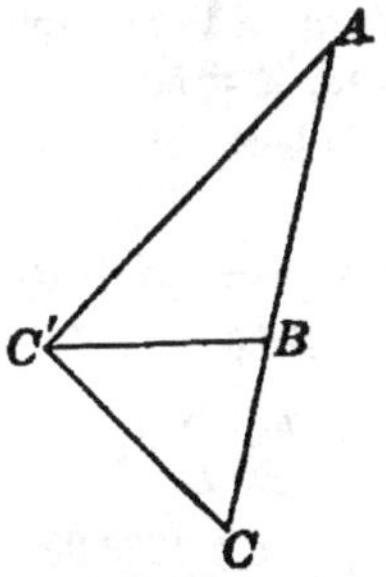

Let A and B be the two rocks, C and C' the first and second positions of the ship. Then given $C = 60°$, $CC'B = 45°$, $CC'A = 90°$, $CC' = 5$; required AB.

$$\begin{aligned} AC &= CC' \sec C \\ &= 5 \times 2 = 10. \\ BC &= CC' \frac{\sin BC'C}{\sin CBC'} \\ &= 5 \frac{\sin 45°}{\sin 75°}. \end{aligned}$$

$$\begin{aligned} \log 5 &= 0.69897 \\ \log \sin 45° &= 9.84949 \\ \text{colog} \sin 75° &= 0.01506 \\ \hline \log BC &= 0.56352 \\ BC &= 3.6603. \\ AB &= AC - BC \\ &= 6.3397. \end{aligned}$$

Distance between rocks, 6.3397 miles.

49. From a window on a level with the bottom of a steeple the elevation of the steeple is 40°, and from a second window 18 ft. higher the elevation is 37° 30′. Find the height of the steeple.

Let A and B be the windows, and C the top of the steeple. Then given $c = 18$, $A = 50°$, $B = 127° 30'$; required height of steeple, $h = b \sin 40°$.

$$C = 180 - (A + B)$$
$$= 2° 30'.$$
$$b = c\,\frac{\sin B}{\sin C}$$
$$= 18\,\frac{\sin 127° 30'}{\sin 2° 30'}.$$
$$h = 18\,\frac{\sin 127° 30' \sin 40°}{\sin 2° 30'}.$$

$$\begin{aligned} \log 18 &= 1.25527 \\ \log \sin 127° 30' &= 9.89947 \\ \log \sin 40° &= 9.80807 \\ \text{colog} \sin 2° 30' &= \underline{1.36032} \\ \log h &= 2.32313 \\ h &= 210.44. \end{aligned}$$

Height of steeple, 210.44 ft.

50. To determine the distance between two inaccessible objects by observing angles at the extremities of a line of known length.

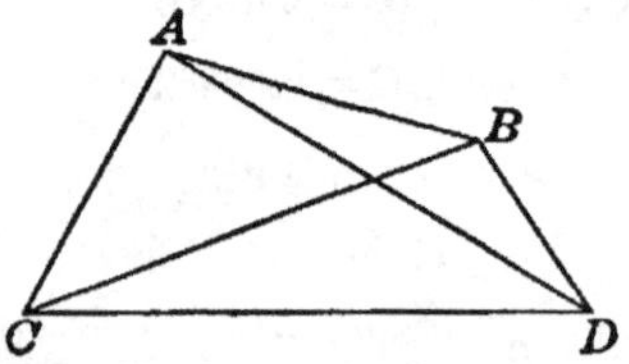

Let A and B be the inaccessible objects, C and D the extremities of the given line. Then, given CD, ACD, BCD, ADC, and BDC; required AB.

$$AC = CD\,\frac{\sin ADC}{\sin CAD}.$$
$$BC = CD\,\frac{\sin BDC}{\sin CBD}.$$

Then, in the triangle CAB, two sides and the included angle are known, and the third side can be computed as usual.

51. Wishing to determine the distance between a church A and a tower B, on the opposite sides of a river, I measure a line CD along the river (C being nearly opposite A), and observe the angle ACB, 58° 20′; ACD, 95° 20′; $ADB =$ 53° 30′, BDC, 98° 45′. CD is 600 feet. What is the distance required?

From the solution of Ex. 51,

$$AC = CD\,\frac{\sin ADC}{\sin CAD}$$
$$= 600\,\frac{\sin 45° 15'}{\sin 39° 25'}.$$
$$BC = CD\,\frac{\sin BDC}{\sin CBD}$$
$$= 600\,\frac{\sin 98° 45'}{\sin 44° 15'}.$$

$$\begin{aligned} \log 600 &= 2.77815 \\ \log \sin 45° 15' &= 9.85137 \\ \text{colog} \sin 39° 25' &= \underline{0.19726} \\ \log AC &= 2.82678 \\ AC &= 671.09. \end{aligned}$$

$$\begin{aligned} \log 600 &= 2.77815 \\ \log \sin 98° 45' &= 9.99492 \\ \text{colog} \sin 44° 15' &= \underline{0.15627} \\ \log BC &= 2.92934 \\ BC &= 849.84. \end{aligned}$$

$\tan \frac{1}{2}(CAB - CBA)$

$$= \frac{BC - AC}{BC + AC} \tan \frac{1}{2}(CAB + CBA)$$

$$= \frac{178.75}{1520.93} \tan 60° 50'.$$

$$\begin{aligned} \log 178.75 &= 2.25224 \\ \text{colog } 1520.93 &= 6.81789-10 \\ \log \tan 60° 50' &= 10.25327 \\ \log \tan \tfrac{1}{2}(CAB - CBA) &= 9.32340 \end{aligned}$$

$$\begin{aligned} \tfrac{1}{2}(CAB - CBA) &= 11° 53' 28'' \\ \tfrac{1}{2}(CAB + CBA) &= 60° 50' \\ CAB &= 72° 43' 28'' \end{aligned}$$

$$AB = BC \frac{\sin ACB}{\sin CAB}$$

$$= 849.84 \frac{\sin 58° 20'}{\sin 72° 43' 28''}.$$

$$\begin{aligned} \log 849.84 &= 2.92934 \\ \log \sin 58° 20' &= 9.92999 \\ \text{colog } \sin 72° 43' 28'' &= 0.02005 \\ \log AB &= 2.87938 \end{aligned}$$

$$AB = 757.50.$$

Required distance, 757.50 ft.

52. Wishing to find the height of a summit A, I measure a horizontal base line CD, 440 yds. At C, the elevation of A is 37° 18′, and the horizontal angle between D and the summit is 76° 18′; at D the horizontal angle between C and the summit is 67° 14′. Find the height.

Let A' be the point directly under A, in the same horizontal plane with CD. Then in the triangle $A'CD$;

$$A'C = CD \frac{\sin D}{\sin A'}$$

$$= 440 \frac{\sin 67° 14'}{\sin 36° 28'}.$$

Also,

$$AA' = A'C \tan CA'A$$

$$= 440 \frac{\sin 67° 14'}{\sin 36° 28'} \tan 37° 18'.$$

$$\begin{aligned} \log 440 &= 2.64345 \\ \log \sin 67° 14' &= 9.96477 \\ \log \tan 37° 18' &= 9.88184 \\ \text{colog } \sin 36° 28' &= 0.22595 \\ \log AA' &= 2.71601 \end{aligned}$$

$$\therefore AA' = 520.01.$$

Height, 520.01 yds.

53. A balloon is observed from two stations 3000 ft. apart. At the first station the horizontal angle of the balloon and the other station is 75° 25′, and the elevation of the balloon is 18°. The horizontal angle of the first station and the balloon, measured at the second station, is 64° 30′. Find the height of the balloon.

Let B be the first station, C the second, A the position of the balloon, and A' the point directly under A, in the same horizontal plane as BC. Then,

$$AA' = A'B \tan A'BA$$

$$= BC \frac{\sin A'CB}{\sin BA'C} \tan A'BA$$

$$= 3000 \frac{\sin 64° 30'}{\sin 40° 5'} \tan 18°.$$

$$\begin{aligned} \log 3000 &= 3.47712 \\ \log \sin 64° 30' &= 9.95549 \\ \log \tan 18° &= 9.51178 \\ \text{colog } \sin 40° 5' &= 0.19118 \\ \log AA' &= 3.13557 \end{aligned}$$

$$AA' = 1366.4.$$

Height of balloon, 1366.4 ft.

54. Two forces, one of 410 pounds, and the other of 320 pounds, make an angle of 51° 37′. Find the intensity and the direction of their resultant.

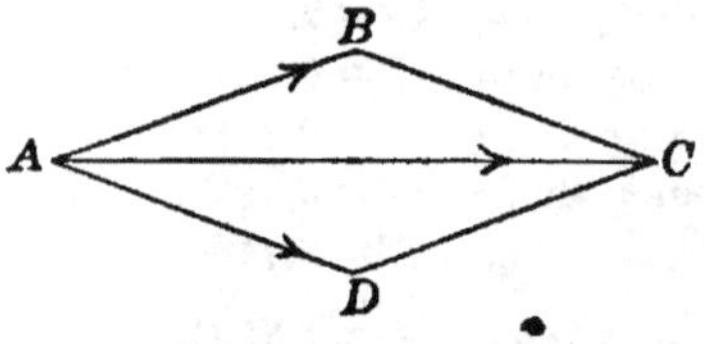

Let AB and AD represent the forces, and AC their resultant. Then, in the triangle ABC, given $c = 410$, $a = 320$, $B = 180° - 51° 37' = 128° 23'$; required b and A.

$$\tan \tfrac{1}{2}(C - A) = \frac{c-a}{c+a} \tan \tfrac{1}{2}(C+A) = \frac{90}{730} \tan 25° 48' 30''.$$

$$\begin{aligned} \log 90 &= 1.95424 \\ \log \tan 25° 48' 30'' &= 9.68448 \\ \text{colog } 730 &= 7.13668-10 \\ \tan \tfrac{1}{2}(C - A) &= 8.77540 \end{aligned}$$

$$\begin{aligned} \tfrac{1}{2}(C - A) &= 3° 24' 43'' \\ \tfrac{1}{2}(C + A) &= 25° 48' 30'' \\ A &= 22° 23' 47'' \end{aligned}$$

$$b = a\frac{\sin B}{\sin A} = 320 \frac{\sin 51° 37'}{\sin 22° 23' 47''}.$$

$$\begin{aligned} \log 320 &= 2.50515 \\ \log \sin 51° 37' &= 9.89425 \\ \text{colog } \sin 22° 23' 47'' &= 0.41906 \\ \log b &= 2.81846 \end{aligned}$$

$$b = 658.36.$$

Intensity of resultant, 658.36 pounds; angle between resultant and first force, 22° 23′ 47″.

55. An unknown force, combined with one of 128 pounds, produces a resultant of 200 pounds, and this resultant makes an angle of 18° 24′ with the known force. Find the intensity and direction of the unknown force.

In the figure for the solution of Ex. 54, given, in the triangle ABC, $c = 128$, $A = 18° 24'$, $b = 200$; required a and B.

$$\tan \tfrac{1}{2}(B - C) = \frac{b-c}{b+c} \tan \tfrac{1}{2}(B + C) = \tfrac{72}{328} \tan 80° 48'.$$

$$\begin{aligned} \log 72 &= 1.85733 \\ \log \tan 80° 48' &= 10.79058 \\ \text{colog } 328 &= 7.48413 - 10 \\ \log \tan \tfrac{1}{2}(B - C) &= 10.13204 \end{aligned}$$

$$\begin{aligned} \tfrac{1}{2}(B - C) &= 53° 34' 44'' \\ \tfrac{1}{2}(B + C) &= 80° 48' \\ B &= 134° 22' 44'' \\ 180 - B &= 45° 37' 16''. \end{aligned}$$

$$a = \frac{b \sin A}{\sin B} = 200 \frac{\sin 18° 24}{\sin 134° 22' 44''}.$$

$$\begin{aligned} \log 200 &= 2.30103 \\ \log \sin 18° 24' &= 9.49920 \\ \text{colog } \sin 134° 22' 44'' &= 0.14586 \\ \log a &= 1.94609 \end{aligned}$$

$$a = 88.326.$$

Intensity of unknown force, 88.326 pounds; angle between known and unknown forces, 45° 37′ 16″.

56. At two stations, the height of a kite subtends the same angle A. The angle which the line joining one station and the kite subtends at the other station is B; and the distance between the two stations is a. Show that the height of the kite is $\frac{1}{2} a \sin A \sec B$.

Let C be the position of the kite, D and E the stations, and C' the point directly under C in the same horizontal plane with DE.

Since the elevation of the kite is the same at D and E, the triangle CDE is isosceles, and

$$CD = CE = \tfrac{1}{2} a \sec B.$$

Also
$$CC' = CD \sin A$$
$$= \tfrac{1}{2} a \sin A \sec B.$$

57. Two towers on a horizontal plane are 120 ft. apart. A person standing successively at their bases observes that the angular elevation of the one is double that of the other; but when he is half way between them, the elevations are complementary. Prove that the heights of the towers are 90 and 40 ft.

Let A and B be the tops of the towers, A' and B' their bases, and C the point half way between them. Then the triangle $AA'C$ and $BB'C$ are similar, and

$$\frac{AA'}{B'C} = \frac{A'C}{BB'}.$$

$$AA' \times BB' = B'C \times A'C$$
$$= 3600.$$

Also, $AB'A' = 2\, BA'B'$.

$$\therefore \tan AB'A' = \frac{2 \tan BA'B'}{1 - \tan^2 BA'B'},$$

or

$$\frac{AA'}{120} = \frac{2\dfrac{BB'}{120}}{1 - \dfrac{BB'^2}{120^2}}$$

$$= \frac{240\, BB'}{120^2 - BB'^2}.$$

$$AA'\,(120^2 - \overline{BB'}^2) = 120 \times 240\, BB'.$$

$$\frac{3600}{BB'}(120^2 - BB'^2) = 120 \times 240\, BB'.$$

$$120^2 - BB'^2 = 8\, BB'^2.$$

$$BB'^2 = 40^2.$$

$$BB' = 40.$$

$$AA' = 90.$$

58. To find the distance of an inaccessible point C from either of two points A and B, having no instruments to measure angles. Prolong CA to a, and CB to b, and join AB, Ab, and Ba. Measure AB, 500; aA, 100; aB, 560; bB, 100; and Ab, 550.

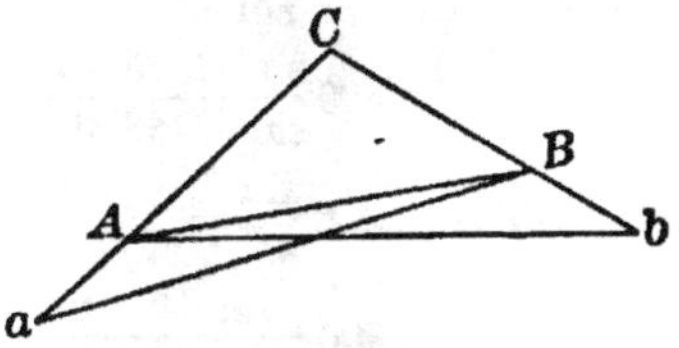

In the triangle aAB,

$$s = \tfrac{1}{2}(500 + 100 + 560)$$
$$= 580.$$

$$\tan \tfrac{1}{2} aAB = \sqrt{\frac{180 \times 480}{580 \times 20}}$$

$$= \sqrt{\frac{96}{29}}.$$

log 96 =	1.98227
colog 29 =	8.53760 − 10
	2) 0.51987
log tan ½ aAB =	10.25993

$$aAB = 122° \; 24' \; 40''.$$

$$CAB = 57° \; 35' \; 20''.$$

In the triangle bAB,

$$s = \tfrac{1}{2}(500 + 550 + 100)$$
$$= 575.$$

$$\tan \tfrac{1}{2} bBA = \sqrt{\frac{75 \times 475}{575 \times 25}}$$
$$= \sqrt{\frac{57}{23}}.$$

$$\begin{aligned} \log 57 &= 1.75587 \\ \text{colog } 23 &= 8.63827 - 10 \\ 2) & \ 0.39414 \\ \log \tan \tfrac{1}{2} bBA &= 10.19707 \\ bBA &= 115° \ 9'. \\ CBA &= 64° \ 51'. \end{aligned}$$

In the triangle ABC,

$$A = 57° \ 35' \ 20'',$$
$$B = 64° \ 51',$$
$$C = 57° \ 33' \ 40''.$$

$$BC = AB \frac{\sin A}{\sin C}$$
$$= 500 \frac{\sin 57° 35' 20''}{\sin 57° 33' 40''}.$$

$$AC = AB \frac{\sin B}{\sin C}$$
$$= 500 \frac{\sin 64° \ 51'}{\sin 57° 33' 40''}.$$

$$\begin{aligned} \log 500 &= 2.69897 \\ \log \sin 57° \ 35' \ 20'' &= 9.92646 \\ \text{colog} \sin 57° \ 33' \ 40'' &= 0.07368 \\ \log BC &= 2.69911 \\ BC &= 500.16. \end{aligned}$$

$$\begin{aligned} \log 500 &= 2.69897 \\ \log \sin 64° \ 51' &= 9.95674 \\ \text{colog} \sin 57° \ 33' \ 40'' &= 0.07368 \\ \log AC &= 2.72939 \\ AC &= 536.27. \end{aligned}$$

Distances of C from A and B, 536.27 ft.; 500.16 ft.

59. Two inaccessible points A and B are visible from D, but no other point can be found whence both are visible. Take some point C, whence A and D can be seen, and measure CD, 200 ft.; ADC, 89°; ACD, 50° 30′. Then take some point E, whence D and B are visible, and measure DE, 200; BDE, 54° 30′; BED, 88° 30′. At D measure ADB, 72° 30′. Compute the distance AB.

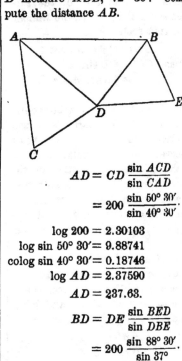

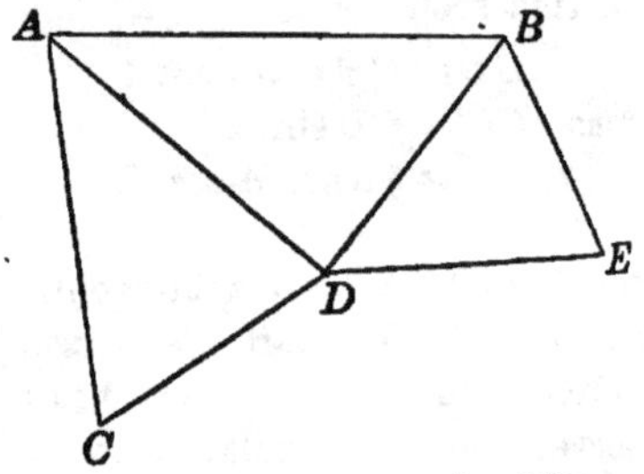

$$AD = CD \frac{\sin ACD}{\sin CAD}$$
$$= 200 \frac{\sin 50° \ 30'}{\sin 40° \ 30'}.$$

$$\begin{aligned} \log 200 &= 2.30103 \\ \log \sin 50° \ 30' &= 9.88741 \\ \text{colog} \sin 40° \ 30' &= 0.18746 \\ \log AD &= 2.37590 \\ AD &= 237.63. \end{aligned}$$

$$BD = DE \frac{\sin BED}{\sin DBE}$$
$$= 200 \frac{\sin 88° \ 30'}{\sin 37°}.$$

$$\begin{aligned} \log 200 &= 2.30103 \\ \log \sin 88° \ 30' &= 9.99985 \\ \text{colog} \sin 37° &= 0.22054 \\ \log BD &= 2.52142 \\ BD &= 332.22. \end{aligned}$$

$$\tan \tfrac{1}{2}(DAB - DBA)$$
$$= \frac{BD - AD}{BD + AD} \tan \tfrac{1}{2}(DAB + DBA)$$
$$= \frac{94.59}{569.85} \tan 53° \ 45'.$$

$$\log 94.59 = 1.97585$$
$$\text{colog } 569.85 = 7.24424$$
$$\log \tan 53^\circ 45' = 10.13476$$
$$\log \tan \tfrac{1}{2}(DAB - DBA) = 9.35485$$

$$\tfrac{1}{2}(DAB - DBA) = 12^\circ 45' 21''$$
$$\tfrac{1}{2}(DAB + DBA) = 53^\circ 45'$$
$$DAB = 66^\circ 30' 21''$$

$$AB = BD \frac{\sin ADB}{\sin DAB}$$
$$= 332.22 \frac{\sin 72^\circ 30'}{\sin 66^\circ 30' 21''}.$$

$$\log 332.22 = 2.52142$$
$$\log \sin 72^\circ 30' = 9.97942$$
$$\text{colog } \sin 66^\circ 30' 21'' = 0.03758$$
$$\log AB = 2.53842$$

$$AB = 345.48.$$

Distance AB, 345.48 ft.

60. To compute the horizontal distance between two inaccessible points A and B, when no point can be found whence both can be seen. Take two points C and D, distant 200 yds., so that A can be seen from C, and B from D. From C measure CF, 200 yds. to F, whence A can be seen; and from D, measure DE, 200 yds. to E, whence B can be seen. Measure AFC, 83°; ACD, 53° 30′; ACF, 54° 31′; BDE, 54° 30′; BDC, 156° 25′; DEB, 88° 30′.

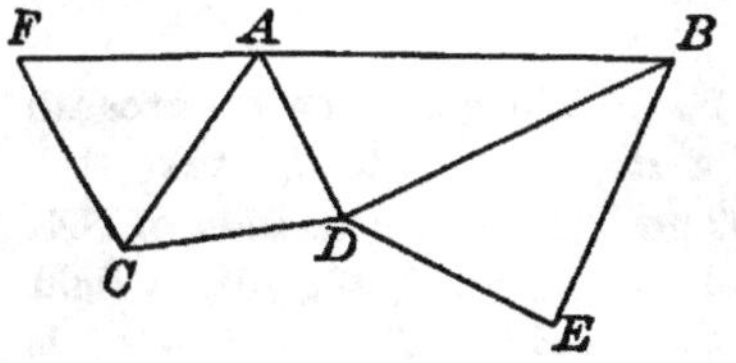

$$AC = CF \frac{\sin AFC}{\sin CAF}$$
$$= 200 \frac{\sin 83^\circ}{\sin 42^\circ 29'}.$$

$$\log 200 = 2.30103$$
$$\log \sin 83^\circ = 9.99675$$
$$\text{colog } \sin 42^\circ 29' = 0.17045$$
$$\log AC = 2.46823$$

$$AC = 293.92.$$

$$BD = DF \frac{\sin BED}{\sin DBE}$$
$$= 200 \frac{\sin 88^\circ 30'}{\sin 37^\circ}$$
$$= 332.22. \quad \text{(cf. Ex. 59).}$$

$$\tan \tfrac{1}{2}(ADC - CAD)$$
$$= \frac{AC - CD}{AC + CD} \tan \tfrac{1}{2}(ADC + CAD)$$
$$= \frac{93.92}{493.92} \tan 63^\circ 15'.$$

$$\log 93.92 = 1.97276$$
$$\text{colog } 493.92 = 7.30634 - 10$$
$$\log \tan 63^\circ 15' = 10.29753$$
$$\log \tan \tfrac{1}{2}(ADC - CAD) = 9.57663$$

$$\tfrac{1}{2}(ADC - CAD) = 20^\circ 40' 8''$$
$$\tfrac{1}{2}(ADC + CAD) = 63^\circ 15'$$
$$ADC = 83^\circ 55' 8''$$

$$BDA = BDC - ADC$$
$$= 156^\circ 25' - 83^\circ 55' 8''$$
$$= 72^\circ 29' 52''.$$

$$AD = AC \frac{\sin ACD}{\sin ADC}$$
$$= 293.92 \frac{\sin 53^\circ 20'}{\sin 83^\circ 55' 8''}.$$

$$\log 293.92 = 2.46823$$
$$\log \sin 53^\circ 20' = 9.90424$$
$$\text{colog } \sin 83^\circ 55' 8'' = 0.00245$$
$$\log AD = 2.37492$$

$$AD = 237.10.$$

$$\tan \tfrac{1}{2}(DAB - DBA)$$
$$= \frac{BD - AD}{BD + AD} \tan \tfrac{1}{2}(DAB + DBA)$$
$$= \frac{95.12}{569.32} \tan 53^\circ\, 45'\, 4''.$$

$$\begin{aligned} \log 95.12 &= 1.97827 \\ \text{colog } 569.32 &= 7.24464 - 10 \\ \log \tan 53^\circ\, 45'\, 4'' &= 10.13478 \\ \log \tan \tfrac{1}{2}(DAB - DBA) &= 9.35769 \end{aligned}$$

$$\begin{aligned} \tfrac{1}{2}(DAB - DBA) &= 12^\circ\, 50'\, 12'' \\ \tfrac{1}{2}(DAB + DBA) &= 53^\circ\, 45'\, 4'' \\ DAB &= 66^\circ\, 35'\, 16'' \end{aligned}$$

$$AB = BD \frac{\sin ADB}{\sin BAD}$$
$$= 332.22 \frac{\sin 72^\circ\, 29'\, 52''}{\sin 66^\circ\, 35'\, 16''}.$$

$$\begin{aligned} \log 332.22 &= 2.52142 \\ \log \sin 72^\circ\, 19'\, 52'' &= 9.97941 \\ \text{colog } \sin 66^\circ\, 35'\, 16'' &= 0.03731 \\ \log AB &= 2.53814 \\ AB &= 345.25. \end{aligned}$$

Distance AB, 345.25 yds.

61. A column in the north temperate zone is east-southeast of an observer, and at noon the extremity of its shadow is northeast of him. The shadow is 80 ft. in length, and the elevation of the column, at the observer's station, is 45°. Find the height of the column.

Let A be the observer's position, B the extremity of the shadow, and C the base of the column. Then, given $A = 67^\circ\, 30'$, $C = 67^\circ\, 30'$, $a = 80$; required b = height of column.

$$b = a \frac{\sin B}{\sin A}$$
$$= 80 \frac{\sin 45^\circ}{\sin 67^\circ\, 30'}.$$

$$\begin{aligned} \log 80 &= 1.90309 \\ \log \sin 45^\circ &= 9.84949 \\ \text{colog } \sin 67^\circ\, 30' &= 0.03438 \\ \log b &= 1.78696 \\ b &= 61.23. \end{aligned}$$

Height of column, 61.23 ft.

62. From the top of a hill the angles of depression of two objects situated in the horizontal plane of the base of the hill are 45° and 30°, and the horizontal angle between the two objects is 30°. Show that the height of the hill equals the distance between the objects.

Let A be the top of the hill, A' the point directly under A in the horizontal plane of the base of the hill, B and C the objects observed.

Then

$$A'B = A'A.$$
$$A'C = A'A \tan 60^\circ$$
$$= \sqrt{3}\, A'A.$$
$$\overline{BC}^2 = \overline{A'B}^2 + \overline{A'C}^2 - 2\, A'B \times A'C \cos BA'C$$
$$= \overline{A'A}^2 + 3\, \overline{A'A}^2 - 3\, \overline{A'A}^2$$
$$= \overline{A'A}^2.$$
$$BC = A'A.$$

63. Wishing to know the breadth of a river from A to B, I take AC, 100 yds. in the prolongation of BA, and then take CD, 200 yds. at right angles to AC. The angle BDA is 37° 18′ 30″. Find AB.

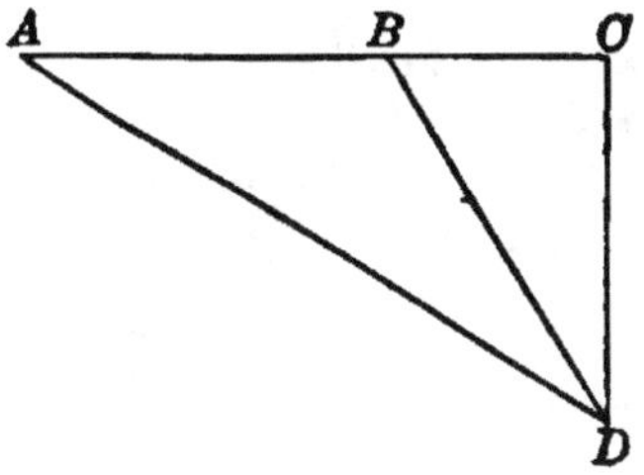

$$\tan BDC = \frac{BC}{CD}$$
$$= \tfrac{1}{2}.$$
$$\log \tan BDC = 9.69897.$$
$$BDC = 26^\circ\ 33'\ 54''.$$
$$ADC = ADB + BDC$$
$$= 63^\circ\ 52'\ 24'.$$
$$AC = CD \tan ADC$$
$$= 200 \tan 63^\circ\ 52'\ 24''.$$
$$\log 200 = \ \ 2.30103$$
$$\log \tan 63^\circ\ 52'\ 24'' = \underline{10.30939}$$
$$\log AC = \ \ 2.61042$$
$$AC = 407.77.$$
$$AB = AC - AB$$
$$= 307.77.$$

64. The sum of the sides of a triangle is 100. The angle at A is double that of B, and the angle at B is double that at C. Determine the sides.

$$B = 2\,C.$$
$$A = 2\,B = 4\,C.$$
$$A + B + C = 7\,C = 180^\circ.$$
$$\therefore C = \ \ 25^\circ\ 42'\ 51\tfrac{3}{7}''.$$
$$B = \ \ 51^\circ\ 25'\ 42\tfrac{6}{7}''.$$
$$A = 102^\circ\ 51'\ 25\tfrac{5}{7}''.$$
$$\frac{a}{c} = \frac{\sin A}{\sin C}.$$
$$\log \sin A = 9.98897$$
$$\log \sin C = \underline{9.63737}$$
$$\log \frac{a}{c} = 0.35160$$
$$\frac{a}{c} = 2.247.$$
$$a = 2.247\,c.$$
$$\frac{b}{c} = \frac{\sin B}{\sin C}.$$
$$\log \sin B = 9.89311$$
$$\log \sin C = \underline{9.63737}$$
$$\log \frac{b}{c} = 0.25574$$
$$\frac{b}{c} = 1.802.$$
$$b = 1.802\,c.$$
$$a + b + c = (2.247 + 1.802 + 1)\,c$$
$$= 5.049\,c.$$
$$\therefore c = \frac{100}{5.049} \quad = 19.806$$
$$a = 2.247\,c \quad = 44.503$$
$$b = 1.802\,c \quad = \underline{35.688}$$
$$a + b + c = 99.997$$

The sides are 19.8, 35.7, 44.5.

65. If $\sin^2 A + 5\cos^2 A = 3$, find A.

$$\sin^2 A + 5\cos^2 A = 3.$$
$$\sin^2 A + 5 - 5\sin^2 A = 3.$$
$$4\sin^2 A = 2.$$
$$\sin^2 A = \tfrac{1}{2}.$$
$$\sin A = \pm\sqrt{\tfrac{1}{2}}.$$
$$\therefore A = \pm 45^\circ,\ \pm 135^\circ.$$

66. If $\sin^2 A = m\cos A - n$, find $\cos A$.

$$\sin^2 A = m\cos A - n.$$
$$1 - \cos^2 A = m\cos A - n.$$
$$\cos^2 A + m\cos A = n + 1.$$
$$\therefore \cos A = \frac{-m \pm \sqrt{m^2 + 4\,(n+1)}}{2}.$$

67. Given $\sin A = m \sin B$, and $\tan A = n \tan B$; find $\sin A$ and $\cos B$.

$$\tan A = n \tan B.$$

$$\frac{\sin A}{\cos A} = n \frac{\sin B}{\cos B}.$$

$$\frac{m \sin B}{\cos A} = \frac{n \sin B}{\cos B}.$$

$$\cos A = \frac{m}{n} \cos B.$$

$$\cos^2 A = \frac{m^2}{n^2} \cos^2 B$$

$$\sin^2 A = m^2 \sin^2 B$$

$$1 = \frac{m^2}{n} \cos^2 B + m^2 \sin^2 B$$

$$\frac{m^2}{n^2} \cos^2 B + m^2 (1 - \cos^2 B) = 1.$$

$$\cos^2 B = \frac{1 - m^2}{\frac{m^2}{n^2} - m^2}$$

$$= \frac{(1 - m^2)\, n^2}{(1 - n^2)\, m^2}.$$

$$\cos B = \frac{n}{m}\sqrt{\frac{1 - m^2}{1 - n^2}}.$$

$$\cos^2 A = \frac{m^2}{n^2} \cos^2 B$$

$$= \frac{1 - m^2}{1 - n^2}.$$

$$\sin^2 A = 1 - \frac{1 - m^2}{1 - n^2}$$

$$= \frac{m^2 - n^2}{1 - n^2}.$$

$$\sin A = \sqrt{\frac{m^2 - n^2}{1 - n^2}}.$$

68. If $\tan^2 A + 4 \sin^2 A = 6$, find A.

$$\tan^2 A + 4 \sin^2 A = 6.$$

$$\frac{\sin^2 A}{1 - \sin^2 A} + 4 \sin^2 A = 6.$$

$$\sin^2 A + 4 \sin^2 A - 4 \sin^4 A$$
$$= 6 - 6 \sin^2 A.$$

$$4 \sin^4 A - 11 \sin^2 A + 6 = 0.$$

$$(4 \sin^2 A - 3)(\sin^2 A - 2) = 0.$$

$$\sin^2 A = \tfrac{3}{4}.$$

$$\sin A = \pm \tfrac{1}{2}\sqrt{3}.$$

$$A = \pm 60°, \pm 120°.$$

69. If $\sin A = \sin 2A$, find A.

$$\sin A = \sin 2A$$
$$= 2 \sin A \cos A.$$

$$\therefore \sin A (1 - 2 \cos A) = 0.$$

$$\therefore \sin A = 0°,$$

or $1 - 2 \cos A = 0°.$

$$A = 0°, 180°, \pm 60°.$$

70. If $\tan 2A = 3 \tan A$, find A.

$$\tan 2A = 3 \tan A.$$

$$\frac{2 \tan A}{1 - \tan^2 A} = 3 \tan A.$$

$$2 \tan A = 3 \tan A - 3 \tan^3 A.$$

$$\tan A (3 \tan^2 A - 1) = 0,$$

$$\tan A = 0,$$

or $3 \tan^2 A - 1 = 0.$

$$A = 0°, 180°, 30°, 210°.$$

71. Prove that $\tan 50° + \cot 50° = 2 \sec 10°$.

$$\tan 50° + \cot 50° = \tan 50° + \frac{1}{\tan 50°}$$

$$= \frac{\tan^2 50° + 1}{\tan 50°}$$

$$= \frac{\sec^2 50°}{\tan 50°}$$

$$= \frac{1}{\sin 50° \cos 50°}$$

$$= \frac{2}{2 \sin 50° \cos 50°}$$

$$= \frac{2}{\sin 100°}$$

$$= 2 \csc 100°$$

$$= 2 \sec 10°.$$

72. Given a regular polygon of n sides, and calling one of them a, find expressions for the radii of the inscribed and circumscribed circles in terms of n and a.

If P, H, D be the sides of a regular inscribed pentagon, hexagon, decagon, prove $P^2 = H^2 + D^2$.

(i.) Angle subtended by each side a at the centre of the circle is $\frac{360^\circ}{n}$. Hence, if r is the radius of the circumscribed circle, and R that of the inscribed circle,

$$\frac{a}{2r} = \sin\frac{180^\circ}{n}.$$

$$\frac{a}{2R} = \tan\frac{180^\circ}{n}.$$

$$\therefore r = \frac{a}{2}\csc\frac{180^\circ}{n}.$$

$$R = \frac{a}{2}\cot\frac{180^\circ}{n}.$$

(ii.) Let $r = 1$; then

$$P = 2\sin 36^\circ.$$
$$H = 2\sin 30^\circ = 1.$$
$$D = 2\sin 18^\circ.$$

To prove $P^2 = H^2 + D^2$,
or, $4\sin^2 36^\circ = 1 + 4\sin^2 18^\circ$.

$$\sin 36^\circ = \cos 54^\circ,$$

or, $\sin 2 \times 18^\circ = \cos 3 \times 18^\circ$.

$$2\sin 18^\circ \cos 18^\circ = 4\cos^3 18^\circ - 3\cos 18^\circ.$$

$$\begin{aligned} 2\sin 18^\circ &= 4\cos^2 18^\circ - 3 \\ &= 4 - 4\sin^2 18^\circ - 3 \\ &= 1 - 4\sin^2 18^\circ. \end{aligned}$$

$$\begin{aligned} \therefore 4\sin^2 18^\circ &= 1 - 2\sin 18^\circ \\ &= 1 - 2\cos 72^\circ. \end{aligned}$$

$$\begin{aligned} 1 + 4\sin^2 18^\circ &= 2 - 2\cos 72^\circ \\ &= 2(1 - \cos 72^\circ) \\ &= 4\sin^2 36^\circ. \end{aligned}$$

73. Obtain the formula for the area of a triangle, given two sides b, c, and the included angle A.

Let p be the length of the perpendicular from B on b. Then

$$\begin{aligned} \text{area} &= \tfrac{1}{2}pb \\ &= \tfrac{1}{2}c\sin A \times b \\ &= \tfrac{1}{2}bc\sin A. \end{aligned}$$

74. Obtain the formula for the area of a triangle, given two angles A, B, and the included side c.

$$a = c\frac{\sin A}{\sin C}.$$

$$b = c\frac{\sin B}{\sin C}.$$

$$\begin{aligned} \text{Area} &= \tfrac{1}{2}ab\sin C \\ &= \tfrac{1}{2}c^2\frac{\sin A\sin B}{\sin C} \\ &= \tfrac{1}{2}c^2\frac{\sin A\sin B}{\sin(A+B)}. \end{aligned}$$

75. Obtain the formula for the area of a triangle, given the three sides.

$$\begin{aligned} \sin B &= 2\sin\tfrac{1}{2}B\cos\tfrac{1}{2}B \\ &= \frac{2}{ac}\sqrt{s(s-a)(s-b)(s-c)}. \end{aligned}$$

(§ 46.)

$$\begin{aligned} \text{Area} &= \tfrac{1}{2}ac\sin B \\ &= \sqrt{s(s-a)(s-b)(s-c)}. \end{aligned}$$

76. If a is a side of an equilateral triangle, its area is $\frac{a^2\sqrt{3}}{4}$.

$$\begin{aligned} \text{Area} &= \tfrac{1}{2}bc\sin A \\ &= \tfrac{1}{2}a^2\sin 60^\circ \\ &= \tfrac{1}{2}a^2 \times \tfrac{1}{2}\sqrt{3} \\ &= \frac{a^2\sqrt{3}}{4}. \end{aligned}$$

77. Two consecutive sides of a rectangle are 52.25 ch. and 38.24 ch. Find its area.

Area = 52.25 × 38.24 sq. ch.

log 52.25 = 1.71809
log 38.24 = 1.58252
log area = 3.30061
Area = 1998 sq. ch.
= 199 A. 3 R. 8 P.

78. Two sides of a parallelogram are 59.8 ch. and 37.05 ch., and the included angle is 72° 10′. Find the area.

Area = 59.8 × 37.05 sin 72° 10′.

log 59.8 = 1.77670
log 37.05 = 1.56879
log sin 72° 10′ = 9.97861
log area = 3.32410
Area = 2109.1 sq. ch.
= 210 A. 3 R. 26 P.

79. Two sides of a parallelogram are 15.36 ch. and 11.46 ch., and the included angle is 47° 30′. Find its area.

Area = 15.36 × 11.46 sin 47° 30′.

log 15.36 = 1.18639
log 11.46 = 1.05918
log sin 47° 30′ = 9.86763
log area = 2.11320
Area = 129.78 sq. ch.
= 12 A. 3 R. 36 P.

80. Two sides of a triangle are 12.38 ch. and 6.78 ch., and the included angle is 46° 24′. Find the area.

Area = $\frac{1}{2}$ × 12.38 × 6.78 sin 46° 24′.

log 6.19 = 0.79169
log 6.78 = 0.83123
log sin 46° 24′ = 9.85984
log area = 1.48276
Area = 30.392 sq. ch.
= 3 A. 0 R. 6 P.

81. Two sides of a triangle are 18.37 ch. and 13.44 ch., and they form a right angle. Find the area.

log 18.37 = 1.26411
log 13.44 = 1.12840
2.39251
2 × area = 246.89 sq. ch.
Area = 123.45 sq. ch.
= 12 A. 1 R. 15 P.

82. Two angles of a triangle are 76° 54′ and 57° 33′ 12″, and the included side is 9 ch. Find the area.

From the solution of Ex. 24,

$$\text{Area} = \tfrac{1}{2}\, 9^2 \frac{\sin 76° 54' \sin 57° 33' 12''}{\sin 134° 27' 12''}.$$

log 40.5 = 1.60746
log sin 76° 54′ = 9.98855
log sin 57° 33′ 12″ = 9.92629
colog sin 134° 27′ 12″ = 0.14641
log area = 1.66871
Area = 46.634 sq. ch.
= 4 A. 2 R. 26 P.

83. Two sides of a triangle are 19.74 ch. and 17.34 ch. The first bears N. 82° 30′ W.; the second S. 24° 15′ E. Find the area.

Included angle = 121° 45′.

log 19.74 = 1.29535
log 17.34 = 1.23905
log sin 121° 45′ = 9.92960
2.46400
2 × area = 291.07 sq. ch.
Area = 145.54 sq. ch.
= 14 A. 2 R. 9 P.

84. The three sides of a triangle are 49 ch., 50.25 ch., and 25.69 ch. Find the area.

From the solution of Ex. 75,

$$\text{Area} = \sqrt{s(s-a)(s-b)(s-c)}.$$

$$s = \tfrac{1}{2}(49 + 50.25 + 25.69)$$
$$= 62.47.$$
$$\therefore s - a = 13.47.$$
$$s - b = 12.22.$$
$$s - c = 36.78.$$

log 62.47 = 1.79567
log 13.47 = 1.12937
log 12.22 = 1.08707
log 36.78 = 1.56561
2) 5.57772
log area = 2.78886
Area = 614.97 sq. ch.
= 61 A. 2 R.

85. The three sides of a triangle are 10.64 ch., 12.28 ch., and 9 ch. Find the area.

$$s = \tfrac{1}{2}(10.64 + 12.28 + 9)$$
$$= 15.96.$$
$$s - a = 5.32.$$
$$s - b = 3.68.$$
$$s - c = 6.96.$$

log 15.96 = 1.20303
log 5.32 = 0.72591
log 3.68 = 0.56585
log 6.96 = 0.84261
2) 3.33740
log area = 1.66870
Area = 46.633 sq. ch.
= 4 A. 2 R. 26 P.

86. The sides of a triangular field, of which the area is 14 acres, are in the ratio of 3, 5, 7. Find the sides.

Let the sides, measured in chains, be $3x$, $5x$, $7x$.

Then $s = \tfrac{1}{2}(3x + 5x + 7x)$
$$= 7.5x.$$
$$s - a = 4.5x.$$
$$s - b = 2.5x.$$
$$s - c = 0.5x.$$

$$140 = \sqrt{7.5x \times 4.5x \times 2.5x \times 0.5x}$$
$$= \frac{x^2}{4}\sqrt{15 \times 9 \times 5}$$
$$= \frac{15x^2}{4}\sqrt{3}.$$
$$\therefore x^2 = \frac{4 \times 140}{15\sqrt{3}} = \frac{112}{3\sqrt{3}}.$$

log 112 = 2.04922
log $3\sqrt{3}$ = 0.71568
2) 1.33354
log x = 0.66677
x = 4.6427.
$3x$ = 13.9281.
$5x$ = 23.2135.
$7x$ = 32.4989.

Sides are 13.93 ch., 23.21 ch., 32.50 ch.

87. In the quadrilateral $ABCD$ we have AB, 17.22 ch.; AD, 7.45 ch.; CD, 14.10 ch.; BC, 5.25 ch.; and the diagonal AC, 15.04 ch. Required the area.

In the triangle, ABC,

$$s = \tfrac{1}{2}(17.22 + 5.25 + 15.04)$$
$$= 18.755.$$
$$s - a = 1.535.$$
$$s - b = 13.505.$$
$$s - c = 3.715.$$

log 18.755 = 1.27311
log 1.535 = 0.18611
log 13.505 = 1.13049
log 3.715 = 0.56996
2) 3.15967
log area = 1.57983
Area = 38.004.

In the triangle ACD,

$$s = \tfrac{1}{2}(15.04 + 14.10 + 7.45)$$
$$= 18.295.$$
$$s - a = 3.255.$$
$$s - b = 4.195.$$
$$s - c = 10.845.$$

$$\begin{aligned} \log 18.295 &= 1.26233 \\ \log 3.255 &= 0.51255 \\ \log 4.195 &= 0.62273 \\ \log 10.895 &= \underline{1.03523} \\ 2)\,&\underline{3.43284} \\ \log \text{area} &= 1.71642 \\ \text{Area} &= 52.050. \end{aligned}$$

$$\begin{aligned} \text{Area } ABC &= 38.004 \\ \text{Area } ACD &= \underline{52.050} \\ \text{Area } ABCD &= 90.054 \text{ sq. ch.} \\ &= 9 \text{ A. } 0 \text{ R. } 1 \text{ P.} \end{aligned}$$

88. The diagonals of a quadrilateral are a and b, and they intersect at an angle D. Show that the area of the quadrilateral is $\frac{1}{2} ab \sin D$.

Let the parts into which the diagonals are divided by their intersection be a_1, a_2, and b_1, b_2, so that $a = a_1 + a_2$ and $b = b_1 + b_2$. Then the areas of the four triangles into which the diagonals divide the quadrilateral are

$$\tfrac{1}{2} a_1 b_1 \sin D, \qquad \tfrac{1}{2} a_2 b_1 \sin D,$$
$$\tfrac{1}{2} a_1 b_2 \sin D, \qquad \tfrac{1}{2} a_2 b_2 \sin D.$$

The area of the quadrilateral is therefore

$$\tfrac{1}{2} a_1 (b_1 + b_2) \sin D + \tfrac{1}{2} a_2 (b_1 + b_2) \sin D$$
$$= \tfrac{1}{2}(a_1 + a_2)(b_1 + b_2) \sin D$$
$$= \tfrac{1}{2} ab \sin D.$$

89. The diagonals of a quadrilateral are 34 and 56, intersecting at an angle of 67°. Find the area.

$$\text{Area} = \tfrac{1}{2} \times 34 \times 56 \times \sin 67°.$$

$$\begin{aligned} \log 17 &= 1.23045 \\ \log 56 &= 1.74819 \\ \log \sin 67° &= \underline{9.96403} \\ \log \text{area} &= 2.94267 \\ \text{Area} &= 876.34. \end{aligned}$$

90. The diagonals of a quadrilateral are 75 and 49, intersecting at an angle of 42°. Find the area.

$$\begin{aligned} \log 75 &= 1.87506 \\ \log 49 &= 1.69020 \\ \log \sin 42° &= \underline{9.82551} \\ &\ 3.39077 \\ 2 \times \text{area} &= 2459. \\ \text{Area} &= 1229.5. \end{aligned}$$

91. Show that the area of a regular polygon of n sides, of which one is a, is $\frac{na^2}{4} \cot \frac{180°}{n}$.

Lines joining the vertices to the centre divide the polygon into n equal isosceles triangles, the bases of which are a, and the vertical angles $\frac{360°}{n}$. The altitudes of the triangles are

$$h = \frac{a}{2} \cot \frac{180°}{n};$$

and their areas are

$$\tfrac{1}{2} ah = \frac{a^2}{4} \cot \frac{180°}{n}.$$

Hence the area of the polygon is

$$\frac{na^2}{4} \cot \frac{180°}{n}.$$

92. One side of a regular pentagon is 25. Find the area.

$$\text{Area} = \frac{5 \times 25^2}{4} \cot \frac{180°}{5}$$
$$= 781.25 \cot 36°.$$

$$\log 781.25 = 2.89279$$
$$\log \cot 36° = \underline{10.13874}$$
$$\log \text{area} = 3.03153$$
$$\text{Area} = 1075.3.$$

93. One side of a regular hexagon is 32. Find the area.

$$\text{Area} = \frac{6 \times 32^2}{4} \cot \frac{180°}{6}$$
$$= 1536 \cot 30°.$$
$$\log 1536 = 3.18639$$
$$\log \cot 30° = \underline{10.23856}$$
$$\log \text{area} = 3.42495$$
$$\text{Area} = 2660.4.$$

94. One side of a regular decagon is 46. Find the area.

$$\text{Area} = \frac{10 \times 46^2}{4} \cot \frac{180°}{10}$$
$$= 5290 \cot 18°.$$
$$\log 5290 = 3.72346$$
$$\log \cot 18° = \underline{10.48822}$$
$$\log \text{area} = 4.21168$$
$$\text{Area} = 16281.$$

95. Find the area of a circle whose circumference is 74 ft.

$$2\pi r = 74.$$
$$r = \frac{37}{\pi}.$$
$$\text{Area} = \pi r^2$$
$$= \frac{37^2}{\pi}.$$
$$\log 37^2 = 3.13640$$
$$\text{colog } \pi = \underline{9.50285 - 10}$$
$$\log \text{area} = 2.63925$$
$$\text{Area} = 435.76 \text{ sq. ft.}$$

96. Find the area of a circle whose radius is 125 ft.

$$\text{Area} = \pi \times 125^2.$$
$$\log 125^2 = 4.19382$$
$$\log \pi = \underline{0.49715}$$
$$\log \text{area} = 4.69097$$
$$\text{Area} = 49088 \text{ sq. ft.}$$

97. In a circle with a diameter of 125 ft. find the area of a sector with an arc of 22°.

Area of sector : area of circle $= 22 : 360$.

$$\therefore \text{area of sector} = \tfrac{22}{360} \pi \left(\tfrac{125}{2}\right)^2$$
$$= \frac{11 \times 125^2}{720} \pi.$$
$$\log 11 = 1.04139$$
$$\log 125^2 = 4.19392$$
$$\text{colog } 720 = 7.14267 - 10$$
$$\log \pi = \underline{0.49715}$$
$$\log \text{area} = 2.87513$$
$$\text{Area} = 750.12 \text{ sq. ft.}$$

98. In a circle with a radius of 44 ft. find the area of sector with an arc of 25°.

$$\text{Area} = \tfrac{25}{360} \pi\, 44^2$$
$$= \frac{1210\pi}{9}.$$
$$\log 1210 = 3.08279$$
$$\log \pi = 0.49715$$
$$\text{colog } 9 = \underline{9.04576}$$
$$\log \text{area} = 2.62570$$
$$\text{Area} = 422.38 \text{ sq. ft.}$$

99. In a circle with a diameter of 50 ft. find the area of a segment with an arc of 280°.

Area of segment = area of sector with same arc + area of triangle with two sides equal to radius, and included angle of 80°.

$$\text{Area of sector} = \tfrac{280}{360}\,\pi\, 25^2$$
$$= \frac{4375\,\pi}{9}.$$

$$\log 4375 = 3.64098$$
$$\log \pi = 0.49715$$
$$\text{colog } 9 = 9.04576 - 10$$
$$\log \text{area} = 3.18389$$

$$\text{Area of sector} = 1527.2.$$

$$\text{Area of triangle} = \tfrac{1}{2}\, 25^2 \sin 80^\circ$$
$$= 312.5 \sin 80^\circ.$$

$$\log 3125 = 2.49485$$
$$\log \sin 80^\circ = 9.99335$$
$$\log \text{area} = 2.48820$$

$$\text{Area of triangle} = 307.75.$$
$$\text{Area of segment} = 1834.95 \text{ sq. ft.}$$

100. Find the area of a segment (less than a semicircle) of which the chord is 20, and the distance of the chord from the middle point of the smaller arc is 2.

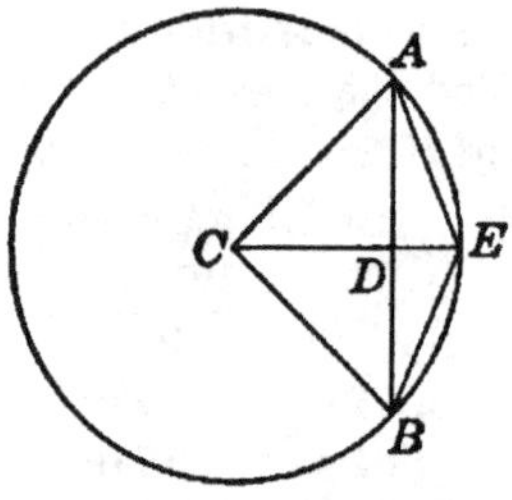

$$\tan AED = \tfrac{10}{2} = 5.$$
$$\log \tan AED = 10.69897.$$
$$AED = 78^\circ\, 41'\, 24''.$$
$$ACD = 180^\circ - 2\,AED$$
$$= 22^\circ\, 37'\, 12''.$$
$$AC = AD \csc ACD$$
$$= 10 \csc 22^\circ 37' 12''.$$

$$\log 10 = 1.00000$$
$$\log \csc 22^\circ 37' 12'' = 0.41497$$
$$\log AC = 1.41497$$
$$AC = 26.$$

$$\text{Area of sector } CAB = \frac{ACB}{360}\,\pi\,\overline{AC}^2.$$
$$ACB = 45^\circ\, 14'\, 24''$$
$$= 162864''.$$
$$360^\circ = 1296000''.$$
$$\text{Area of sector} = \tfrac{162864}{1296000}\,\pi\, 26^2$$
$$= \tfrac{377}{3000}\,\pi\, 26^2.$$

$$\log 377 = 2.57634$$
$$\log \pi = 0.49715$$
$$\log 26^2 = 2.82994$$
$$\text{colog } 3000 = 6.52288 - 10$$
$$\log \text{area} = 2.42631$$

$$\text{Area of sector} = 266.87.$$

$$\text{Area of triangle } CAB$$
$$= AD \times CD$$
$$= 10\,(26 - 2)$$
$$= 240.$$
$$\text{Area of segment} = 26.87.$$

101. If r is the radius of a circle, the area of a regular circumscribed polygon of n sides is $nr^2 \tan \dfrac{180^\circ}{n}$.

The area of a regular inscribed polygon is $\dfrac{n}{2} r^2 \sin \dfrac{360^\circ}{n}$.

Lines drawn from the vertices to the centre divide the polygon into n equal isosceles triangles, the bases of which are the sides of the polygon and the vertical angles $\dfrac{360^\circ}{n}$.

In the circumscribed polygon, each side $= 2\,r \tan \dfrac{180^\circ}{n}$, and the altitude of each triangle is r. Hence the area of each triangle is $r^2 \tan \dfrac{180^\circ}{n}$, and the area of the polygon $nr^2 \tan \dfrac{180^\circ}{n}$.

In the inscribed polygon, each side $= 2r\sin\frac{180^\circ}{n}$, and the altitude of each triangle is $r\csc\frac{180^\circ}{n}$. Hence the area of each triangle is $r^2\sin\frac{180^\circ}{n}\csc\frac{180^\circ}{n} = \frac{r^2}{2}\sin\frac{360^\circ}{n}$, and the area of the polygon is $\frac{nr^2}{2}\sin\frac{360^\circ}{n}$.

102. If a is a side of a regular polygon of n sides, the area of the inscribed circle is $\frac{\pi a^2}{4}\cot^2\frac{180^\circ}{n}$.

The area of the circumscribed circle is $\frac{\pi a^2}{4}\csc^2\frac{180^\circ}{n}$.

If r is the radius of the inscribed circle,
$$a = 2r\tan\frac{180^\circ}{n}.$$
$$\therefore r = \frac{a}{2}\cot\frac{180^\circ}{n}.$$
$$\pi r^2 = \frac{\pi a^2}{2}\cot^2\frac{180^\circ}{n}.$$

If R is the radius of the circumscribed circle,
$$a = 2R\sin\frac{180^\circ}{n}.$$
$$\therefore R = \frac{a}{2}\csc\frac{180^\circ}{n}.$$
$$\pi R^2 = \frac{\pi a^2}{4}\csc^2\frac{180^\circ}{n}.$$

103. The area of a regular polygon inscribed in a circle is to that of the circumscribed polygon of the same number of sides as 3 to 4. Find the number of sides.

$$\frac{n}{2}r^2\sin\frac{360^\circ}{n} : nr^2\tan\frac{180^\circ}{n} = 3:4.$$
$$2nr^2\sin\frac{360^\circ}{n} = 3nr^2\tan\frac{180^\circ}{n}.$$
$$2\sin\frac{360^\circ}{n} = 3\tan\frac{180^\circ}{n}.$$
$$4\sin\frac{180^\circ}{n}\cos\frac{180^\circ}{n} = 3\,\frac{\sin\frac{180^\circ}{n}}{\cos\frac{180^\circ}{n}}.$$
$$4\cos^2\frac{180^\circ}{n} = 3.$$
$$\cos\frac{180^\circ}{n} = \tfrac{1}{2}\sqrt{3}.$$
$$\frac{180^\circ}{n} = 30^\circ.$$
$$n = 6.$$

104. The area of a regular polygon inscribed in a circle is a geometric mean between the areas of an inscribed and a circumscribed regular polygon of half the number of sides.

Area of inscribed polygon of $2n$ sides $= nr^2\sin\frac{180^\circ}{n}$.

Area of inscribed polygon of n sides $= \frac{n}{2}r^2\sin\frac{360^\circ}{n}$.

Area of circumscribed polygon of n sides $= nr^2\tan\frac{180^\circ}{n}$.

$$\frac{n}{2}r^2\sin\frac{360^\circ}{n}\times nr^2\tan\frac{180^\circ}{n}$$
$$= \frac{n^2r^4}{2}\sin\frac{360^\circ}{n}\tan\frac{180^\circ}{n}$$
$$= n^2r^4\sin\frac{180^\circ}{n}\cos\frac{180^\circ}{n}\,\frac{\sin\frac{180^\circ}{n}}{\cos\frac{180^\circ}{n}}$$
$$= n^2r^4\sin^2\frac{180^\circ}{n}$$
$$= \left(nr^2\sin\frac{180^\circ}{n}\right)^2.$$

105. The area of a circumscribed regular polygon is an harmonic mean between the areas of an inscribed regular polygon of the same number of sides and of a circumscribed regular polygon of half that number.

Area of circumscribed polygon of $2n$ sides

$$= a = 2nr^2 \tan\frac{90^\circ}{n}.$$

Area of inscribed polygon of $2n$ sides

$$= b = nr^2 \sin\frac{180^\circ}{n}.$$

Area of circumscribed polygon of n sides

$$= c = nr^2 \tan\frac{180^\circ}{n}.$$

To prove

$$\frac{2}{a} = \frac{1}{b} + \frac{1}{c}.$$

$$\frac{1}{b} + \frac{1}{c} = \frac{1}{nr^2 \sin\frac{180^\circ}{n}} + \frac{1}{nr^2 \tan\frac{180^\circ}{n}}$$

$$= \frac{1 + \cos\frac{180^\circ}{n}}{nr^2 \sin\frac{180^\circ}{n}}$$

$$= \frac{2\cos^2\frac{90^\circ}{n}}{2nr^2 \sin\frac{90^\circ}{n}\cos\frac{90^\circ}{n}}$$

$$= \frac{\cot\frac{90^\circ}{n}}{nr^2}$$

$$= \frac{2}{2nr^2 \tan\frac{90^\circ}{n}}$$

$$= \frac{2}{a}.$$

106. The perimeter of a circumscribed regular triangle is double that of the inscribed regular triangle.

Each side of circumscribed triangle $= 2r\tan 60^\circ = 2\sqrt{3}\,r.$

Each side of inscribed triangle

$$= 2r\sin 60^\circ = \sqrt{3}\,r.$$

107. The square described about a circle is four-thirds the inscribed dodecagon.

Area of square $= 4r^2.$

Area of dodecagon $= \frac{12}{2} r^2 \sin\frac{360^\circ}{12}$

$$= 6r^2 \sin 30^\circ$$

$$= 3r^2.$$

108. Two sides of a triangle are 3 and 12, and the included angle is 30°. Find the hypotenuse of an isosceles right triangle of equal area.

Area of given triangle

$$= \tfrac{1}{2} \times 3 \times 12 \sin 30^\circ$$

$$= 9.$$

Side of required triangle

$$= \sqrt{2 \times 9}$$

$$= 3\sqrt{2}.$$

Hypotenuse of required triangle,

$$= \sqrt{2\,(3\sqrt{2})^2}$$

$$= \sqrt{36}$$

$$= 6.$$

Required hypotenuse, 6.

110. Taking the earth's equatorial diameter to be 7925.6 miles, find the length in feet of the arc of one minute of a great circle.

Circumference of great circle

$$= \pi \times 7925.6.$$

Length of arc of 1′, in feet

$$= \frac{\pi \times 7925.6 \times 5280}{360 \times 60}$$

$$= \frac{7925.6 \times 5280\,\pi}{21600}.$$

log 7925.6 = 3.89903
log 5280 = 3.72263
log π = 0.49715
colog 21600 = 5.66555 − 10
3.78436

Arc of 1′, 6086.4 ft.

111. A ship sails from latitude 43° 45′ S., on a course N. by E., 2345 miles. Find the latitude reached and the departure made.

Course, 11° 15′ E.

Diff. in lat. = 2345 cos 11° 15′.
Depart. = 2345 sin 11° 15′.

log 2345 = 3.37014
log cos 11° 15′ = 9.99157
log diff. lat. = 3.36171

Diff. lat. = 2299.9′
= 38° 20′.

log 2345 = 3.37014
log sin 11° 15′ = 9.29024
log depart. = 2.66038

Depart. = 457.49.

Latitude reached, 5° 25′ S.; departure, 457.5 miles.

112. A ship sails from latitude 1° 45′ N., on a course S.E. by E., and reaches latitude 2° 31′ S. Find the distance and the departure.

Course, 56° 15′.

Diff. in lat. = 4° 16′ = 256 miles.
Dist. = 256 sec 56° 15′.
Depart. = 256 tan 56° 15′.

log 256 = 2.40824
log sec 56° 15′ = 0.25526
log dist. = 2.66350

Dist. = 460.79.

log 256 = 2.40824
log tan 56° 15′ = 10.17511
log depart. = 2.58335

Depart. = 383.13.

Distance, 460.8 miles; departure, 383.1 miles.

113. A ship sails from latitude 13° 17′ S., on a course N.E. by E. ¾ E., until the departure is 207 miles. Find the distance, and the latitude reached.

Course, 64° 41′ 15″.
Depart., 207 miles.

Dist. = 207 csc 64° 41′ 15″.
Diff. in lat. = 207 cot 64° 41′ 15″.

log 207 = 2.31597
log csc 64° 41′ 15″ = 0.04385
log dist. = 2.35982

Dist. = 228.99.

log 207 = 2.31597
log cot 64° 41′ 15″ = 9.67483
log diff. lat. = 1.99080

Diff. lat. = 97.904′
= 1° 38′.
13° 17′ − 1° 38′ = 11° 39′.

Distance, 229 miles; latitude reached, 11° 39′ S.

114. A ship sails on a course between S. and E., 244 miles, leaving latitude 2° 52′ S., and reaching latitude 5° 8′ S. Find the course and the departure.

Diff. in lat., 2° 16′ = 136 miles.
Dist., 244 miles.

$$\cos(\text{course}) = \tfrac{136}{244}.$$

$$\text{Depart.} = \sqrt{244^2 - 136^2}.$$

$$\begin{aligned} \log 136 &= 2.13354 \\ \text{colog } 244 &= 7.61261 - 10 \\ \log\cos(\text{course}) &= 9.74615 \end{aligned}$$

$$\text{Course} = 56° \; 7' \; 32''.$$

$$\sqrt{244^2 - 136^2} = \sqrt{108 \times 380}.$$

$$\begin{aligned} \log 108 &= 2.03342 \\ \log 380 &= 2.57978 \\ 2)\, &4.61320 \\ \log \text{depart.} &= 2.30660 \end{aligned}$$

$$\text{Depart.} = 202.58.$$

Course, S. 56° 7′ 30″ E.; departure, 202.6 miles.

115. A ship sails from latitude 32° 18′ N., on a course between N. and W., making a distance of 344 miles, and a departure of 103 miles. Find the course, and the latitude reached.

Dist., 344 miles.
Depart., 103 miles.

$$\sin(\text{course}) = \tfrac{103}{344}.$$

$$\text{diff. in lat.} = \sqrt{344^2 - 103^2}.$$

$$\begin{aligned} \log 103 &= 2.01284 \\ \text{colog } 344 &= 7.46344 - 10 \\ \log\sin(\text{course}) &= 9.47628 \end{aligned}$$

$$\text{Course} = 17° \; 25' \; 22''.$$

$$\sqrt{344^2 - 103^2} = \sqrt{241 \times 447}.$$

$$\begin{aligned} \log 241 &= 2.38202 \\ \log 447 &= 2.65031 \\ 2)\, &5.03233 \\ \log(\text{diff. in lat.}) &= 2.51616 \end{aligned}$$

$$\text{Diff. in lat.} = 328.22' = 5° \; 28'.$$

$$32° \; 18' + 5° \; 28' = 37° \; 46'.$$

Course, N. 17° 25′ W.; latitude reached, 37° 46′ N.

116. A ship sails on a course between S. and E., making a difference of latitude 136 miles, and a departure 203 miles. Find the distance and the course.

Diff. in. lat., 136 miles.
Depart., 203 miles.

$$\tan(\text{course}) = \tfrac{203}{136}.$$

$$\begin{aligned} \log 203 &= 2.30750 \\ \text{colog } 136 &= 7.86646 - 10 \\ \log\tan(\text{course}) &= 10.17396 \end{aligned}$$

$$\text{Course} = 56° \; 10' \; 49''.$$

$$\text{Dist.} = 203 \csc 56° \; 10' \; 49''.$$

$$\begin{aligned} \log 203 &= 2.30750 \\ \log\csc 56° \; 10' \; 49'' &= 0.08051 \\ \log \text{dist.} &= 2.38801 \end{aligned}$$

$$\text{Dist.} = 244.35.$$

Course S. 56° 11′ E.; distance, 244.3 miles.

117. A ship sails due north 15 *statute* miles an hour for one day. What is the distance in a straight line from the point left to the point reached? (Take earth's radius, 3962.8 statute miles.)

Distance sailed in one day

$$\begin{aligned} &= 15 \times 24 = 360 \text{ miles} \\ &= \frac{360}{2\pi \times 3962.8} \times 360° \\ &= \frac{129600°}{2\pi \times 3962.8} \\ &= \frac{64800°}{3962.8\,\pi}. \end{aligned}$$

$$\log 64800 = 4.81158$$
$$\text{colog } 3962.8 = 6.40200 - 10$$
$$\text{colog } \pi = \underline{9.50285 - 10}$$
$$0.71643$$

Distance sailed

$= 5.2051°$

$= 5° 12' 18''$.

Chord of arc sailed

$= 2 \times 3962.8 \sin 2° 36' 9''$

$= 7925.6 \sin 2° 36' 9''$.

$$\log 7925.6 = 3.89903$$
$$\log \sin 2° 36' 9'' = \underline{8.65712}$$
$$\log \text{chord} = 2.55615$$

Chord = 359.87.

Required distance, 359.87 miles.

121. A ship in latitude 42° 16′ N., longitude 72° 16′ W., sails due east a distance of 149 miles. What is the position of the point reached?

Diff. long. = 149 sec 42° 16′.

$$\log 149 = 2.17319$$
$$\log \sec 42° 16' = \underline{0.13076}$$
$$\log \text{diff. long.} = 2.30395$$

Diff. long. = 201.35′

= 3° 21′.

Longitude of position reached, 68° 55′ W.

122. A ship in latitude 44° 49′ S., longitude 119° 42′ E., sails due west until it reaches longitude 117° 16′ E. Find the distance made.

Diff. long. = 2° 26′

= 146 miles.

Depart. = 146 cos 44° 49′.

$$\log 146 = 2.16435$$
$$\log \cos 44° 49' = \underline{9.85087}$$
$$\log \text{depart.} = 2.01522$$

Depart. = 103.57.

Distance made, 103.6 miles.

124. A ship leaves latitude 31° 14′ N., longitude 42° 19′ W., and sails E.N.E. 325 miles. Find the position reached.

Course, 67° 30′.

Diff. lat. = 325 cos 67° 30′.

$$\log 325 = 2.51188$$
$$\log \cos 67° 30' = \underline{9.58284}$$
$$\log \text{diff. lat.} = 2.09472$$

Diff. lat. = 124.37′

= 2° 4′.

Mid. lat. = 32° 16′.

Depart. = 325 sin 67° 30′.

Diff. long. = 325 sin 67° 30′ sec 32° 16′

$$\log 325 = 2.51188$$
$$\log \sin 67° 30' = 9.96562$$
$$\log \sec 32° 16' = \underline{0.07285}$$
$$\log \text{diff. long.} = 2.55035$$

Diff. long. = 355.10′

= 5° 55′.

Latitude of position reached, 33° 18′ N.; longitude, 36° 24′ W.

125. Find the bearing and distance of Cape Cod from Havana. (Cape Cod, 42° 2′ N., 70° 3′ W.; Havana, 23° 9′ N., 82° 22′ W.)

Diff. long. = 12° 19′ = 739 miles.

Diff. lat. = 18° 53′ = 1133 miles.

Mid. lat. = 32° 35′ 30″.

Depart. = diff. long. × cos mid. lat.

= 739 cos 32° 35′ 30″.

$$\tan(\text{course}) = \frac{\text{depart.}}{\text{diff. lat.}}$$
$$= \frac{739 \cos 32° 35' 30''}{1133}.$$

$$\log 739 = 2.86864$$
$$\log \cos 32° 35' 30'' = 9.92559$$
$$\text{colog } 1133 = \underline{6.94577}$$
$$\log \tan(\text{course}) = 9.74000$$

Course = 28° 47′ 26″.

Dist. = diff. lat. × sec (course)
= 1133 sec 28° 47′ 26″.

$$\log 1133 = 3.05423$$
$$\log \sec 28° 47' 26'' = 0.05730$$
$$\log \text{dist.} = 3.11153$$

Dist. = 1292.8.

Bearing, N. 28° 47′ E.; distance, 1293 miles.

126. Leaving latitude 49° 57′ N., longitude 15° 16′ W., a ship sails between S. and W. till the departure is 194 miles and the latitude is 47° 18′ N. Find the course, distance, and longitude reached.

Diff. lat. = 2° 39′ = 159 miles.
Mid. lat. = 48° 37′ 30″.
Depart. = 194 miles.
Diff. long. = 194 sec 48° 37′ 30″.

$$\log 194 = 2.28780$$
$$\log \sec 48° 37' 30'' = 0.17981$$
$$\log \text{diff. long.} = 2.46761$$

Diff. long. = 293.50′
= 4° 53′.

$$\tan (\text{course}) = \tfrac{194}{159}.$$

$$\log 194 = 2.28780$$
$$\log 159 = 2.20140$$
$$\log \tan (\text{course}) = 10.08640$$

Course = 50° 39′ 44″.

dist. = 159 sec 50° 39′ 44″.

$$\log 159 = 2.20140$$
$$\log \sec 50° 39' 44'' = 0.19799$$
$$\log \text{dist.} = 2.39939$$

Dist. = 250.83.

Course S. 50° 40′ W.; distance, 250.8 miles; longitude reached, 20° 9′ W.

127. Leaving latitude 42° 30′ N., longitude 58° 51′ W., a ship sails S.E. by S. 300 miles. Find the position reached.

Course, 33° 45′.
Diff. lat. = 300 cos. 33° 45′.

$$\log 300 = 2.47712$$
$$\log \cos 33° 45' = 9.91985$$
$$\log \text{diff. lat.} = 2.39697$$

Diff. lat. = 249.44′
= 4° 9′.

Mid. lat. = 40° 25′ 30″.
Depart. = 300 sin 33° 45′.
Diff. long. = 300 sin 33° 45′ sec 40° 25′ 30″.

$$\log 300 = 2.47712$$
$$\log \sin 33° 45' = 9.74474$$
$$\log \sec 40° 25' 30'' = 0.11847$$
$$\log \text{diff. long.} = 2.34033$$

Diff. long. = 218.94′
= 3° 39′.

Latitude of position reached, 38° 21′ N.; longitude, 55° 12′ W.

128. Leaving latitude 49° 57′ N., longitude 30° W., a ship sails S. 39° W., and reaches latitude 47° 44′ N. Find the distance and longitude reached.

Course, 39°.

Diff. lat. = 2° 13′ = 133 miles.
Mid. lat. = 48° 50′ 30″.
Dist. = 133 sec 39°.

$$\log 133 = 2.12385$$
$$\log \sec 39° = 0.10950$$
$$\log \text{dist.} = 2.23335$$

Dist. = 171.14.

Depart. $= 133 \tan 39°$.
Diff. long. $= 133 \tan 39°$
$\sec 48° 50' 30''$.

$$\log 133 = 2.12385$$
$$\log \tan 39° = 9.90837$$
$$\log \sec 48° 50' 30'' = 0.18168$$
$$\log \text{diff. long.} = 2.21390$$
$$\text{Diff. long.} = 163.64'$$
$$= 2° 44'.$$

Distance, 171 miles; longitude reached, 32° 44′ W.

129. Leaving latitude 37° N., longitude 32° 16′ W., a ship sails between N. and W. 300 miles, and reaches latitude 41° N. Find the course, and longitude reached.

Diff. lat. $= 4° = 240$ miles.
Mid. lat. $= 39°$.
Dist. $= 300$.

$$\cos(\text{course}) = \tfrac{240}{300}.$$
$$\log 240 = 2.38021$$
$$\log 300 = 2.47712$$
$$\log \cos(\text{course}) = 9.90309$$
$$\text{Course} = 36° 52' 12''.$$
$$\text{Depart.} = \sqrt{300^2 - 240^2}$$
$$= \sqrt{60 \times 540}$$
$$= 180.$$
$$\text{Diff. long.} = 180 \sec 39°.$$
$$\log 180 = 2.25527$$
$$\log \sec 39° = 0.10950$$
$$\log \text{diff. long.} = 2.36477$$
$$\text{Diff. long.} = 231.62'$$
$$= 3° 52'.$$

Course, N. 36° 52′ W.; longitude reached, 36° 8′ W.

130. Leaving latitude 50° 10′ S., longitude 30° E., a ship sails E.S.E., making 160 miles departure. Find the distance and position reached.

Course, 67° 30′.

Depart. $= 160$ miles.
Dist. $= 160 \csc 67° 30'$.
Diff. lat. $= 160 \cot 67° 30'$.

$$\log 160 = 2.20412$$
$$\log \csc 67° 30' = 0.03438$$
$$\log \text{dist.} = 2.23850$$
$$\text{Dist.} = 173.18.$$

$$\log 160 = 2.20412$$
$$\log \cot 67° 30' = 9.61722$$
$$\log \text{diff. lat.} = 1.82134$$
$$\text{Diff. lat.} = 66.273'$$
$$= 1° 6'.$$

Lat. reached $= 51° 16'$.
Mid. lat. $= 50° 43'$.
Diff. long. $= 160 \sec 50° 43'$.

$$\log 160 = 2.20412$$
$$\log \sec 50° 43' = 0.19849$$
$$\log \text{diff. long.} = 2.40261$$
$$\text{Diff. long.} = 252.70'$$
$$= 4° 13'.$$

Distance, 173 miles; latitude of position reached, 51° 16′ S.; longitude, 34° 13′ E.

131. Leaving latitude 49° 30′ N., longitude 25° W., a ship sails between S. and E. 215 miles, making a departure of 167 miles. Find the course, and position reached.

$$\sin(\text{course}) = \tfrac{167}{215}.$$
$$\log 167 = 2.22272$$
$$\log 215 = 2.33244$$
$$\log \sin(\text{course}) = 9.89028$$
$$\text{Course} = 50° 57' 48''.$$

$$\text{Diff. lat.} = \sqrt{215^2 - 167^2}$$
$$= \sqrt{48 \times 382}$$

$$\log 48 = 1.68124$$
$$\log 382 = \underline{2.58206}$$
$$2)\,\underline{4.26330}$$
$$\text{log diff. lat.} = 2.13165$$

$$\text{Diff. lat.} = 135.41'$$
$$= 2° 15'.$$

$$\text{Mid. lat.} = 48° 22' 30''.$$
$$\text{Diff. long.} = 167 \sec 48° 22' 30''.$$

$$\log 167 = 2.22272$$
$$\log \sec 48° 22' 30'' = \underline{0.17767}$$
$$\text{log diff. long.} = 2.40039$$

$$\text{Diff. long.} = 251.41'$$
$$= 4° 11''.$$

Course, S. 50° 58′ E.; latitude of position reached, 47° 15′ N., longitude, 20° 49′ W.

132. Leaving latitude 43° S., longitude 21° W., a ship sails 273 miles, and reaches latitude 40° 17′ S. What are the *two* courses and longitudes, either one of which will satisfy the data?

The two courses make equal angles with the meridian on opposite sides.

$$\text{Diff. lat.} = 2° 43' = 163 \text{ miles.}$$
$$\text{Dist.} = 273 \text{ miles.}$$

$$\cos(\text{course}) = \tfrac{163}{273}.$$

$$\log 163 = 2.21219$$
$$\log 273 = \underline{2.43616}$$
$$\log \cos(\text{course}) = 9.77603$$

$$\text{Course} = 53° 20' 21''.$$

$$\text{Depart.} = \sqrt{273^2 - 163^2}$$
$$= \sqrt{110 \times 436}.$$
$$\text{Mid. lat.} = 41° 38' 30''.$$
$$\text{Diff. long.} = \sqrt{110 \times 436} \sec 41° 38'' 30''.$$

$$\log \sqrt{110} = 1.02069$$
$$\log \sqrt{436} = 1.31975$$
$$\log \sec 41° 38' 30'' = \underline{0.12649}$$
$$\text{log diff. long.} = 2.46693$$

$$\text{Diff. long.} = 293.04'$$
$$= 4° 53'.$$

(i.) Course N. 53° 20′ E.; longitude of position reached, 16° 7′ W.

(ii.) Course N. 53° 20′ W.; longitude of position reached, 25° 53′ W.

133. Leaving latitude 17° N., longitude 119° E., a ship sails 219 miles, making a departure of 162 miles. What four sets of answers do we get?

The four courses all make the same angle with the meridian.

$$\sin(\text{course}) = \tfrac{162}{219}.$$

$$\log 162 = 2.20952$$
$$\log 219 = \underline{2.34044}$$
$$\log \sin(\text{course}) = 9.86908$$

$$\text{Course} = 47° 42' 33''.$$

$$\text{Diff. lat.} = \sqrt{219^2 - 162^2}$$
$$= \sqrt{57 \times 381}.$$

$$\log 57 = 1.75587$$
$$\log 381 = \underline{2.58092}$$
$$2)\,\underline{4.33679}$$
$$\text{log diff. lat.} = 2.16839$$

$$\text{Diff. lat.} = 147.36'$$
$$= 2° 27'.$$

(i.) $\text{Mid. lat.} = 18° 13' 30''.$

$$\text{Diff. long.} = 162 \sec 18° 13' 30''.$$

$$\log 162 = 2.20952$$
$$\log \csc 18° 13' 30'' = \underline{0.02235}$$
$$\text{log diff. long.} = 2.23187$$

$$\text{Diff. long.} = 170.56'$$
$$= 2° 51'.$$

(ii.) Mid. lat. = 15° 46′ 30″.

Diff. long. = 162 sec 15° 46′ 30″.

log 162 = 2.20952
log sec 15° 46′ 30″ = 0.01667
log diff. long. = 2.22619
Diff. long. = 168.34′
= 2° 48′.

(i.) Course N. 47° 42.5′ E.; latitude of position reached, 19° 27′ N., longitude 121° 51′ E.

Course N. 47° 42.5′ W.; latitude of position reached, 19° 27′ N., longitude 116° 9′ E.

(ii.) Course S. 47° 42.5′ E.; latitude of position reached, 14° 33′ N., longitude 121° 48′.

Course S. 47° 42.5′ W.; latitude of position reached, 14° 33′ N., longitude 116° 12′ E.

134. A ship in latitude 30° sails due east 360 statute miles. What is the shortest distance from the point left to the point reached?

Solve the same problem for latitudes 45°, 60°, etc.

Radius of parallel
= 3962.8 cos lat.

Arc sailed, in degrees

$$= \frac{360 \times 360°}{2\pi \times 3962.8 \text{ cos lat.}}.$$

log 360^2 = 5.11260
colog 2π = 9.20182 − 10
colog 3962.8 = 6.40200 − 10
0.71642

Arc sailed, in degrees
= 5.205° sec lat.

Arc sailed, in minutes
= 312.30′ sec. lat.

Chord of arc
= 2 rad. of parallel sin (½ arc)
= 2 × 3962.8 cos lat. sin (156.15′ sec lat.)
= 7925.6 cos lat. sin (156.15′ sec lat.).

(i.) lat. = 30°.

log 156.15′ = 2.19354
log sec 30° = 0.06247
log (½ arc) = 2.25601
½ arc = 180.30′
= 3° 0′ 18″.

log 7925.6 = 3.89903
log sin 3° 0′ 18″ = 8.71952
log cos 30° = 9.93753
log chord = 2.55608
Chord = 359.82.

(ii.) lat. = 45°.

log 156.15 = 2.19354
log sec 45° = 0.15051
log (½ arc) = 2.34405
½ arc = 220.82′
= 3° 40′ 49″.

log 7925.6 = 3.89903
log sin 3° 40′ 49″ = 8.80746
log cos 45° = 9.84949
log chord = 2.55598
Chord = 359.73.

(iii.) lat. = 60°.

sec lat. = 2.
½ arc = 312.30′
= 5° 12′ 18″.

log 7925.6 = 3.89903
log sin 5° 12′ 18″ = 8.95782
log cos 60° = 9.69897
log chord = 2.55582
Chord = 359.60.

Shortest distance, in lat. 30°, 359.82 miles; in lat. 45°, 359.73 miles; in lat. 60°, 359.60 miles; in general 7925.6 cos lat. × sin (156.15′ sec. lat.).

137. A ship leaves Cape Cod (Ex. 125), and sails S.E. by S. 114 miles, N. by E. 94 miles, W.N.W. 42 miles. Solve as in Ex. 136.

First course, 33° 45′.

Diff. lat. = 114 cos 33° 45′.
Depart. = 114 sin 33° 45′.

$$\begin{aligned} \log 114 &= 2.05690 \\ \log \cos 33^\circ 45' &= \underline{9.91985} \\ \log \text{diff. lat.} &= 1.97675 \\ \text{Diff. lat.} &= 94.787 \text{ S.} \end{aligned}$$

$$\begin{aligned} \log 114 &= 2.05690 \\ \log \sin 33^\circ 45' &= \underline{9.74474} \\ \log \text{depart.} &= 1.80164 \\ \text{Depart.} &= 63.334 \text{ E.} \end{aligned}$$

Second course, 11° 15′.

Diff. lat. = 94 cos 11° 15′.
Depart. = 94 sin 11° 15′.

$$\begin{aligned} \log 94 &= 1.97313 \\ \log \cos 11^\circ 15' &= \underline{9.99157} \\ \log \text{diff. lat.} &= 1.96470 \\ \text{Diff. lat.} &= 92.194 \text{ N.} \end{aligned}$$

$$\begin{aligned} \log 94 &= 1.97313 \\ \log \sin 11^\circ 15' &= \underline{9.29024} \\ \log \text{depart.} &= 1.26337 \\ \text{Depart.} &= 18.339 \text{ E.} \end{aligned}$$

Third course, 67° 30′.

Diff. lat. = 42 cos 67° 30′.
Depart. = 42 sin 67° 30′.

$$\begin{aligned} \log 42 &= 1.62325 \\ \log \cos 67^\circ 30' &= \underline{9.58284} \\ \log \text{diff. lat.} &= 1.20609 \\ \text{Diff. lat.} &= 16.073 \text{ N.} \end{aligned}$$

$$\begin{aligned} \log 42 &= 1.62325 \\ \log \sin 67^\circ 30' &= \underline{9.96562} \\ \log \text{depart.} &= 1.58887 \\ \text{Depart.} &= 38.804 \text{ W.} \end{aligned}$$

Total diff. lat. = 13.48′ N.
= 13′ 29″ N.
Lat. of C. Cod = 42° 2′.
Lat. reached = 42° 15′ N.
Mid. lat. = 42° 8′ 44′.
Total depart. = 42.869 E.
Diff. long. = 42.869 sec 42° 8′ 44″

$$\begin{aligned} \log 42.867 &= 1.63214 \\ \log \sec 42^\circ 8' 44'' &= \underline{0.12992} \\ \log \text{diff. long.} &= 1.76206 \\ \text{Diff. long.} &= 57.817' \text{ E.} \\ &= 58' \text{ E.} \end{aligned}$$

Long. of Cape Cod = 70° 3′ W.
Long. reached = 69° 5′ W.

$$\tan(\text{course}) = \frac{42.869}{13.48}.$$

$$\begin{aligned} \log 42.869 &= 1.63214 \\ \log 13.48 &= \underline{1.12969} \\ \log \tan(\text{course}) &= 10.50245 \\ \text{Course} &= 72^\circ 32' 40''. \end{aligned}$$

Dist. = 13.48 sec 72° 32′ 40″.

$$\begin{aligned} \log 13.48 &= 1.12969 \\ \log \sec 72^\circ 32' 40'' &= \underline{0.52293} \\ \log \text{dist.} &= 1.65262 \\ \text{Dist.} &= 44.939. \end{aligned}$$

Course N. 72° 33′ E.; distance, 45 miles; latitude reached, 42° 15′ N., longitude 69° 5′ W.

138. A ship leaves Cape of Good Hope (latitude 34° 22′ S., longitude 18° 30′ E.) and sails N.W. 126 miles, N. by E. 84 miles, W.S.W. 217 miles. Solve as in Ex. 136.

First course, 45°.

Diff. lat. = 126 cos 45°.
Depart. = 126 sin 45°.

log 126 = 2.10037
log cos 45° = 9.84949

log diff. lat. = 1.94986

Diff. lat. = 89.096 N.
Depart. = 89.096 W.

Second course, 11° 15′.

Diff. lat. = 84 cos 11° 15′.
Depart. = 84 sin 11° 15′.

log 84 = 1.92428
log cos 11° 15′ = 9.99157

log diff. lat. = 1.91585

Diff. lat. = 82.386 N.

log 84 = 1.92421
log sin 11° 15′ = 9.29024

log depart. = 1.21445

Depart. = 16.389 E.

Third course, 67° 30′.

Diff. lat. = 217 cos 67° 30″.
Depart. = 217 sin 67° 30″.

log 217 = 2.33646
log cos 67° 30′ = 9.58284

log. diff. lat. = 1.91930

Diff. lat. = 83.042 S.

log 217 = 2.33646
log sin 67° 30′ = 9.96562

log depart. = 2.30208

Depart. = 200.49 W.

Total diff. lat. = 88.440′ N.
= 1° 28′ 26″ N.
Lat. reached = 32° 53′ 34″ S.
Mid. lat. = 33° 37′ 47″.
Total depart. = 273.197 W.
Diff. long. = 273.20 sec 33° 37′ 47″ W.

log 273.20 = 2.43648
log sec 33° 37′ 47″ = 0.07954

log diff. long. = 2.51602

Diff. long. = 328.11′
= 5° 28′.
Long. reached = 13° 2′ E.

$$\tan(\text{course}) = \frac{273.20}{88.44}.$$

log 273.20 = 2.43648
log 88.44 = 1.94665

log tan (course) = 10.48983

Course = 72° 3′ 43″
Dist. = 88.44 sec 72° 3′ 43″.

log 88.44 = 1.94665
log sec 72° 3′ 43″ = 0.51147

log dist. = 2.45812

Dist. = 287.16.

Course, N. 72° 4′ W.; distance, 287 miles; latitude reached, 32° 54′ S., longitude 13° 2′ E.

Problems in Goniometry. Page 99.

1. Prove that $\sin x + \cos x = \sqrt{2}\cos(x - \frac{1}{4}\pi)$.

$$\sin x + \cos x = \sqrt{2}\left(\frac{1}{\sqrt{2}}\sin x + \frac{1}{\sqrt{2}}\cos x\right)$$
$$= \sqrt{2}(\sin\tfrac{1}{4}\pi\sin x + \cos\tfrac{1}{4}\pi\cos x)$$
$$= \sqrt{2}\cos(x - \tfrac{1}{4}\pi).$$

2. Prove that $\sin x - \cos x = -\sqrt{2}\cos(x + \frac{1}{4}\pi)$.

$$\sin x - \cos x = \sqrt{2}\left(\frac{1}{\sqrt{2}}\sin x - \frac{1}{\sqrt{2}}\cos x\right)$$
$$= \sqrt{2}(\sin\tfrac{1}{4}\pi\sin x - \cos\tfrac{1}{4}\pi\cos x)$$
$$= -\sqrt{2}\cos(x + \tfrac{1}{4}\pi).$$

3. Prove that $\sin x + \sqrt{3}\cos x = 2\sin(x + \frac{1}{3}\pi)$.

$$\sin x + \sqrt{3}\cos x = 2\left(\tfrac{1}{2}\sin x + \frac{\sqrt{3}}{2}\cos x\right)$$
$$= 2(\cos\tfrac{1}{3}\pi\sin x + \sin\tfrac{1}{3}\pi\cos x)$$
$$= 2\sin(x + \tfrac{1}{3}\pi).$$

4. Prove that $\sin(x + \frac{1}{3}\pi) + \sin(x - \frac{1}{3}\pi) = \sin x$.

$$\sin(x + \tfrac{1}{3}\pi) = \sin x\cos\tfrac{1}{3}\pi + \cos x\sin\tfrac{1}{3}\pi$$
$$= \tfrac{1}{2}\sin x + \frac{\sqrt{3}}{2}\cos x.$$
$$\sin(x - \tfrac{1}{3}\pi) = \tfrac{1}{2}\sin x - \frac{\sqrt{3}}{2}\cos x.$$
$$\sin(x + \tfrac{1}{3}\pi) + \sin(x - \tfrac{1}{3}\pi) = \sin x.$$

5. Prove that $\cos(x + \frac{1}{6}\pi) + \cos(x - \frac{1}{6}\pi) = \sqrt{3}\cos x$.

$$\cos(x + \tfrac{1}{6}\pi) = \frac{\sqrt{3}}{2}\cos x - \tfrac{1}{2}\sin x.$$
$$\cos(x - \tfrac{1}{6}\pi) = \frac{\sqrt{3}}{2}\cos x + \tfrac{1}{2}\sin x.$$
$$\cos(x + \tfrac{1}{6}\pi) + \cos(x - \tfrac{1}{6}\pi) = \sqrt{3}\cos x.$$

6. Prove that $\tan x + \sec x = \tan(\frac{1}{2}x + \frac{1}{4}\pi)$.

$$\tan x + \sec x = \frac{\sin x}{\cos x} + \frac{1}{\cos x}$$
$$= \frac{\sin x + 1}{\cos x}$$

$$= \frac{1 - \cos(x + \frac{1}{2}\pi)}{\sin(x + \frac{1}{2}\pi)}$$
$$= \frac{2 \sin^2 \frac{1}{2}(x + \frac{1}{2}\pi)}{2 \sin \frac{1}{2}(x + \frac{1}{2}\pi) \cos \frac{1}{2}(x + \frac{1}{2}\pi)}$$
$$= \frac{\sin(\frac{1}{2}x + \frac{1}{4}\pi)}{\cos(\frac{1}{2}x + \frac{1}{4}\pi)}$$
$$= \tan(\tfrac{1}{2}x + \tfrac{1}{4}\pi).$$

7. Prove that $\tan x + \sec x = \dfrac{1}{\sec x - \tan x}$.

$$\sec^2 x = 1 + \tan^2 x.$$
$$\sec^2 x - \tan^2 x = 1.$$
$$\sec x + \tan x = \frac{1}{\sec x - \tan x}.$$

8. Prove that $\dfrac{1 - \tan x}{1 + \tan x} = \dfrac{\cot x - 1}{\cot x + 1}$.

$$\frac{\cot x - 1}{\cot x + 1} = \frac{\dfrac{1}{\tan x} - 1}{\dfrac{1}{\tan x} + 1}$$
$$= \frac{1 - \tan x}{1 + \tan x}.$$

9. Prove that $\dfrac{\sin x}{1 + \cos x} + \dfrac{1 + \cos x}{\sin x} = 2 \csc x$.

$$\frac{\sin x}{1 + \cos x} + \frac{1 + \cos x}{\sin x} = \frac{\sin^2 x + (1 + \cos x)^2}{\sin x (1 + \cos x)}$$
$$= \frac{\sin^2 x + \cos^2 x + 2 \cos x + 1}{\sin x (1 + \cos x)}$$
$$= \frac{1 + 2 \cos x + 1}{\sin x (1 + \cos x)}$$
$$= \frac{2(1 + \cos x)}{\sin x (1 + \cos x)}$$
$$= \frac{2}{\sin x}$$
$$= 2 \csc x.$$

10. Prove that

$$\tan x + \cot x = 2 \csc 2x.$$

$$\tan x + \cot x = \frac{\sin x}{\cos x} + \frac{\cos x}{\sin x}$$
$$= \frac{\sin^2 x + \cos^2 x}{\sin x \cos x}$$
$$= \frac{1}{\sin x \cos x}$$
$$= \frac{2}{2 \sin x \cos x}$$
$$= \frac{2}{\sin 2x}$$
$$= 2 \csc 2x.$$

11. Prove that

$$\cot x - \tan x = 2\cot 2x.$$

$$\cot x - \tan x = \frac{\cos x}{\sin x} - \frac{\sin x}{\cos x}$$
$$= \frac{\cos^2 x - \sin^2 x}{\sin x \cos x}$$
$$= \frac{2\cos 2x}{\sin 2x}$$
$$= 2\cot 2x.$$

12. Prove that

$$1 + \tan x \tan 2x = \sec 2x.$$

$$1 + \tan x \tan 2x$$
$$= 1 + \frac{\sin x \sin 2x}{\cos x \cos 2x}$$
$$= 1 + \frac{2\sin^2 x \cos x}{\cos x(1 - 2\sin^2 x)}$$
$$= 1 + \frac{2\sin^2 x}{1 - 2\sin^2 x}$$
$$= \frac{1}{1 - 2\sin^2 x}$$
$$= \frac{1}{\cos 2x}$$
$$= \sec 2x.$$

13. Prove that

$$\sec 2x = \frac{\sec^2 x}{2 - \sec^2 x}.$$

$$\sec 2x = \frac{1}{\cos 2x}$$
$$= \frac{1}{2\cos^2 x - 1}$$
$$= \frac{\frac{1}{\cos^2 x}}{2 - \frac{1}{\cos^2 x}}$$
$$= \frac{\sec^2 x}{2 - \sec^2 x}.$$

14. Prove that

$$2\sec 2x = \sec(x + 45^\circ)\sec(x - 45^\circ).$$

$$2\sec 2x$$
$$= \frac{2}{\cos 2x}$$
$$= \frac{2}{\cos^2 x - \sin^2 x}$$
$$= \frac{2}{(\cos x - \sin x)(\cos x + \sin x)}$$

(Exs. 1 and 2):

$$= \frac{2}{2\cos(x + 45^\circ)\cos(x - 45^\circ)}$$
$$= \sec(x + 45^\circ)\sec(x - 45^\circ).$$

15. Prove that

$$\tan 2x + \sec 2x = \frac{\cos x + \sin x}{\cos x - \sin x}.$$

$$\tan 2x + \sec 2x$$
$$= \frac{\sin 2x}{\cos 2x} + \frac{1}{\cos 2x}$$
$$= \frac{\sin 2x + 1}{\cos 2x}$$
$$= \frac{1 - \cos(2x + 90^\circ)}{\sin(2x + 90^\circ)}$$
$$= \frac{2\sin^2(x + 45^\circ)}{2\sin(x + 45^\circ)\cos(x + 45^\circ)}$$
$$= \frac{\sin(x + 45^\circ)}{\cos(x + 45^\circ)}$$
$$= \frac{\sqrt{\tfrac{1}{2}}\sin x + \sqrt{\tfrac{1}{2}}\cos x}{\sqrt{\tfrac{1}{2}}\sin x - \sqrt{\tfrac{1}{2}}\cos x}$$
$$= \frac{\sin x + \cos x}{\sin x - \cos x}.$$

16. Prove that $\sin 2x = \dfrac{2\tan x}{1 + \tan^2 x}$.

$$\frac{2\tan x}{1 + \tan^2 x} = \frac{2\tan x}{\sec^2 x}$$
$$= 2\tan x \cos^2 x$$
$$= 2\sin x \cos x$$
$$= \sin 2x.$$

17. Prove that $2\sin x+\sin 2x=\dfrac{2\sin^3 x}{1-\cos x}$.

$$2\sin x+\sin 2x=2\sin x+2\sin x\cos x$$
$$=2\sin x(1+\cos x)$$

But
$$1-\cos^2 x=\sin^2 x.$$
$$\therefore 1+\cos x=\frac{\sin^2 x}{1-\cos x}.$$
$$2\sin x+\sin 2x=2\sin x\frac{\sin^2 x}{1-\cos x}$$
$$=\frac{2\sin^3 x}{1-\cos x}.$$

18. Prove that $\sin 3x=\dfrac{\sin^2 2x-\sin^2 x}{\sin x}$.

By [20], $$\sin 2x+\sin x=2\sin\tfrac{3}{2}x\cos\tfrac{1}{2}x.$$
By [21], $$\sin 2x-\sin x=2\cos\tfrac{3}{2}x\sin\tfrac{1}{2}x.$$
$$\therefore \sin^2 2x-\sin^2 x=2\sin\tfrac{3}{2}x\cos\tfrac{3}{2}x\times 2\sin\tfrac{1}{2}x\cos\tfrac{1}{2}x$$
$$=\sin 3x\sin x.$$
$$\sin 3x=\frac{\sin^2 2x-\sin^2 x}{\sin x}.$$

19. Prove that $\tan 3x=\dfrac{3\tan x-\tan^3 x}{1-3\tan^2 x}$.

$$\tan 3x=\tan(2x+x)$$
$$=\frac{\tan 2x+\tan x}{1-\tan 2x\tan x}$$

[14], $$=\frac{\dfrac{2\tan x}{1-\tan^2 x}+\tan x}{1-\dfrac{2\tan x}{1-\tan^2 x}\tan x}$$
$$=\frac{3\tan x-\tan^3 x}{1-3\tan^2 x}.$$

20. Prove that $\dfrac{\tan 2x+\tan x}{\tan 2x-\tan x}=\dfrac{\sin 3x}{\sin x}$.

By [24], $$\frac{\sin A+\sin B}{\sin A-\sin B}=\frac{\tan\frac{1}{2}(A+B)}{\tan\frac{1}{2}(A-B)}.$$
$$\therefore \frac{\sin A}{\sin B}=\frac{\tan\frac{1}{2}(A+B)+\tan\frac{1}{2}(A-B)}{\tan\frac{1}{2}(A+B)-\tan\frac{1}{2}(A-B)}.$$

Let $A=3x$, $B=x$; then
$$\frac{\sin 3x}{\sin x}=\frac{\tan 2x+\tan x}{\tan 2x-\tan x}.$$

21. Prove that $\sin(x+y)+\cos(x-y)=2\sin(x+\frac{1}{4}\pi)\sin(y+\frac{1}{4}\pi)$.

$$\sin(x+y)=\sin x\cos y+\cos x\sin y.$$
$$\cos(x-y)=\cos x\cos y+\sin x\sin y.$$
$$\sin(x+y)+\cos(x-y)=(\sin x+\cos x)\cos y+(\cos x+\sin x)\sin y$$
$$=(\sin x+\cos x)(\sin y+\cos y).$$

But $\sin x+\cos x=\sqrt{2}\left(\frac{1}{\sqrt{2}}\sin x+\frac{1}{\sqrt{2}}\cos y\right)$

$$=\sqrt{2}\sin(x+\tfrac{1}{4}\pi).$$

Similarly, $\sin y+\cos y=\sqrt{2}\sin(y+\frac{1}{4}\pi)$.

$\therefore \sin(x+y)+\cos(x-y)=2\sin(x+\frac{1}{4}\pi)\sin(y+\frac{1}{4}\pi)$.

22. Prove that $\sin(x+y)-\cos(x-y)=-2\sin(x-\frac{1}{4}\pi)\sin(y-\frac{1}{4}\pi)$.

$$\sin(x+y)=\sin x\cos y+\cos x\sin y.$$
$$\cos(x-y)=\cos x\cos y+\sin x\sin y.$$
$$\sin(x+y)-\cos(x-y)=(\sin x-\cos x)\cos y+(\cos x-\sin x)\sin y$$
$$=(\sin x-\cos x)(\cos y-\sin y)$$
$$=-2\sin(x-\tfrac{1}{4}\pi)\sin(y-\tfrac{1}{4}\pi).$$

23. Prove that $\tan x+\tan y=\frac{\sin(x+y)}{\cos x\cos y}$.

$$\tan x+\tan y=\frac{\sin x}{\cos x}+\frac{\sin y}{\cos y}$$
$$=\frac{\sin x\cos y+\cos x\sin y}{\cos x\cos y}$$
$$=\frac{\sin(x+y)}{\cos x\cos y}.$$

24. Prove that $\tan(x+y)=\frac{\sin 2x+\sin 2y}{\cos 2x+\cos 2y}$.

By [20], $\sin 2x+\sin 2y=2\sin(x+y)\cos(x-y)$.

By [22], $\cos 2x+\cos 2y=2\cos(x+y)\cos(x-y)$.

$$\therefore \frac{\sin 2x+\sin 2y}{\cos 2x+\cos 2y}=\frac{2\sin(x+y)\cos(x-y)}{2\cos(x+y)\cos(x-y)}$$
$$=\tan(x+y).$$

25. Prove that $\frac{\sin x+\cos y}{\sin x-\cos y}=\frac{\tan[\frac{1}{2}(x+y)+45^\circ]}{\tan[\frac{1}{2}(x-y)-45^\circ]}$.

$$\sin x+\cos y=\sin x+\sin(y+90^\circ)$$

[20], $=2\sin\frac{1}{2}(x+y+90^\circ)\cos\frac{1}{2}(x-y-90^\circ)$.

$$\sin x-\cos y=2\cos\tfrac{1}{2}(x+y+90^\circ)\sin\tfrac{1}{2}(x-y-90^\circ).$$
$$\therefore \frac{\sin x+\cos y}{\sin x-\cos y}=\frac{\tan\frac{1}{2}(x+y+90^\circ)}{\tan\frac{1}{2}(x-y-90^\circ)}$$
$$=\frac{\tan[\frac{1}{2}(x+y)+45^\circ]}{\tan[\frac{1}{2}(x-y)-45^\circ]}.$$

26. Prove that $\sin 2x + \sin 4x = 2 \sin 3x \cos x$.

By [20], $\sin A + \sin B = 2 \sin \frac{1}{2}(A + B) \cos \frac{1}{2}(A - B)$.

$\therefore \sin 2x + \sin 4x = 2 \sin 3x \cos x$.

27. Prove that
$$\begin{aligned}\sin 4x &= 4 \sin x \cos x - 8 \sin^3 x \cos x\\ &= 8 \cos^3 x \sin x - 4 \sin x \cos x.\end{aligned}$$

$$\begin{aligned}\sin 4x &= 2 \sin 2x \cos 2x\\ &= 4 \sin x \cos x (1 - 2 \sin^2 x)\\ &= 4 \sin x \cos x - 8 \sin^3 x \cos x;\\ &= 4 \sin x \cos x (2 \cos^2 x - 1)\\ &= 8 \cos^3 x \sin x - 4 \sin x \cos x.\end{aligned}$$

28. Prove that
$$\begin{aligned}\cos 4x &= 1 - 8 \cos^2 x + 8 \cos^4 x\\ &= 1 - 8 \sin^2 x + 8 \sin^4 x.\end{aligned}$$

$$\begin{aligned}\cos 4x &= 2 \cos^2 2x - 1\\ &= 2(2 \cos^2 x - 1)^2 - 1\\ &= 8 \cos^4 x - 8 \cos^2 x + 2 - 1\\ &= 1 - 8 \cos^2 x + 8 \cos^4 x;\\ &= 1 - 2 \sin^2 2x\\ &= 1 - 2(4 \sin^2 x \cos^2 x)\\ &= 1 - 8 \sin^2 x (1 - \sin^2 x)\\ &= 1 - 8 \sin^2 x + 8 \sin^4 x.\end{aligned}$$

29. Prove that $\cos 2x + \cos 4x = 2 \cos 3x \cos 2x$.

By [22], $\cos A + \cos B = 2 \cos \frac{1}{2}(A + B) \cos \frac{1}{2}(A - B)$.

$\therefore \cos 2x + \cos 4x = 2 \cos 3x \cos 2x$.

30. Prove that $\sin 3x - \sin x = 2 \cos 2x \sin x$.

By [21], $\sin A - \sin B = 2 \cos \frac{1}{2}(A + B) \sin \frac{1}{2}(A - B)$.

$\therefore \sin 3x - \sin x = 2 \cos 2x \sin x$.

31. Prove that $\sin^3 x \sin 3x + \cos^3 x \cos 3x = \cos^3 2x$.

$$\begin{aligned}\sin^3 x \sin 3x &= \sin x \sin^2 x \sin 3x\\ &= \sin x (1 - \cos^2 x) \sin 3x\\ &= \sin x \sin 3x - \sin x \cos^2 x \sin 3x.\end{aligned}$$

$$\cos^3 x \cos 3x = \cos x \cos 3x - \cos x \sin^2 x \cos 3x.$$

$\therefore \sin^3 x \sin 3x + \cos^3 x \cos 3x$
$$\begin{aligned}&= \sin x \sin 3x + \cos x \cos 3x\\ &\qquad - \sin x \cos^2 x \sin 3x - \cos x \sin^2 x \cos 3x\\ &= \cos 2x - \sin x \cos x (\cos x \sin 3x + \sin x \cos 3x)\end{aligned}$$

$$= \cos 2x - \sin x \cos x \sin 4x$$
$$= \cos 2x - \tfrac{1}{2} \sin 2x \sin 4x$$
$$= \cos 2x - \sin^2 2x \cos 2x$$
$$= \cos 2x (1 - \sin^2 2x)$$
$$= \cos^3 2x.$$

32. Prove that $\cos^4 x - \sin^4 x = \cos 2x$,

$$\cos^4 x - \sin^4 x = (\cos^2 + \sin^2 x)(\cos^2 x - \sin^2 x)$$
$$= 1 \times \cos 2x$$
$$= \cos 2x.$$

33. Prove that $\cos^4 x + \sin^4 x = 1 - \tfrac{1}{2} \sin^2 2x$.

$$\cos^4 x + \sin^4 x = (\cos^2 x + \sin^2 x)^2 - 2 \sin^2 x \cos^2 x$$
$$= 1 - 2 \sin^2 x \cos^2 x$$
$$= 1 - \tfrac{1}{2} \sin^2 2x.$$

34. Prove that $\cos^6 x - \sin^6 x = \cos 2x (1 - \sin^2 x \cos^2 x)$.

$$\cos^6 x - \sin^6 x = (\cos^2 x - \sin^2 x)(\cos^4 x + \cos^2 x \sin^2 x + \sin^4 x)$$
$$= \cos 2x [(\cos^2 x + \sin^2 x)^2 - \cos^2 x \sin^2 x]$$
$$= \cos 2x (1 - \cos^2 x \sin^2 x).$$

35. Prove that $\cos^6 x + \sin^6 x = 1 - 3 \sin^2 x \cos^2 x$.

$$\cos^6 x + \sin^6 x = (\cos^2 x + \sin^2 x)(\cos^4 x - \cos^2 x \sin^2 x + \sin^4 x)$$
$$= \cos^4 x - \cos^2 x \sin^2 x + \sin^4 x$$
$$= (\cos^2 x + \sin^2 x)^2 - 3 \cos^2 x \sin^2 x$$
$$= 1 - 3 \cos^2 x \sin^2 x.$$

36. Prove that $\dfrac{\sin 3x + \sin 5x}{\cos 3x - \cos 5x} = \cot x$.

By [20], $\sin 3x + \sin 5x = 2 \sin 4x \cos x$.

By [23], $\cos 3x - \cos 5x = 2 \sin 4x \sin x$.

$$\therefore \frac{\sin 3x + \sin 5x}{\cos 3x - \cos 5x} = \frac{\cos x}{\sin x} = \cot x.$$

37. Prove that $\dfrac{\sin 3x + \sin 5x}{\sin x + \sin 3x} = 2 \cos 2x$.

By [20], $\sin 3x + \sin 5x = 2 \sin 4x \cos x$.

$\sin x + \sin 3x = 2 \sin 2x \cos x$.

$$\therefore \frac{\sin 3x + \sin 5x}{\sin x + \sin 3x} = \frac{\sin 4x}{\sin 2x}$$
$$= \frac{2 \sin 2x \cos 2x}{\sin 2x}$$
$$= 2 \cos 2x.$$

38. Prove that $\csc x - 2 \cot 2x \cos x = 2 \sin x$.

$$\begin{aligned}\csc x - 2 \cot 2x \cos x &= \csc x - 2\frac{\cos 2x}{\sin 2x}\cos x\\ &= \csc x - \frac{\cos 2x}{\sin x}\\ &= \frac{1}{\sin x} - \frac{\cos 2x}{\sin x}\\ &= \frac{1 - \cos 2x}{\sin x}\\ &= \frac{2 \sin^2 x}{\sin x}\\ &= 2 \sin x.\end{aligned}$$

39. Prove that $(\sin 2x - \sin 2y) \tan (x + y) = 2(\sin^2 x - \sin^2 y)$.

$$\begin{aligned}\sin 2x - \sin 2y &= 2 \cos (x + y) \sin (x - y).\\ (\sin 2x - \sin 2y) \tan (x + y) &= 2 \sin (x + y) \sin (x - y).\\ \sin x + \sin y &= 2 \sin \tfrac{1}{2}(x + y) \cos \tfrac{1}{2}(x - y).\\ \sin x - \sin y &= 2 \cos \tfrac{1}{2}(x + y) \sin \tfrac{1}{2}(x - y).\\ \therefore \sin^2 x - \sin^2 y &= 4 \sin \tfrac{1}{2}(x + y) \cos \tfrac{1}{2}(x + y)\\ &\quad \sin \tfrac{1}{2}(x - y) \cos \tfrac{1}{2}(x - y)\\ &= \sin (x + y) \sin (x - y).\\ 2(\sin^2 x - \sin^2 y) &= 2 \sin (x + y) \sin (x - y)\\ &= (\sin 2x - \sin 2y) \tan (x + y).\end{aligned}$$

40. Prove that $(1 + \cot x + \tan x)(\sin x - \cos x) = \frac{\sec x}{\csc^2 x} - \frac{\csc x}{\sec^2 x}$.

$$\begin{aligned}(1 + \cot x + \tan x)(\sin x - \cos x) &= \sin x - \cos x + \cos x - \frac{\cos^2 x}{\sin x}\\ &\quad + \frac{\sin^2 x}{\cos x} - \sin x\\ &= \frac{\sin^2 x}{\cos x} - \frac{\cos^2 x}{\sin x}\\ &= \frac{\sec x}{\csc^2 x} - \frac{\csc x}{\sec^2 x}.\end{aligned}$$

41. Prove that $\sin x + \sin 3x + \sin 5x = \frac{\sin^2 3x}{\sin x}$.

By [20], $\quad \sin x + \sin 5x = 2 \sin 3x \cos 2x$.

$$\begin{aligned}\therefore \sin x + \sin 3x + \sin 5x &= \sin 3x + 2 \sin 3x \cos 2x\\ &= \sin 3x (1 + 2 \cos 2x).\end{aligned}$$

Also, $\sin 3x - \sin x = 2\cos 2x \sin x.$

$$\frac{\sin 3x}{\sin x} - 1 = 2\cos 2x.$$

$$1 + 2\cos 2x = \frac{\sin 3x}{\sin x}.$$

$$\therefore \sin x + \sin 3x + \sin 5x = \sin 3x \frac{\sin 3x}{\sin x}$$

$$= \frac{\sin^2 3x}{\sin x}.$$

42. Prove that $\dfrac{3\cos x + \cos 3x}{3\sin x - \sin 3x} = \cot^3 x.$

By [22], $3\cos x + \cos 3x = 2\cos x + (\cos x + \cos 3x)$

$$= 2\cos x + 2\cos x \cos 2x$$

$$= 2\cos x(1 + \cos 2x)$$

$$= 4\cos^3 x.$$

By [21], $3\sin x - \sin 3x = 2\sin x + (\sin x - \sin 3x)$

$$= 2\sin x - 2\sin x \cos 2x$$

$$= 2\sin x(1 - \cos 2x)$$

$$= 4\sin^3 x.$$

$$\therefore \frac{3\cos x + \cos 3x}{3\sin x - \sin 3x} = \frac{4\cos^3 x}{4\sin^3 x}$$

$$= \cot^3 x.$$

43. Prove that $\sin 3x = 4\sin x \sin(60° + x)\sin(60° - x).$

$$\sin(60° + x) = \tfrac{1}{2}\sqrt{3}\cos x + \tfrac{1}{2}\sin x.$$

$$\sin(60° - x) = \tfrac{1}{2}\sqrt{3}\cos x - \tfrac{1}{2}\sin x.$$

$$\sin(60° + x)\sin(60° - x) = \tfrac{3}{4}\cos^2 x - \tfrac{1}{4}\sin^2 x$$

$$= \frac{3(1 - \sin^2 x) - \sin^2 x}{4}$$

$$= \frac{3 - 4\sin^2 x}{4}.$$

$$4\sin x \sin(60° + x)\sin(60° - x) = \sin x(3 - 4\sin^2 x)$$

$$= 3\sin x - 4\sin^3 x$$

$$= \sin 3x.$$

44. Prove that $\sin 4x = 2\sin x \cos 3x + \sin 2x.$

By [21], $\sin 4x - \sin 2x = 2\cos 3x \sin x.$

$$\therefore \sin 4x = 2\cos 3x \sin x + \sin 2x.$$

45. Prove that $\sin x + \sin(x - \tfrac{2}{3}\pi) + \sin(\tfrac{1}{3}\pi - x) = 0.$

By [20], $\sin(x - \tfrac{2}{3}\pi) + \sin(\tfrac{1}{3}\pi - x) = 2\sin(-\tfrac{1}{6}\pi)\cos(x - \tfrac{1}{2}\pi)$

$$= -\sin x.$$

$$\therefore \sin x + \sin(x - \tfrac{2}{3}\pi) + \sin(\tfrac{1}{3}\pi - x) = 0.$$

46. Prove that $\cos x \sin(y-z)+\cos y \sin(z-x)+\cos z \sin(x-y)=0$.

$$\cos x \sin(y-z) = \cos x \sin y \cos z - \cos x \cos y \sin z.$$
$$\cos y \sin(z-x) = \cos y \sin z \cos x - \cos y \cos z \sin x.$$
$$\cos z \sin(x-y) = \cos z \sin x \cos y - \cos z \cos x \sin y.$$
$$\therefore \cos x \sin(y-z) + \cos y \sin(z-x) + \cos z \sin(x-y) = 0.$$

47. Prove that

$$\cos(x+y)\sin y - \cos(x+z)\sin z = \sin(x+y)\cos y - \sin(x+z)\cos z.$$

$$\sin(x+y)\cos y - \cos(x+y)\sin y = \sin x.$$
$$\sin(x+z)\cos z - \cos(x+z)\sin z = \sin x.$$
$$\therefore \sin(x+y)\cos y - \cos(x+y)\sin y$$
$$= \sin(x+z)\cos z - \cos(x+z)\sin z$$
$$\cos(x+y)\sin y - \cos(x+z)\sin z$$
$$= \sin(x+y)\cos z - \cos(x+z)\sin z.$$

48. Prove that

$$\cos(x+y+z) + \cos(x+y-z) + \cos(x-y+z) + \cos(y+z-x)$$
$$= 4\cos x \cos y \cos z.$$

By [22], $\cos[(x+y)+z] + \cos[(x+y)-z] = 2\cos(x+y)\cos z.$

$$\cos[z+(x-y)] + \cos[z-(x-y)] = 2\cos z \cos(x-y).$$
$$\therefore \cos(x+y+z) + \cos(x+y-z) + \cos(x-y+z)$$
$$+ \cos(y+z-x) = 2\cos(x+y)\cos z + 2\cos(x-y)\cos z$$
$$= 2\cos z[\cos(x+y) + \cos(x-y)]$$
$$= 2\cos z(2\cos x \cos y)$$
$$= 4\cos x \cos y \cos z.$$

49. Prove that $\sin(x+y)\cos(x-y) + \sin(y+z)\cos(y-z)$
$+ \sin(z+x)\cos(z-x) = \sin 2x + \sin 2y + \sin 2z.$

By [20], $\sin(x+y)\cos(x-y) = \frac{1}{2}(\sin 2x + \sin 2y).$

$$\sin(y+z)\sin(y-z) = \tfrac{1}{2}(\sin 2y + \sin 2z).$$
$$\sin(z+x)\sin(z-x) = \tfrac{1}{2}(\sin 2z + \sin 2x).$$
$$\therefore \sin(x+y)\cos(x-y) + \sin(y+z)\cos(y-z) + \sin(z+x)\sin(z-x)$$
$$= \sin 2x + \sin 2y + \sin 2z.$$

50. Prove that $\dfrac{\sin 75^\circ + \sin 15^\circ}{\sin 75^\circ - \sin 15^\circ} = \tan 60^\circ.$

By [20], $\sin 75^\circ + \sin 15^\circ = 2\sin 45^\circ \cos 30^\circ.$

By [21], $\sin 75^\circ - \sin 15^\circ = 2\cos 45^\circ \sin 30^\circ.$

$$\frac{\sin 75^\circ + \sin 15^\circ}{\sin 75^\circ - \sin 15^\circ} = \frac{2\sin 45^\circ \cos 30^\circ}{2\cos 45^\circ \sin 30^\circ}$$
$$= \tan 45^\circ \cot 30^\circ$$
$$= \tan 60^\circ.$$

51. Prove that $\cos 20^\circ + \cos 100^\circ + \cos 140^\circ = 0$.

By [22], $\cos 20^\circ + \cos 100^\circ = 2\cos 60^\circ \cos 40^\circ$

$= \cos 40^\circ.$

Also $\cos 140^\circ = \cos(180 - 40^\circ)$

$= -\cos 40^\circ.$

$\therefore \cos 20^\circ + \cos 100^\circ + \cos 140^\circ = 0.$

52. Prove that $\cos 36^\circ + \sin 36^\circ = \sqrt{2}\cos 9^\circ$.

$$\cos 36^\circ + \sin 36^\circ = \sqrt{2}\left(\frac{1}{\sqrt{2}}\cos 36^\circ + \frac{1}{\sqrt{2}}\sin 36^\circ\right)$$

$$= \sqrt{2}(\cos 45^\circ \cos 36^\circ + \sin 45^\circ \sin 36^\circ)$$

$$= \sqrt{2}\cos(45^\circ - 36^\circ)$$

$$= \sqrt{2}\cos 9^\circ.$$

53. Prove that $\tan 11^\circ 15' + 2\tan 22^\circ 30' + 4\tan 45^\circ = \cot 11^\circ 15'$.

By Ex. 11, $\cot 11^\circ 15' - \tan 11^\circ 15' = 2\cot 22^\circ 30'$.

$2\cot 22^\circ 30' - 2\tan 22^\circ 30' = 4\cot 45^\circ.$

$\therefore \cot 11^\circ 15' - \tan 11^\circ 15' - 2\tan 22^\circ 30' = 4\cot 45^\circ = 4\tan 45^\circ.$

$\tan 11^\circ 15' + 2\tan 22^\circ 30' + 4\tan 45^\circ = \cot 11^\circ 15'.$

54. If A, B, C are the angles of a plane triangle, prove that

$$\sin 2A + \sin 2B + \sin 2C = 4\sin A \sin B \sin C.$$

$A + B + C = 180^\circ.$

By [20], $\sin 2A + \sin 2B = 2\sin(A+B)\cos(A-B)$

$= 2\sin C\cos(A-B).$

$\therefore \sin 2A + \sin 2B + \sin 2C = 2\sin C\cos(A-B) - 2\sin C\cos C$

$= 2\sin C[\cos(A-B) + \cos(A+B)]$

$= 4\sin C\sin A\sin B.$

55. If A, B, C are the angles of a plane triangle, prove that

$$\cos 2A + \cos 2B + \cos 2C = -1 - 4\cos A\cos B\cos C.$$

By [22], $\cos 2A + \cos 2B = 2\cos(A+B)\cos(A-B)$

$= -2\cos C\cos(A-B).$

$\therefore \cos 2A + \cos 2B + \cos 2C = -2\cos C\cos(A-B) + \cos 2C$

$= -2\cos C\cos(A-B) + (1 + \cos 2C) - 1$

$= -2\cos C\cos(A-B) + 2\cos^2 C - 1$

$= 2\cos C[\cos C - \cos(A-B)] - 1$

$= 2\cos C[-\cos(A+B) - \cos(A-B)] - 1$

$= 2\cos C(-2\cos A\cos B) - 1$

$= -4\cos A\cos B\cos C - 1.$

56. If A, B, C are the angles of a plane triangle, prove that

$$\sin 3A + \sin 3B + \sin 3C = -4 \cos \frac{3A}{2} \cos \frac{3B}{2} \cos \frac{3C}{2}.$$

By [20],
$$\begin{aligned}\sin 3A + \sin 3B &= 2 \sin \tfrac{3}{2}(A+B) \cos \tfrac{3}{2}(A-B)\\ &= 2 \sin \tfrac{3}{2}(180 - C) \cos \tfrac{3}{2}(A-B)\\ &= -2 \cos \tfrac{3}{2}C \cos \tfrac{3}{2}(A-B).\end{aligned}$$

$$\begin{aligned}\sin 3C &= 2 \sin \tfrac{3}{2}C \cos \tfrac{3}{2}C\\ &= 2 \sin \tfrac{3}{2}(180 - A - B) \cos \tfrac{3}{2}C\\ &= 2 \cos \tfrac{3}{2}(A+B) \cos \tfrac{3}{2}C.\end{aligned}$$

$$\begin{aligned}\therefore \sin 3A + \sin 3B + \sin 3C &= -2 \cos \tfrac{3}{2}C\,[\cos \tfrac{3}{2}(A-B) + \cos \tfrac{3}{2}(A+B)]\\ &= -2 \cos \tfrac{3}{2}C\,(2 \cos \tfrac{3}{2}A \cos \tfrac{3}{2}B)\\ &= -4 \cos \tfrac{3}{2}A \cos \tfrac{3}{2}B \cos \tfrac{3}{2}C.\end{aligned}$$

57. If A, B, C are the angles of a plane triangle, prove that

$$\cos^2 A + \cos^2 B + \cos^2 C = 1 - 2 \cos A \cos B \cos C.$$

$$\cos^2 A = \frac{1 + \cos 2A}{2}.$$

$$\cos^2 B = \frac{1 + \cos 2B}{2}.$$

$$\cos^2 C = \frac{1 + \cos 2C}{2}.$$

$$\cos^2 A + \cos^2 B + \cos^2 C = \frac{3 + \cos 2A + \cos 2B + \cos 2C}{2}.$$

But, Ex. 55,

$$\cos 2A + \cos 2B + \cos 2C = -1 - 4 \cos A \cos B \cos C.$$

$$\begin{aligned}\therefore \cos^2 A + \cos^2 B + \cos^2 C &= \frac{3 - 1 - 4 \cos A \cos B \cos C}{2}\\ &= 1 - 2 \cos A \cos B \cos C.\end{aligned}$$

58. If $A + B + C = 90°$, prove that

$$\tan A \tan B + \tan B \tan C + \tan C \tan A = 1.$$

$$\begin{aligned}&\tan A \tan B + \tan B \tan C + \tan C \tan A\\ &= \tan A \tan B + (\tan A + \tan B) \tan C\\ &= \tan A \tan B + \frac{\tan A + \tan B}{\tan (A+B)}\\ &= \tan A \tan B + \frac{\tan A + \tan B}{\dfrac{\tan A + \tan B}{1 - \tan A \tan B}}\\ &= \tan A \tan B + 1 - \tan A \tan B\\ &= 1.\end{aligned}$$

59. If $A+B+C=90°$, prove that

$$\sin^2 A+\sin^2 B+\sin^2 C=1-2\sin A\sin B\sin C.$$

$$\begin{aligned}\sin C&=\cos(A+B)\\&=\cos A\cos B-\sin A\sin B.\\ \sin C+\sin A\sin B&=\cos A\cos B.\\ \sin^2 C+2\sin A\sin B\sin C+\sin^2 A\sin^2 B&\\&=\cos^2 A\cos^2 B.\\ \sin^2 C+2\sin A\sin B\sin C&=\cos^2 A\cos^2 B-\sin^2 A\sin^2 B\\&=(1-\sin^2 A)(1-\sin^2 B)-\sin^2 A\sin^2 B.\\&=1-\sin^2 A-\sin^2 B.\\ \therefore \sin^2 A+\sin^2 B+\sin^2 C&=1-2\sin A\sin B\sin C.\end{aligned}$$

60. If $A+B+C=90°$, prove that

$$\sin 2A+\sin 2B+\sin 2C=4\cos A\cos B\cos C.$$

By [20], $\sin 2A+\sin 2B=2\sin(A+B)\cos(A-B)$

$$\begin{aligned}&=2\cos C\cos(A-B).\\ \therefore \sin 2A+\sin 2B+\sin 2C&=2\cos C\cos(A-B)+2\sin C\cos C\\&=2\cos C[\cos(A-B)+\sin C]\\&=2\cos C[\cos(A-B)+\cos(A+B)]\\&=4\cos A\cos B\cos C.\end{aligned}$$

61. Prove that $\sin(\sin^{-1}x+\sin^{-1}y)=x\sqrt{1-y^2}+y\sqrt{1-x^2}$.

$$\begin{aligned}\sin(\sin^{-1}x+\sin^{-1}y)&=\sin(\sin^{-1}x)\cos(\sin^{-1}y)\\&\quad+\cos(\sin^{-1}x)\sin(\sin^{-1}y)\\&=x\sqrt{1-y^2}+y\sqrt{1-x^2}.\end{aligned}$$

62. Prove that $\tan(\tan^{-1}x+\tan^{-1}y)=\dfrac{x+y}{1-xy}$.

By [6], $\tan(\tan^{-1}x+\tan^{-1}y)=\dfrac{\tan(\tan^{-1}x)+\tan(\tan^{-1}y)}{1-\tan(\tan^{-1}x)\tan(\tan^{-1}y)}$

$$=\frac{x+y}{1-xy}.$$

63. Prove that $2\tan^{-1}x=\tan^{-1}\dfrac{2x}{1-x^2}$.

By [14], $\tan(2\tan^{-1}x)=\dfrac{2\tan(\tan^{-1}x)}{1-\tan^2(\tan^{-1}x)}$

$$=\frac{2x}{1-x^2}.$$

$$\therefore 2\tan^{-1}x=\tan^{-1}\frac{2x}{1-x^2}.$$

64. Prove that $\quad 2\sin^{-1}x = \sin^{-1}(2x\sqrt{1-x^2})$.

$$\sin(2\sin^{-1}x) = 2\sin(\sin^{-1}x)\cos(\sin^{-1}x)$$
$$= 2x\sqrt{1-x^2}.$$
$$\therefore 2\sin^{-1}x = \sin^{-1}2x\sqrt{1-x^2}.$$

65. Prove that $\quad 2\cos^{-1}x = \cos^{-1}(2x^2-1)$.

$$\cos(2\cos^{-1}x) = 2\cos^2(\cos^{-1}x) - 1$$
$$= 2x^2 - 1.$$
$$\therefore 2\cos^{-1}x = \cos^{-1}(2x^2-1).$$

66. Prove that $\quad 3\tan^{-1}x = \tan^{-1}\dfrac{3x-x^3}{1-3x^2}$.

$$\tan(3\tan^{-1}x) = \frac{\tan(\tan^{-1}x) + \tan(2\tan^{-1}x)}{1-\tan(\tan^{-1}x)\tan(2\tan^{-1}x)}$$
$$= \frac{x+\tan(2\tan^{-1}x)}{1-x\tan 2\tan^{-1}x}$$

(Ex. 63),
$$= \frac{x+\dfrac{2x}{1-x^2}}{1-x\dfrac{2x}{1-x^2}}$$
$$= \frac{3x-x^3}{1-3x^2}.$$
$$\therefore 3\tan^{-1}x = \tan^{-1}\frac{3x-x^3}{1-3x^2}.$$

67. Prove that $\quad \sin^{-1}\sqrt{\dfrac{x}{y}} = \tan^{-1}\sqrt{\dfrac{x}{y-x}}$.

Let $\quad \sin^{-1}\sqrt{\dfrac{x}{y}} = n.$

Then $\quad \sqrt{\dfrac{x}{y}} = \sin n.$

$$\sqrt{\frac{y-x}{y}} = \cos n.$$
$$\sqrt{\frac{x}{y-x}} = \tan n.$$
$$\therefore n = \tan^{-1}\sqrt{\frac{x}{y-x}}.$$
$$\sin^{-1}\sqrt{\frac{x}{y}} = \tan^{-1}\sqrt{\frac{y}{y-x}}.$$

68. Prove that $\sin^{-1}\sqrt{\frac{x-y}{x-z}} = \tan^{-1}\sqrt{\frac{x-y}{y-z}}$.

Let $\sin^{-1}\sqrt{\frac{x-y}{x-z}} = n$.

Then $\sqrt{\frac{x-y}{x-z}} = \sin n$.

$$\sqrt{\frac{y-z}{x-z}} = \cos n.$$

$$\sqrt{\frac{x-y}{y-z}} = \tan n.$$

$$n = \tan^{-1}\sqrt{\frac{x-y}{y-z}}.$$

$$\sin^{-1}\sqrt{\frac{x-y}{x-z}} = \tan^{-1}\sqrt{\frac{x-y}{y-z}}.$$

69. Prove that $\tan^{-1}\frac{1}{1-2x+4x^2} + \tan^{-1}\frac{1}{1+2x+4x^2} = \tan^{-1}\frac{1}{2x^2}$.

$$\tan\left(\tan^{-1}\frac{1}{1-2x+4x^2} + \tan^{-1}\frac{1}{1+2x+4x^2}\right)$$

$$= \frac{\frac{1}{1-2x+4x^2} + \frac{1}{1+2x+4x^2}}{1 - \frac{1}{(1-2x+4x^2)(1+2x+4x^2)}}$$

$$= \frac{1+2x+4x^2+1-2x+4x^2}{(1-2x+4x^2)(1+2x+4x^2)-1}$$

$$= \frac{2+8x^2}{4x^2+16x^4}$$

$$= \frac{1}{2x^2}.$$

$$\therefore \tan^{-1}\frac{1}{1-2x+4x^2} + \tan^{-1}\frac{1}{1+2x+4x^2} = \tan^{-1}\frac{1}{2x^2}$$

70. Prove that

$$\sin^{-1}x = \sec^{-1}\frac{1}{\sqrt{1-x^2}}.$$

Let $\sin^{-1}x = n$.

Then $x = \sin n$.

$$\sqrt{1-x^2} = \cos n.$$

$$\frac{1}{\sqrt{1-x^2}} = \sec n.$$

$$n = \sec^{-1}\frac{1}{\sqrt{1-x^2}}.$$

$$\sin^{-1}x = \sec^{-1}\frac{1}{\sqrt{1-x^2}}.$$

71. Prove that

$$2\sec^{-1}x=\tan^{-1}\frac{2\sqrt{x^2-1}}{2-x^2}.$$

Let $2\sec^{-1}x=n$.

Then $x=\sec\frac{1}{2}n$.

$$\frac{1}{x}=\cos\frac{1}{2}n.$$

$$2\left(\frac{1}{x}\right)^2-1=\cos n.$$

$$\frac{2-x^2}{x^2}=\cos n.$$

$$\frac{x^2}{2-x^2}=\sec n.$$

$$\left(\frac{x^2}{2-x^2}\right)^2-1=\tan^2 n.$$

$$\frac{4x^2-4}{(2-x^2)^2}=\tan^2 n.$$

$$\tan n=\frac{2\sqrt{x^2-1}}{2-x^2}.$$

$$n=\tan^{-1}\frac{2\sqrt{x^2-1}}{2-x^2}.$$

72. Prove that $\tan^{-1}\frac{1}{2}+\tan^{-1}\frac{1}{3}=45°$.

$$\tan(\tan^{-1}\tfrac{1}{2}+\tan^{-1}\tfrac{1}{3})=\frac{\frac{1}{2}+\frac{1}{3}}{1-\frac{1}{2}\times\frac{1}{3}}$$

$$=1.$$

$$\therefore\ \tan^{-1}\tfrac{1}{2}+\tan^{-1}\tfrac{1}{3}=\tan^{-1}1$$

$$=45°.$$

73. Prove that $\tan^{-1}\frac{1}{3}+\tan^{-1}\frac{1}{5}=\tan^{-1}\frac{4}{7}$.

$$\tan(\tan^{-1}\tfrac{1}{3}+\tan^{-1}\tfrac{1}{5})=\frac{\frac{1}{3}+\frac{1}{5}}{1-\frac{1}{3}\times\frac{1}{5}}$$

$$=\tfrac{4}{7}.$$

$$\therefore\ \tan^{-1}\tfrac{1}{3}+\tan^{-1}\tfrac{1}{5}=\tan^{-1}\tfrac{4}{7}.$$

74. Prove that $\sin^{-1}\frac{3}{5}+\sin^{-1}\frac{12}{13}=\sin^{-1}\frac{63}{65}$.

$$\sin(\sin^{-1}\tfrac{3}{5}+\sin^{-1}\tfrac{12}{13})=\tfrac{3}{5}\times\tfrac{5}{13}+\tfrac{4}{5}\times\tfrac{12}{13}$$

$$=\tfrac{63}{65}.$$

$$\therefore\ \sin^{-1}\tfrac{3}{5}+\sin^{-1}\tfrac{12}{13})=\sin^{-1}\tfrac{63}{65}.$$

75. Prove that $\sin^{-1}\frac{1}{\sqrt{82}}+\sin^{-1}\frac{4}{\sqrt{41}}=45°$.

$$\sin\left(\sin^{-1}\frac{1}{\sqrt{82}}+\sin^{-1}\frac{4}{\sqrt{41}}\right)=\frac{1}{\sqrt{82}}\times\frac{5}{\sqrt{41}}+\frac{9}{\sqrt{82}}\times\frac{4}{\sqrt{41}}$$

$$=\frac{5}{41\sqrt{2}}+\frac{36}{41\sqrt{2}}$$

$$=\frac{1}{\sqrt{2}}.$$

$$\therefore\ \sin^{-1}\frac{1}{\sqrt{82}}+\sin^{-1}\frac{4}{\sqrt{41}}=\sin^{-1}\frac{1}{\sqrt{2}}.$$

$$=45°.$$

76. Prove that $\sec^{-1}\frac{5}{3} + \sec^{-1}\frac{13}{12} = \sec^{-1}\frac{65}{16}$.

$$\sec^{-1}\tfrac{5}{3} + \sec^{-1}\tfrac{13}{12} = \cos^{-1}\tfrac{3}{5} + \cos^{-1}\tfrac{12}{13}.$$

$$\cos\left(\cos^{-1}\tfrac{3}{5} + \cos^{-1}\tfrac{12}{13}\right) = \tfrac{3}{5} \times \tfrac{12}{13} - \tfrac{4}{5} \times \tfrac{5}{13}$$
$$= \tfrac{16}{65}.$$

$$\therefore \cos^{-1}\tfrac{3}{5} + \cos^{-1}\tfrac{12}{13} = \cos^{-1}\tfrac{16}{65}.$$
$$\sec^{-1}\tfrac{5}{3} + \sec^{-1}\tfrac{13}{12} = \sec^{-1}\tfrac{65}{16}.$$

77. Prove that $\tan^{-1}(2 + \sqrt{3}) + \tan^{-1}(2 - \sqrt{3}) = \sec^{-1}2$.

Let $\tan^{-1}(2 + \sqrt{3}) - \tan^{-1}(2 - \sqrt{3}) = n$.

Then
$$\tan n = \frac{(2 + \sqrt{3}) - (2 - \sqrt{3})}{1 + (2 + \sqrt{3})(2 - \sqrt{3})}$$
$$= \frac{2\sqrt{3}}{2}$$
$$= \sqrt{3}.$$
$$\therefore n = 60^\circ.$$
$$\sec n = 2.$$
$$\therefore \tan^{-1}(2 + \sqrt{3}) - \tan^{-1}(2 - \sqrt{3}) = \sec^{-1}2.$$

78. Prove that $\tan^{-1}\frac{1}{3} + \tan^{-1}\frac{1}{5} + \tan^{-1}\frac{1}{7} + \tan^{-1}\frac{1}{8} = 45^\circ$.

Let
$$\tan^{-1}\tfrac{1}{3} + \tan^{-1}\tfrac{1}{5} = n,$$
and
$$\tan^{-1}\tfrac{1}{7} + \tan^{-1}\tfrac{1}{8} = \nu.$$

Then
$$\tan n = \frac{\frac{1}{3} + \frac{1}{5}}{1 - \frac{1}{3} \times \frac{1}{5}}$$
$$= \tfrac{4}{7}.$$
$$\tan \nu = \frac{\frac{1}{7} + \frac{1}{8}}{1 - \frac{1}{7} \times \frac{1}{8}}$$
$$= \tfrac{3}{11}.$$
$$\tan(n + \nu) = \frac{\frac{4}{7} + \frac{3}{11}}{1 - \frac{4}{7} \times \frac{3}{11}}$$
$$= 1.$$
$$n + \nu = \tan^{-1}1$$
$$= 45^\circ.$$
$$\tan^{-1}\tfrac{1}{3} + \tan^{-1}\tfrac{1}{5} + \tan^{-1}\tfrac{1}{7} + \tan^{-1}\tfrac{1}{8} = 45^\circ.$$

79. Given $\cos x = \frac{3}{5}$, find $\sin \frac{1}{2}x$ and $\cos \frac{1}{2}x$.

$$\cos x = 2\cos^2\frac{x}{2} - 1.$$
$$\therefore 2\cos^2\frac{x}{2} = 1 + \tfrac{3}{5}.$$
$$\cos^2\frac{x}{2} = \tfrac{4}{5}.$$
$$\cos\frac{x}{2} = \pm\frac{2}{\sqrt{5}}.$$
$$\sin^2\frac{x}{2} = \tfrac{1}{5}.$$
$$\sin\frac{x}{2} = \pm\frac{1}{\sqrt{5}}.$$
$$\sin\tfrac{1}{2}x = \pm\frac{1}{\sqrt{5}};\quad \cos\tfrac{1}{2}x = \pm\frac{2}{\sqrt{5}}.$$

80. Given $\tan x = \frac{1}{2}$, find $\tan \frac{1}{2}x$.

$$\tan x = \frac{2\tan\frac{1}{2}x}{1-\tan^2\frac{1}{2}x}.$$

$$\frac{1}{2} = \frac{2\tan\frac{1}{2}x}{1-\tan^2\frac{1}{2}x}.$$

$$1-\tan^2\tfrac{1}{2}x = 4\tan\tfrac{1}{2}x.$$

$$\tan^2\tfrac{1}{2}x + 4\tan\tfrac{1}{2}x - 1 = 0.$$

$$\therefore \tan\tfrac{1}{2}x = \pm\sqrt{5}-2.$$

81. Given $\sin x + \cos x = \sqrt{\frac{1}{2}}$, find $\cos 2x$.

$$\sin x + \cos x = \sqrt{\tfrac{1}{2}}$$

$$\sin^2 x + 2\sin x\cos x + \cos^2 x = \tfrac{1}{2}.$$

$$1 + 2\sin x\cos x = \tfrac{1}{2}.$$

$$2\sin x\cos x = -\tfrac{1}{2}.$$

$$\sin 2x = -\tfrac{1}{2}.$$

$$\therefore \cos 2x = \pm\tfrac{1}{2}\sqrt{3}.$$

82. Given $\tan 2x = \frac{24}{7}$, find $\sin x$.

$$\tan 2x = \tfrac{24}{7}.$$

$$\sec^2 2x = 1 + \tan^2 2x = \tfrac{625}{49}.$$

$$\cos^2 2x = \tfrac{49}{625}.$$

$$\cos 2x = \pm\tfrac{7}{25}.$$

$$1 - 2\sin^2 x = \pm\tfrac{7}{25}.$$

$$\sin^2 x = \tfrac{9}{25} \text{ or } \tfrac{16}{25}.$$

$$\sin x = \pm\tfrac{3}{5} \text{ or } \pm\tfrac{4}{5}.$$

83. Given $\cos 3x = \frac{23}{27}$, find $\tan 2x$.

$$\cos 3x = 4\cos^3 x - 3\cos x.$$

$$4\cos^3 x - 3\cos x = \tfrac{23}{27}.$$

By trial one solution is

$$\cos x = -\tfrac{1}{3}.$$

$$4\cos^3 x - 3\cos x - \tfrac{23}{27} = (\cos x + \tfrac{1}{3})(4\cos^2 x - \tfrac{4}{3}\cos x - \tfrac{23}{9}).$$

From $4\cos^2 x - \frac{4}{3}\cos x - \frac{23}{9} = 0$

we have $\cos x = \frac{1}{6} \pm \frac{1}{3}\sqrt{6}.$

But $\frac{1}{6} + \frac{1}{3}\sqrt{6} > 1,$

and $\frac{1}{6} - \frac{1}{3}\sqrt{6} < -1.$

Hence the only solution is

$$\cos x = -\tfrac{1}{3}.$$

Then
$$\sin x = \pm\tfrac{2}{3}\sqrt{2}.$$

$$\tan x = \pm 2\sqrt{2}.$$

$$\tan 2x = \frac{2\tan x}{1-\tan^2 x} = \frac{\pm 4\sqrt{2}}{1-8} = \pm\tfrac{4}{7}\sqrt{2}.$$

84. Given $2\csc x - \cot x = \sqrt{3}$, find $\sin\frac{1}{2}x$.

$$2\csc x - \cot x = \sqrt{3}.$$

$$\frac{2}{\sin x} - \frac{\cos x}{\sin x} = \sqrt{3}.$$

$$2 - \cos x = \sqrt{3}\sin x.$$

$$4 - 4\cos x + \cos^2 x = 3\sin^2 x = 3 - 3\cos^2 x.$$

$$4\cos^2 x - 4\cos x + 1 = 0.$$

$$\cos x = \tfrac{1}{2}.$$

$$1 - 2\sin^2\tfrac{1}{2}x = \tfrac{1}{2}.$$

$$\sin^2\tfrac{1}{2}x = \tfrac{1}{4}.$$

$$\sin\tfrac{1}{2}x = \tfrac{1}{2}.$$

85. Find $\sin 18^\circ$; $\cos 36^\circ$.

(i.)
$$54^\circ = 90^\circ - 36^\circ.$$

$$3\times 18^\circ = 90^\circ - 2\times 18^\circ.$$

$$\cos(3\times 18^\circ) = \sin(2\times 18^\circ).$$

$$4\cos^3 18^\circ - 3\cos 18^\circ = 2\sin 18^\circ\cos 18^\circ.$$

$$4\cos^2 18^\circ - 3 = 2\sin 18^\circ.$$

$$4 - 4\sin^2 18^\circ - 3 = 2\sin 18^\circ.$$

$$4\sin^2 18^\circ - 2\sin 18^\circ - 1 = 0.$$

$$\therefore \sin 18^\circ = \frac{\sqrt{5}-1}{4}.$$

(ii.)
$$\cos 36^\circ = 1 - 2\sin^2 18^\circ = 1 - 2\left(\frac{\sqrt{5}-1}{4}\right)^2 = \frac{\sqrt{5}+1}{4}.$$

86. Solve the equation

$$\sin x = 2\sin(\tfrac{1}{3}\pi + x).$$

$$\sin x = 2\sin(\tfrac{1}{3}\pi + x),$$
$$= \sqrt{3}\cos x + \sin x.$$
$$\sqrt{3}\cos x = 0.$$
$$\cos x = 0.$$
$$\therefore x = \tfrac{1}{2}\pi \text{ or } \tfrac{3}{2}\pi.$$

87. Solve the equation

$$\sin 2x = 2\cos x.$$

$$\sin 2x = 2\cos x.$$
$$2\sin x\cos x = 2\cos x.$$
$$2\cos x(\sin x - 1) = 0.$$

(i.) $\cos x = 0.$
$$x = 90°, 270°.$$

(ii.) $\sin x = 1.$
$$x = 90°.$$
$$\therefore x = 90° \text{ or } 270°.$$

88. Solve the equation

$$\cos 2x = 2\sin x.$$

$$\cos 2x = 2\sin x.$$
$$1 - 2\sin^2 x = 2\sin x.$$
$$2\sin^2 x + 2\sin x - 1 = 0.$$
$$\sin x = \frac{-1 \pm \sqrt{3}}{2}.$$
$$\sin x = \frac{-1 + \sqrt{3}}{2}.$$
$$x = \sin^{-1}\frac{\sqrt{3} - 1}{2}.$$

89. Solve the equation

$$\sin x + \cos x = 1.$$

$$\sin x + \cos x = 1.$$
$$\sin^2 x + 2\sin x\cos x + \cos^2 x = 1.$$
$$2\sin x\cos x = 0.$$

(i.) $\sin x = 0.$
$$x = 0°, 180°.$$

(ii.) $\cos x = 0.$
$$x = 90°, 270°.$$
$$\therefore x = 0°, 90°, 180° \text{ or } 270°.$$

But $x = 180°, 270°$ do not satisfy the given equation.

Hence $x = 0°$ or $90°$.

90. Solve the equation

$$\sin x + \cos 2x = 4\sin^2 x.$$

$$\sin x + \cos 2x = 4\sin^2 x.$$
$$\sin x + 1 - 2\sin^2 x = 4\sin^2 x.$$
$$6\sin^2 x - \sin x - 1 = 0.$$
$$\sin x = \tfrac{1}{2} \text{ or } -\tfrac{1}{3}.$$
$$x = 30° \text{ or } \sin^{-1}(-\tfrac{1}{3}).$$

91. Solve the equation

$$4\cos 2x + 3\cos x = 1.$$

$$4\cos 2x + 3\cos x = 1.$$
$$8\cos^2 x - 4 + 3\cos x = 1.$$
$$8\cos^2 x + 3\cos x - 5 = 0.$$
$$\cos x = -1 \text{ or } \tfrac{5}{8}.$$
$$x = 180° \text{ or } \cos^{-1}\tfrac{5}{8}.$$

92. Solve the equation

$$\sin x + \sin 2x = \sin 3x.$$

$$\sin x + \sin 2x = \sin 3x.$$
$$\sin x + 2\sin x\cos x = 3\sin x - 4\sin^3 x.$$
$$4\sin^3 x - 2\sin x + 2\sin x\cos x = 0.$$
$$\sin x(2\sin^2 x - 1 + \cos x) = 0.$$
$$\sin x(1 - 2\cos^2 x + \cos x) = 0.$$

(i.) $\sin x = 0.$
$$x = 0, 180°.$$

(ii.) $2\cos^2 x - \cos x - 1 = 0.$
$$\cos x = 1, -\tfrac{1}{2}.$$
$$x = 0°, 120°, 240°.$$
$$\therefore x = 0°, 120°, 180°, \text{ or } 240°.$$

93. Solve the equation $\sin 2x = 3\sin^2 x - \cos^2 x.$

$$\sin 2x = 3\sin^2 x - \cos^2 x.$$
$$2\sin x\cos x = 3\sin^2 x - \cos^2 x.$$
$$3\sin^2 x - 2\sin x\cos x - \cos^2 x = 0.$$
$$(3\sin x + \cos x)(\sin x - \cos x) = 0.$$

(i.)
$$3\sin x + \cos x = 0.$$
$$3\tan x + 1 = 0.$$
$$\tan x = -\tfrac{1}{3}.$$

(ii.)
$$\sin x - \cos x = 0.$$
$$\tan x = 1.$$
$$x = 45^\circ \text{ or } 225^\circ.$$
$$\therefore x = 45^\circ,\ 225^\circ, \text{ or } \tan^{-1}(-\tfrac{1}{3}).$$

94. Solve the equation $\tan x + \tan 2x = \tan 3x.$

$$\tan x + \tan 2x = \tan 3x.$$
$$\tan x + \frac{2\tan x}{1-\tan^2 x} = \frac{3\tan x - \tan^3 x}{1 - 3\tan^2 x}.$$
$$\tan x\left(1 + \frac{2}{1-\tan^2 x} - \frac{3-\tan^2 x}{1-3\tan^2 x}\right) = 0.$$

(i.)
$$\tan x = 0.$$
$$x = 0^\circ,\ 180^\circ.$$

(ii.)
$$1 + \frac{2}{1-\tan^2 x} - \frac{3-\tan^2 x}{1-3\tan^2 x} = 0.$$
$$(1-\tan^2 x)(1-3\tan^2 x) + 2(1-3\tan^2 x) - (1-\tan^2 x)(3-\tan^2 x) = 0.$$
$$-6\tan^2 x + 2\tan^4 x = 0.$$
$$\tan^2 x(\tan^2 x - 3) = 0.$$
$$\tan x = 0 \text{ or } \pm\sqrt{3}.$$
$$x = 0^\circ,\ 180^\circ;\ \pm 60^\circ,\ \pm 120^\circ.$$
$$\therefore x = 0^\circ,\ \pm 60^\circ,\ \pm 120^\circ, \text{ or } 180^\circ.$$

95. Solve the equation $\cot x - \tan x = \sin x + \cos x.$

$$\cot x - \tan x = \sin x + \cos x.$$
$$\frac{\cos x}{\sin x} - \frac{\sin x}{\cos x} = \sin x + \cos x.$$
$$\frac{\cos^2 x - \sin^2 x}{\sin x\cos x} = \sin x + \cos x.$$
$$\cos^2 x - \sin^2 x = \sin x\cos x(\sin x + \cos x).$$
$$(\sin x + \cos x)(\cos x - \sin x - \sin x\cos x) = 0.$$

(i.) $\sin x + \cos x = 0.$

$$\tan x = -1.$$

$$x = 135^\circ,\ -45^\circ.$$

(ii.) $\cos x - \sin x - \sin x \cos x = 0.$

$$\cos x - \sin x = \sin x \cos x.$$

$$\cos^2 x - 2\sin x \cos x + \sin^2 x = \sin^2 x \cos^2 x.$$

$$1 - 2\sin x \cos x = \sin^2 x \cos^2 x.$$

$$\sin^2 x \cos^2 x + 2\sin x \cos x - 1 = 0.$$

$$\sin x \cos x = -1 \pm \sqrt{2}.$$

$$\sin 2x = -2 + 2\sqrt{2}.$$

$$2x = \sin^{-1}(2\sqrt{2} - 2).$$

$$x = \tfrac{1}{2}\sin^{-1}(2\sqrt{2} - 2).$$

$$\therefore x = -45^\circ,\ 135^\circ, \text{ or } \tfrac{1}{2}\sin^{-1}(2\sqrt{2} - 2).$$

96. Solve the equation $\tan^2 x = \sin 2x.$

$$\tan^2 x = \sin 2x.$$

$$\tan^2 x = 2\sin x \cos x$$

$$= 2\tan x \cos^2 x.$$

$$= \frac{2\tan x}{\sec^2 x}$$

$$= \frac{2\tan x}{1 + \tan^2 x}.$$

$$\tan^2 x + \tan^4 x = 2\tan x.$$

$$\tan x(\tan^3 x + \tan x - 2) = 0.$$

(i.) $\tan x = 0.$

$$x = 0^\circ,\ 180^\circ.$$

(ii.) $\tan^3 x + \tan x - 2 = 0.$

$$(\tan x - 1)(\tan^2 x + \tan x + 2) = 0.$$

$$\tan x = 1.$$

$$x = 45^\circ \text{ or } 225^\circ.$$

$$\therefore x = 0^\circ,\ 45^\circ,\ 180^\circ,\ 225^\circ.$$

97. Solve the equation $\tan x + \cot x = \tan 2x.$

$$\tan x + \cot x = \tan 2x.$$

$$\tan x + \frac{1}{\tan x} = \frac{2\tan x}{1 - \tan^2 x}.$$

$$\frac{\tan^2 x + 1}{\tan x} = \frac{2\tan x}{1 - \tan^2 x}.$$

$$1 - \tan^4 x = 2\tan^2 x.$$

$$\tan^4 x + 2\tan^2 x - 1 = 0.$$

$$\tan^2 x = -1 \pm \sqrt{2}$$

$$= -1 + \sqrt{2}.$$

$$\sec^2 x = 1 + \tan^2 x$$
$$= \sqrt{2}.$$
$$\cos^2 x = \frac{1}{\sqrt{2}}\cdot$$
$$\cos x = \pm\sqrt{\frac{1}{\sqrt{2}}}\cdot$$
$$\therefore x = \cos^{-1}\left(\pm\sqrt{\frac{1}{\sqrt{2}}}\right)\cdot$$

98. Solve the equation $\dfrac{1-\tan x}{1+\tan x} = \cos 2x.$

$$\frac{1-\tan x}{1+\tan x} = \cos 2x.$$
$$\frac{\cos x - \sin x}{\cos x + \sin x} = \cos^2 x - \sin^2 x$$
$$= (\cos x - \sin x)(\cos x + \sin x).$$
$$\cos x - \sin x = (\cos x - \sin x)(\cos x + \sin x)^2.$$
$$(\cos x - \sin x)[1 - (\cos x + \sin x)^2] = 0.$$

(i.) $\cos x - \sin x = 0.$
$$\tan x = 1.$$
$$x = 45°,\ 225°.$$

(ii.) $1 - (\cos x + \sin x)^2 = 0.$
$$1 - (\cos^2 x + 2\sin x\cos x + \sin^2 x) = 0.$$
$$1 - (1 + 2\sin x\cos x) = 0.$$
$$\sin x\cos x = 0.$$
$$x = 0°,\ 90°,\ 180°,\ 270°.$$
$$\therefore x = 0°,\ 45°,\ 90°,\ 180°,\ 225°,\ 270°.$$

99. Solve the equation $\sin x + \sin 2x = 1 - \cos 2x.$

$$\sin x + \sin 2x = 1 - \cos 2x.$$
$$\sin x + 2\sin x\cos x = 2\sin^2 x.$$
$$\sin x(1 + 2\cos x - 2\sin x) = 0.$$

(i.) $\sin x = 0.$
$$x = 0,\ 180°.$$

(ii.) $1 + 2\cos x - 2\sin x = 0.$
$$\sin x - \cos x = \tfrac{1}{2}.$$
$$\sin^2 x - 2\sin x\cos x + \cos^2 x = \tfrac{1}{4}.$$
$$2\sin x\cos x = \tfrac{3}{4}.$$
$$\sin 2x = \tfrac{3}{4}.$$
$$x = \tfrac{1}{2}\sin^{-1}\tfrac{3}{4}.$$
$$\therefore x = 0°,\ 180°,\ \text{or}\ \tfrac{1}{2}\sin^{-1}\tfrac{3}{4}.$$

100. Solve the equation $\sec 2x + 1 = 2\cos x$.

$$\sec 2x + 1 = 2\cos x.$$
$$\frac{1}{\cos 2x} + 1 = 2\cos x.$$
$$1 + \cos 2x = 2\cos x \cos 2x.$$
$$2\cos^2 x = 2\cos x \cos 2x.$$
$$\cos x(\cos x - \cos 2x) = 0.$$

(i.) $$\cos x = 0.$$
$$x = 90°, \ 270°.$$

(ii.) $$\cos x - \cos 2x = 0.$$
$$\cos x = \cos 2x.$$
$$x = \pm 2x + n\,360.$$
$$x = 0°, \ 120°, \ 240°.$$
$$\therefore x = 0°, \ \pm 90°, \ \pm 120°.$$

101. Solve the equation $\tan 2x + \tan 3x = 0$.

$$\tan 2x + \tan 3x = 0.$$
$$\tan 2x = -\tan 3x = \tan(-3x).$$
$$2x = -3x \text{ or } 180° - 3x.$$

(i.) $$5x = 0° + n\,360.$$
$$x = 0°, \ 72°, \ 144°, \ 216°, \ 288°.$$

(ii.) $$5x = 180° + n\,360°.$$
$$x = 36°, \ 108°, \ 180°, \ 252°, \ 334°.$$
$$\therefore x = 0°, \ \pm 36°, \ \pm 72°, \ \pm 108°, \ \pm 144°, \ 180°.$$

102. Solve the equation $\tan(\frac{1}{4}\pi + x) + \tan(\frac{1}{4}\pi - x) = 4$.

$$\tan(\tfrac{1}{4}\pi + x) + \tan(\tfrac{1}{4}\pi - x) = 4.$$
$$\frac{1 + \tan x}{1 - \tan x} + \frac{1 - \tan x}{1 + \tan x} = 4.$$
$$(1 + \tan x)^2 + (1 - \tan x)^2 = 4(1 - \tan^2 x).$$
$$2 + 2\tan^2 x = 4 - 4\tan^2 x.$$
$$6\tan^2 x = 2.$$
$$\tan x = \pm\frac{1}{\sqrt{3}}.$$
$$x = \pm\tfrac{1}{6}\pi, \ \pm\tfrac{5}{6}\pi.$$

103. Solve the equation $\sqrt{1 + \sin x} - \sqrt{1 - \sin x} = 2\cos x$.

$$\sqrt{1 + \sin x} - \sqrt{1 - \sin x} = 2\cos x.$$
$$1 + \sin x - 2\sqrt{1 - \sin^2 x} + 1 - \sin x = 4\cos^2 x.$$
$$2 \pm 2\cos x = 4\cos^2 x.$$
$$2\cos^2 x \pm \cos x - 1 = 0.$$
$$\cos x = \pm\tfrac{1}{2} \text{ or } \pm 1.$$
$$x = \pm 60°, \ \pm 120°, \ 0°, \ 180°.$$

104. Solve the equation $\tan x \tan 3x = -\frac{2}{5}$.

$$\tan x \tan 3x = -\tfrac{2}{5}.$$

By Ex. 19, $$\tan 3x = \frac{3\tan x - \tan^3 x}{1 - 3\tan^2 x}.$$

$$\therefore \frac{3\tan^2 x - \tan^4 x}{1 - 3\tan^2 x} = -\tfrac{2}{5}.$$

$$15\tan^2 x - 5\tan^4 x = -2 + 6\tan^2 x.$$

$$5\tan^4 x - 9\tan^2 x - 2 = 0.$$

$$\tan^2 x = 2,\ -\tfrac{1}{5}.$$

$$\tan^2 x = 2.$$

$$\tan x = \pm\sqrt{2}.$$

$$x = \tan^{-1}\sqrt{2}.$$

105. Solve the equation $\sin(45° + x) + \cos(45° - x) = 1$.

$$\sin(45° + x) + \cos(45° - x) = 1.$$

$$2\cos(45° - x) = 1.$$

$$\cos(45° - x) = \tfrac{1}{2}.$$

$$45° - x = \pm 60°.$$

$$x = 105°,\ -15°.$$

106. Solve the equation $\tan x + \sec x = a$.

$$\tan x + \sec x = a.$$

$$\sec x = a - \tan x.$$

$$\sec^2 x = a^2 - 2a\tan x + \tan^2 x.$$

$$1 + \tan^2 x = a^2 - 2a\tan x + \tan^2 x.$$

$$1 = a^2 - 2a\tan x.$$

$$\tan x = \frac{a^2 - 1}{2a}.$$

$$-\cot x = \frac{2a}{a^2 - 1}$$

$$= \cot(2\cot^{-1} a).$$

$$\therefore x = -2\cot^{-1} a.$$

107. Solve the equation $\cos 2x = a(1 - \cos x)$.

$$\cos 2x = a(1 - \cos x).$$

$$2\cos^2 x - 1 = a - a\cos x.$$

$$2\cos^2 x + a\cos x = a + 1.$$

$$\cos x = \frac{-a \pm \sqrt{a^2 + 8a + 8}}{4}.$$

$$x = \cos^{-1}\left(\frac{-a \pm \sqrt{a^2 + 8a + 8}}{4}\right).$$

108. Solve the equation $\cos 2x(1-\tan x) = a(1+\tan x)$.

$$\cos 2x(1-\tan x) = a(1+\tan x).$$

$$\cos 2x = a\frac{1+\tan x}{1-\tan x}.$$

$$\cos^2 x - \sin^2 x = a\frac{\cos x + \sin x}{\cos x - \sin x}.$$

(i.) $\cos x + \sin x = 0.$

$$\tan x = -1.$$

$$x = 135^\circ,\ -45^\circ.$$

(ii.) $\cos x - \sin x = \dfrac{a}{\cos x - \sin x}.$

$$\cos^2 x - 2\sin x\cos x + \sin^2 x = a.$$

$$2\sin x\cos x = 1 - a.$$

$$\sin 2x = 1 - a.$$

$$x = \tfrac{1}{2}\sin^{-1}(1-a).$$

$$\therefore x = 135^\circ,\ -45^\circ,\ \text{or } \tfrac{1}{2}\sin^{-1}(1-a).$$

109. Solve the equation $\sin^6 x + \cos^6 x = \frac{7}{12}\sin^2 2x$.

$$\sin^6 x + \cos^6 x = \tfrac{7}{12}\sin^2 2x.$$

$$\begin{aligned}\sin^6 x + \cos^6 x &= (\sin^2 x + \cos^2 x)(\sin^4 x - \sin^2 x\cos^2 x + \cos^4 x)\\ &= \sin^4 x - \sin^2 x\cos^2 x + \cos^4 x\\ &= (\sin^2 x + \cos^2 x)^2 - 3\sin^2 x\cos^2 x\\ &= 1 - 3\sin^2 x\cos^2 x.\end{aligned}$$

$$\begin{aligned}1 - 3\sin^2 x\cos^2 x &= \tfrac{7}{12}\sin^2 2x\\ &= \tfrac{7}{3}\sin^2 x\cos^2 x.\end{aligned}$$

$$\tfrac{16}{3}\sin^2 x\cos^2 x = 1$$

$$4\sin^2 x\cos^2 x = \tfrac{3}{4}.$$

$$\sin^2 2x = \tfrac{3}{4}.$$

$$\sin 2x = \pm\tfrac{1}{2}\sqrt{3}.$$

$$2x = \pm 60^\circ,\ \pm 120^\circ.$$

$$\begin{aligned}x &= \pm 30^\circ,\ \pm 210^\circ,\ \pm 60^\circ,\ \pm 240^\circ.\\ &= \pm 30^\circ,\ \pm 150^\circ,\ \pm 60^\circ,\ \pm 120^\circ.\end{aligned}$$

110. Solve the equation

$$\cos 3x + 8\cos^3 x = 0.$$

$$\cos 3x + 8\cos^3 x = 0.$$

$$4\cos^3 x - 3\cos x + 8\cos^3 x = 0.$$

$$12\cos^3 x - 3\cos x = 0.$$

$$\cos x(4\cos^2 x - 1) = 0.$$

(i.) $\cos x = 0.$

$$x = 90^\circ,\ 270^\circ.$$

(ii.) $4\cos^2 x - 1 = 0.$

$$\cos x = \pm\tfrac{1}{2}.$$

$$x = \pm 60,\ \pm 120^\circ$$

$$\therefore x = \pm 60^\circ,\ \pm 90^\circ,\ \text{or } \pm 120^\circ.$$

111. Solve the equation $\sec(x+120^\circ)+\sec(x-120^\circ)=2\cos x$.

$$\sec(x+120^\circ)+\sec(x-120^\circ)=2\cos x.$$

$$\frac{1}{\cos(x+120^\circ)}+\frac{1}{\cos(x-120^\circ)}=2\cos x.$$

$$\frac{\cos(x+120^\circ)+\cos(x-120^\circ)}{\cos(x+120^\circ)\cos(x-120)}=2\cos x.$$

[22], $$\frac{2\cos x\cos 120^\circ}{\cos^2 x\cos^2 120-\sin^2 x\sin^2 120}=2\cos x.$$

$$\frac{-\cos x}{\frac{1}{4}(\cos^2 x-3\sin^2 x)}=2\cos x.$$

$$2\cos x+\cos x(\cos^2 x-3\sin^2 x)=0.$$

$$2\cos x+\cos x(4\cos^2 x-3)=0.$$

$$\cos x(4\cos^2 x-1)=0.$$

(i.) $$\cos x=0.$$

$$x=90^\circ,\ 270^\circ.$$

(ii.) $$4\cos^2 x-1=0.$$

$$\cos x=\pm\tfrac{1}{2}.$$

$$x=\pm 60^\circ,\ \pm 120^\circ.$$

$$\therefore x=\pm 60^\circ,\ \pm 90^\circ,\ \text{or} \pm 120^\circ.$$

112. Solve the equation $\csc x=\cot x+\sqrt{3}$.

$$\csc x=\cot x+\sqrt{3}.$$

$$\csc^2 x=\cot^2 x+2\sqrt{3}\cot x+3.$$

$$1+\cot^2 x=\cot^2 x+2\sqrt{3}\cot x+3.$$

$$2\sqrt{3}\cot x=-2.$$

$$\cot x=-\frac{1}{\sqrt{3}}.$$

$$x=-60^\circ,\ 120^\circ.$$

$x=-60^\circ$ does not satisfy the given equation.

$$\therefore x=120^\circ.$$

113. Solve the equation

$$4\cos 2x+6\sin x=5.$$

$$4\cos 2x+6\sin x=5.$$

$$4(1-2\sin^2 x)+6\sin x=5.$$

$$8\sin^2 x-6\sin x+1=0.$$

$$\sin x=\tfrac{1}{2},\ \tfrac{1}{4}.$$

$$x=30^\circ,\ 150^\circ,\ \sin^{-1}\tfrac{1}{4}.$$

114. Solve the equation

$$\cos x-\cos 2x=1.$$

$$\cos x-\cos 2x=1.$$

$$\cos x-(2\cos^2 x-1)=1.$$

$$2\cos^2 x-\cos x=0.$$

$$\cos x=0,\ \tfrac{1}{2}.$$

$$x=\pm 90^\circ, \pm 60^\circ.$$

115. Solve the equation $\sin 4x - \sin 2x = \sin x$.

$$\sin 4x - \sin 2x = \sin x.$$

[20], $$2 \cos 3x \sin x = \sin x.$$

$$\sin x (2 \cos 3x - 1) = 0.$$

(i.) $$\sin x = 0.$$

$$x = 0°, 180°.$$

(ii.) $$2 \cos 3x - 1 = 0.$$

$$\cos 3x = \tfrac{1}{2}.$$

$$3x = \pm 60 + n\, 360°.$$

$$x = \pm 20°, \pm 140°, \pm 260°$$

$$= \pm 20°, \pm 140°, \pm 100°.$$

$$\therefore x = 0°, \pm 20°, \pm 100°, \pm 140°, \pm 180°.$$

116. Solve the equation $2 \sin^2 x + \sin^2 2x = 2$.

$$2 \sin^2 x + \sin^2 2x = 2.$$

$$2 \sin^2 x + 4 \sin^2 x \cos^2 x = 2.$$

$$\sin^2 x + 2 \sin^2 x (1 - \sin^2 x) = 1.$$

$$2 \sin^4 x - 3 \sin^2 x + 1 = 0.$$

$$\sin^2 x = 1, \tfrac{1}{2}.$$

$$\sin x = \pm 1, \pm \sqrt{\tfrac{1}{2}}.$$

$$x = \pm 90°, \pm 45°, \pm 135°.$$

117. Solve the equation $\cos 5x + \cos 3x + \cos x = 0$.

$$\cos 5x + \cos 3x + \cos x = 0.$$

[22], $$2 \cos 4x \cos x + \cos x = 0.$$

(i.) $$\cos x = 0.$$

$$x = 90°, 270°.$$

(ii.) $$2 \cos 4x + 1 = 0.$$

$$\cos 4x = -\tfrac{1}{2}.$$

$$4x = \pm 120° + n\, 360°.$$

$$x = \pm 30°, \pm 120°, \pm 210°, \pm 300°$$

$$= \pm 30°, \pm 120°, \pm 150°, \pm 60°.$$

$$\therefore x = \pm 30°, \pm 60°, \pm 90°, \pm 120°, \text{ or } \pm 150°.$$

118. Solve the equation $\sec x - \cot x = \csc x - \tan x$.

$$\sec x - \cot x = \csc x - \tan x.$$

$$\frac{1}{\cos x} - \frac{\cos x}{\sin x} = \frac{1}{\sin x} - \frac{\sin x}{\cos x}.$$

$$\sin x - \cos^2 x = \cos x - \sin^2 x.$$

$$(\sin x - \cos x)(1 + \sin x + \cos x) = 0.$$

(i.)
$$\sin x - \cos x = 0.$$
$$\tan x = 1.$$
$$x = 45°,\ 225°.$$

(ii.)
$$1 + \sin x + \cos x = 0.$$
$$\sin x + \cos x = -1.$$
$$\sin^2 x + 2\sin x\cos x + \cos^2 x = 1.$$
$$2\sin x\cos x = 0.$$
$$\sin 2x = 0.$$
$$2x = 0,\ 180°.$$
$$x = 0°,\ 180°,\ 90°,\ 270°.$$
$$\therefore x = 0°,\ 45°,\ \pm 90°, \text{ or } 225°.$$

119. Solve the equation $\tan^2 x + \cot^2 x = \frac{10}{3}$.

$$\tan^2 x + \cot^2 x = \tfrac{10}{3}.$$
$$\tan^2 x + \frac{1}{\tan^2 x} = \tfrac{10}{3}.$$
$$\tan^4 x - \tfrac{10}{3}\tan^2 x + 1 = 0.$$
$$\tan^2 x = 3,\ \tfrac{1}{3}.$$
$$\tan x = \pm\sqrt{3},\ \pm\frac{1}{\sqrt{3}}.$$
$$x = \pm 60°,\ \pm 120°,\ \pm 30°, \pm 150°.$$

120. Solve the equation $\sin 4x - \cos 3x = \sin 2x$.

$$\sin 4x - \cos 3x = \sin 2x.$$
$$\sin 4x - \sin 2x = \cos 3x.$$
[21],
$$2\cos 3x\sin x = \cos 3x.$$
$$\cos 3x\,(2\sin x - 1) = 0.$$

(i.)
$$\cos 3x = 0.$$
$$3x = \pm 90° + n\,360°.$$
$$x = \pm 30°,\ \pm 150°,\ \pm 90°.$$

(ii.)
$$2\sin x - 1 = 0.$$
$$\sin x = \tfrac{1}{2}.$$
$$x = 30°,\ 150°.$$
$$\therefore x = \pm 30,\ \pm 90°,\ \pm 150°.$$

121. Solve the equation $\sin x + \cos x = \sec x$.

$$\sin x + \cos x = \sec x.$$
$$\sin x + \cos x = \frac{1}{\cos x}.$$
$$\cos x\sin x + \cos^2 x = 1.$$
$$\cos x\sin x = 1 - \cos^2 x$$
$$= \sin^2 x.$$
$$\sin x\,(\cos x - \sin x) = 0.$$

(i.) $\sin x = 0.$

$$x = 0°,\ 180°.$$

(ii.) $\cos x - \sin x = 0.$

$$\tan x = 1.$$
$$x = 45°,\ 225°.$$
$$\therefore x = 0°,\ 45°,\ 180°,\ 225°.$$

122. Solve the equation $2 \cos x \cos 3x + 1 = 0.$

$$2 \cos x \cos 3x + 1 = 0.$$
$$2 \cos x (4 \cos^3 x - 3 \cos x) + 1 = 0.$$
$$8 \cos^4 x - 6 \cos^2 x + 1 = 0.$$
$$\cos^2 x = \tfrac{1}{2},\ \tfrac{1}{4}.$$
$$\cos x = \pm \sqrt{\tfrac{1}{2}},\ \pm \tfrac{1}{2}.$$
$$x = \pm 45°,\ \pm 135°,\ \pm 60°,\ \pm 120°.$$

123. Solve the equation $\cos 3x - 2 \cos 2x + \cos x = 0.$

$$\cos 3x - 2 \cos 2x + \cos x = 0.$$
$$\cos 3x + \cos x = 2 \cos 2x.$$

[22], $2 \cos 2x \cos x = 2 \cos 2x.$

$$\cos 2x (\cos x - 1) = 0.$$

(i.) $\cos 2x = 0.$

$$2x = \pm 90° + n\, 360°.$$
$$x = \pm 45°,\ \pm 135°.$$

(ii.) $\cos x - 1 = 0.$

$$\cos x = 1.$$
$$x = 0°.$$
$$\therefore x = 0°,\ \pm 45°,\ \pm 135°.$$

124. Solve the equation $\tan 2x \tan x = 1.$

$$\tan 2x \tan x = 1.$$
$$\tan 2x = \text{ctn}\ x.$$
$$2x = \pm 90° - x.$$
$$3x = \pm 90° + n\, 360°.$$
$$x = \pm 30°,\ \pm 150°,\ \pm 270°.$$
$$\therefore x = \pm 30°,\ \pm 90°,\ \pm 150.$$

125. Solve the equation $\sin (x + 12°) + \sin (x - 8°) = \sin 20°.$

$$\sin (x + 12°) + \sin (x - 8°) = \sin 20°.$$

[20], $2 \sin (x + 2°) \cos 10° = \sin 20°$

$$= 2 \sin 10° \cos 10°.$$
$$\sin (x + 2°) = \sin 10°.$$
$$x + 2° = 10°,\ 170°.$$
$$x = 8° \text{ or } 168°.$$

126. Solve the equation $\tan(60° + x)\tan(60° - x) = -2$.

$$\tan(60 + x)\tan(60° - x) = -2.$$

[6],
$$\frac{\sqrt{3} + \tan x}{1 - \sqrt{3}\tan x} \times \frac{\sqrt{3} - \tan x}{1 + \sqrt{3}\tan x} = -2.$$

$$\frac{3 - \tan^2 x}{1 - 3\tan^2 x} = -2.$$

$$3 - \tan^2 x = -2 + 6\tan^2 x.$$

$$7\tan^2 x = 5.$$

$$\tan x = \sqrt{\tfrac{5}{7}}.$$

$$x = \tan^{-1}\sqrt{\tfrac{5}{7}}.$$

127. Solve the equation $\sin(x + 120°) + \sin(x + 60°) = \frac{3}{2}$.

$$\sin(x + 120°) + \sin(x + 60°) = \tfrac{3}{2}.$$

[20],
$$2\sin(x + 90°)\cos 30° = \tfrac{3}{2}.$$

$$2\cos x \times \frac{\sqrt{3}}{2} = \tfrac{3}{2}.$$

$$\cos x = \tfrac{1}{2}\sqrt{3}.$$

$$x = \pm 30°.$$

128. Solve the equation $\sin(x + 30°)\sin(x - 30°) = \frac{1}{2}$.

$$\sin(x + 30°)\sin(x - 30°) = \tfrac{1}{2}.$$

[23],
$$-\tfrac{1}{2}(\cos 2x - \cos 60°) = \tfrac{1}{2}.$$

$$\cos 2x - \cos 60° = -1.$$

$$\cos 2x = -\tfrac{1}{2}.$$

$$2x = \pm 120° + n\,360°.$$

$$x = \pm 60°,\ \pm 240°$$

$$= \pm 60°,\ \pm 120°.$$

129. Solve the equation $\sin^4 x + \cos^4 x = \frac{5}{8}$.

$$\sin^4 x + \cos^4 x = \tfrac{5}{8}.$$

$$\sin^4 x + 2\sin^2 x\cos^2 x + \cos^4 x = 2\sin^2 x\cos^2 x + \tfrac{5}{8}.$$

$$(\sin^2 x + \cos^2 x)^2 = 2\sin^2 x\cos^2 x + \tfrac{5}{8}.$$

$$1 = 2\sin^2 x\cos^2 x + \tfrac{5}{8}.$$

$$2\sin^2 x\cos^2 x = \tfrac{3}{8}.$$

$$4\sin^2 x\cos^2 x = \tfrac{3}{4}.$$

$$\sin^2 2x = \tfrac{3}{4}.$$

$$\sin 2x = \pm\tfrac{1}{2}\sqrt{3}.$$

$$2x = \pm 60°,\ \pm 120°.$$

$$x = \pm 30°,\ \pm 60°,\ \pm 120°, \text{ or } \pm 150°.$$

130 Solve the equation $\sin^4 x - \cos^4 x = \frac{7}{25}$.

$$\sin^4 x - \cos^4 x = \tfrac{7}{25}.$$
$$(\sin^2 x + \cos^2 x)(\sin^2 x - \cos^2 x) = \tfrac{7}{25}.$$
$$\sin^2 x - \cos^2 x = \tfrac{7}{25}.$$
$$2\sin^2 x - 1 = \tfrac{7}{25}.$$
$$\sin^2 x = \tfrac{16}{25}.$$
$$\sin x = \pm \tfrac{4}{5}.$$
$$x = \pm \sin^{-1} \tfrac{4}{5}.$$

131. Solve the equation $\tan(x + 30°) = 2\cos x$.

Let $$x + 30° = y.$$
Then $$\tan y = 2\cos(y - 30°).$$
$$\frac{\sin y}{\cos y} = \sqrt{3}\cos y + \sin y.$$
$$\sin y = \sqrt{3}\cos^2 y + \sin y \cos y.$$
$$\sin y(1 - \cos y) = \sqrt{3}\cos^2 y.$$
$$\sin^2 y(1 - \cos y)^2 = 3\cos^4 y.$$
$$(1 - \cos^2 y)(1 - 2\cos y + \cos^2 y) = 3\cos^4 y.$$
$$1 - 2\cos y + 2\cos^3 y - \cos^4 y = 3\cos^4 y.$$
$$4\cos^4 y - 2\cos^3 y + 2\cos y - 1 = 0.$$
$$(2\cos y - 1)(2\cos^3 y + 1) = 0.$$

(i.) $$2\cos y - 1 = 0.$$
$$\cos y = \tfrac{1}{2}.$$
$$y = \pm 60°.$$
$$x = 30°, \ -90°.$$

(ii.) $$2\cos^3 y + 1 = 0.$$
$$\cos^3 y = -\tfrac{1}{2}.$$
$$\cos y = -\sqrt[3]{\tfrac{1}{2}}.$$
$$y = \cos^{-1}\left(-\frac{1}{\sqrt[3]{2}}\right).$$
$$x = \cos^{-1}\left(-\frac{1}{\sqrt[3]{2}}\right) - 30°$$
$$= 150° - \cos^{-1}\frac{1}{\sqrt[3]{2}}.$$
$$\therefore x = 30°, \ -90°, \text{ or } 150° - \cos^{-1}\frac{1}{\sqrt[3]{2}}.$$

But $x = -90°$ does not satisfy the original equation.

$$\therefore x = 30°, \text{ or } 150° - \cos^{-1}\left(\frac{1}{\sqrt[3]{2}}\right).$$

132. Solve the equation $\sec x = 2 \tan x + \frac{1}{4}$.

$$\sec x = 2 \tan x + \tfrac{1}{4}.$$
$$\sec^2 x = 4 \tan^2 x + \tan x + \tfrac{1}{16}.$$
$$1 + \tan^2 x = 4 \tan^2 x + 1 = x + \tfrac{1}{16}.$$
$$3 \tan^2 x + \tan x - \tfrac{15}{16} = 0.$$
$$\tan x = \tfrac{5}{12},\ -\tfrac{3}{4}.$$
$$x = \tan^{-1} \tfrac{5}{12},\ \text{or}\ -\tan^{-1} \tfrac{3}{4}.$$

133. Solve the equations $\sin(x - y) = \cos x$, $\cos(x + y) = \sin x$.

$$\sin(x - y) = \cos x.$$
$$x - y = 90^\circ - x,\ \text{or}\ 90^\circ + x.$$
$$y = 2x - 90^\circ,\ \text{or}\ -90^\circ.$$
$$\cos(x + y) = \sin x.$$
$$x + y = 90^\circ - x,\ \text{or}\ x - 90^\circ.$$
$$y = 90^\circ - 2x,\ \text{or}\ -90^\circ.$$

(i.) $y = -90^\circ$, x indeterminate.

(ii.) $y = 2x - 90^\circ = 90^\circ - 2x$.

$$\therefore x = 45^\circ,\ y = 0^\circ,\ \text{or}\ x = 135^\circ,\ y = 180^\circ,$$
$$\text{or}\ x = 225^\circ,\ y = 0^\circ,\ \text{or}\ x = 315^\circ,\ y = 180^\circ.$$
$$\therefore x = 45^\circ,\ y = 0^\circ;\ x = 135^\circ,\ y = 180^\circ;\ x = -45^\circ,\ y = 180^\circ;$$
$$x = -135^\circ,\ y = 0^\circ,\ \text{or}\ y = -90^\circ,\ x\ \text{indeterminate}.$$

134. Solve the equations $\tan x + \tan y = a$, $\cot x + \cot y = b$.

$$\tan x + \tan y = a.$$
$$\cot x + \cot y = b.$$
$$\tan x = a - \tan y.$$
$$\cot x = b - \cot y.$$
$$(a - \tan y)(b - \cot y) = 1.$$
$$ab - b \tan y - a \cot y + 1 = 1.$$
$$b \tan y + a \cot y = ab.$$
$$b \tan^2 y + a = ab \tan y.$$
$$b \tan^2 y - ab \tan y + a = 0.$$
$$\tan y = \frac{ab \pm \sqrt{a^2b^2 - 4ab}}{2b}.$$
$$\tan x = a - \tan y$$
$$= \frac{ab \mp \sqrt{a^2b^2 - 4ab}}{2b}.$$
$$\therefore x = \tan^{-1}\left(\frac{ab \mp \sqrt{a^2b^2 - 4ab}}{2b}\right).$$
$$y = \tan^{-1}\left(\frac{ab \pm \sqrt{a^2b^2 - 4ab}}{2b}\right).$$

135. Solve the equation $\sin(x + 12°)\cos(x - 12°) = \cos 33° \sin 57°$.

$$\sin(x + 12°)\cos(x - 12°) = \cos 33° \sin 57°.$$

[20], $$\tfrac{1}{2}(\sin 2x + \sin 24°) = \tfrac{1}{2}(\sin 90° + \sin 24°).$$

$$\sin 2x = \sin 90°.$$

$$2x = 90° + n\,360°.$$

$$x = 45° \text{ or } 225°.$$

136. Solve the equation $\sin^{-1} x + \sin^{-1} \tfrac{1}{2}x = 120°$.

$$\sin^{-1} x + \sin^{-1} \tfrac{1}{2}x = 120°.$$

$$\sin(\sin^{-1} x + \sin^{-1} \tfrac{1}{2}x) = \tfrac{1}{2}\sqrt{3}.$$

$$x\sqrt{1 - \tfrac{1}{4}x^2} + \tfrac{1}{2}x\sqrt{1 - x^2} = \tfrac{1}{2}\sqrt{3}.$$

$$x\sqrt{4 - x^2} + x\sqrt{1 - x^2} = \sqrt{3}.$$

$$x^2(4 - x^2) = 3 - 2\sqrt{3}\,x\sqrt{1 - x^2} + x^2(1 - x^2).$$

$$3x^2 - 3 = -2\sqrt{3}\,x\sqrt{1 - x^2}.$$

$$9x^4 - 18x^2 + 9 = 12x^2(1 - x^2).$$

$$21x^4 - 30x^2 + 9 = 0.$$

$$7x^4 - 10x^2 + 3 = 0.$$

$$x^2 = 1,\ \tfrac{3}{7}.$$

$$x = \pm 1,\ \pm\sqrt{\tfrac{3}{7}}.$$

137. Solve the equation $\tan^{-1} x + \tan^{-1} 2x = \tan^{-1} 3\sqrt{3}$.

$$\tan^{-1} x + \tan^{-1} 2x = \tan^{-1} 3\sqrt{3}.$$

$$\tan(\tan^{-1} x + \tan^{-1} 2x) = 3\sqrt{3}.$$

$$\frac{x + 2x}{1 - 2x^2} = 3\sqrt{3}.$$

$$3x = 3\sqrt{3}(1 - 2x^2).$$

$$x = \sqrt{3}(1 - 2x^2).$$

$$2\sqrt{3}x^2 + x - \sqrt{3} = 0.$$

$$x = \frac{1}{\sqrt{3}},\ -\tfrac{1}{2}\sqrt{3}.$$

138. Solve the equation $\sin^{-1} x + 2\cos^{-1} x = \tfrac{2}{3}\pi$.

$$\sin^{-1} x + 2\cos^{-1} x = \tfrac{2}{3}\pi.$$

$$\sin(\sin^{-1} x + 2\cos^{-1} x) = \tfrac{1}{2}\sqrt{3}.$$

$$x\cos(2\cos^{-1} x) + \sqrt{1 - x^2}\sin(2\cos^{-1} x) = \tfrac{1}{2}\sqrt{3}.$$

$$x(2x^2 - 1) + \sqrt{1 - x^2} \times 2x\sqrt{1 - x^2} = \tfrac{1}{2}\sqrt{3}.$$

$$x(2x^2 - 1) + 2x(1 - x^2) = \tfrac{1}{2}\sqrt{3}.$$

$$x = \tfrac{1}{2}\sqrt{3}.$$

139. Solve the equation $\sin^{-1}x + 3\cos^{-1}x = 210°$.

$$\sin^{-1}x + 3\cos^{-1}x = 210°.$$
$$\sin(\sin^{-1}x + 3\cos^{-1}x) = -\tfrac{1}{2}.$$
$$x\cos(3\cos^{-1}x) + \sqrt{1-x^2}\,\sin(3\cos^{-1}x) = -\tfrac{1}{2}.$$
$$x(4x^3 - 3x) + \sqrt{1-x^2}\,[3\sqrt{1-x^2} - 4(1-x^2)^{\frac{3}{2}}] = -\tfrac{1}{2}.$$
$$x(4x^3 - 3x) + (1-x^2)[3 - 4(1-x^2)] = -\tfrac{1}{2}.$$
$$4x^4 - 3x^2 - 1 + 5x^2 - 4x^4 = -\tfrac{1}{2}.$$
$$2x^2 = \tfrac{1}{2}.$$
$$x = \pm\tfrac{1}{2}.$$

140. Solve the equation $\tan^{-1}x + 2\cot^{-1}x = 135°$.

$$\tan^{-1}x + 2\cot^{-1}x = 135°.$$
$$\tan(\tan^{-1}x + 2\cot^{-1}x) = -1.$$
$$\tan\left(\tan^{-1}x + 2\tan^{-1}\frac{1}{x}\right) = -1.$$
$$\frac{x + \dfrac{\frac{2}{x}}{1-\frac{1}{x^2}}}{1 - x\dfrac{\frac{2}{x}}{1-\frac{1}{x^2}}} = -1.$$
$$\frac{x - \frac{1}{x} + \frac{2}{x}}{1 - \frac{1}{x^2} - 2} = -1.$$
$$\frac{x^3 + x}{-x^2 - 1} = -1.$$
$$x^3 + x = 1 + x^2.$$
$$x^3 - x^2 + x - 1 = 0.$$
$$(x-1)(x^2+1) = 0.$$
$$\therefore x = 1.$$

141. Solve the equation $\tan^{-1}(x+1) + \tan^{-1}(x-1) = \tan^{-1}2x$.

$$\tan^{-1}(x+1) + \tan^{-1}(x-1) = \tan^{-1}2x.$$
$$\tan[\tan^{-1}(x+1) + \tan^{-1}(x-1)] = 2x.$$
$$\frac{x+1+x-1}{1-(x^2-1)} = 2x.$$
$$\frac{2x}{2-x^2} = 2x.$$
$$x = 2x - x^3.$$
$$x^3 - x = 0.$$
$$x = 0,\ 1,\ -1.$$

142. Solve the equation $\tan^{-1}\dfrac{x+2}{x+1} + \tan^{-1}\dfrac{x-2}{x-1} = \tfrac{3}{4}\pi$.

$$\tan^{-1}\frac{x+2}{x+1} + \tan^{-1}\frac{x-2}{x-1} = \tfrac{3}{4}\pi.$$
$$\tan\left(\tan^{-1}\frac{x+2}{x+1} + \tan^{-1}\frac{x-2}{x-1}\right) = -1.$$
$$\frac{\dfrac{x+2}{x+1} + \dfrac{x-2}{x-1}}{1 - \dfrac{(x+2)(x-2)}{(x+1)(x-1)}} = -1.$$
$$\frac{(x-1)(x+2) + (x+1)(x-2)}{(x+1)(x-1) - (x+2)(x-2)} = -1.$$
$$\frac{2x^2 - 4}{3} = -1.$$
$$x^2 = \tfrac{1}{2}.$$
$$x = \pm\sqrt{\tfrac{1}{2}}.$$

143. Solve the equation

$$\tan^{-1}\frac{2x}{1-x^2}=60°.$$

$$\tan^{-1}\frac{2x}{1-x^2}=60°.$$

$$2\tan^{-1}x=60°.$$

$$\tan^{-1}x=30°,\ 210°.$$

$$x=\tan 30°,\ \tan 210°$$

$$=\frac{1}{\sqrt{3}}.$$

144. Find the value of $a\sec x+b\csc x$, when $\tan x=\sqrt[3]{\frac{b}{a}}.$

$$\tan x=\frac{b^{\frac{1}{3}}}{a^{\frac{1}{3}}}.$$

$$\sec^2x=1+\frac{b^{\frac{2}{3}}}{a^{\frac{2}{3}}}$$

$$=\frac{a^{\frac{2}{3}}+b^{\frac{2}{3}}}{a^{\frac{2}{3}}}.$$

$$\cot x=\frac{a^{\frac{1}{3}}}{b^{\frac{1}{3}}}.$$

$$\csc^2x=\frac{a^{\frac{2}{3}}+b^{\frac{2}{3}}}{b^{\frac{2}{3}}}.$$

$$\therefore \sec x=\frac{(a^{\frac{2}{3}}+b^{\frac{2}{3}})^{\frac{1}{2}}}{a^{\frac{1}{3}}}.$$

$$\csc x=\frac{(a^{\frac{2}{3}}+b^{\frac{2}{3}})^{\frac{1}{2}}}{b^{\frac{1}{3}}}.$$

$$a\sec x+b\csc x$$

$$=a^{\frac{2}{3}}(a^{\frac{2}{3}}+b^{\frac{2}{3}})^{\frac{1}{2}}+b^{\frac{2}{3}}(a^{\frac{2}{3}}+b^{\frac{2}{3}})^{\frac{1}{2}}$$

$$=(a^{\frac{2}{3}}+b^{\frac{2}{3}})(a^{\frac{2}{3}}+b^{\frac{2}{3}})^{\frac{1}{2}}$$

$$=(a^{\frac{2}{3}}+b^{\frac{2}{3}})^{\frac{3}{2}}.$$

145. Find the value of $\sin 3x$, when $\sin 2x=\sqrt{1-m^2}$.

$$\sin 2x=\sqrt{1-m^2}.$$

$$\cos^2 2x=m^2.$$

$$\cos 2x=\pm m.$$

$$1-2\sin^2x=\pm m.$$

$$2\sin^2x=1\pm m.$$

$$\sin x=\sqrt{\frac{1\pm m}{2}}.$$

$$\sin 3x=3\sin x-4\sin^3x$$

$$=3\left(\frac{1\pm m}{2}\right)^{\frac{1}{2}}-4\left(\frac{1\pm m}{2}\right)^{\frac{3}{2}}$$

$$=\left(\frac{1\pm m}{2}\right)^{\frac{1}{2}}\left(3-4\frac{(1\pm m)}{2}\right)$$

$$=\left(\frac{1\pm m}{2}\right)^{\frac{1}{2}}(1\mp 2m).$$

146. Find the value of

$\frac{\csc^2x-\sec^2x}{\csc^2x+\sec^2x}$, when $\tan x=\sqrt{\tfrac{1}{7}}$.

$$\tan x=\sqrt{\tfrac{1}{7}}.$$

$$\sec^2x=1+\tfrac{1}{7}.$$

$$=\tfrac{8}{7}.$$

$$\cot x=\sqrt{7}.$$

$$\csc^2x=1+7$$

$$=8.$$

$$\therefore \frac{\csc^2x-\sec^2x}{\csc^2x+\sec^2x}=\frac{8-\frac{8}{7}}{8+\frac{8}{7}}.$$

$$=\tfrac{3}{4}.$$

147. Find the value of $\sin x$, when $\tan^2x+3\cot^2x=4$.

$$\tan^2x+3\cot^2x=4.$$

$$\tan^2x+\frac{3}{\tan^2x}=4.$$

$$\tan^4x-4\tan^2x+3=0.$$

$$\tan^2x=1,\ 3.$$

$$\cot^2x=1,\ \tfrac{1}{3}.$$

$$\csc^2x=2,\ \tfrac{4}{3}.$$

$$\sin^2x=\tfrac{1}{2},\ \tfrac{3}{4}.$$

$$\sin x=\sqrt{\tfrac{1}{2}},\ \tfrac{1}{2}\sqrt{3}.$$

148. Find the value of $\cos x$, when $5\tan x + \sec x = 5$.

$$5\tan x + \sec x = 5.$$
$$5\sin x + 1 = 5\cos x.$$
$$5\sin x = 5\cos x - 1.$$
$$25(1-\cos^2 x) = 25\cos^2 x - 10\cos x + 1.$$
$$50\cos^2 x - 10\cos x - 24 = 0.$$
$$\cos x = \tfrac{4}{5},\ -\tfrac{3}{5}.$$

149. Find the value of $\sec x$, when $\tan x = \dfrac{a}{\sqrt{2a+1}}$.

$$\tan x = \frac{a}{\sqrt{2a+1}}.$$
$$\sec^2 x = 1 + \frac{a^2}{2a+1}$$
$$= \frac{a^2+2a+1}{2a+1}.$$
$$\sec x = \frac{a+1}{\sqrt{2a+1}}.$$

150. Simplify the expression $\dfrac{(\cos x + \cos y)^2 + (\sin x + \sin y)^2}{\cos^2 \frac{1}{2}(x-y)}$.

$$\frac{(\cos x + \cos y)^2 + (\sin x + \sin y)^2}{\cos^2 \frac{1}{2}(x-y)}$$
$$= \frac{[2\cos\frac{1}{2}(x+y)\cos\frac{1}{2}(x-y)]^2 + [2\sin\frac{1}{2}(x+y)\cos\frac{1}{2}(x-y)]^2}{\cos^2\frac{1}{2}(x-y)}$$
$$= 4\cos^2\tfrac{1}{2}(x+y) + 4\sin^2\tfrac{1}{2}(x+y) = 4.$$

151. Simplify the expression $\dfrac{\sin(x+2y) - 2\sin(x+y) + \sin x}{\cos(x+2y) - 2\cos(x+y) + \cos x}$.

$$\frac{\sin(x+2y) - 2\sin(x+y) + \sin x}{\cos(x+2y) - 2\cos(x+y) + \cos x}$$
$$= \frac{[\sin(x+2y) + \sin x] - 2\sin(x+y)}{[\cos(x+2y) + \cos x] - 2\cos(x+y)}$$
$$= \frac{2\sin(x+y)\cos y - 2\sin(x+y)}{2\cos(x+y)\cos y - 2\cos(x+y)}$$
$$= \frac{\sin(x+y)(\cos y - 1)}{\cos(x+y)(\cos y - 1)}$$
$$= \frac{\sin(x+y)}{\cos(x+y)} = \tan(x+y).$$

152. Simplify the expression $\dfrac{\sin(x-z) + 2\sin x + \sin(x+z)}{\sin(y-z) + 2\sin y + \sin(y+z)}$.

$$\frac{\sin(x-z) + 2\sin x + \sin(x+z)}{\sin(y-z) + 2\sin y + \sin(y+z)}$$
$$= \frac{[\sin(x-z) + \sin(x+z)] + 2\sin x}{[\sin(y-z) + \sin(y+z)] + 2\sin y}$$
$$= \frac{2\sin x\cos z + 2\sin x}{2\sin y\cos z + 2\sin y}$$
$$= \frac{\sin x(\cos z + 1)}{\sin y(\cos z + 1)} = \frac{\sin x}{\sin y}.$$

153. Simplify the expression $\frac{\cos 6x - \cos 4x}{\sin 6x + \sin 4x}$.

$$\frac{\cos 6x - \cos 4x}{\sin 6x + \sin 4x} = \frac{2\cos 5x \cos x}{2\sin 5x \cos x}$$
$$= \frac{\cos 5x}{\sin 5x}$$
$$= \cot 5x.$$

154. Simplify the expression $\tan^{-1}(2x+1) + \tan^{-1}(2x-1)$.

$$\tan\left[\tan^{-1}(2x+1) + \tan^{-1}(2x-1)\right] = \frac{2x+1+2x-1}{1-(2x+1)(2x-1)}.$$
$$= \frac{4x}{2-4x^2}$$
$$= \frac{2x}{1-2x^2}.$$
$$\therefore \tan^{-1}(2x+1) + \tan^{-1}(2x-1) = \tan^{-1}\frac{2x}{1-2x^2}.$$

155. Simplify the expression
$$\frac{1}{1+\sin^2x} + \frac{1}{1+\cos^2x} + \frac{1}{1+\sec^2x} + \frac{1}{1+\csc^2x}.$$

$$\frac{1}{1+\sin^2x} + \frac{1}{1+\cos^2x} + \frac{1}{1+\sec^2x} + \frac{1}{1+\csc^2x}$$
$$= \left(\frac{1}{1+\sin^2x} + \frac{1}{1+\csc^2x}\right) + \left(\frac{1}{1+\cos^2x} + \frac{1}{1+\sec^2x}\right)$$
$$= \left(\frac{1}{1+\sin^2x} + \frac{\sin^2x}{1+\sin^2x}\right) + \left(\frac{1}{1+\cos^2x} + \frac{\cos^2x}{1+\cos^2x}\right)$$
$$= \frac{1+\sin^2x}{1+\sin^2x} + \frac{1+\cos^2x}{1+\cos^2x}$$
$$= 1 + 1.$$
$$= 2.$$

156. Simplify the expression $2\sec^2x - \sec^4x - 2\csc^2x + \csc^4x$.

$$2\sec^2x - \sec^4x - 2\csc^2x + \csc^4x$$
$$= -1 + 2\sec^2x - \sec^4x + 1 - 2\csc^2x + \csc^4x$$
$$= -(\sec^2x - 1)^2 + (\csc^2x - 1)^2$$
$$= -\tan^2x + \cot^2x.$$

ENTRANCE EXAMINATION PAPERS. PAGE 106.

I.

1. Prove that

$$\cos \text{co-}\theta = \sin \theta;$$
$$\sec (\tfrac{1}{2}\pi + \theta) = -\csc \theta;$$
$$\tan (-\theta) = -\tan \theta;$$
$$\csc (\pi - \theta) = \csc \theta.$$

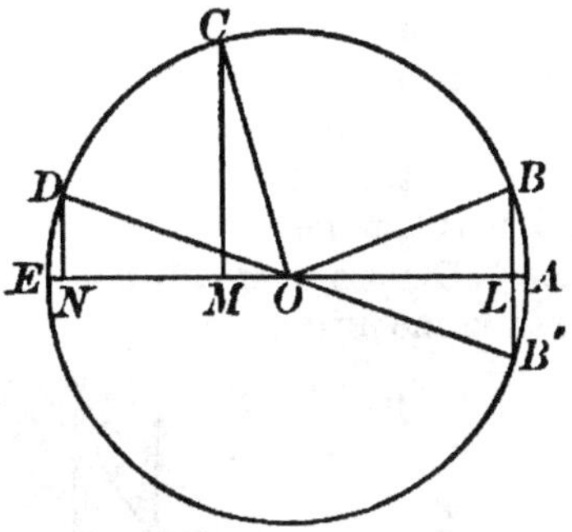

In the figure let

$$AOB = \theta,$$
$$AOB' = -\theta,$$
$$AOC = 90° + \theta,$$
$$AOD = 180° - \theta,$$
$$OB' = OB = OC = OD,$$
$$BB',\ CM,\ ND \perp OA.$$

Then the triangles LOB, LOB', MCO, and NOD are equal, and

(i.) $$\cos OBL = \frac{BL}{OB} = \sin LOB.$$
$$\therefore \cos \text{co-}\theta = \sin \theta.$$

(ii.) $$\sec AOC = -\frac{OC}{OM} = -\frac{OB}{BL} = -\csc AOB.$$
$$\therefore \sec (\tfrac{1}{2}\pi + \theta) = -\csc \theta.$$

(iii.) $$\tan AOB' = -\frac{LB'}{OL} = -\frac{LB}{OL} = -\tan AOB.$$
$$\therefore \tan (-\theta) = \tan \theta.$$

(iv.) $$\csc AOD = \frac{OD}{DN} = \frac{OB}{LB} = \csc AOB.$$
$$\csc (\pi - \theta) = \csc \theta.$$

2. Draw the curve of tangents, and show the changes in the value of this function as the arc increases from 0° to 60°.

Suppose $y = \tan x$.

As x increases from 0° to 90° and from 90° to 180°, y increases from 0 to $+\infty$ and from $-\infty$ to 0, and as x continues to increase from 180° to 360°, y takes the same series of values again, since $\tan (x + 180°) = \tan x$. The initial value may be any negative or positive integral multiple of 180°, instead of 0°, without change in the series of values of y. Also, at equal angular distances in opposite directions from any quadrantal angle $\tan x$ has equal values with opposite signs. For the first quadrant we have for corresponding values of x and y

$$x = 0,\ 30°,\ 45°,\ 60°,\ 90°.$$
$$y = 0,\ \frac{1}{\sqrt{3}},\ 1,\ \sqrt{3},\ \infty.$$

Suppose now that the values of x are laid off along a given horizontal straight line from any initial point 0, and at each point a perpendicular is erected of a length equal to y, the perpendicular being drawn upward if y is positive, and downward if y is negative. The extremities of these perpendiculars form the required curve. The shape of this curve is shown in the figure.

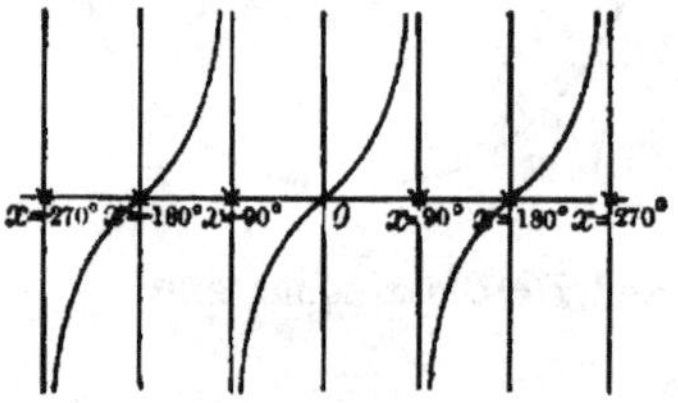

This curve consists of an infinite series of parallel branches, whose horizontal distance from each other is 180°. Each branch consists of equal upper and lower halves, the one of which can be obtained by rotating the other through 180° about the middle point of the branch.

3. In terms of positive angles less than 45°, express the values of $\sin - 250°$, $\csc \frac{13}{12}\pi$, $\tan - \frac{16}{3}\pi$. Also find all the values of θ in terms of α when $\cos\theta = \sqrt{\sin^2\alpha}$.

(i.) $\sin - 250° = \sin(110° - 360°)$
$= \sin 110°$
$= \sin(90° + 20°)$
$= \cos 20°.$

(ii.) $\csc \frac{13}{12}\pi = \csc \frac{13}{12} 180°$
$= \csc(180° + \frac{1}{12} 180°)$
$= -\csc \frac{180°}{12}$
$= -\csc 15°.$

(iii.) $\tan - \frac{16}{3}\pi = \tan(\frac{2}{3}\pi - 6\pi)$
$= \tan \frac{2}{3}\pi$
$= \tan 120°.$
$= \tan(90° + 30)$
$= -\cot 30°.$

(iv.) $\cos\theta = \sqrt{\sin^2\alpha}$
$= \pm \sin\alpha$
$= \pm \cos(90° - \alpha).$

If $\cos\theta = +\cos(90° - \alpha)$
$\theta = \pm(90° - \alpha) + n\,360°$
where n is any integer.

If $\cos\theta = -\cos(90° - \alpha)$
$\theta = 180° \pm (90° - \alpha) + n\,360°.$

$\therefore \theta = 90° - \alpha + n\,360°,$
$\alpha - 90° + n\,360°,$
$270° - \alpha + n\,360°,$
$90° + \alpha + n\,360°.$

Or, $\theta = 90° - \alpha + n\,360°,$
$270° + \alpha + n\,360°,$
$270° - \alpha + n\,360°,$
$90° + \alpha + n\,360°.$

Or, $\theta = (2n + 1)\,90° \pm \alpha.$

4. (*a*) Given $\cos x = 0.5$, find $\cos 2x$ and $\tan 2x$.

(*b*) Prove that vers $(180° - A +)$ vers $(360° - A) = 2$.

(*a*) $\cos x = 0.5$
$\sin x = \sqrt{1 - \cos^2 x}$
$= \pm\sqrt{0.75}$
$= \pm 0.5\sqrt{3};$

$\tan x = \frac{\sin x}{\cos x}$
$= \pm\sqrt{3}.$

$\cos 2x = \cos^2 x - \sin^2 x$
$= 0.25 - 0.75$
$= -\frac{1}{2}.$

$\tan 2x = \frac{2\tan x}{1 - \tan^2 x}$
$= \frac{\pm 2\sqrt{3}}{1 - 3}$
$= \pm\sqrt{3}.$

(b)
$$\text{vers}\,(180^\circ - A) = 1 - \cos(180^\circ - A)$$
$$= 1 + \cos A.$$
$$\text{vers}\,(360^\circ - A) = 1 - \cos(360^\circ - A)$$
$$= 1 - \cos A.$$
$$\therefore \text{vers}\,(180^\circ - A) + \text{vers}\,(360^\circ - A)$$
$$= 1 + \cos A + 1 - \cos A$$
$$= 2.$$

5. Prove the check formulæ:
$$a + b : c = \cos \tfrac{1}{2}(A - B) : \sin \tfrac{1}{2} C;$$
$$a - b : c = \sin \tfrac{1}{2}(A - B) : \cos \tfrac{1}{2} C.$$

By the Law of Signs (§ 33),
$$a : b : c = \sin A : \sin B : \sin C$$
$$\therefore a + b : c = (\sin A + \sin B) : \sin C$$
$$a - b : c = (\sin A - \sin B) : \sin C$$

By [20] and [21],
$$\sin A + \sin B = 2 \sin \tfrac{1}{2}(A + B) \cos \tfrac{1}{2}(A - B)$$
$$= 2 \sin \tfrac{1}{2}(180^\circ - C) \cos \tfrac{1}{2}(A - B)$$
$$= 2 \sin (90^\circ - \tfrac{1}{2} C) \cos \tfrac{1}{2}(A - B)$$
$$= 2 \cos \tfrac{1}{2} C \cos \tfrac{1}{2}(A - B).$$
$$\sin A - \sin B = 2 \cos \tfrac{1}{2}(A + B) \sin \tfrac{1}{2}(A - B)$$
$$= 2 \sin \tfrac{1}{2} C \sin \tfrac{1}{2}(A - B).$$
$$\therefore a + b : c = 2 \cos \tfrac{1}{2} C \sin \tfrac{1}{2}(A - B) : \sin C$$
$$= 2 \cos \tfrac{1}{2} C \sin \tfrac{1}{2}(A - B) : 2 \sin \tfrac{1}{2} C \cos \tfrac{1}{2} C$$
$$= \sin \tfrac{1}{2}(A - B) : \sin \tfrac{1}{2} C.$$
$$a + b : c = 2 \sin \tfrac{1}{2} C \cos \tfrac{1}{2}(A - B) : \sin C$$
$$= 2 \sin \tfrac{1}{2} C \cos \tfrac{1}{2}(A - B) : 2 \sin \tfrac{1}{2} C \cos \tfrac{1}{2} C$$
$$= \cos \tfrac{1}{2}(A - B) : \cos \tfrac{1}{2} C.$$

6. In a right triangle, r (the hypotenuse) is given, and one acute angle is n times the other; find the sides about the right angle in terms of r and n.

Let A and B be the acute angles of the triangle, a and b the sides opposite A and B respectively.

Then $B = nA.$
$$A + B = 90^\circ.$$
$$\therefore A = \frac{90^\circ}{n + 1}.$$
$$B = \frac{n\,90^\circ}{n + 1}.$$
$$a = r \sin A.$$
$$= r \sin \frac{90^\circ}{n + 1}.$$
$$b = r \sin B$$
$$= r \sin \frac{n\,90^\circ}{n + 1}$$
$$= r \cos \frac{90^\circ}{n + 1}.$$

7. The tower of McGraw Hall is 125 feet high, and from its summit the angles of depression of the bases of two trees on the campus, which stand on the same level as the Hall, are respectively 57° 44′ and 16° 59′, and the angle subtended by the line joining the trees is 99° 30′. Find the distance between the trees.

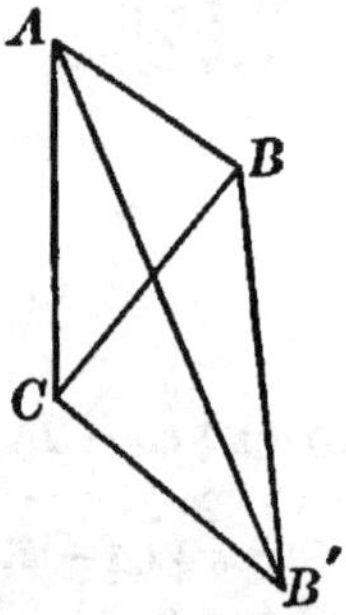

Let A and C be the summit and base of the tower, B and B' the bases of the trees. Then in the triangle ABC,

$$C = 90°,$$
$$AC = 125,$$
$$A = 90° - 57° 44'$$
$$= 32° 16'.$$

$$AB = AC \sec A$$
$$= 125 \sec 32° 16'.$$

Similarly in the right triangle $AB'C$,

$$AB' = AC \sec A$$
$$= 125 \sec 73° 1'.$$

$$\log 125 = 2.09691$$
$$\log \sec 32° 16' = 0.07285$$
$$\log AB = 2.16976$$
$$AB = 147.83.$$
$$\log 125 = 2.09691$$
$$\log \sec 73° 1' = 0.53448$$
$$\log AB' = 2.63139$$
$$AB' = 427.95.$$

In the triangle ABB',

$$A = 99° 30'.$$
$$b' = 147.83.$$
$$b = 427.95.$$

$$\tan \tfrac{1}{2}(B - B') = \frac{b - b'}{b + b'} \tan \tfrac{1}{2}(B + B')$$
$$= \frac{280.12}{575.78} \tan 40° 15'.$$

$$\log 280.12 = 2.44734$$
$$\text{colog } 575.78 = 7.23974$$
$$\log \tan 40° 15' = 9.92766$$
$$\log \tan \tfrac{1}{2}(B - B') = 9.61474$$
$$\tfrac{1}{2}(B - B') = 22° 23' 3''$$
$$\tfrac{1}{2}(B + B') = 40° 15'$$
$$B = 62° 38' 3''$$

$$BB' = AB' \frac{\sin A}{\sin B}$$
$$= 427.95 \frac{\sin 99° 30'}{\sin 62° 38' 3''}.$$

$$\log 427.95 = 2.63139$$
$$\log \sin 99° 30' = 9.99400$$
$$\text{colog } \sin 62° 38' 3'' = 0.05155$$
$$\log BB' = 2.67694$$
$$BB' = 475.27.$$

Required distance, 475.27 ft.

II.

1. Prove that

$$\cot(-\theta) = -\cot\theta;$$
$$\csc\overline{\pi-\theta} = \csc\theta;$$
$$\sin(\pi+\theta) = -\sin\theta;$$
$$\sec \text{co-}\theta = \csc\theta;$$
$$\cos(\tfrac{1}{2}\pi+\theta) = -\sin\theta.$$

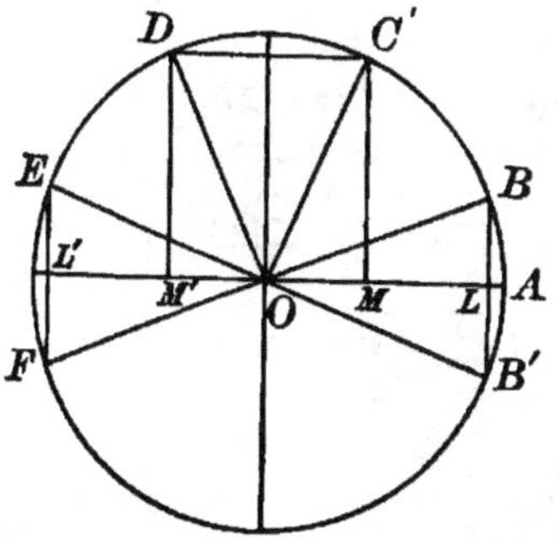

Let $AOB = \theta.$

$AOB' = -\theta.$

$AOC = 90° - \theta.$

$AOD = 90° + \theta.$

$AOE = 180° - \theta.$

$AOF = 180° + \theta.$

$OA = OB = OB' = OC$
$= OD = OE = OF.$

$BL, CM, DM'\ EL' \perp OA.$

Then the triangles LOB, LOB', MCO, $M'DO$, $L'OE$, $L'OF$ are equal.

(i.) $\cot AOB' = -\dfrac{OL}{LB'} = -\dfrac{OL}{LB} = -\cot AOB.$

$\therefore \cot(-\theta) = -\cot\theta.$

(ii.) $\csc AOE = \dfrac{OE}{L'E} = \dfrac{OB}{LB} = \csc AOB.$

$\therefore \csc\overline{\pi-\theta} = \csc\theta.$

(iii.) $\sin AOF = -\dfrac{L'F}{OF} = -\dfrac{AB}{OB} = -\sin AOB.$

$\sin(\pi+\theta) = -\sin\theta.$

(iv.) $\sec AOC = \dfrac{OC}{OM} = \dfrac{OB}{AB} = \csc AOB.$

$\therefore \sec \text{co-}\theta = \csc\theta.$

(v.) $\cos AOD = -\dfrac{OM'}{OD} = -\dfrac{AB}{OB} = -\sin AOB.$

$\therefore \cos(\tfrac{1}{2}\pi+\theta) = -\sin\theta.$

2. Show that in any plane triangle $\sin\tfrac{1}{2}A = \sqrt{\dfrac{(s-b)(s-c)}{bc}}.$

By the Law of Cosines (§ 34),

$$a^2 = b^2 + c^2 - 2bc\cos A.$$

$$\therefore \cos A = \frac{b^2+c^2-a^2}{2bc}.$$

$$1 - \cos A = \frac{2bc - b^2 - c^2 + a^2}{2bc}.$$

$$= \frac{a^2 - (b-c)^2}{2bc}.$$

By [16], $1 - \cos A = 2 \sin^2 \frac{1}{2} A.$

$$\therefore 2 \sin^2 \tfrac{1}{2} A = \frac{a^2 - (b-c)^2}{2bc}.$$

$$\sin \tfrac{1}{2} A = \tfrac{1}{2} \sqrt{\frac{a^2 - (b-c)^2}{2bc}}.$$

But $a^2 - (b-c)^2 = (a+b-c)(a-b+c),$

or if $s = \frac{1}{2}(a+b+c).$

$$a^2 - (b-c)^2 = (2s - 2c)(2s - 2b)$$
$$= 4(s-b)(s-c).$$

$$\therefore \sin \tfrac{1}{2} A = \tfrac{1}{2} \sqrt{\frac{4(s-b)(s-c)}{bc}}$$

$$= \sqrt{\frac{(s-b)(s-c)}{bc}}.$$

3. Find the value of $\sin(\theta \pm \theta')$ in terms of $\sin \theta$, $\cos \theta$, $\sin \theta'$, and $\cos \theta'$.

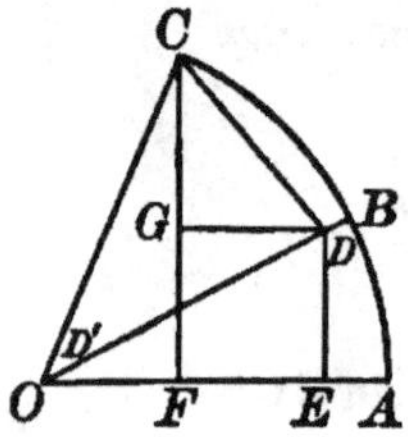

Let $AOB = \theta$, $BOC = \theta'$,
$OA = OB = OC$,
$CD \perp OB$, CF, $DE \perp OA$, $DG \perp CF$.

Then $AOC = \theta + \theta'$.
$DCG = \theta$.

$$\sin(\theta + \theta') = \frac{CF}{OC}$$
$$= \frac{CG + DE}{OC}$$
$$= \frac{DE}{OC} + \frac{CG}{OC}$$
$$= \frac{DE}{OD} \times \frac{OD}{OC} + \frac{CG}{CD} \times \frac{CD}{OC}$$
$$= \sin \theta \cos \theta' + \cos \theta \sin \theta'.$$

4. Given $\tan 45° = 1$; find all the functions of $22° 30'$.

$$\tan 45° = 1,$$
$$\sec^2 45° = 1 + \tan^2 45°$$
$$= 2.$$
$$\sec 45° = \sqrt{2}.$$
$$\cos 45° = \frac{1}{\sqrt{2}}.$$
$$\sin 45° = \tan 45° \cos 45°$$
$$= \frac{1}{\sqrt{2}}.$$

By [17],

$$\sin 22° 30' = \sqrt{\frac{1 - \cos 45°}{2}}$$
$$= \sqrt{\frac{1 - \sqrt{\frac{1}{2}}}{2}}$$
$$= \sqrt{\frac{\sqrt{2} - 1}{2\sqrt{2}}}$$
$$= \tfrac{1}{2}\sqrt{2 - \sqrt{2}},$$

and
$$\cos 22° 30' = \sqrt{\frac{1 + \cos 45°}{2}}$$
$$= \tfrac{1}{2}\sqrt{2 + \sqrt{2}}.$$
$$\tan 22° 30' = \frac{\sin 22° 30'}{\cos 22° 30'}$$
$$= \frac{\sqrt{2 - \sqrt{2}}}{\sqrt{2 + \sqrt{2}}}$$

$$= \frac{2-\sqrt{2}}{\sqrt{2^2-(\sqrt{2})^2}}$$
$$= \frac{2-\sqrt{2}}{\sqrt{2}}$$
$$= \sqrt{2}-1.$$

$$\cot 22° 30' = \frac{1}{\tan 22° 30'}$$
$$= \frac{1}{\sqrt{2}-1}$$
$$= \sqrt{2}+1.$$

$$\sec 22° 30' = \sqrt{1+\tan^2 22° 30'}$$
$$= \sqrt{4-2\sqrt{2}}.$$

$$\csc 22° 30' = \sqrt{1+\cot^2 22° 30'}$$
$$= \sqrt{4+2\sqrt{2}}.$$

5. Determine the number of solutions of each of the triangles:

$a = 13.4$, $b = 11.46$, $A = 77° 20'$;
$c = 58$, $a = 75$, $C = 60°$;
$b = 109$, $a = 94$, $A = 92° 10'$;
$c = 309$, $b = 360$, $C = 21° 14' 25''$.

(i.) $a = 13.4$, $b = 11.46$, $A = 77° 20'$.

$a > b$

$\therefore$ one solution.

(ii.) $c = 58$, $a = 75$, $C = 60°$.

$$a \sin C = 75 \times \tfrac{1}{2}\sqrt{3}$$
$$= 75 \times 0.866 +$$
$$= 64. +.$$

$c < a \sin C$

$\therefore$ no solution.

(iii.) $b = 109$, $a = 94$, $A = 92° 10'$.

$A > 90°$, $a < b$,

$\therefore$ no solution.

(iv.) $c = 309$, $b = 360$, $C = 21° 14' 25''$.

$$b \sin C < b \sin 30°$$
$$< \tfrac{1}{2} b$$
$$< 180.$$

$\therefore b > c > b \sin C.$

$\therefore$ two solutions.

6. In a parallelogram, given side a, diagonal d, and the angle A formed by the diagonals; find the other diagonal and the other side.

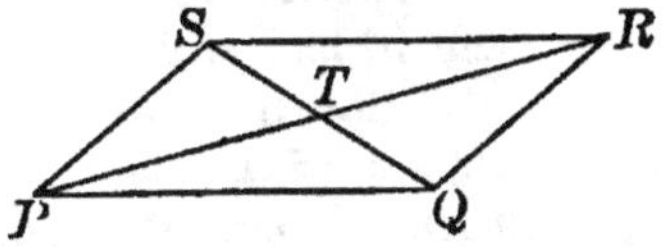

In the parallelogram $PQRS$, given $PQ = a$, $QS = d$, $PTQ = A$, required PR and PS.

$$TQ = \tfrac{1}{2} QS$$
$$= \tfrac{1}{2} d.$$

$$\sin TPQ = \frac{TQ}{PQ} \sin PTQ$$
$$= \tfrac{1}{2}\frac{d}{a} \sin A.$$

From this formula, TPQ must be determined by logarithms.

$$TQP = 180° - TPQ - A.$$

$$PT = PQ \frac{\sin TQP}{\sin PTQ}$$
$$= a\frac{\sin(180° - TPQ - A)}{\sin A}$$
$$= a\frac{\sin(TPQ + A)}{\sin A}.$$

$$PR = 2\,PT$$
$$= 2a\frac{\sin(TPQ + A)}{\sin A}.$$

From this formula PR must be determined by logarithms.

$$\tan \tfrac{1}{2}(QSP - QPS)$$
$$= \frac{PQ - QS}{PQ + PS} \tan \tfrac{1}{2}(QSP + QPS)$$
$$= \frac{a-d}{a+d} \tan (90° - \tfrac{1}{2} PQS)$$
$$= \frac{a-d}{a+d} \cot \tfrac{1}{2} PQS,$$

from which QSP and QPS must be determined.

$$PS = PQ \frac{\sin PQS}{\sin QSP},$$

from which PS may be determined.

7. A and B are two objects whose distance, on account of intervening obstacles, cannot be directly measured. At the summit of a hill, whose height above the common horizontal plane of the objects is known to be 517.3 yds., angle ACB is found to be 15° 13′ 15″. The angles of elevation of C viewed from A and B are 21° 9′ 18″ and 23° 15′ 34″ respectively. Find the distance from A to B.

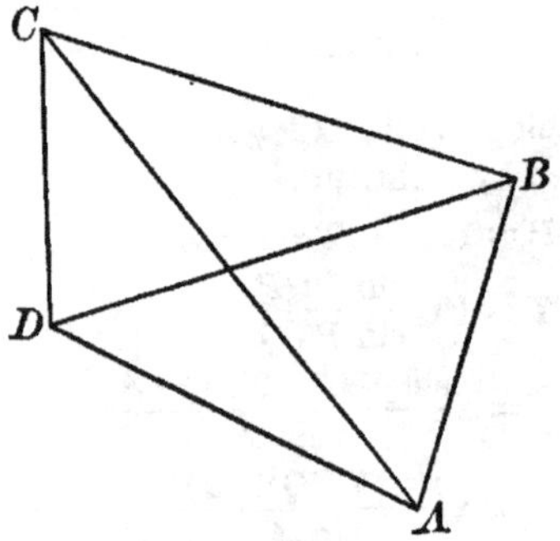

Let D be the point directly under C in the same horizontal plane with A and B. Then

$$AC = CD \csc CAD$$
$$= 517.3 \csc 21° 9' 18''.$$

$$\log 517.3 = 2.71374$$
$$\log \csc 21° 9' 18'' = 0.44262$$
$$\log AC = 3.15636$$
$$AC = 1433.4.$$

$$BC = CD \csc CBD$$
$$= 517.3 \csc 23° 15' 34''.$$

$$\log 517.3 = 2.71374$$
$$\log \csc 23° 15' 34'' = 0.40352$$
$$\log BC = 3.11726$$
$$BC = 1310.0.$$

$$\tan \tfrac{1}{2}(ABC - BAC)$$
$$= \frac{AC - BC}{AC + BC} \tan \tfrac{1}{2}(ABC + BAC)$$
$$= \frac{123.4}{2743.4} \tan 82° 23' 22''.$$

$$\log 123.4 = 2.09132$$
$$\text{colog } 2743.4 = 6.56171\text{-}10$$
$$\log \tan 82° 23' 22'' = 10.87414$$
$$\log \tan \tfrac{1}{2}(ABC - BAC) = 9.52717$$
$$\tfrac{1}{2}(ABC - BAC) = 18° 36' 20''$$
$$\tfrac{1}{2}(ABC + BAC) = 82° 23' 22''$$
$$ABC = 100° 59' 42''$$

$$AB = AC \frac{\sin ACB}{\sin ABC}$$
$$= 1433.4 \frac{\sin 15° 13' 15''}{\sin 100° 59' 42''}.$$

$$\log 1433.4 = 3.15636$$
$$\log \sin 15° 13' 15'' = 9.41919$$
$$\text{colog} \sin 100° 59' 42'' = 0.00804$$
$$\log AB = 2.58359$$
$$AB = 383.35$$

Distance AB is 383.35 yds.

III.

1. Trace the value of $\tan \theta$ and that of $\csc \theta$, as θ increases from 0° to 360°.

(i.) When $\theta = 0°$, $\tan \theta = 0$; and, as θ increases from 0° to 90°, $\tan \theta$ increases from 0 to ∞. As θ increases from 90° to 180° and from 180° to 270°, $\tan \theta$ increases from $-\infty$ to 0 and from 0 to $+\infty$; and, as θ increases from 270° to 360°, $\tan \theta$ increases from $-\infty$ to 0. Since $\tan (180° + \theta) = \tan \theta$, the succession of values of $\tan \theta$ is the same from $\theta = 180°$ to $\theta = 360°$ as from $\theta = 0°$ to $\theta = 180°$; and, since $\tan (180° - \theta) = -\tan \theta$, $\tan \theta$ takes the same series of values in the second quadrant as in the first, but in the opposite order and with the opposite sign. In the first quadrant the values of tan θ for several angles are given in the following table:

$$\theta = 0^\circ,\ 30^\circ,\ 45^\circ,\ 60^\circ,\ 90^\circ.$$

$$\tan\theta = 0,\ \sqrt{2},\ 1,\ \frac{2}{\sqrt{3}},\ \infty.$$

(ii.) When $\theta = 0^\circ$, $\csc\theta = \infty$; and as θ increases from 0° to 90°, $\csc\theta$ decreases from ∞ to 1. As θ increases from 90° to 180° and from 180° to 270°, $\csc\theta$ increases from 1 to ∞ and from $-\infty$ to -1; and as θ increases from 270° to 360°, $\csc\theta$ decreases from -1 to $-\infty$. Since $\csc(180^\circ + \theta) = -\csc\theta$, $\csc\theta$ takes the same succession of values from $\theta = 180^\circ$ to $\theta = 360^\circ$ as from $\theta = 0^\circ$ to $\theta = 180^\circ$, but with the opposite sign; and since $\csc(180^\circ - \theta) = \csc\theta$, $\csc\theta$ takes the same series of values in the second quadrant as in the first, but in opposite order. In the first quadrant the values of $\csc\theta$ for several angles are given in the following table:

$$\theta = 0^\circ,\ 30^\circ,\ 45^\circ,\ 60^\circ,\ 90^\circ.$$

$$\csc\theta = \infty,\ 2,\ \sqrt{2},\ \frac{2}{\sqrt{3}},\ 1.$$

2. (*a*) Find the remaining function of θ when $\cos\theta = -\frac{1}{2}\sqrt{3}$.

(*b*) Determine all the values of θ that will satisfy the relation $\cot\theta = 2\cos\theta$.

(*a*) $\cos\theta = -\frac{1}{2}\sqrt{3}$.

$$\sin\theta = \sqrt{1-\cos^2\theta} = \pm\tfrac{1}{2}.$$

$$\tan\theta = \frac{\sin\theta}{\cos\theta} = \mp\frac{1}{\sqrt{3}}.$$

$$\cot\theta = \frac{1}{\tan\theta} = \mp\sqrt{3}.$$

$$\sec\theta = \frac{1}{\cos\theta} = -\frac{2}{\sqrt{3}}.$$

$$\csc\theta = \frac{1}{\sin\theta} = \pm 2.$$

(*b*) $\cot\theta = 2\cos\theta$.

$$\frac{\cos\theta}{\sin\theta} = 2\cos\theta.$$

$$\cos\theta = 2\sin\theta\cos\theta.$$

$$\cos\theta = \sin 2\theta.$$

$$\theta = 90^\circ - 2\theta \text{ or } 2\theta - 90^\circ.$$

(i.) $3\theta = 90^\circ + n\,360^\circ$.

$$\theta = 30^\circ + n\,120^\circ$$

$$= 30^\circ,\ 150^\circ,\ 270^\circ.$$

(ii.) $\theta = 90^\circ$.

$$\therefore \theta = 30^\circ,\ 90^\circ,\ 150^\circ,\ 270^\circ.$$

3. Prove the identity

$$\tan A - \cot A = \frac{\sin^2 A - \cos^2 A}{\sin A\cos A} = -2\cot 2A.$$

$$\tan A - \cot A = \frac{\sin A}{\cos A} - \frac{\cos A}{\sin A} = \frac{\sin^2 A - \cos^2 A}{\sin A\cos A}.$$

But $\cos^2 A - \sin^2 A = \cos 2A$

$$2\sin A\cos A = \sin 2A.$$

$$\therefore \tan A - \cot A = \frac{-\cos 2A}{\frac{1}{2}\sin 2A} = -2\cot 2A.$$

4. Derive an expression for the sine of half an angle in a triangle in terms of the sides of the triangle. See II., Ex. 2.

5. Construct a figure and explain fully (giving formulæ) how you would find the height above its base, and the distance from the observer, of an inaccessible vertical object that is visible from two points whose distance apart is known, and which can be seen from one another.

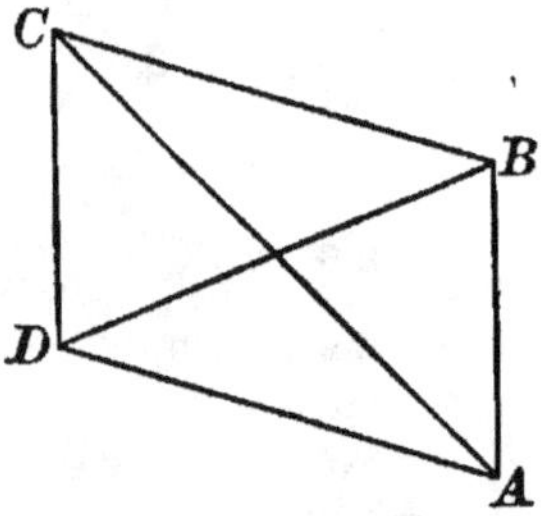

Let CD be the vertical object, A and B the two points of observation.

Measure the angles CAB, CBA, and CAD. Then

$$AC = AB\frac{\sin CBA}{\sin ACB}$$

$$= AB\frac{\sin CBA}{\sin(180^\circ - CAB - CBA)}$$

$$= AB\frac{\sin CBA}{\sin(CAB + CBA)}.$$

$$CB = AB\frac{\sin CAB}{\sin(CAB + CBA)}.$$

$$CD = AC\sin CAD$$

$$= AB\frac{\sin CBA \sin CAD}{\sin(CAB + CBA)}.$$

6. Given two sides of a plane triangle equal respectively to 121.34 and 216.7, and the included angle 47° 21′ 11″, to find the remaining parts of the triangle.

Let $b = 121.34$.

$c = 216.7$.

$A = 47^\circ\ 21'\ 11''$.

Then $\tan\frac{1}{2}(C - B)$

$$= \frac{c - b}{c + b}\tan\tfrac{1}{2}(C + B)$$

$$= \frac{95.36}{338.04}\tan 66^\circ\ 19'\ 24''.$$

$$\begin{aligned}
\log 95.36 &= 1.97937\\
\text{colog } 338.04 &= 7.47103-10\\
\log\tan 66^\circ\ 19'\ 24'' &= \underline{10.35805}\\
\log\tan\tfrac{1}{2}(C - B) &= 9.80845\\
\tfrac{1}{2}(C - B) &= 32^\circ\ 45''\ 19''\\
\tfrac{1}{2}(C + B) &= \underline{66^\circ\ 19'\ 24''}\\
C &= 99^\circ\ \ 4'\ 43''.\\
B &= 33^\circ\ 34'\ \ 5''.
\end{aligned}$$

$$a = b\frac{\sin A}{\sin B}$$

$$= 121.34\frac{\sin 47^\circ\ 21'\ 11''}{\sin 33^\circ\ 34'\ \ 5''}.$$

$$\begin{aligned}
\log 121.34 &= 2.08400\\
\log\sin 47^\circ\ 21'\ 11'' &= 9.86661\\
\text{colog}\sin 33^\circ\ 34'\ \ 5'' &= \underline{0.25733}\\
\log a &= 2.20794\\
a &= 161.42.
\end{aligned}$$

7. In a right triangle, if the difference of the base and the perpendicular is 12 yds., and the angle at the base is 38° 1′ 8″, what is the length of the hypotenuse?

$$\begin{aligned}
a - b &= 12,\\
B &= 38^\circ\ \ 1'\ \ 8''.\\
A &= 51^\circ\ 58'\ 52''.
\end{aligned}$$

$$a + b = (a - b)\frac{\tan\frac{1}{2}(A + B)}{\tan\frac{1}{2}(A - B)}$$

$$= 12\frac{\tan 45^\circ}{\tan 6^\circ\ 58'\ 52''}$$

$$= 12\cot 6^\circ\ 58'\ 52''.$$

$$\begin{aligned}
\log 12 &= 1.07918\\
\log\cot 6^\circ\ 58'\ 52'' &= \underline{10.91204}\\
\log(a + b) &= 1.99122\\
a + b &= 97.998\\
a - b &= \underline{12}\\
a &= 54.999\\
c &= a\csc A\\
&= 54.999\csc 51^\circ\ 58'\ 52''.
\end{aligned}$$

$$\log 54.999 = 1.74035$$
$$\log \csc 51^\circ 58' 52'' = \underline{0.10358}$$
$$\log c = 1.84393$$
$$c = 69.812.$$

Length of hypotenuse, 69.812 yds.

IV.

1. By means of an equilateral triangle, one of whose angles is bisected, find the numerical values of the functions of 30° and 60°.

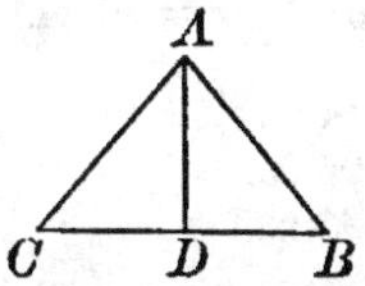

Let ABC be an equilateral triangle; AD the perpendicular from A on BC; and let the length of each side of the triangle be 1. Then

$$BAD = 30^\circ,\ DBA = 60^\circ.$$
$$AD = \sqrt{AB^2 - BD^2}$$
$$= \sqrt{1 - \tfrac{1}{4}}$$
$$= \tfrac{1}{2}\sqrt{3}.$$

$$\sin BAD = \cos ABD = \frac{BD}{AB}$$
$$\sin 30^\circ = \cos 60^\circ \quad = \tfrac{1}{2};$$
$$\cos BAD = \sin ABD = \frac{AD}{AB}$$
$$\cos 30^\circ = \sin 60^\circ \quad = \tfrac{1}{2}\sqrt{3};$$
$$\tan BAB = \cot ABD = \frac{BD}{AD}$$
$$\tan 30^\circ = \cot 60^\circ \quad = \frac{1}{\sqrt{3}};$$
$$\cot BAD = \tan ABD = \frac{AD}{BD}$$
$$\cot 30^\circ = \tan 60^\circ \quad = \sqrt{3};$$
$$\sec BAD = \csc ABD = \frac{AB}{AD}$$
$$\sec 30^\circ = \csc 60^\circ \quad = \frac{2}{\sqrt{3}};$$
$$\csc BAD = \sec ABD = \frac{AB}{BD}$$
$$\csc 30^\circ = \sec 60^\circ \quad = 2.$$

2. If θ be any angle, prove that

$$\sin\theta = \tan\theta : \sqrt{1 + \tan^2\theta},$$
$$\cos\theta = \sqrt{\csc^2\theta - 1} : \csc\theta.$$

(i.) $$\sqrt{1 + \tan^2\theta} = \sqrt{\sec^2\theta}$$
$$= \sec\theta.$$
$$\therefore \tan\theta : \sqrt{1 + \tan^2\theta} = \tan\theta : \sec\theta$$
$$= \frac{\sin\theta}{\cos\theta} : \frac{1}{\cos\theta}$$
$$= \sin\theta.$$

(ii.) $$\sqrt{\csc^2\theta - 1} = \sqrt{\cot^2\theta}$$
$$= \cot\theta.$$
$$\therefore \sqrt{\csc^2\theta - 1} : \csc\theta = \cot\theta : \csc\theta$$
$$= \frac{\cos\theta}{\sin\theta} : \frac{1}{\sin\theta}$$
$$= \cos\theta.$$

3. Prove that $\dfrac{\sin\theta + \sin\theta'}{\cos\theta - \cos\theta'} = -\cot\tfrac{1}{2}(\theta - \theta')$, when θ and θ' are any angles.

$$\sin\theta = \sin[\tfrac{1}{2}(\theta + \theta') + \tfrac{1}{2}(\theta - \theta')]$$
$$= \sin\tfrac{1}{2}(\theta + \theta')\cos\tfrac{1}{2}(\theta - \theta') + \cos\tfrac{1}{2}(\theta + \theta')\sin\tfrac{1}{2}(\theta - \theta').$$
$$\sin\theta' = \sin[\tfrac{1}{2}(\theta + \theta') - \tfrac{1}{2}(\theta - \theta')]$$
$$= \sin\tfrac{1}{2}(\theta + \theta')\cos\tfrac{1}{2}(\theta - \theta') - \cos\tfrac{1}{2}(\theta + \theta')\sin\tfrac{1}{2}(\theta - \theta').$$
$$\therefore \sin\theta + \sin\theta' = 2\sin\tfrac{1}{2}(\theta + \theta')\cos\tfrac{1}{2}(\theta - \theta').$$
$$\cos\theta = \cos[\tfrac{1}{2}(\theta + \theta') + \tfrac{1}{2}(\theta - \theta')]$$
$$= \cos\tfrac{1}{2}(\theta + \theta')\cos\tfrac{1}{2}(\theta - \theta') - \sin\tfrac{1}{2}(\theta + \theta')\sin\tfrac{1}{2}(\theta - \theta').$$

$$\cos\theta' = \cos[\tfrac{1}{2}(\theta+\theta') - \tfrac{1}{2}(\theta-\theta')]$$
$$= \cos\tfrac{1}{2}(\theta+\theta')\cos\tfrac{1}{2}(\theta-\theta') + \sin\tfrac{1}{2}(\theta+\theta')\sin\tfrac{1}{2}(\theta-\theta').$$

$$\therefore \cos\theta - \cos\theta' = -2\sin\tfrac{1}{2}(\theta+\theta')\sin\tfrac{1}{2}(\theta-\theta').$$

$$\frac{\sin\theta+\sin\theta'}{\cos\theta-\cos\theta'} = \frac{2\sin\frac{1}{2}(\theta+\theta')\cos\frac{1}{2}(\theta-\theta')}{-2\sin\frac{1}{2}(\theta+\theta')\sin\frac{1}{2}(\theta-\theta')}$$
$$= -\frac{\cos\frac{1}{2}(\theta-\theta')}{\sin\frac{1}{2}(\theta-\theta')}$$
$$= -\cot\tfrac{1}{2}(\theta-\theta').$$

4. Find $\sin 2\theta$, $\cos 2\theta$, and $\tan 2\theta$, in terms of functions of 2θ.

$$\sin 2\theta = \sin(\theta+\theta)$$
$$= \sin\theta\cos\theta + \cos\theta\sin\theta$$
$$= 2\sin\theta\cos\theta.$$

$$\cos 2\theta = \cos(\theta+\theta)$$
$$= \cos\theta\cos\theta - \sin\theta\sin\theta$$
$$= \cos^2\theta - \sin^2\theta.$$

$$\tan^2\theta = \tan(\theta+\theta)$$
$$= \frac{\tan\theta+\tan\theta}{1-\tan\theta\tan\theta}$$
$$= \frac{2\tan\theta}{1-\tan^2\theta}.$$

5. Assuming the law of sines for a plane triangle, prove that

$(a+b) : c = \cos\frac{1}{2}(A-B) : \sin\frac{1}{2}C,$
$(a-b) : c = \sin\frac{1}{2}(A-B) : \cos\frac{1}{2}C.$

See I., Ex. 5.

6. At 120 ft. distance, and on a level with the foot of a steeple, the angle of elevation of its top is 62° 27′; find the height.

$A = 62° 27'$, $b = 120$.

$$a = b\tan A$$
$$= 120\tan 62° 27'.$$

$$\log 120 = 2.07918$$
$$\log\tan 62° 27' = 10.28260$$
$$\log a = 2.36178$$
$$a = 230.03.$$

Height of steeple, 230.03 ft.

7. Solve the plane triangle given the three sides,

$a = 48.76$, $b = 62.92$, $c = 80.24$.

$$s = \tfrac{1}{2}(a+b+c)$$
$$= 95.96.$$
$$s-a = 47.20.$$
$$s-b = 33.04.$$
$$s-c = 15.72.$$

$$r = \sqrt{\frac{(s-a)(s-b)(s-c)}{s}}$$
$$= \sqrt{\frac{47.20\times 33.04\times 15.72}{95.96}}.$$

$$\log 47.20 = 1.67394$$
$$\log 33.04 = 1.51904$$
$$\log 15.72 = 1.19645$$
$$\text{colog } 95.96 = 8.01791-10$$
$$2)\,2.40734$$
$$\log r = 1.20367.$$

$$\log\tan\tfrac{1}{2}A = 9.52973$$
$$\log\tan\tfrac{1}{2}B = 9.68463$$
$$\log\tan\tfrac{1}{2}C = 10.00722$$

$$\tfrac{1}{2}A = 18° 42' 29''$$
$$\tfrac{1}{2}B = 25° 48' 56''$$
$$\tfrac{1}{2}C = 45° 28' 35''$$

$$A = 37° 24' 58''$$
$$B = 51° 37' 52''$$
$$C = 90° 57' 10''$$

$$A+B+C = 180°\ 0'\ 0''.$$

V.

1. In how many years will a sum of money double itself at 4 per cent, interest being compounded semi-annually?

Let the sum be S. Then the amount at the end of n years is

$$(1.02)^{2n}S,$$

and $(1.02)^{2n}S = 2\,S.$

$$(1.02)^{2n} = 2.$$
$$2\,n \log 1.02 = \log 2.$$
$$n = \tfrac{1}{2}\,\frac{\log 2}{\log 1.02}.$$
$$\log 2 = 0.30103.$$
$$\log 1.02 = 0.00860.$$
$$\tfrac{1}{2}\,\frac{\log 2}{\log 1.02} = 17.5.$$
$$\therefore n = 17.5.$$

The sum will double itself in $17\tfrac{1}{2}$ years.

2. Given $\sin^2 x = \dfrac{1+\sqrt{1-m^2}}{2}$, find $\sin 2\,x$ and $\tan 2\,x$.

$$\sin^2 x = \frac{1+\sqrt{1-m^2}}{2}.$$
$$\cos 2\,x = 1 - 2\sin^2 x$$
$$= -\sqrt{1-m^2}.$$
$$\sin 2\,x = \sqrt{1-\cos^2 2\,x}$$
$$= \pm m.$$
$$\tan 2\,x = \frac{\sin 2\,x}{\cos 2\,x}$$
$$= \pm\frac{m}{\sqrt{1-m^2}}.$$

3. Find all values of x under 360°, which satisfy the equation $\sqrt{8}\cos 2\,x = 1 - 2\sin x$.

$$\sqrt{8\cos 2\,x} = 1 - 2\sin x.$$
$$8\cos 2\,x = 1 - 4\sin x + 4\sin^2 x.$$
$$8\,(1-2\sin^2 x) = 1 - 4\sin x + 4\sin^2 x.$$
$$20\sin^2 x - 4\sin x - 7 = 0.$$
$$\sin x = \tfrac{7}{10} \text{ or } -\tfrac{1}{2}.$$

(i.) $\sin x = \tfrac{7}{10}.$

$x = \sin^{-1}\tfrac{7}{10}.$

(ii.) $\sin x = -\tfrac{1}{2}.$

$x = 330° \text{ or } 210°.$

4. What is always the value of

$$2\sin^2 x\sin^2 y + 2\cos^2 x\cos^2 y - \cos 2\,x\cos 2\,y\,?$$

$2\sin^2 x\sin^2 y + 2\cos^2 x\cos^2 y - \cos 2\,x\cos 2\,y$

$= 2\,(\sin x\sin y + \cos x\cos y)^2 - 4\sin x\sin y\cos x\cos y - \cos 2\,x\cos 2\,y$

$= 2\cos^2(x-y) - (\sin 2\,x\sin 2\,y + \cos 2\,x\cos 2\,y)$

$= 2\cos^2(x-y) - \cos(2\,x - 2\,y)$

$= 2\cos^2(x-y) - \cos 2\,(x-y)$

$= 2\cos^2(x-y) - [2\cos^2(x-y) - 1]$

$= 1.$

5. Find the area of a parallelogram, if its diagonals are 2 and 3, intersecting each other at an angle of 35°.

Area of each of the four triangles into which the diagonals divide the parallelogram:

$$= \tfrac{1}{2}\times\tfrac{2}{2}\times\tfrac{3}{2}\sin 35°$$
$$= \tfrac{3}{4}\sin 35°.$$
$$\therefore \text{Whole area} = 3\sin 35°.$$
$$\sin 35° = 0.5736.$$
$$\therefore \text{Whole area} = 1.7208.$$

6. Find the bearing and distance from Cape Horn (55° 55′ S., 67° 40′ W.) to Falkland Islands (51° 40′ S., 59° W.).

Diff. lat. $= 4° 15' = 255'$.
Mid. lat. $= 53° 47' 30''$.
Diff. long. $= 8° 40' = 520'$.

Depart. = Diff. long. × cos mid. lat.
$= 520 \cos 53° 47' 30''$.

$$\tan \text{course} = \frac{\text{Depart.}}{\text{Diff. lat.}} = \frac{520}{255} \cos 53° 47' 30''.$$

log 520 = 2.71600
colog 255 = 7.59346−10
log cos 53° 47′ 30″ = 9.77139
log tan course = 10.08085
course = N. 50° 18′ 9″ E.

Dist. = diff. lat. × sec course
= 255 sec 50° 18′ 9″.

log 255 = 2.40654
log sec 50° 18′ 9″ = 0.19468
log dist. = 2.60122
dist. = 399.23.

Bearing, N. 50° 18′ E.; distance, 399 miles.

VI.

1. In a certain system of logarithms $\bar{1}.25$ is the logarithm of $\frac{1}{8}$. What is the base?

[Be careful to remember what $\bar{1}.25$ means.]

Let the base $= b$,

Then $b^{\bar{1}.25} = \frac{1}{8}$.
$b^{-0.75} = \frac{1}{8}$.
$b^{-\frac{3}{4}} = \frac{1}{8}$.
$b = 8^{\frac{4}{3}}$
$= 16$.

The base is 16.

2. Find the tangent of $3x$ in terms of the tangent of x.

$$\tan 3x = \tan(2x + x) = \frac{\tan 2x + \tan x}{1 - \tan 2x \tan x} = \frac{\dfrac{2\tan x}{1-\tan^2 x} + \tan x}{1 - \dfrac{2 \tan x}{1 - \tan^2 x}\tan x} = \frac{3\tan x - \tan^3 x}{1 - 3\tan^2 x}.$$

3. One angle of a triangle is 35°, and one of the sides including the angle is 24. What are the smallest values the other sides can have?

The smallest value of the side opposite the given angle is 24 sin 35°. The third side may have any value from 0 to ∞.

log 24 = 1.38021
log sin 35° = 9.75859
log (24 sin 35°) = 1.13880
24 sin 35° = 13.766.

4. Find all the values of x, under 360°, which satisfy the equation
$$\tan 2x (\tan^2 x - 1) = 2 \sec^2 x - 6.$$

$$\tan 2x(\tan^2 x - 1) = 2\sec^2 x - 6.$$

$$\frac{2\tan x}{1 - \tan^2 x}(\tan^2 x - 1) = 2(1 + \tan^2 x) - 6.$$

$$-2\tan x = 2\tan^2 x - 4.$$

$$\tan^2 x + \tan x = 2.$$

$$\tan x = 1 \text{ or } -2.$$

(i.) $\tan x = 1$.
$x = 45°$ or $225°$.

(ii.) $\tan x = -2$.
$x = \tan^{-1}(-2)$.

5. Two ships leave Cape Cod (42° N. 70° W.), one sailing E., the other sailing N.E. How many miles must each sail to reach longitude 65° W.?

Diff. long. = 5° = 300′.

(i.) For the first ship,

$$\begin{aligned}\text{Dist.} &= \text{diff. long.} \times \cos \text{lat.}\\ &= 300 \cos 42°.\end{aligned}$$

$$\begin{aligned}\log 300 &= 2.47712\\ \log \cos 42° &= \underline{9.87107}\\ \log \text{dist.} &= 2.34819\\ \text{Dist.} &= 222.94.\end{aligned}$$

(ii.) For the second ship,

Course = 45°.

∴ depart. = diff. lat.

Let diff. lat. = d.

Then

$$\text{diff. long.} = \text{depart.} \times \sec \text{mid. lat.}$$

$$300 = d \sec (42° + \tfrac{1}{2} d).$$

By trial $d = 3°\ 36'\ 30''$

$= 216.5'$,

$$\begin{aligned}\text{Dist.} &= \text{diff. lat.} \times \sec \text{course}\\ &= 216.5' \sec 45°.\end{aligned}$$

$$\begin{aligned}\log 216.5 &= 2.33546\\ \log \sec 45° &= \underline{0.15051}\\ \log \text{dist.} &= 2.48597\\ \text{Dist.} &= 306.17.\end{aligned}$$

First ship must sail 223 miles; second ship 306 miles.

6. If $A + B + C = 180°$, find the value of

$$\tan A + \tan B + \tan C - \tan A \tan B \tan C.$$

$$\begin{aligned}&\tan A + \tan B + \tan C - \tan A \tan B \tan C\\ &\quad = \tan A + \tan B + \tan C\,(1 - \tan A \tan B)\\ &\quad = \tan A + \tan B + \tan\,[180° - (A + B)]\ (1 - \tan A \tan B)\\ &\quad = \tan A + \tan B - \tan\,(A + B)\ (1 - \tan A \tan B)\\ &\quad = \tan A + \tan B - \frac{\tan A + \tan B}{1 - \tan A \tan B}(1 - \tan A \tan B)\\ &\quad = 0.\end{aligned}$$

VII.

1. What is the base, when $\log 0.008 = -1.5$?

If the base $= b$,

$$\begin{aligned}b^{-1.5} &= 0.008.\\ b^{-\frac{3}{2}} &= \tfrac{8}{1000}.\\ b^{\frac{3}{2}} &= \tfrac{1000}{8}\\ &= 125.\\ \therefore b &= 125^{\frac{2}{3}}\\ &= 25.\end{aligned}$$

The base is 25.

2. If $\cos (a - b) = 3 \cos (a + b)$, find the value of $\dfrac{\sec(a + b)}{\sec a \sec b}$.

$$\cos (a - b) = 3 \cos (a + b).$$

$$\begin{aligned}&\cos a \cos b + \sin a \sin b\\ &\qquad = 3 \cos a \cos b - 3 \sin a \sin b.\end{aligned}$$

$$2 \cos a \cos b = 4 \sin a \sin b.$$

$$\cos a \cos b = 2 \sin a \sin b.$$

$$\begin{aligned}\cos (a + b) &= \cos a \cos b - \sin a \sin b\\ &= \tfrac{1}{2} \cos a \cos b.\end{aligned}$$

$$\begin{aligned}\sec (a + b) &= \frac{2}{\cos a \cos b}\\ &= 2 \sec a \sec b.\end{aligned}$$

$$\frac{\sec (a + b)}{\sec a \sec b} = 2.$$

3. The area of an oblique-angled triangle is 50. One angle is 30°, and a side adjacent to that angle is 12. Solve the triangle.

$$\text{Area} = \tfrac{1}{2}bc \sin A.$$

$$50 = \frac{c}{2} \times 12 \sin 30^\circ$$

$$= 3c.$$

$$\therefore c = \tfrac{50}{3}.$$

$$\tan \tfrac{1}{2}(C - B) = \frac{c-b}{c+b} \tan \tfrac{1}{2}(C+B)$$

$$= \frac{\frac{14}{3}}{\frac{86}{3}} \tan \tfrac{1}{2}(180^\circ - 30^\circ)$$

$$= \tfrac{7}{43} \tan 75^\circ.$$

$$\begin{aligned}
\log 7 &= 0.84510 \\
\text{colog } 43 &= 8.36653 \\
\log \tan 75^\circ &= 10.57195 \\
\log \tan \tfrac{1}{2}(C - B) &= 9.78358 \\
\tfrac{1}{2}(C - B) &= 31^\circ\, 16'\, 50'' \\
\tfrac{1}{2}(C + B) &= 75^\circ \\
C &= 106^\circ\, 16'\, 50'' \\
B &= 43^\circ\, 43'\, 10''.
\end{aligned}$$

$$a = b\frac{\sin A}{\sin B}$$

$$= 12 \frac{\sin 30^\circ}{\sin 43^\circ 43' 10''}$$

$$= \frac{6}{\sin 43^\circ\, 43'\, 10''}.$$

$$\begin{aligned}
\log 6 &= 0.77815 \\
\text{colog } \sin 43^\circ\, 43'\, 10'' &= 0.16044 \\
\log a &= 0.93859 \\
a &= 8.6814.
\end{aligned}$$

4. Find all values of x, less than 360°, which satisfy the equation

$$\sin 2x - \cos x = \cos^2 x.$$

$$\sin 2x - \cos x = \cos^2 x.$$

$$2 \sin x \cos x - \cos x = \cos^2 x.$$

$$\cos x\,(2 \sin x - 1 - \cos x) = 0.$$

(i.) $\cos x = 0.$

$$x = 90^\circ \text{ or } 270^\circ.$$

(ii.) $2 \sin x - 1 - \cos x = 0.$

$$2 \sin x - 1 = \cos x.$$

$$4 \sin^2 x - 4 \sin x + 1 = \cos^2 x$$

$$= 1 - \sin^2 x.$$

$$5 \sin^2 x - 4 \sin x = 0.$$

$$\sin x = 0 \text{ or } \tfrac{4}{5}.$$

$$x = 0^\circ,\ 180^\circ,\ 53^\circ\, 7'\, 48'', \text{ or } 126^\circ\, 52'\, 12''.$$

$$\therefore x = 0^\circ,\ 90^\circ,\ 180^\circ,\ 270^\circ,\ 53^\circ\, 7'\, 48'', \text{ or } 126^\circ\, 52'\, 12''.$$

Of these values, however, only the following satisfy the original equation:

$$x = 90^\circ,\ 180^\circ,\ 270^\circ,\ 53^\circ\, 7'\, 48''.$$

5. Find, by Middle Latitude Sailing, the course and distance from Cape Cod (Lat. 42° 2′ N., Long. 70° 4′ W.) to Fayal (Lat. 38° 32′ N., Long. 28° 39′ W.).

Diff. lat. = 3° 30′ = 210′.
Mid. lat. = 40° 17′.
Diff. long. = 41° 25′ = 2485′.
Depart. = diff. long. × cos mid. lat.
= 2485 cos 40° 17′.

$$\text{tan course} = \frac{\text{depart.}}{\text{diff. lat.}}$$

$$= \frac{2485 \cos 40^\circ 17'}{210}.$$

$$\begin{aligned}
\log 2485 &= 3.39533 \\
\text{colog } 210 &= 7.67778 \\
\log \cos 40^\circ\, 17' &= 9.88244 \\
\log \text{tan course} &= 10.95555
\end{aligned}$$

Course = S. 83° 40′ 43″ E.

Dist. = diff. lat. × sec course
= 210 sec 83° 40′ 43″.

$$\begin{aligned}
\log 210 &= 2.32222 \\
\log \sec 83^\circ\, 40'\, 43'' &= 0.95819 \\
\log \text{dist.} &= 3.28041 \\
\text{Dist.} &= 1907.3
\end{aligned}$$

Course, S. 83° 41′; distance, 1907 miles.

6. In any triangle ABC, prove

$$\tan \tfrac{1}{2}A \tan \tfrac{1}{2}B + \tan \tfrac{1}{2}A \tan \tfrac{1}{2}C + \tan \tfrac{1}{2}B \tan \tfrac{1}{2}C = 1.$$

$$\begin{aligned}
&\tan \tfrac{1}{2}A \tan \tfrac{1}{2}B + \tan \tfrac{1}{2}A \tan \tfrac{1}{2}C + \tan \tfrac{1}{2}B \tan \tfrac{1}{2}C \\
&= \tan \tfrac{1}{2}A \tan \tfrac{1}{2}B + \tan \tfrac{1}{2}C (\tan \tfrac{1}{2}A + \tan \tfrac{1}{2}B) \\
&= \tan \tfrac{1}{2}A \tan \tfrac{1}{2}B + \tan [90^\circ - \tfrac{1}{2}(A + B)] (\tan \tfrac{1}{2}A + \tan \tfrac{1}{2}B) \\
&= \tan \tfrac{1}{2}A \tan \tfrac{1}{2}B + \cot \tfrac{1}{2}(A + B) (\tan \tfrac{1}{2}A + \tan \tfrac{1}{2}B) \\
&= \tan \tfrac{1}{2}A \tan \tfrac{1}{2}B + \frac{1 - \tan \tfrac{1}{2}A \tan \tfrac{1}{2}B}{\tan \tfrac{1}{2}A + \tan \tfrac{1}{2}B} (\tan \tfrac{1}{2}A + \tan \tfrac{1}{2}B) \\
&= 1.
\end{aligned}$$

VIII.

1. What is the base of a system of logarithms in which

$$\log \tfrac{1}{243} = \bar{2}.33\tfrac{1}{3}?$$

Let the base $= b$.

Then

$$\begin{aligned}
b^{\bar{2}.33\frac{1}{3}} &= \tfrac{1}{243}. \\
b^{-1.66\frac{2}{3}} &= \tfrac{1}{243}. \\
b^{-\frac{5}{3}} &= 3^{-5}. \\
b &= (3^{-5})^{-\frac{3}{5}} \\
&= 3^3 \\
&= 27.
\end{aligned}$$

The base is 27.

2. Given the area of a right triangle, and the smallest angle, find the legs of the triangle in terms of the data.

Let area $= F$

given angle $= A$.

Then,

$$\begin{aligned}
ab &= 2F. \\
\frac{a}{b} &= \tan A. \\
a^2 &= 2F \tan A. \\
b^2 &= 2F \cot A. \\
a &= \sqrt{2F \tan A}. \\
b &= \sqrt{2F \cot A}.
\end{aligned}$$

3. Find a and b, given $\dfrac{\sin a}{\sin b} = \sqrt{2}$, and $\dfrac{\tan a}{\tan b} = \sqrt{3}$.

$$\begin{aligned}
\frac{\sin a}{\sin b} &= \sqrt{2}. \\
\frac{\tan a}{\tan b} &= \sqrt{3}. \\
\sin a &= \sqrt{2} \sin b. \\
\tan a &= \sqrt{3} \tan b. \\
\frac{\sin a}{\cos a} &= \sqrt{3} \frac{\sin b}{\cos b}. \\
\frac{\sin a}{\sin b} &= \sqrt{3} \frac{\cos a}{\cos b}. \\
\sqrt{2} &= \sqrt{3} \frac{\cos a}{\cos b}. \\
\cos a &= \sqrt{\tfrac{2}{3}} \cos b. \\
\sin^2 a &= 2 \sin^2 b. \\
\cos^2 a &= \tfrac{2}{3} \cos^2 b. \\
\sin^2 a + \cos^2 a &= 2 \sin^2 b + \tfrac{2}{3} \cos^2 b. \\
2 \sin^2 b + \tfrac{2}{3} \cos^2 b &= 1. \\
2 \sin^2 b + \tfrac{2}{3} (1 - \sin^2 b) &= 1. \\
\sin^2 b &= \tfrac{1}{4}. \\
\sin b &= \pm \tfrac{1}{2}. \\
b &= \pm 30^\circ \text{ or } \pm 150^\circ. \\
\sin a &= \sqrt{2} \sin b \\
&= \pm \frac{1}{\sqrt{2}}. \\
a &= \pm 45^\circ \text{ or } \pm 135^\circ.
\end{aligned}$$

4. One angle of an oblique-angled triangle is 45°, and an adjacent side is $\sqrt{2}$. What is the smallest value the opposite side can have? Solve the triangle when the opposite side is $\frac{5}{4}$.

(i.) Smallest value of opposite side is

$$\sqrt{2}\sin 45° = 1.$$

(ii.) $a = \sqrt{2}$, $b = \frac{5}{4}$, $B = 45°$.

$a > b > a \sin B$. $\therefore$ two solutions.

$$\sin A = \frac{a}{b}\sin B$$
$$= \sqrt{2} \times \tfrac{4}{5}\sin 45°$$
$$= \tfrac{4}{5}.$$

$$\log 4 = 0.60206$$
$$\text{colog } 5 = \underline{9.30103}$$
$$\log \sin A = 9.90309$$

$A = 53° 7' 48''$ or $126° 52' 12''$.

$C = 180 - (A + B)$

$= 81° 52' 12''$ or $8° 7' 48''$

$$c = b\frac{\sin C}{\sin B}.$$

$\log b =$	0.09691	0.09691
$\log \sin C =$	9.99561	9.15051
$\text{colog} \sin B =$	0.15051	0.15051
$\log c =$	0.24303	9.39793 − 10
$c =$	1.75	2.50

5. A ship leaves Cape Cod (42° 2′ N., 70° 4′ W.) and sails 200 knots on a course S. 40° E. Find the latitude and longitude reached.

Diff. lat. = distance × cos course
= 200 cos 40°.

$$\log 200 = 2.30103$$
$$\log \cos 40° = \underline{9.88425}$$
$$\log \text{diff. lat.} = 2.18528$$
$$\text{diff. lat.} = 153.21'$$
$$= 2° 33' 13''.$$

Mid. lat. = 40° 45′ 24″.

Diff. long. = depart. × sec. mid. lat.

Depart. = distance × sin course
= 200 sin 40°.

Diff. long. = 200 sin 40° sec 40° 45′ 24″.

$$\log 200 = 2.30103$$
$$\log \sin 40° = 9.80807$$
$$\log \sec 40° 45' 24'' = \underline{0.12062}$$
$$\log \text{diff. long.} = 2.22972$$
$$\text{diff. long.} = 169.72'$$
$$= 2° 49' 43''.$$

Latitude reached, 39° 29′ N.; longitude, 67° 14′ W.

6. If $2 \tan 2a = \tan 2b \sin 2b$, find the relation between the tangents of a and b.

$$2 \tan 2a = \tan 2b \sin 2b.$$

$$\frac{4 \tan a}{1 - \tan^2 a} = \frac{2 \tan b}{1 - \tan^2 b} \times 2 \sin b \cos b$$
$$= \frac{4 \tan b}{1 - \tan^2 b} \times \tan b \cos^2 b$$
$$= \frac{4 \tan^2 b}{1 - \tan^2 b} \times \frac{1}{\sec^2 b}$$
$$= \frac{4 \tan^2 b}{(1 - \tan^2 b)(1 + \tan^2 b)}$$
$$= \frac{4 \tan^2 b}{1 - \tan^4 b}.$$

$$\frac{\tan a}{1 - \tan^2 a} = \frac{\tan^2 b}{1 - \tan^4 b}.$$

$$\tan a\,(1-\tan^4 b)=\tan^2 b\,(1-\tan^2 a).$$

$$\tan^2 b\tan^2 a+(1-\tan^4 b)\tan a-\tan^2 b=0.$$

$$\therefore \tan a=\tan^2 b \text{ or } -\frac{1}{\tan^2 b}$$

$$=\tan^2 b \text{ or } -\cot^2 b.$$

IX.

1. What is the base of the system of logarithms when $\log 3 = 0.3976$?

Let the base $= b$,

Then $b^{0.3976} = 3.$

$$0.3976 \log b = \log 3$$

$$\log b=\frac{\log 3}{0.3976}$$

$$=\frac{0.47712}{0.3976}$$

$$\log 47712 = 4.67863$$

$$\log 39760 = 4.59945$$

$$\log(\log b) = .07918$$

$$\log b = 1.2000$$

$$b = 15.849.$$

The base is 15.849.

2. Solve the right-angled triangle in which one angle is 30°, and the difference of the legs is 4.

$$a-b=4,\ B=30°,\ A=60°.$$

$$\frac{b}{a}=\tan B$$

$$=\frac{1}{\sqrt{3}}.$$

$$b=\frac{1}{\sqrt{3}}a.$$

$$a-b=a-\frac{a}{\sqrt{3}}$$

$$=4.$$

$$a=\frac{4\sqrt{3}}{\sqrt{3}-1}$$

$$=2(3+\sqrt{3}).$$

$$b=\frac{a}{\sqrt{3}}$$

$$=2(\sqrt{3}+1).$$

$$c^2=a^2+b^2$$

$$=a^2+\frac{a^2}{3}$$

$$=\frac{4a^2}{3}$$

$$c=\frac{2a}{\sqrt{3}}$$

$$=4(\sqrt{3}+1).$$

3. Find x, given $\sec x = 2\tan x + 2$.

$$\sec x = 2\tan x + 2.$$

$$\sec^2 x = 4\tan^2 x + 8\tan x + 4.$$

$$1+\tan^2 x = 4\tan^2 x + 8\tan x + 4.$$

$$3\tan^2 x + 8\tan x + 3 = 0.$$

$$\tan x=\frac{-4\pm\sqrt{7}}{3}$$

$$=\frac{-4\pm 2.6459}{3}$$

$$=1.3541 \text{ or } -6.6459.$$

$$\therefore x = 126°\,26'\,44'', -53°\,33'\,16'', 98°\,33'\,31'', \text{ or } -81°\,26'\,29''.$$

4. One angle of a triangle is double another angle. The side opposite the first angle is three-halves of the side opposite the second angle. Find the angles.

$$A=2B,\ a=\tfrac{3}{2}b.$$

$$\sin A=\frac{a}{b}\sin B$$

$$=\tfrac{3}{2}\sin\tfrac{1}{2}A.$$

$$2 \sin \tfrac{1}{2} A \cos \tfrac{1}{2} A = \tfrac{3}{2} \sin \tfrac{1}{2} A.$$
$$\cos \tfrac{1}{2} A = \tfrac{3}{4}.$$
$$\cos B = \tfrac{3}{4}.$$
$$\cos A = 2 \cos^2 \tfrac{1}{2} A - 1$$
$$= \tfrac{1}{8}.$$
$$\log \cos A = 9.09691$$
$$A = 82° 49' 9''.$$
$$\log \cos B = 9.87506.$$
$$B = 41° 24' 35''.$$
$$C = 180° - (A + B)$$
$$= 55° 46' 16''.$$

5. Find, by Middle Latitude Sailing, the course and distance from Funchal [32° 38′ N., 16° 54′ W.] to Gibraltar [36° 7′ N., 5° 21′ W.].

Diff. lat. $= 3° 29' = 209'$.
Mid. lat. $= 34° 22' 30''$.
Diff. long. $= 11° 33' = 693'$.

Depart. = diff. long. × sec. mid. lat.
$= 693 \sec 34° 22' 30''$.

$$\tan \text{course} = \frac{\text{depart.}}{\text{diff. lat.}}$$
$$= \frac{693}{209} \sec 34° 22' 30''.$$

$$\log 693 = 2.84073$$
$$\text{colog } 209 = 7.67985$$
$$\log \sec 34° 22' 30'' = \underline{0.08335}$$
$$\log \tan \text{course} = 10.60393$$
$$\text{Course} = \text{N. } 76° 2' 18'' \text{ E.}$$

Dist. = diff. lat. × sec course
$= 209 \sec 76° 2' 18''$.

$$\log 209 = 2\ 32015$$
$$\log \sec 76° 2' 18'' = \underline{0.61749}$$
$$\log \text{dist.} = 2.93764$$
$$\text{Dist.} = 866.24.$$

Course, N. 76° 2′ E.; distance, 866 miles.

6. Reduce to its simplest form $\cos 2x \tan (45° + x) - \sin 2x$.

$$\cos 2x \tan (45° + x) - \sin 2x$$
$$= (\cos^2 x - \sin^2 x) \frac{1 + \tan x}{1 - \tan x} - 2 \sin x \cos x$$
$$= (\cos^2 x - \sin^2 x) \frac{\cos x + \sin x}{\cos x - \sin x} - 2 \sin x \cos x$$
$$= (\cos x + \sin x)^2 - 2 \sin x \cos x$$
$$= \cos^2 x + \sin^2 x$$
$$= 1.$$

X.

1. If the base of our system of logarithms were 20 instead of 10, what would be the logarithm of one-tenth?

$$\log_{20} \tfrac{1}{10} = \log_{20} 10 \times \log_{10} \tfrac{1}{10}$$
$$= \frac{1}{\log_{10} 20} \times \log_{10} \tfrac{1}{10}$$
$$= -\frac{1}{\log_{10} 20}$$
$$= -\frac{1}{1.30103}$$
$$= \bar{1}.23138.$$

2. The area of a right triangle is 6, and the sum of the three sides is 12. Solve the triangle.

$$ab = 12.$$

$$a+b+\sqrt{a^2+b^2} = 12.$$

$$a+b = 12-\sqrt{a^2+b^2}.$$

$$a^2+2ab+b^2 = 144-24\sqrt{a^2+b^2}+a^2+b^2.$$

$$ab = 72-12\sqrt{a^2+b^2}.$$

$$12 = 72-12\sqrt{a^2+b^2}.$$

$$\sqrt{a^2+b^2} = 5.$$

$$a^2+b^2 = 25.$$

$$a^2+2ab+b^2 = 25+24 = 49.$$

$$a+b = 7.$$

$$a^2-2ab+b^2 = 25-24 = 1.$$

$$a-b = 1.$$

$$\therefore a = 4.$$

$$b = 3.$$

$$c = 5.$$

$$\tan A = \frac{a}{b} = \tfrac{4}{3}.$$

$$\log\tan A = 10.12494.$$

$$A = 53° 7' 48''.$$

$$B = 36° 52' 12''.$$

3. Reduce to its simplest form

$$\cos^2 B + \sin^2 B \cos 2A - \sin^2 A \cos 2B.$$

$$\begin{aligned}
&\cos^2 B + \sin^2 B \cos 2A - \sin^2 A \cos 2B.\\
&= 1 - \sin^2 B + \sin^2 B \cos 2A - \sin^2 A \cos 2B\\
&= 1 + \sin^2 B(\cos 2A - 1) - \sin^2 A \cos 2B\\
&= 1 + \sin^2 B(2\sin^2 A) - \sin^2 A \cos 2B\\
&= 1 + \sin^2 A(2\sin^2 B - \cos 2B)\\
&= 1 + \sin^2 A(2\sin^2 B - 1 + 2\sin^2 B)\\
&= 1 - \sin^2 A + 4\sin^2 A\sin^2 B\\
&= \cos^2 A + 4\sin^2 A\sin^2 B.
\end{aligned}$$

4. Two angles of a triangle are 40° 14′ and 60° 37′. The sum of the two opposite sides is 10. Find these sides.

$$a-b = (a+b)\frac{\tan\frac{1}{2}(A-B)}{\tan\frac{1}{2}(A+B)} = 10\frac{\tan 10° 11' 30''}{\tan 50° 25' 30''}.$$

$$\begin{aligned}
\log 10 &= 1.00000\\
\log\tan 10° 11' 30'' &= 9.25473\\
\text{colog}\tan 50° 25' 30'' &= 9.91726\\
\log(a-b) &= 0.17199\\
a-b &= 1.4859\\
a+b &= 10\\
a &= 5.743\\
b &= 4.257.
\end{aligned}$$

The sides are 5.743 and 4.257.

5. A ship leaves Cape of Good Hope (34° 22′ S., 18° 30′ E.) and sails N. 40° W. to latitude 30° S. Find, by Middle Latitude Sailing, the longitude reached and the distance sailed.

Diff. lat. $= 4° 22' = 262'$.
Mid. lat. $= 32° 11'$.
Course $= 40°$.
Depart. $=$ diff. lat. × tan course $= 262 \tan 40°$.
Diff. long. $=$ depart. × sec mid. lat. $= 262 \tan 40° \sec 32° 11'$.

$$\log 262 = 2.41830$$
$$\log \tan 40° = 9.92381$$
$$\log \sec 32° 11' = 0.07245$$
$$\log \text{diff. long.} = 2.41456$$
$$\text{Diff. long.} = 259.75'$$
$$= 4° 19' 45''.$$

Dist. $=$ diff. lat. × sec course $= 262 \sec 40°$.

$$\log 262 = 2.41830$$
$$\log \sec 40° = 0.11575$$
$$\log \text{dist.} = 2.53405$$
$$\text{dist.} = 342.02.$$

Longitude reached, 14° 10′ E.; distance sailed, 342 miles.

6. The base angles of a triangle are 22° 30′ and 112° 30′. Find the ratio between the base and the height of the triangle.

Let $A = 22° 30'$, $B = 112° 30'$, $C =$ base, $p =$ altitude.

$$C = 180° - (A + B) = 45°.$$
$$p = b \sin A.$$
$$c = b \frac{\sin C}{\sin B}.$$
$$\frac{c}{p} = \frac{\sin C}{\sin A \sin B} = \frac{\sin 45°}{\sin 22° 30' \sin 112° 30'} = \frac{2 \sin 22° 30' \cos 22° 30'}{\sin 22° 30' \cos 22° 30'} = 2.$$

The base is twice the altitude.

XI.

1. What is meant by the logarithm of a number n in the system whose base is 8? What will be the logarithm of 4 in this system?

(i.) The logarithm of a number n in the system whose base is 8 is the power to which 8 must be raised to produce n.

(ii.) The logarithm of 4 in this system is $\frac{2}{3}$.

2. Establish the formula:

$$\sin \tfrac{3}{2} x = \pm (1 + 2 \cos x) \sqrt{\frac{1 - \cos x}{2}}.$$

Which sign should be used when x lies in the first quadrant? When x lies in the second quadrant?

$$\begin{aligned}\sin \tfrac{3}{2} x &= \sin (x + \tfrac{1}{2} x) \\ &= \sin x \cos \tfrac{1}{2} x + \cos x \sin \tfrac{1}{2} x \\ &= 2 \sin \tfrac{1}{2} x \cos^2 \tfrac{1}{2} x + (1 - 2 \sin^2 \tfrac{1}{2} x) \sin \tfrac{1}{2} x \\ &= \sin \tfrac{1}{2} x (2 \cos^2 \tfrac{1}{2} x - 2 \sin^2 \tfrac{1}{2} x + 1) \\ &= \sin \tfrac{1}{2} x (2 \cos x + 1).\end{aligned}$$

But $1 - \cos x = 2\sin^2 \frac{1}{2}x.$

$$\therefore \sin \tfrac{1}{2}x = \pm\sqrt{\frac{1-\cos x}{2}}.$$

$$\sin \tfrac{3}{2}x = \pm(1 + 2\cos x)\sqrt{\frac{1-\cos x}{2}}.$$

If x lies in the first quadrant, $\sin \frac{3}{2}x$ and $\cos x$ are positive, and the positive sign must be used.

If x lies in the second quadrant and is $< 120°$, $\sin \frac{3}{2}x$ and $1 + 2\cos x$ are positive; and if x is $> 120°$, $\sin \frac{3}{2}x$ and $1 + 2\cos x$ are negative. Hence the positive sign should be used in both cases.

3. In a triangle two angles are equal to 32° 47′ and 49° 28′ respectively, and the length of the included side is .072. Solve the triangle.

$A = 32° 47'$, $B = 49° 28'$, $c = 0.72.$

$$C = 180° - (A + B)$$
$$= 97° 45'.$$

$$a = c\frac{\sin A}{\sin C}$$
$$= 0.72\frac{\sin 32° 47'}{\sin 97° 45'}.$$

$\log .072 = 8.85733 - 10$
$\log \sin 32° 47' = 9.73357$
$\text{colog} \sin 97° 45' = 0.00399$
$\log a = 8.59489 - 10$
$a = 0.039345.$

$$b = c\frac{\sin B}{\sin C}$$
$$= 0.072\frac{\sin 49° 28'}{\sin 97° 45'}$$

$\log .072 = 8.85733 - 10$
$\log \sin 49° 28' = 9.88083$
$\text{colog} \sin 97° 45' = 0.00399$
$\log b = 8.74215 - 10$
$b = 0.055226.$

4. A circular tent 30 ft. in diameter subtends at a certain point an angle of 15°. Find the distance of this point from the centre of the tent.

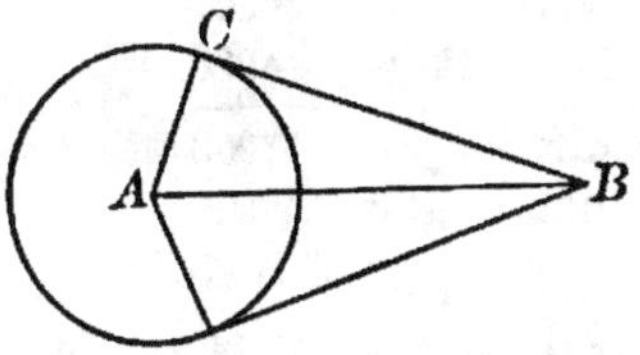

Let A be the centre of the tent, B the point of observation, and BC the tangent from B to the circle representing the tent. Then

$AC = 15$, $B = 7° 30'$, $C = 90°.$

$$AB = AC \csc B$$
$$= 15 \csc 7° 30'.$$

$\log 15 = 1.17609$
$\log \csc 7° 30' = 0.88430$
$\log AB = 2.06039$
$AB = 114.92.$

Distance of point of observation from centre of tent, 115 ft.

5. A ship leaves Lat. 42° 2′ N., Long. 70° 3′ W., and sails N. 40° E., a distance of 420 miles. Find, by Middle Latitude Sailing, the position reached.

Diff. lat. = dist. × cos course
$= 420 \cos 40°.$

$\log 420 = 2.62325$
$\log \cos 40° = 9.88425$

log diff. lat. = 2.50750
Diff. lat. = 321.74′
= 5° 21′ 44″.

Mid. lat. = 44° 42′ 52″.

Depart. = dist. × sin course
$= 420 \sin 40°.$

Diff. long. = depart. × sec. mid. lat.
$= 420 \sin 40 \sec 44° 42' 52''.$

$\log 420 = 2.62325$
$\log \sin 40° = 9.80807$
$\log \sec 44° 42' 52'' = 0.14836$

log. diff. long. = 2.57968
diff. long. = 379.91′
= 6° 19′ 55″.

Latitude reached, 47° 24′ N.; longitude, 63° 43′ W.

XII.

1. Express an angle of 60° in radians.

$$360° = 2\pi \text{ radians}.$$

$$\therefore 60° = \frac{\pi}{3} \text{ radians}.$$

2. Represent geometrically the different trigonometric functions of an angle. State the signs of each function for each quadrant.

See *Trigonometry*, § 21.

3. Express $\tan \phi$ and $\sec \phi$ in terms of $\sin \phi$.

$$\tan \phi = \frac{\sin \phi}{\cos \phi}$$

$$= \frac{\sin \phi}{\sqrt{1 - \sin^2 \phi}}.$$

$$\sec \phi = \frac{1}{\cos \phi}$$

$$= \frac{1}{\sqrt{1 - \sin^2 \phi}}.$$

4. Derive the formula

$$\sin \alpha + \sin \beta = 2 \sin \tfrac{1}{2}(\alpha + \beta) \cos \tfrac{1}{2}(\alpha - \beta).$$

$$\sin \alpha = \sin [\tfrac{1}{2}(\alpha + \beta) + \tfrac{1}{2}(\alpha - \beta)]$$

$$= \sin \tfrac{1}{2}(\alpha + \beta) \cos \tfrac{1}{2}(\alpha - \beta) + \cos \tfrac{1}{2}(\alpha + \beta) \sin \tfrac{1}{2}(\alpha - \beta).$$

$$\sin \beta = \sin [\tfrac{1}{2}(\alpha + \beta) - \tfrac{1}{2}(\alpha - \beta)]$$

$$= \sin \tfrac{1}{2}(\alpha + \beta) \cos \tfrac{1}{2}(\alpha - \beta) - \cos \tfrac{1}{2}(\alpha + \beta) \sin \tfrac{1}{2}(\alpha - \beta).$$

$$\therefore \sin \alpha + \sin \beta = 2 \sin \tfrac{1}{2}(\alpha + \beta) \cos \tfrac{1}{2}(\alpha - \beta)$$

5. Show that if a, b, and c are the sides of a triangle and A is the angle opposite the side a,

then $a^2 = b^2 + c^2 - 2bc \cos A.$

See *Trigonometry*, § 34.

6. Given $\cos 2x = 2 \sin x$, find the value of $\sin x$.

$$\cos 2x = 2 \sin x.$$

$$1 - 2 \sin^2 x = 2 \sin x.$$

$$2 \sin^2 x + 2 \sin x - 1 = 0.$$

$$\sin x = \frac{-1 \pm \sqrt{3}}{2}.$$

$$\sin x = \frac{\sqrt{3} - 1}{2}.$$

7. Given two sides of a triangle $a = 450.2$, $b = 425.4$, and their included angle $C = 62° 8'$; find the remaining parts of the triangle.

$$\tan \tfrac{1}{2}(A - B) = \frac{a-b}{a+b} \tan \tfrac{1}{2}(A+B)$$
$$= \frac{24.8}{875.6} \tan 58° 56'.$$

$$\begin{aligned} \log 24.8 &= 1.39445 \\ \text{colog } 875.6 &= 7.05769 - 10 \\ \log \tan 58° 56' &= \underline{10.22008} \\ \log \tan \tfrac{1}{2}(A - B) &= 8.67222 \\ \tfrac{1}{2}(A - B) &= 2° 41' 30'' \\ \tfrac{1}{2}(A + B) &= \underline{58° 56'} \\ A &= 61° 37' 30''. \\ B &= 56° 14' 30''. \end{aligned}$$

$$c = a \frac{\sin C}{\sin A}$$
$$= 450.2 \frac{\sin 62° 8'}{\sin 61° 37' 30''}.$$

$$\begin{aligned} \log 450.2 &= 2.65341 \\ \log \sin 62° 8' &= 9.94647 \\ \text{colog } \sin 61° 37' 30'' &= \underline{0.05559} \\ \log c &= 2.65547 \\ c &= 452.34. \end{aligned}$$

XIII.

1. Express an angle of 15° in radians.

$$360° = 2\pi \text{ radians.}$$
$$\therefore 15° = \frac{\pi}{12} \text{ radians.}$$

2. Write the simplest equivalents for $\sin(\pi + \phi)$, $\tan(2\pi - \phi)$, $\cos(\frac{3}{2}\pi - \phi)$, $\sec(\pi + \phi)$.

$$\begin{aligned} \sin(\pi + \phi) &= -\sin \phi. \\ \tan(2\pi - \phi) &= -\tan \phi. \\ \cos(\tfrac{3}{2}\pi - \phi) &= -\sin \phi. \\ \sec(\pi + \phi) &= -\sec \phi. \end{aligned}$$

3. Express (*a*) $\tan \phi$ in terms of $\sin \phi$, $\cos \phi$, and $\cot \phi$, respectively, and (*b*) $\cos \phi$ in terms of $\tan \phi$, $\sec \phi$, and $\operatorname{cosec} \phi$, respectively.

(*a*)
$$\tan \phi = \frac{\sin \phi}{\cos \phi} = \frac{\sin \phi}{\sqrt{1 - \sin^2 \phi}}; \quad = \frac{\sqrt{1 - \cos^2 \phi}}{\cos \phi}; \quad = \frac{1}{\cot \phi}.$$

(*b*)
$$\cos \phi = \frac{1}{\sec \phi} = \frac{1}{\sqrt{1 + \tan^2 \phi}}; \quad = \sqrt{1 - \sin^2 \phi} = \sqrt{1 - \frac{1}{\csc^2 \phi}} = \frac{\sqrt{\csc^2 \phi - 1}}{\csc \phi}.$$

4. Show (*a*) that $\sin(\alpha + \beta) + \sin(\alpha - \beta) = 2 \sin \alpha \cos \beta$;
(*b*) that $\cos(\alpha + \beta) + \cos(\alpha - \beta) = 2 \cos \alpha \cos \beta$.

(*a*)
$$\begin{aligned} \sin(\alpha + \beta) &= \sin \alpha \cos \beta + \cos \alpha \sin \beta. \\ \sin(\alpha - \beta) &= \sin \alpha \cos \beta - \cos \alpha \sin \beta. \\ \therefore \sin(\alpha + \beta) + \sin(\alpha - \beta) &= 2 \sin \alpha \cos \beta. \end{aligned}$$

(*b*)
$$\begin{aligned} \cos(\alpha + \beta) &= \cos \alpha \cos \beta - \sin \alpha \sin \beta. \\ \cos(\alpha - \beta) &= \cos \alpha \cos \beta + \sin \alpha \sin \beta. \\ \therefore \cos(\alpha + \beta) + \cos(\alpha - \beta) &= 2 \cos \alpha \cos \beta. \end{aligned}$$

5. Assume the formula $\cos\alpha = \frac{b^2+c^2-a^2}{2bc}$, and show that $\sin^2\frac{1}{2}\alpha = \frac{(s-b)(s-c)}{bc}$, when $s=\frac{1}{2}(a+b+c)$.

$$\cos\alpha = \frac{b^2+c^2-a^2}{2bc}.$$

$$1-2\sin^2\tfrac{1}{2}\alpha = \frac{b^2+c^2-a^2}{2bc}.$$

$$2\sin^2\tfrac{1}{2}\alpha = 1-\frac{b^2+c^2-a^2}{2bc}$$

$$= \frac{a^2-(b^2-2bc+c^2)}{2bc}$$

$$= \frac{[a+(b-c)][a-(b-c)]}{2bc}$$

$$= \frac{(a+b-c)(a-b+c)}{2bc}$$

$$= \frac{(2s-2c)(2s-2b)}{2bc}$$

$$\sin^2\tfrac{1}{2}\alpha = \frac{(s-b)(s-c)}{bc}.$$

6. Obtain a formula for $\tan\frac{1}{2}\alpha$ in terms of $\cos\alpha$.

$$\cos\alpha = 1-2\sin^2\tfrac{1}{2}\alpha$$

$$= 2\cos^2\tfrac{1}{2}\alpha - 1.$$

$$\therefore \sin^2\tfrac{1}{2}\alpha = \frac{1-\cos\alpha}{2}.$$

$$\cos^2\tfrac{1}{2}\alpha = \frac{1+\cos\alpha}{2}.$$

$$\tan^2\tfrac{1}{2}\alpha = \frac{\sin^2\frac{1}{2}\alpha}{\cos^2\frac{1}{2}\alpha}$$

$$= \frac{1-\cos\alpha}{1+\cos\alpha}.$$

$$\tan\tfrac{1}{2}\alpha = \sqrt{\frac{1-\cos\alpha}{1+\cos\alpha}}.$$

7. The base of a triangle $c = 556.7$, and the two adjacent angles $\alpha = 65°\ 20'.2$, $\beta = 70°\ 00'.5$; calculate the area of the triangle.

$$\gamma = 180° - (\alpha+\beta)$$

$$= 44°\ 39'.3.$$

$$a = c\frac{\sin\alpha}{\sin\gamma}.$$

$$\text{Area} = \tfrac{1}{2}ac\sin\beta$$

$$= \tfrac{1}{2}c^2\frac{\sin\alpha\sin\beta}{\sin\gamma}$$

$$= \tfrac{1}{2}(556.7)^2\frac{\sin 65°\ 20'.2 \sin 70°\ 00'.5}{\sin 44°\ 39'.3}.$$

$\log (556.7)^2 = 5.49124$
$\text{colog } 2 = 9.69897-10$
$\log\sin 65°\ 20'.2 = 9.95845$
$\log\sin 70°\ 00'.5 = 9.97301$
$\text{colog}\sin 44°\ 39'.3 = 0.15314$
$\log \text{area} = 5.27481$
$\text{Area} = 188280.$

8. Given $0 < \alpha < 90°$, and $\log\cos\alpha = \bar{1}.85254$, to determine α.

$$\alpha = 45°\ 24'\ 20''.$$

XIV.

1. Reduce an angle of 3.5 radians to degrees.

$$1 \text{ radian} = \frac{180°}{\pi}.$$

$$\therefore 3\tfrac{1}{2} \text{ radians} = \frac{630°}{\pi}$$

$$= \frac{630°}{3.14159}.$$

$\log 630 = 2.79934$
$\log 3.14159 = 0.49715$
2.30219
$3\frac{1}{2}$ radians $= 200.54°$
$= 200°\ 32'\ 24''.$

More accurately,

1 radian $= 57.29578°.$
$3\frac{1}{2}$ radians $= 200.535°$
$= 200°\ 32'\ 6''.$

2. Define the different trigonometrical functions of an angle, and give their algebraic signs for an angle in each quadrant.

See *Trigonometry*, § 21.

3. Write simple equivalents for the following functions: $\sin(-\alpha)$; $\cos(-\alpha)$; $\tan(\frac{1}{2}\pi+\alpha)$; $\sec(\frac{3}{2}\pi-\alpha)$.

$$\sin(-\alpha) = -\sin\alpha.$$
$$\cos(-\alpha) = \cos\alpha.$$
$$\tan\left(\frac{\pi}{2}+\alpha\right) = -\cot\alpha.$$
$$\sec(\tfrac{3}{2}\pi-\alpha) = -\csc\alpha.$$

4. Express cosec α in terms, respectively, of $\sin\alpha$, $\cos\alpha$, $\tan\alpha$, $\cot\alpha$, $\sec\alpha$.

$$\text{cosec}\,\alpha = \frac{1}{\sin\alpha}$$
$$= \frac{1}{\sqrt{1-\cos^2\alpha}}$$
$$= \sqrt{1+\frac{1}{\tan^2\alpha}}$$
$$= \sqrt{1+\cot^2\alpha}$$
$$= \frac{\sec\alpha}{\sqrt{\sec^2\alpha-1}}.$$

5. Reduce $(\cos\alpha\cos\beta-\sin\alpha\sin\beta)^2+(\sin\alpha\cos\beta+\cos\alpha\sin\beta)^2$ to its simplest equivalent.

$$(\cos\alpha\cos\beta-\sin\alpha\sin\beta)^2+(\sin\alpha\cos\beta+\cos\alpha\sin\beta)^2$$
$$= \cos^2(\alpha+\beta)+\sin^2(\alpha+\beta)$$
$$= 1.$$

6. Show that $\tan\left(\frac{\pi}{4}-\alpha\right) = \frac{1-\tan\alpha}{1+\tan\alpha}$.

$$\tan(45^\circ-\alpha) = \frac{\tan 45^\circ - \tan\alpha}{\tan 45^\circ + \tan\alpha}$$
$$= \frac{1-\tan\alpha}{1+\tan\alpha}.$$

7. The sum of two sides, a and b, of a triangle is 546.7 ft., the sum of the opposite angles, α and β, is 124°, and the ratio $\sin\alpha : \sin\beta = 1.003$; determine the angles and sides of the triangle.

$$\gamma = 180^\circ - (\alpha+\beta)$$
$$= 56^\circ.$$
$$a+b = 546.7.$$
$$\frac{a}{b} = \frac{\sin\alpha}{\sin\beta}$$
$$= 1.003.$$

$$a = 1.003\,b.$$
$$a+b = 2.003\,b.$$
$$2.003\,b = 546.7$$
$$b = \frac{546.7}{2.003}$$
$$= 272.94.$$
$$a = 1.003\,b$$
$$= 273.76.$$
$$\tan\tfrac{1}{2}(\alpha-\beta) = \frac{a-b}{a+b}\tan\tfrac{1}{2}(\alpha+\beta)$$
$$= \frac{0.81882}{546.7}\tan 62^\circ.$$

$$\log 0.81882 = 9.91319$$
$$\text{colog } 546.7 = 7.26225 - 10$$
$$\log \tan 62^\circ = 10.27433$$
$$\log \tan \tfrac{1}{2}(\alpha - \beta) = 7.44977$$
$$\tfrac{1}{2}(\alpha - \beta) = 9' \, 41''$$
$$\tfrac{1}{2}(\alpha + \beta) = 62^\circ$$
$$\alpha = 62^\circ \, 9' \, 41''$$
$$\beta = 61^\circ \, 50' \, 19''.$$

$$c = b \frac{\sin \gamma}{\sin \beta}$$
$$= 272.94 \frac{\sin 56^\circ}{\sin 61^\circ \, 50' \, 19''}.$$

$$\log 274.94 = 2.43606$$
$$\log \sin 56^\circ = 9.91857$$
$$\text{colog } \sin 61^\circ \, 50' \, 19'' = 0.05472$$
$$\log c = 2.40935$$
$$c = 256.65.$$

8. Given $0 < \alpha < 90^\circ$, and $\log \cot \alpha = 0.03293$, to determine α.

$$\alpha = 47^\circ \, 10' \, 12''.$$

XV.

1. Express (*a*) an angle of 2 radians in degrees; (*b*) an angle of 30° in radians.

(*a*) 1 radian = 57.29578°.

$$\therefore 2 \text{ radians} = 114.59156^\circ$$
$$= 114^\circ \, 35' \, 30''.$$

(*b*) $360^\circ = 2\pi$ radians.

$$\therefore 30^\circ = \frac{\pi}{6} \text{ radians.}$$

2. Give simple equivalents for the following functions:

$\tan(-x)$, $\operatorname{cosec}(-x)$, $\sin\left(x + \frac{\pi}{2}\right)$, $\sin\left(x - \frac{\pi}{2}\right)$, $\tan\left(\frac{3\pi}{2} - x\right)$, $\sin(2\pi - x)$.

$$\tan(-x) = -\tan x.$$
$$\csc(-x) = -\csc x.$$
$$\sin\left(x + \frac{\pi}{2}\right) = \cos x.$$
$$\sin\left(x - \frac{\pi}{2}\right) = -\cos x.$$
$$\tan\left(\frac{3\pi}{2} - x\right) = \cot x.$$
$$\sin(2\pi - x) = -\sin x.$$

3. Given $\tan x = \frac{a}{b}$ to express $\sin x$, $\cos x$, $\cot x$, $\sec x$, and $\operatorname{cosec} x$ in terms of a and b.

$$\sin x = \frac{1}{\csc x}$$
$$= \frac{1}{\sqrt{1 + \cot^2 x}}$$
$$= \frac{a}{\sqrt{a^2 + b^2}}.$$
$$\cos x = \sqrt{1 - \sin^2 x}$$
$$= \frac{b}{\sqrt{a^2 + b^2}}.$$
$$\cot x = \frac{1}{\tan x}$$
$$= \frac{b}{a}.$$
$$\sec x = \frac{1}{\cos x}$$
$$= \frac{\sqrt{a^2 + b^2}}{b}.$$
$$\csc x = \frac{1}{\sin x}$$
$$= \frac{\sqrt{a^2 + b^2}}{a}.$$

4. Show that

$$\tan a \pm \tan b = \frac{\sin(a \pm b)}{\cos a \cos b}.$$

$$\sin(a \pm b) = \sin a \cos b \pm \cos a \sin b.$$
$$\frac{\sin(a \pm b)}{\cos a \cos b} = \frac{\sin a \cos b}{\cos a \cos b} \pm \frac{\cos a \sin b}{\cos a \cos b}$$
$$= \tan a \pm \tan b.$$

5. Derive the formulæ

$$\cos \tfrac{1}{2} a = \pm \sqrt{\frac{1 + \cos a}{2}}, \ \sin \tfrac{1}{2} a = \pm \sqrt{\frac{1 - \cos a}{2}}.$$

$$\cos a = 2 \cos^2 \tfrac{1}{2} a - 1$$
$$= 1 - 2 \sin^2 \tfrac{1}{2} a.$$
$$\therefore \cos \tfrac{1}{2} a = \pm \sqrt{\frac{1 + \cos a}{2}}.$$
$$\sin \tfrac{1}{2} a = \pm \sqrt{\frac{1 - \cos a}{2}}.$$

6. Given $180° < \phi < 270°$, and $\log \cot \phi = 0.3232$, to determine ϕ.

$$\phi = 222° \ 52' \ 12''.$$

7. The sides of a triangle are $a = 32.5$ ft., $b = 33.1$ ft., $c = 32.4$ ft. Calculate the area of the triangle and the angle c opposite the side C, using the following formulæ:

$$S = \sqrt{p(p-a)(p-b)(p-c)} = \tfrac{1}{2} ab \sin C,$$

in which S denotes the area of the triangle, and $p = \frac{1}{2}(a + b + c)$.

$$p = 49, \qquad \log p = 1.69020$$
$$p - a = 16.5, \qquad \log (p - a) = 1.21748$$
$$p - b = 15.9, \qquad \log (p - b) = 1.20140$$
$$p - c = 16.6, \qquad \log (p - c) = 1.22011$$
$$\log S^2 = 5.32919$$
$$\log S = 2.66459.$$
$$S = 461.94.$$
$$\sin c = \frac{2S}{ab}$$
$$= \frac{2 \times 461.94}{32.5 \times 33.1}.$$
$$\log 2 = 0.30103$$
$$\log 461.94 = 2.66459$$
$$\text{colog } 32.5 = 8.48812$$
$$\text{colog } 33.1 = 8.48017$$
$$\log \sin C = 9.93391$$
$$C = 59° \ 11' \ 8''.$$

Exercise XXIII. Page 119.

1. Given $\log_{10}2 = 0.30103$, $\log_{10}3 = 0.47712$, $\log_{10}7 = 0.84510$; find $\log_{10}6$, $\log_{10}14$, $\log_{10}21$, $\log_{10}4$, $\log_{10}12$, $\log_{10}5$, $\log_{10}\frac{1}{2}$, $\log_{10}\frac{1}{4}$, $\log_{10}\frac{7}{9}$, $\log_{10}\frac{21}{20}$.

$$\begin{aligned}\log_{10}6 &= \log_{10}2 + \log_{10}3\\ \log_{10}2 &= 0.30103\\ \log_{10}3 &= \underline{0.47712}\\ \therefore \log_{10}6 &= 0.77815\end{aligned}$$

$$\begin{aligned}\log_{10}14 &= \log_{10}2 + \log_{10}7\\ \log_{10}2 &= 0.30103\\ \log_{10}7 &= \underline{0.84510}\\ \therefore \log_{10}14 &= 1.14613\end{aligned}$$

$$\begin{aligned}\log_{10}21 &= \log_{10}3 + \log_{10}7\\ \log_{10}3 &= 0.47712\\ \log_{10}7 &= \underline{0.84510}\\ \therefore \log_{10}21 &= 1.32222\end{aligned}$$

$$\begin{aligned}\log_{10}4 &= 2\log_{10}2\\ \log_{10}2 &= 0.30103\\ &\underline{\qquad\qquad 2}\\ \therefore \log_{10}4 &= 0.60206\end{aligned}$$

$$\begin{aligned}\log_{10}12 &= \log_{10}3 + \log_{10}4\\ \log_{10}3 &= 0.47712\\ \log_{10}4 &= \underline{0.60206}\\ \therefore \log_{10}12 &= 1.07918\end{aligned}$$

$$\begin{aligned}\log_{10}5 &= \log_{10}10 - \log_{10}2\\ \log_{10}10 &= 1.00000\\ \log_{10}2 &= \underline{0.30103}\\ \therefore \log_{10}5 &= 0.69897\end{aligned}$$

$$\begin{aligned}\log_{10}\tfrac{1}{2} &= \log_{10}1 - \log_{10}2\\ \log_{10}1 &= 0.00000\\ \log_{10}2 &= \underline{0.30103}\\ \therefore \log_{10}\tfrac{1}{2} &= \bar{1}.69897\end{aligned}$$

$$\begin{aligned}\log_{10}\tfrac{1}{4} &= 2\log_{10}\tfrac{1}{2}\\ \log_{10}\tfrac{1}{2} &= \bar{1}.69897\\ &\underline{\qquad\qquad 2}\\ \therefore \log_{10}\tfrac{1}{4} &= \bar{1}.39794\end{aligned}$$

$$\begin{aligned}\log_{10}\tfrac{7}{9} &= \log_{10}7 - \log_{10}9\\ \log_{10}7 &= 0.87510\\ \log_{10}3^2 &= \underline{0.95424}\\ \therefore \log_{10}\tfrac{7}{9} &= \bar{1}.89086\end{aligned}$$

$$\begin{aligned}\log_{10}\tfrac{21}{20} &= \log_{10}21 - \log_{10}20\\ \log_{10}21 &= 1.32222\\ \log_{10}(10\times 2) &= \underline{1.30103}\\ \therefore \log_{10}\tfrac{21}{20} &= 0.02119\end{aligned}$$

2. With the data of example 1; find

$$\log_2 10,\ \log_2 5,\ \log_3 5,\ \log_7 \tfrac{1}{2},\ \log_5 \tfrac{9}{175}.$$

$$\log_2 10 = \frac{1}{\log_{10}2} = \frac{1}{0.30103} = 3.3226.$$

$$\log_2 5 = \frac{\log_{10}5}{\log_{10}2} = \frac{0.69897}{0.30103} = 2.3224.$$

$$\log_3 5 = \frac{\log_{10}5}{\log_{10}3} = \frac{0.69897}{0.47712} = 1.4650.$$

$$\log_7 \tfrac{1}{2} = \frac{\log_{10}\frac{1}{2}}{\log_{10}7} = \frac{-0.30103}{9.87510} = -0.3562.$$

$$\log_5 \tfrac{9}{343} = 2 \log_5 3 - 3 \log_5 7$$
$$= \frac{2 \log_{10} 3 - 3 \log_{10} 7}{\log_{10} 5}$$
$$= \frac{1.39794 - 2.53530}{0.47712}$$
$$= -2.3838.$$

3. Given $\log_{10} e = 0.43429$; find

$\log_e 2$, $\log_e 3$, $\log_e 5$, $\log_e 7$, $\log_e 8$, $\log_e 9$, $\log_e \tfrac{2}{3}$, $\log_e \tfrac{4}{5}$, $\log_e \tfrac{35}{27}$, $\log_e \tfrac{7}{60}$.

$$\log_e 2 = \frac{\log_{10} 2}{\log_{10} e} = \frac{0.30103}{0.43429} = 0.69315.$$
$$\log_e 3 = \frac{0.47712}{0.43429} = 1.09861.$$
$$\log_e 5 = \frac{0.69897}{0.43429} = 1.60944.$$
$$\log_e 7 = \frac{0.84510}{0.43429} = 1.94591.$$
$$\log_e 8 = 3 \log_e 2 = 2.07944.$$
$$\log_e 9 = 2 \log_e 3 = 2.19722.$$
$$\log_e \tfrac{2}{3} = \log_e 2 - \log_e 3 = -0.40546.$$
$$\log_e \tfrac{4}{5} = 2 \log_e 2 - \log_e 5 = -0.22314.$$
$$\log_e \tfrac{35}{27} = \log_e 5 + \log_e 7 - 3 \log_e 3 = 0.25952.$$
$$\log_e \tfrac{7}{60} = \log_e 7 - (\log_e 5 + \log_e 3 + 2 \log_e 2)$$
$$= -2.14843.$$

4. Find x from the equations $5^x = 12$, $16^x = 10$, $27^x = 4$.

$$5^x = 12. \quad \therefore x \log_{10} 5 = \log_{10} 12.$$
$$x = \frac{\log_{10} 12}{\log_{10} 5} = \frac{1.07918}{0.69897} = 1.5439.$$
$$16^x = 10. \quad \therefore x \log_{10} 16 = \log_{10} 10.$$
$$x = \frac{\log_{10} 10}{\log_{10} 16} = \frac{1.00000}{1.20412} = 0.83048.$$
$$27^x = 4. \quad \therefore x \log_{10} 27 = \log_{10} 4.$$
$$x = \frac{\log_{10} 4}{\log_{10} 27} = \frac{0.60206}{1.43136} = 0.42061.$$

Exercise XXIV. Page 125.

1. Calculate to five places of decimals $\log_e 3$, $\log_e 5$, $\log_e 7$.

In the case of $\log_e 3$ the calculation is carried out below to ten places, for use in the next example.

In the formula

$$\log_e \frac{z+1}{z} = 2\left(\frac{1}{2z+1} + \frac{1}{3(2z+1)^3} + \frac{1}{5(2z+1)^5} + \cdots\right),$$

let $z = \frac{1}{2}.$ Then $\frac{z+1}{z} = 3,\ 2z+1 = 2,$

and $\log_e 3 = 2\left(\frac{1}{2} + \frac{1}{3\times 2^3} + \frac{1}{5\times 2^5} + \cdots\right)$

2	2.00000000000		
4	1.00000000000	÷ 1 =	1.00000000000
4	0.25000000000	÷ 3 =	0.08333333333
4	0.06250000000	÷ 5 =	0.01250000000
4	0.01562500000	÷ 7 =	0.00223214286
4	0.00390625000	÷ 9 =	0.00043402778
4	0.00097656250	÷ 11 =	0.00008877841
4	0.00024414062	÷ 13 =	0.00001878005
4	0.00006103515	÷ 15 =	0.00000406901
4	0.00001525879	÷ 17 =	0.00000089758
4	0.00000381470	÷ 19 =	0.00000020077
4	0.00000095367	÷ 21 =	0.00000004541
4	0.00000023842	÷ 23 =	0.00000001037
4	0.00000005960	÷ 25 =	0.00000000238
4	0.00000001490	÷ 27 =	0.00000000055
4	0.00000000372	÷ 29 =	0.00000000013
4	0.00000000093	÷ 31 =	0.00000000003
	0.00000000023	÷ 33 =	0.00000000001
			1.09861228867

$$\therefore \log_e 3 = 1.0986122886.$$

Again, let $z = 4.$ Then $z+1 = 5,\ 2z+1 = 9,$ and

$$\log_e \frac{5}{4} = 2\left(\frac{1}{9} + \frac{1}{3\times 9^3} + \frac{1}{5\times 9^5} + \cdots\right).$$

9	2.000000		
9	0.222222	÷ 1 =	0.222222
9	0.024691		
9	0.002743	÷ 3 =	0.000914
9	0.000305		
	0.000034	÷ 5 =	0.000007
			0.223143

$$\therefore \log_e \tfrac{5}{4} = 0.22314.$$
$$\log_e 5 = 0.22314 + \log_e 4$$
$$= 0.22314 + 2 \times 0.69315$$
$$= 1.60944.$$

Again, let $z = 6$. Then $z + 1 = 7$, $2z + 1 = 13$, and

$$\log \frac{7}{6} = 2\left(\frac{1}{13} + \frac{1}{3 \times 13^3} + \frac{1}{5 \times 13^5} + \cdots\cdots\right).$$

$$\begin{array}{r|l}
13 & 2.000000 \\
13 & 0.153846 \div 1 = 0.153846 \\
13 & 0.011834 \\
13 & 0.000910 \div 3 = 0.000303 \\
13 & 0.000070 \\
 & 0.000005 \div 5 = \underline{0.000001} \\
 & \qquad\qquad\qquad\;\; 0.154150
\end{array}$$

$$\begin{aligned}
\therefore \log_e \tfrac{7}{6} &= 0.15415. \\
\log 7 &= 0.15415 + \log_e 6 \\
&= 0.15415 + \log_e 2 + \log_e 3 \\
&= 1.94591.
\end{aligned}$$

2. Calculate to ten places of decimals $\log_e 10$.

Let $z = 9$. Then $z + 1 = 10$, $2z + 1 = 19$,

and
$$\log_e \frac{10}{9} = 2\left(\frac{1}{19} + \frac{1}{3 \times 19^3} + \frac{1}{5 \times 19^5} + \cdots\cdots\right).$$

$$\begin{array}{r|l}
19 & 2.00000000000 \\
19 & 0.10526315789 \div 1 = 0.10526315789 \\
19 & 0.00554016621 \\
19 & 0.00029158769 \div 3 = 0.00009719590 \\
19 & 0.00001534672 \\
19 & 0.00000080772 \div 5 = 0.00000016154 \\
19 & 0.00000004251 \\
 & 0.00000000224 \div 7 = \underline{0.00000000032} \\
 & \qquad\qquad\qquad\qquad\;\; 0.10536051565
\end{array}$$

$$\begin{aligned}
\therefore \log_e \tfrac{10}{9} &= 0.1053605156. \\
\log_e 10 &= 0.1053605156 + 2 \log_e 3 \\
&= 2.3025850930.
\end{aligned}$$

3. Calculate to five places of decimals $\log_{10} 2$, $\log_{10} e$, $\log_{10} 11$.

$$\log_{10} 2 = \frac{\log_e 2}{\log_e 10} = \frac{0.693147}{2.30258} = 0.30103.$$

$$\log_{10} e = \frac{1}{\log_e 10} = \frac{1}{2.30258} = 0.43429.$$

To calculate $\log_{10} 11$, let $z = 10$. Then $z + 1 = 11$, $2z + 1 = 21$, and

$$\log_{10} \tfrac{11}{10} = 2 \log_{10} e \left(\frac{1}{21} + \frac{1}{3 \times 21^3} + \frac{1}{5 \times 215} + \cdots\cdots\right).$$

$$
\begin{array}{r|l}
21 & 2.000000 \\
\hline
21 & 0.095238 \div 1 = 0.095238 \\
\hline
21 & 0.004535 \\
\hline
 & 0.000216 \div 3 = \underline{0.000072} \\
 & \qquad\qquad\qquad\quad 0.095310
\end{array}
$$

$$
\begin{aligned}
\therefore \log_{10} \tfrac{11}{10} &= 0.09531 \times \log_{10} e \\
&= 0.09531 \times 0.43429 \\
&= 0.04139. \\
\log_{10} 11 &= 0.04139 + \log_{10} 10 \\
&= 1.04139.
\end{aligned}
$$

Exercise XXV. Page 126.

1. Given $\pi = 3.1415926$, compute $\sin 1'$, $\cos 1'$, and $\tan 1'$ to as many decimal places as possible.

The circular measure of $1'$ is

$$\frac{\pi}{10800} = \frac{3.1415926}{10800} = 0.0002908882 +,$$

the next figure being 0 or 1.

Again, taking the value of $\sin 1'$ as computed in the text-book, $0.00029088 +$, we have

$$
\begin{aligned}
\cos 1' &> \sqrt{1 - (0.00029089)^2} \\
&> \sqrt{1 - 0.000000084617} \\
&> \sqrt{0.999999915383} \\
&> 0.999999957691.
\end{aligned}
$$

Also,

$$
\begin{aligned}
\cos 1' &< \sqrt{1 - (0.00029088)^2} \\
&< \sqrt{1 - 0.000000084612} \\
&< \sqrt{0.999999915388} \\
&< 0.999999957694.
\end{aligned}
$$

Hence, $\cos 1' = 0.99999995769$, correct to eleven decimal places.

But $\sin x > x \cos x$.

$$
\begin{aligned}
\therefore \sin 1' &> 0.0002908882 \times 0.99999995769 \\
&> 0.0002908882\,(1 - 0.00000004231) \\
&> 0.0002908882 - 0.00000000002 \\
&> 0.00029088818.
\end{aligned}
$$

Therefore sin 1′ lies between 0.00029088818 and 0.00029088821. That is, correct to nine places of decimals,

$$\sin 1' = 0.000290888,$$

the next two figures being 18, 19, 20 or 21.

Repeating the process, beginning with the last value of sin 1′, the computation can be carried still further. To eleven places

$$\sin 1' = 0.00029088820 +.$$

From the values of sin 1′ and cos 1′ we have

$$\begin{aligned}\tan 1' &= \frac{\sin 1'}{\cos 1'}\\ &= \frac{0.000290888}{0.99999995769}\\ &= 0.000290888012 +.\end{aligned}$$

2. Compute sin 2′ by the same method, and also by the formula $\sin 2x = 2 \sin x \cos x$. To how many places do the two results agree?

The circular measure of 2′ is

$$\frac{\pi}{5400} = \frac{3.1415926}{5400} = 0.0005817764 +.$$

Hence sin 2′ lies between 0 and 0.0005817765,

and

$$\begin{aligned}\cos 2' &> \sqrt{1 - (0.00058178)^2}\\ &> \sqrt{1 - 0.0000003502}\\ &> \sqrt{0.9999996498}\\ &> 0.9999998249.\end{aligned}$$

But

$$\sin x > x \cos x,$$

hence

$$\begin{aligned}\sin 2' &> 0.0005817765 \times 0.9999998249\\ &> 0.0005817765\,(1 - 0.00000002)\\ &> 0.0005817765 - 0.00000000012\\ &> 0.0005817763.\end{aligned}$$

Hence, sin 2′ = 0.000581776, correct to nine decimal places.

Again,

$$\begin{aligned}\sin 2' &= 2 \sin 1' \cos 1'\\ &= 2 \times 0.000290888 \times 0.99999995769\\ &= 0.000581776\,(1 - 0.00000004231)\\ &= 0.000581776 - 0.000000000025\\ &= 0.000581776 +.\end{aligned}$$

The two methods, therefore, both carry the calculation to the same number of places.

3. Compute sin 1° to four places of decimals.

The circular measure of 1° is

$$\frac{\pi}{180} = \frac{3.1415926}{180} = 0.01745329.$$

$$\text{Hence, } \cos 1° > \sqrt{1-(0.018)^2}$$
$$> \sqrt{0.999676}$$
$$> 0.999.$$

$$\text{And} \quad \sin 1° > 0.01745\,(1-0.001)$$
$$> 0.01745 - 0.000$$
$$> 0.0174.$$

Hence, to four decimal places, $\sin 1° = 0.0174$.

4. From the formula $\cos x = 1 - 2\sin^2\frac{x}{2}$ show that $\cos x > 1 - \frac{x^2}{2}$.

$$\text{Since} \quad \sin x < x,$$
$$\text{we have} \quad \sin\frac{x}{2} < \frac{x}{2}.$$
$$\sin^2\frac{x}{2} < \frac{x^2}{4}.$$
$$1 - 2\sin^2\frac{x}{2} > 1 - \frac{x^2}{2}.$$
$$\therefore \cos x > 1 - \frac{x^2}{2}.$$

5. Show by aid of a table of natural sines that $\sin x$ and x agree to four decimal places for all angles less than 4° 40′.

The circular measure of 4° 40′, or 280′, is

$$\frac{280\,\pi}{10800} = \frac{7\,\pi}{270}$$
$$= \frac{7 \times 3.1415926}{270}$$
$$= 0.08145.$$

The circular measure of 4° 41′ is

$$0.08145 + 0.00029 = 0.08174.$$

From a table,

$$\sin 4° 40' = 0.0814.$$
$$\sin 4° 41' = 0.0816.$$

Hence $\sin x$ and the circular measure of x agree for 4° 40′, and therefore for all smaller angles to four decimal places; but they differ for larger angles.

6. If the values of $\log x$ and $\log \sin x$ agree to five decimal places, find from a table the greatest value x can have.

Let x be expressed in minutes. Then its circular measure is

$$\frac{\pi x}{10.800 \times 60}$$

and its logarithm is

$$\log x'' + (\log \pi - \log 648000)$$
$$= \log x'' + (0.49715 - 5.81158)$$
$$= \log x'' - 5.31443$$
$$= \log x'' + 4.68557 - 10.$$

But from the explanation preceding Table IV., if we remember that log sines are given in the Table increased by 10, we have

$$\log \sin x + 10 = \log x'' + S$$
$$\log \sin x = \log x'' + S - 10$$

Hence, if, for five places, $\log \sin x = \log x$, we have

$$\log x'' + 4.68557 - 10 = \log x'' + S - 10$$
$$\therefore S = 4.68557.$$

But, the greatest angle for which this value of S can be used is given in the Table as 2409″. Hence, the greatest angle for which $\log x$ and $\log \sin x$ agree to five decimal places is

$$2409'' = 40'\,9''.$$

EXERCISE XXVI. PAGE 128.

1. Compute the sine and cosine of 6′ to seven decimal places.

From Example 2, Exercise XXV,

$$\sin 2' = 0.000581776+.$$

Also, from Example 1 of the same Exercise,

$$\cos 1' = 0.999999958.$$

Hence,
$$\begin{aligned}\sin 3' &= 2 \sin 2' \cos 1' - \sin 1' \\ &= 2 \times 0.000581776\ (1 - 0.0000001) - 0.000290888 \\ &= 2 \times 0.00058178 - 0.00029089 \\ &= 0.00087266.\end{aligned}$$

Also,
$$\begin{aligned}\cos 2' &= 2 \cos^2 1' - 1 \\ &= 2\ (0.999999958)^2 - 1 \\ &= 2 \times 0.99999992 - 1 \\ &= 0.999999840+.\end{aligned}$$

$$\begin{aligned}\cos 3' &= 2 \cos 2' \cos 1' - \cos 1' \\ &= 2 \times 0.99999984 \times 0.9999999 - 0.99999996 \\ &= 2 \times 0.99999974 - 0.99999996 \\ &\doteq 0.99999978+.\end{aligned}$$

Finally,
$$\begin{aligned}\sin 6' &= 2 \sin 3' \cos 3' \\ &= 2 \times 0.00087266 \times 0.99999978 \\ &= 0.00174532.\end{aligned}$$

$$\begin{aligned}\cos 6' &= 2 \cos^2 3' - 1 \\ &= 2\ (0.9999998)^2 - 1 \\ &= 2 \times 0.9999996 - 1 \\ &= 0.9999992+.\end{aligned}$$

2. In the formula (1) let $y = 1°$. Assuming $\sin 1° = 0.017454+$, $\cos 1° = 0.999848+$, compute the sines and cosines from degree to degree as far as 4°.

$$\begin{aligned}\sin 2° &= 2 \sin 1° \cos 1° \\ &= 2 \times 0.017454 \times 0.999848 \\ &= 0.034902.\end{aligned}$$

$$\begin{aligned}\sin 3° &= 2 \sin 2° \cos 1° - \sin 1° \\ &= 2 \times 0.034902 \times 0.999848 - 0.017454 \\ &= 0.052340.\end{aligned}$$

$$\begin{aligned}\sin 4° &= 2 \sin 3° \cos 1° - \sin 2° \\ &= 2 \times 0.052340 \times 0.999898 - 0.034902 \\ &= 0.069762.\end{aligned}$$

$$\cos 2^\circ = 2\cos^2 1^\circ - 1$$
$$= 2\,(0.999848)^2 - 1$$
$$= 2 \times 0.999696 - 1$$
$$= 0.999392.$$

$$\cos 3^\circ = (2\cos 2^\circ - 1)\cos 1^\circ$$
$$= 0.998784 \times 0.999848$$
$$= 0.998632.$$

$$\cos 4^\circ = 2\cos 3^\circ \cos 1^\circ - \cos 2^\circ$$
$$= 2 \times 0.998632 \times 0.999848 - 0.999392$$
$$= 1.996960 - 0.999392$$
$$= 0.997568.$$

Exercise XXVII. Page 132.

1. Find the six 6th roots of -1; of $+1$.

$-1 = \cos 180^\circ + i \sin 180^\circ.$
$+1 = \cos 0^\circ + 1 \sin 0^\circ.$

Hence, the six 6th roots of -1 are

$$\cos 30^\circ + i\sin 30^\circ = \frac{\sqrt{3}+i}{2}.$$
$$\cos 90^\circ + i\sin 90^\circ = i.$$
$$\cos 150^\circ + i\sin 150^\circ = \frac{-\sqrt{3}+i}{2}.$$
$$\cos 210^\circ + i\sin 210^\circ = \frac{-\sqrt{3}-i}{2}.$$
$$\cos 270^\circ + i\sin 270^\circ = -i.$$
$$\cos 330^\circ + i\sin 330^\circ = \frac{\sqrt{3}-i}{2}.$$

The six 6th roots of $+1$ are

$$\cos 0^\circ + i\sin 0^\circ = +1.$$
$$\cos 60^\circ + i\sin 60^\circ = \frac{1+\sqrt{-3}}{2}.$$
$$\cos 120^\circ + i\sin 120^\circ = \frac{-1+\sqrt{-3}}{2}.$$
$$\cos 180^\circ + i\sin 180^\circ = -1.$$
$$\cos 240^\circ + i\sin 240^\circ = \frac{-1-\sqrt{-3}}{2}.$$
$$\cos 300^\circ + i\sin 300^\circ = \frac{1-\sqrt{-3}}{2}.$$

2. Find the three cube roots of i.

$i = \cos 90^\circ + i\sin 90^\circ.$

Hence the three cube roots of i are

$$\cos 30^\circ + i\sin 30^\circ = \frac{\sqrt{3}+i}{2}.$$
$$\cos 150^\circ + i\sin 150^\circ = \frac{-\sqrt{3}+i}{2}.$$
$$\cos 270^\circ + i\sin 270^\circ = -i.$$

3. Find the four 4th roots of $-i$.

$-i = \cos 270^\circ + i\sin 270^\circ.$

Hence the four 4th roots of $-i$ are

$$\cos 67\tfrac{1}{2}^\circ + i\sin 67\tfrac{1}{2}^\circ.$$
$$\cos 157\tfrac{1}{2}^\circ + i\sin 157\tfrac{1}{2}^\circ.$$
$$\cos 247\tfrac{1}{2}^\circ + i\sin 247\tfrac{1}{2}^\circ.$$
$$\cos 337\tfrac{1}{2}^\circ + i\sin 337\tfrac{1}{2}^\circ.$$

4. Express $\sin 4\theta$ and $\cos 4\theta$ in terms of $\sin\theta$ and $\cos\theta$.

$$\sin 4\theta = 4\cos^3\theta\sin\theta - \frac{4\times3\times2}{\lfloor 3}\cos\theta\sin^3\theta$$

$$= 4\cos^3\theta\sin\theta - 4\cos\theta\sin^3\theta.$$

$$\cos 4\theta = \cos^4\theta - \frac{4\times3}{\lfloor 2}\cos^2\theta\sin^2\theta + \frac{4\times3\times2\times1}{\lfloor 4}\sin^4\theta$$

$$= \cos^4\theta - 6\cos^2\theta\sin^2\theta + \sin^4\theta.$$

Exercise XXVIII. Page 134.

1. Verify by the series for $\sin x$ and $\cos x$ that $\sin^2 x + \cos^2 x = 1$.

$$\sin x = x - \frac{x^3}{6} + \frac{x^5}{120} - \frac{x^7}{5040} + \cdots\cdots.$$

$$\cos x = 1 - \frac{x^2}{2} + \frac{x^4}{24} - \frac{x^6}{720} + \cdots\cdots.$$

$$\therefore \sin^2 x = x^2 - \frac{x^4}{3} + \frac{2x^6}{45} - \frac{x^8}{315} + \cdots\cdots.$$

$$\cos^2 x = 1 - x^2 + \frac{x^4}{3} - \frac{2x^6}{45} + \frac{x^8}{315} + \cdots\cdots.$$

$$\sin^2 x + \cos^2 x = 1.$$

2. Verify by the series that $\sin(-x) = -\sin x$ and $\cos(-x) = \cos x$.

The series for the sine consists entirely of odd powers of x and, therefore, changes its sign with x; while the series for the cosine consists entirely of even powers, and is unchanged when x changes its sign.

3. Verify from the series that $\sin 2x = 2\sin x\cos x$.

$$\sin 2x = 2x - \frac{(2x)^3}{6} + \frac{(2x)^5}{120} - \frac{(2x)^7}{5040} + \cdots\cdots$$

$$= 2x - \frac{4x^3}{3} + \frac{4x^5}{15} - \frac{8x^7}{315} + \cdots\cdots$$

$$= 2\left(x - \frac{2x^3}{3} + \frac{2x^5}{15} - \frac{4x^7}{315} + \cdots\cdots\right).$$

Also, $$\sin x\cos x = x - \frac{2x^3}{3} + \frac{2x^5}{5} - \frac{4x^7}{315} + \cdots\cdots.$$

$$\therefore \sin 2x = 2\sin x\cos x.$$

4. Verify from the series that $\cos 2x = 1 - 2\sin^2 x$.

$$\sin^2 x = x^2 - \frac{x^4}{3} + \frac{2x^6}{45} - \frac{x^8}{315} + \cdots\cdots.$$

$$1 - 2\sin^2 x = 1 - 2x^2 + \frac{2x^4}{3} - \frac{4x^6}{45} + \frac{2x^8}{315} - \cdots\cdots$$

$$= 1 - \frac{(2x)^2}{2} + \frac{(2x)^4}{24} - \frac{(2x)^6}{720} + \frac{(2x)^8}{40320} - \cdots\cdots$$

$$= \cos 2x.$$

5. Find the series for sec x as far as the term containing the 6th power of x.

$$\sec x = \frac{1}{\cos x} = 1 \div \left(1 - \frac{x^2}{2} + \frac{x^4}{24} - \frac{x^6}{720} + \cdots\cdots\right)$$
$$= 1 + \frac{x^2}{2} + \frac{5x^4}{24} + \frac{61x^6}{720} + \cdots\cdots.$$

6. Find the series for $x \cot x$, noting that $x \cot = \dfrac{x}{\sin x} \cos x$.

$$x \cos x = x - \frac{x^3}{2} + \frac{x^5}{24} - \frac{x^7}{720} + \cdots\cdots.$$
$$\sin x = x - \frac{x^3}{6} + \frac{x^5}{120} - \frac{x^7}{5040} + \cdots\cdots.$$
$$\frac{x \cos x}{\sin x} = 1 - \frac{x^2}{3} - \frac{2x^4}{45} - \frac{11x^6}{1890} - \cdots\cdots.$$

7. Calculate sin 10° and cos 10° to 6 places of decimals.

The circular measure of 10° is $\dfrac{\pi}{18}$.

Hence
$$\sin 10° = \frac{\pi}{18} - \frac{1}{6}\left(\frac{\pi}{18}\right)^3 + \frac{1}{120}\left(\frac{\pi}{18}\right)^5 - \cdots\cdots.$$
$$\cos 10° = 1 - \frac{1}{2}\left(\frac{\pi}{18}\right)^2 + \frac{1}{24}\left(\frac{\pi}{18}\right)^4 - \frac{1}{720}\left(\frac{\pi}{18}\right)^6 + \cdots\cdots.$$

Taking $\pi = 3.141592$, we find

$$\pi = 3.141592 \qquad \frac{\pi}{18} = 0.174533$$
$$\pi^2 = 9.869604 \qquad \frac{\pi^2}{2 \times 18^2} = 0.015231$$
$$\pi^3 = 31.006269 \qquad \frac{\pi^3}{6 \times 18^3} = 0.000886$$
$$\pi^4 = 97.409083 \qquad \frac{\pi^4}{24 \times 18^4} = 0.000039$$
$$\pi^5 = 306.019597 \qquad \frac{\pi^5}{120 \times 18^5} = 0.000001$$

$$\therefore \sin 10° = 0.174533 - 0.000886 + 0.000001$$
$$= 0.173648.$$
$$\cos 10° = 1 - 0.015231 + 0.000039$$
$$= 0.984808.$$

NOTE. The powers of π need be computed only once, and can then be used for finding the functions of all angles.

8. Calculate tan 15° to 6 places of decimals.

The circular measure of 15° is $\frac{\pi}{12}$.

Hence $$\tan 15° = \frac{\pi}{12} + \frac{1}{3}\left(\frac{\pi}{12}\right)^3 + \frac{2}{15}\left(\frac{\pi}{12}\right)^5 + \frac{17}{315}\left(\frac{\pi}{12}\right)^7 + \cdots\cdots.$$

$$\pi = 3.141592 \qquad \frac{\pi}{12} = 0.261799$$

$$\pi^3 = 31.006269 \qquad \frac{\pi^3}{3 \times 12^3} = 0.005981$$

$$\pi^5 = 306.019597 \qquad \frac{2\pi^5}{15 \times 12^5} = 0.000164.$$

$$\therefore \tan 15° = \overline{0.267944}$$

9. From the exponential values of $\cos x$ show that $\cos 3x = 4\cos^3 x - 3\cos x$.

$$\cos x = \tfrac{1}{2}(e^{xi} + e^{-xi}).$$
$$\therefore \cos 3x = \tfrac{1}{2}(e^{3xi} + e^{-3xi})$$
$$= \tfrac{1}{2}(e^{xi} + e^{-xi})(e^{2xi} - 1 + e^{-2xi})$$
$$= \cos x\,[\{\tfrac{1}{2}(e^{xi} + e^{-xi})\}^2 4 - 3]$$
$$= \cos x\,(4\cos^2 x - 3)$$
$$= 4\cos^3 x - 3\cos x.$$

10. From the exponential value of $\sin x$ show that $\sin 3x = 3\sin x - 4\sin^3 x$.

$$\sin 3x = \frac{1}{2i}(e^{3xi} - e^{-3xi})$$
$$= \frac{1}{2i}(e^{xi} - e^{-xi})(e^{2xi} + 1 + e^{-2xi})$$
$$= \sin x\left[\left\{\frac{1}{2i}\left(e^{xi} - e^{-xi}\right)\right\}^2(-4) + 3\right]$$
$$= \sin x\,(-4\sin^2 x + 3)$$
$$= 3\sin x - 4\sin^3 x.$$

SPHERICAL TRIGONOMETRY.

Exercise XXIX. Page 137.

1. The angles of a triangle are 70°, 80°, and 100°; find the sides of the polar triangle.

Given $A = 70°$, $B = 80°$, $C = 100°$; to find a', b', c'.

$$a' = 180° - 70° = 110°.$$
$$b' = 180° - 80° = 100°.$$
$$c' = 180° - 100° = 80°.$$

2. The sides of a triangle are 40°, 90°, and 125°; find the angles of the polar triangle.

Given $a = 40°$, $b = 90°$, $c = 125°$; required A', B', C'.

$$A' = 180° - 40° = 140°.$$
$$B' = 180° - 90° = 90°.$$
$$C' = 180° - 125° = 55°.$$

3. Prove that the polar of a quadrantal triangle is a right triangle.

Let the triangle ABC be a quadrantal triangle.

Then $b = 90°$.

Let $A'B'C'$ be the polar triangle.

$$B' + b = 180°.$$

But $b = 90°$.

$$\therefore B' = 90°.$$

$\therefore$ triangle $A'B'C'$ is a right triangle.

4. Prove that, if a triangle have three right angles, the sides of the triangle are quadrants.

Every vertex is the pole of the opposite side. Every side is, therefore, 90°.

5. Prove that, if a triangle have two right angles, the sides opposite these angles are quadrants, and the third angle is measured by the number of degrees in the opposite side.

Let ABC be the triangle, and

$$B = C = 90°.$$

Then A is the pole of a. Therefore b and c are quadrants, and the angle A is equal to the side BC measured in degrees.

6. How can the sides of a spherical triangle, given in degrees, be found in units of length, when the length of the radius of the sphere is known?

Since the sides of the triangle are arcs of great circles, every degree of arc is $\frac{1}{360}$ of the circumference of a great circle or $\frac{2\pi r}{360}$, where r is the radius of the sphere. Hence, to find the length of a side, multiply its measure in degrees by $\frac{2\pi r}{360}$ or $\frac{\pi r}{180}$.

7. Find the length of the sides of the triangle in Example 2, if the radius of the sphere is 4 feet.

The sides are

$$a = 40° = 40 \times \frac{\pi \times 4}{180}\text{ ft.} = \frac{8\pi}{9}\text{ ft.}$$

$$b = 90° = 90 \times \frac{\pi \times 4}{180}\text{ ft.} = 2\pi\text{ ft.}$$

$$c = 125° = 125 \times \frac{\pi \times 4}{180}\text{ ft.} = \frac{25\pi}{9}\text{ ft.}$$

Exercise XXX. Page 140.

1. Prove, by aid of Formula [38], that the hypotenuse of a right triangle is *less than* or *greater than* 90°, according as the two legs are *alike* or *unlike* in kind.

By Formula [37],

$$\cos c = \cos a \cos b.$$

If a and b are both $< 90°$ or both $> 90°$, $\cos a$ and $\cos b$ have the same sign. Hence $\cos c$ is positive, and $c < 90°$.

But if a and b are unlike in kind, $\cos a$ and $\cos b$ have opposite signs. Hence $\cos c$ is negative, and $c > 90°$.

2. Prove, by aid of Formula [41], that in a right spherical triangle each leg and the opposite angle are always alike in kind.

Formula [41],

$$\cos A = \cos a \times \sin B.$$

$B < 180°$. $\therefore \sin B$ is positive.

Hence sign of $\cos A$ is same as sign of $\cos a$, and both must be either greater or less than 90°; that is, alike in kind.

3. What inferences may be drawn respecting the values of the other parts: (i.) if $c = 90°$; (ii.) if $a = 90°$; (iii.) if $c = 90°$ and $a = 90°$; (iv.) if $a = 90°$ and $b = 90°$?

(i.) If $c = 90°$,

$$0 = \cos a \times \cos b$$

$$\cos a \text{ or } \cos b = 0.$$

$$\therefore a \text{ or } b = 90°.$$

If $a = 90°$,

$$\cos A = 0 \times \sin B = 0.$$

$$\therefore A = 90°.$$

Hence, from Ex. 5, § 42,

$$B = b.$$

(ii.) If $a = 90°$,

$$\cos A = 0 \times \sin B.$$

$$\therefore A = 90°.$$

$$\therefore c = 90°.$$

$$B = b.$$

(iii.) If $c = 90°$
and $a = 90°$,
from (i.) and (ii.)

$$A = 90°.$$

$$B = b.$$

(iv.) If $a = 90°$
and $b = 90°$,
from (ii.) $c = 90°$.

$$B = b = 90°.$$

4. Deduce from [38]-[43] and [18]-[23] the formula

$$\tan^2 \tfrac{1}{2} b = \tan \tfrac{1}{2}(c-a) \tan \tfrac{1}{2}(c+a).$$

From [38], page 138, we have

$$\cos b = \frac{\cos c}{\cos a};$$

$$\text{whence } 1 - \cos b = \frac{\cos a - \cos c}{\cos a}.$$

$$1 + \cos b = \frac{\cos a + \cos c}{\cos a}.$$

$$\therefore \frac{1-\cos b}{1+\cos b} = \frac{\cos a - \cos c}{\cos a + \cos c}.$$

But by [18], page 55,

$$\frac{1-\cos b}{1+\cos b} = \tan^2 \tfrac{1}{2} b.$$

And if in [23] and [22], page 56, we write a and c in place of A and B and divide [23] by [22], we get

$$\frac{\cos a - \cos c}{\cos a + \cos c} = -\tan \tfrac{1}{2}(a+c) \tan \tfrac{1}{2}(a-c) = \tan \tfrac{1}{2}(c+a) \tan \tfrac{1}{2}(c-a).$$

$$\therefore \tan^2 \tfrac{1}{2} b = \tan \tfrac{1}{2}(c+a) \tan \tfrac{1}{2}(c-a).$$

5. Deduce from [38]-[43] and [18]-[23] the formula

$$\tan^2 (45° - \tfrac{1}{2} A) = \tan \tfrac{1}{2}(c-a) \cot \tfrac{1}{2}(c+a).$$

From [39],

$$\sin A = \frac{\sin a}{\sin c};$$

whence, operating as in Example 4, we have

$$\frac{1-\sin A}{1+\sin A} = \frac{\sin c - \sin a}{\sin c + \sin a}.$$

If in [19], page 55, we substitute $90° + A$ for z, and remember that $\cos(90° + A) = -\sin A$, [19] reduces to the form

$$\frac{1-\sin A}{1+\sin A} = \cot^2 (45° + \tfrac{1}{2} A) = \tan^2 (45° - \tfrac{1}{2} A),$$

(since $45° + \tfrac{1}{2} A$ and $45° - \tfrac{1}{2} A$ are complementary angles).

And by dividing [21] by [20], page 56, and writing c for A and a for B, we have

$$\frac{\sin c - \sin a}{\sin c + \sin a} = \tan \tfrac{1}{2}(c-a) \cot \tfrac{1}{2}(c+a).$$

$$\therefore \tan^2 (45° - \tfrac{1}{2} A) = \tan \tfrac{1}{2}(c-a) \cot \tfrac{1}{2}(c+a).$$

6. Deduce from [38]-[43] and [18]-[23] the formula

$$\tan^2 \tfrac{1}{2} B = \sin(c-a) \csc(c+a).$$

From [40], by operating as before,

$$\frac{1-\cos B}{1+\cos B} = \frac{\tan c - \tan a}{\tan c + \tan a}.$$

$$\text{But } \frac{\tan c - \tan a}{\tan c + \tan a} = \frac{\dfrac{\sin c}{\cos c} - \dfrac{\sin a}{\cos a}}{\dfrac{\sin c}{\cos c} + \dfrac{\sin a}{\cos a}} = \frac{\sin c \cos a - \cos c \sin a}{\sin c \cos a + \cos c \sin a} = \frac{\sin(c-a)}{\sin(c+a)}.$$

And by [18], page 55,

$$\frac{1-\cos B}{1+\cos B} = \tan^2 \tfrac{1}{2} B.$$

$$\therefore \tan^2 \tfrac{1}{2} B = \frac{\sin(c-a)}{\sin(c+a)} = \sin(c-a)\csc(c+a).$$

7. Deduce from [38]-[43] and [18]-[23] the formula

$$\tan^2 \tfrac{1}{2} c = -\cos(A+B)\sec(A-B).$$

By [43], $\cos c = \cot A \cot B = \dfrac{\cot A}{\tan B};$

whence, as before,

$$\frac{1-\cos c}{1+\cos c}=\frac{\tan B-\cot A}{\tan B+\cot A},$$

or $$\tan^2 \tfrac{1}{2} c=\frac{-\cos(A+B)}{\cos(A-B)},$$

$$=-\cos(A+B)\sec(A-B).$$

8. Deduce from [38]-[43] and [18]-[23] the formula

$$\tan^2 \tfrac{1}{2} A=\tan[\tfrac{1}{2}(A+B)-45°]\tan[\tfrac{1}{2}(A-B)+45°].$$

From [41], $$\cos a=\cos A \csc B=\frac{\cos A}{\sin B},$$

whence, as before,

$$\frac{1-\cos a}{1+\cos a}=\frac{\sin B-\cos A}{\sin B+\cos A},$$

or, $$\tan^2 \tfrac{1}{2} c=\frac{\sin B+\sin(A-90°)}{\sin B-\sin(A-90°)}.$$

If in [20] and [21] page 56 we substitute B for A and $A-90°$ for B, and divide [20] by [21], we obtain

$$\frac{\sin B+\sin(A-90°)}{\sin B-\sin(A-90°)}=\tan\tfrac{1}{2}(A+B-90°)\cot\tfrac{1}{2}(B-A+90°)$$

$$=\tan[\tfrac{1}{2}(A+B)-45°]\cot[\tfrac{1}{2}(B-A)+45°]$$

$$=\tan[\tfrac{1}{2}(A+B)-45°]\tan[\tfrac{1}{2}(A-B)+45°],$$

since $\tfrac{1}{2}(A-B)+45°=90°-[\tfrac{1}{2}(B-A)+45°]$.

$$\therefore \tan^2\tfrac{1}{2}A=\tan[\tfrac{1}{2}(A+B)-45°]\tan[\tfrac{1}{2}(A-B)+45°].$$

9. Deduce from [38]-[43] and [18]-[23] the formula

$$\tan^2(45°-\tfrac{1}{2}c)=\tan\tfrac{1}{2}(A-a)\cot\tfrac{1}{2}(A+a).$$

By [39], $$\sin c=\frac{\sin a}{\sin A}.$$

$$\therefore \cos(90°-c)=\frac{\sin a}{\sin A},$$

$$\tan^2\tfrac{1}{2}(90°-c)=\frac{\sin A-\sin a}{\sin A+\sin a},$$

or, by [20], [21], $$\tan^2(45°-\tfrac{1}{2}c)=\tan\tfrac{1}{2}(A+a)\cot\tfrac{1}{2}(A-a).$$

10. Deduce from [38]-[43] and [18]-[23] the formula

$$\tan^2(45°-\tfrac{1}{2}b)=\sin(A-a)\csc(A+a).$$

By [42], $$\sin b=\frac{\tan a}{\tan A}.$$

$$\therefore \cos(90°-b)=\frac{\tan a}{\tan A},$$

whence $\tan^2 \frac{1}{2}(90° - b) = \frac{\tan A - \tan a}{\tan A + \tan a}$,

or, by Ex. 6, $\tan^2 (45° - \frac{1}{2}b) = \frac{\sin (A - a)}{\sin (A + a)}$

$= \sin (A - a) \csc (A + a)$.

11. Deduce from [38]–[43] and [19]–[23] the formula

$$\tan^2 (45° - \tfrac{1}{2}B) = \tan \tfrac{1}{2}(A - a) \tan \tfrac{1}{2}(A + a).$$

By [41], $\sin B = \frac{\cos A}{\cos a}$,

whence $\tan^2 (45° - \frac{1}{2}B) = \frac{\cos a - \cos A}{\cos a + \cos A}$,

or, by [22], [23], $\tan^2 (45° - \frac{1}{2}B) = -\tan \frac{1}{2}(a + A) \tan \frac{1}{2}(a - A)$

$= \tan \frac{1}{2}(A + a) \tan \frac{1}{2}(A - a)$.

Exercise XXXI. Page 142.

1. Show that Napier's Rules lead to the equations contained in Formulas [38], [39], [40], and [41].

$\cos c = \cos a \cos b$.

$\sin a = \cos (\text{co. } c) \cos (\text{co. } A)$
$= \sin c \sin A$.

$\sin b = \cos (\text{co. } c) \cos (\text{co. } B)$
$= \sin c \sin B$.

$\sin (\text{co. } B) = \tan a \tan (\text{co. } c)$.
$\cos B = \tan a \cot c$.
$\sin (\text{co. } A) = \tan b \tan (\text{co. } c)$.
$\cos A = \tan b \cot c$.

$\sin (\text{co. } A) = \cos a \cos (\text{co. } B)$.
$\cos A = \cos a \sin B$.
$\sin (\text{co. } B) = \cos b \cos (\text{co. } A)$.
$\cos B = \cos b \sin A$.

$\sin a = \tan (\text{co. } B) \tan b$
$= \cot B \tan b$.

$\sin b = \tan a \tan (\text{co. } A)$
$= \tan a \cot A$.

$\cos c = \tan (\text{co. } A) \tan (\text{co. } B)$
$= \cot A \cot B$.

2. What will Napier's Rules become, if we take as the five parts of the triangle, the hypotenuse, the two oblique angles, and the *complements* of the two legs?

Every part being replaced by its complement, every function is replaced by the complementary function. Napier's Rules therefore become:

(i.) Cosine of middle part equals product of cotangents of adjacent parts.

(ii.) Cosine of middle part equals product of sines of opposite parts.

Exercise XXXII. Page 146.

1. Solve the right triangle, given $a = 36° 27'$, $b = 43° 32' 31''$.

Taking co. c as the middle part, by Rule II.,

$$\cos c = \cos a \cos b.$$
$$\log \cos a = 9.90546$$
$$\log \cos b = 9.86026$$
$$\log \cos c = 9.76572$$
$$c = 54° 20'.$$

Taking a as the middle part, we have, by Rule I.,

$$\sin a = \tan b \cot B,$$

whence $\tan b = \sin a \tan B$,

and $\tan B = \dfrac{\tan b}{\sin a}$

$$\log \tan b = 9.97989$$
$$\text{colog} \sin a = 0.22613$$
$$\log \tan B = 10.20402$$
$$B = 57° 59' 19.3''.$$

Taking b as the middle part, by Rule I.,

$$\sin b = \tan a \cot A.$$
$$\tan a = \sin b \tan A.$$
$$\tan A = \frac{\tan a}{\sin b}.$$
$$\log \tan a = 9.86842$$
$$\text{colog} \sin b = 0.16185$$
$$\log \tan A = 10.03027$$
$$A = 46° 59' 43.2''.$$

2. Solve the right triangle, given $a = 86° 40'$, $b = 32° 40'$.

$$\cos c = \cos a \cos b.$$
$$\tan A = \tan a \csc b.$$
$$\tan B = \tan b \csc a.$$
$$\log \cos a = 8.76451$$
$$\log \cos b = 9.92522$$
$$\log \cos c = 8.68973$$
$$c = 87° 11' 39.8''.$$
$$\log \tan a = 11.23475$$
$$\log \csc b = 0.26781$$
$$\log \tan A = 11.50256$$
$$A = 88° 11' 57.8''.$$
$$\log \tan b = 9.80697$$
$$\log \csc a = 0.00074$$
$$\log \tan B = 9.80771$$
$$B = 32° 42' 38.7''.$$

3. Solve the right triangle, given $a = 50°$, $b = 36° 54' 49''$.

$$\cos c = \cos a \cos b.$$
$$\tan A = \tan a \csc b.$$
$$\tan B = \tan b \csc a.$$
$$\log \cos a = 9.80807$$
$$\log \cos b = 9.90284$$
$$\log \cos c = 9.71091$$
$$c = 59° 4' 25.7''.$$
$$\log \tan a = 10.07619$$
$$\log \csc b = 0.22141$$
$$\log \tan A = 10.29760$$
$$A = 63° 15' 13.13''.$$
$$\log \tan b = 9.87575$$
$$\log \csc a = 0.11575$$
$$\log \tan B = 9.99150$$
$$B = 44° 26' 21.6''.$$

4. Solve the right angle, given $a = 120° 10'$, $b = 150° 59' 44''$.

$$\cos c = \cos a \cos b.$$
$$\tan A = \tan a \csc b.$$
$$\tan B = \tan b \csc a.$$
$$\log \cos a = 9.70115$$
$$\log \cos b = 9.94180$$
$$\log \cos c = 9.64295$$
$$c = 63° 55' 43.3''.$$
$$\log \tan a = 10.23565$$
$$\log \csc b = 0.31437$$
$$\log \tan A = 10.55002$$
$$A = 105° 44' 21.25''.$$

$$\log \tan b = 9.74383$$
$$\log \csc a = 0.06320$$
$$\log \tan B = 9.80703$$
$$B = 147° 19' 47.14''.$$

5. Solve the right triangle, given $c = 55° 9' 32''$, $a = 22° 15' 7''$.

$$\tan b = \sin a \tan B.$$
$$\cos A = \tan b \cot c.$$
$$\cos B = \tan a \cot c.$$

$$\log \sin a = 9.57828$$
$$\log \tan B = 10.52709$$
$$\log \tan b = 10.10537$$
$$b = 51° 53'.$$

$$\log \tan b = 10.10537$$
$$\log \cot c = 9.84266$$
$$\log \cos A = 9.94803$$
$$A = 27° 28' 25.71''.$$

$$\log \tan a = 9.61188$$
$$\log \cot c = 9.84266$$
$$\log \cos B = 9.45454$$
$$B = 73° 27' 11.16''.$$

6. Solve the right triangle, given $c = 23° 49' 51''$, $a = 14° 16' 35''$.

$$\cos b = \cos c \sec a.$$
$$\sin A = \sin a \csc c.$$
$$\cos B = \tan a \cot c.$$

$$\log \cos c = 9.96130$$
$$\log \sec a = 0.01362$$
$$\log \cos b = 9.97492$$
$$b = 19° 17'.$$

$$\log \sin a = 9.39199$$
$$\log \csc c = 0.39358$$
$$\log \sin A = 9.78557$$
$$A = 37° 36' 49.4''.$$

$$\log \tan a = 9.40522$$
$$\log \cot c = 10.33488$$
$$\log \cos B = 9.76050$$
$$B = 54° 49' 23.3''.$$

7. Solve the right triangle, given $c = 44° 33' 17''$, $a = 32° 9' 17''$.

$$\cos b = \cos c \sec a.$$
$$\sin A = \sin a \csc c.$$
$$\cos B = \tan a \cot c.$$

$$\log \cos c = 9.85283$$
$$\log \sec a = 0.07231$$
$$\log \cos b = 9.92514$$
$$b = 32° 41'.$$

$$\log \sin a = 9.72608$$
$$\log \csc c = 0.15391$$
$$\log \sin A = 9.87999$$
$$A = 49° 20' 16.3''.$$

$$\log \tan a = 9.79840$$
$$\log \cot c = 10.00675$$
$$\log \cos B = 9.80515$$
$$B = 50° 19' 16''.$$

8. Solve the right triangle, given $c = 97° 13' 4''$, $a = 132° 14' 12''$.

$$\cos b = \cos c \sec a.$$
$$\sin A = \sin a \csc c.$$
$$\cos B = \tan a \cot c.$$

$$\log \cos c = 9.09914$$
$$\log \sec a = 0.17250$$
$$\log \cos b = 9.27164$$
$$b = 79° 13' 38.18''.$$

$$\log \sin a = 9.86945$$
$$\log \csc c = 0.00345$$
$$\log \sin A = 9.87290$$
$$A = 48° 16' 10''.$$

But A and a must be of the same kind,

$$\therefore A = 131° 43' 50''.$$

$$\log \tan a = 10.04196$$
$$\log \cot c = 9.10259$$
$$\log \cos B = 9.14455$$
$$B = 81° 58' 53.3''.$$

9. Solve the right triangle, given $a = 77° 21' 50''$, $A = 83° 56' 40''$.

$\sin c = \sin a \csc A.$
$\sin b = \tan a \cot A.$
$\sin B = \sec a \cos A.$

$\log \sin a = 9.98935$
$\log \csc A = 0.00243$
$\log \sin c = 9.99178$
$c = 78° 53' 20''.$

Since c is found from its sine, it may have two values which are supplements of each other.

Hence also $c = 101° 6' 40''.$

$\log \tan a = 10.64939$
$\log \cot A = 9.02565$
$\log \sin b = 9.67504$
$b = 28° 14' 31.3'',$
or $= 151° 45' 28.7''.$

$\log \sec a = 0.66004$
$\log \cos A = 9.02323$
$\log \sin B = 9.68327$
$B = 28° 49' 57.4'',$
or $= 151° 10' 2.6''.$

10. Solve the right triangle, given $a = 77° 21' 50''$, $A = 40° 40' 40''$.

$\sin c = \sin a \csc A.$

But $\sin A < \sin a.$

$\therefore \sin c > 1$, which is impossible.

11. Solve the right triangle, given $a = 92° 47' 32''$, $B = 50° 2' 1''$.

$\tan c = \tan a \sec B.$
$\tan b = \sin a \tan B.$
$\cos A = \cos a \sin B.$

$\log \tan a = 11.31183$
$\log \sec B = 0.19223$
$\log \tan c = 31.50406$
$c = 91° 47' 40''.$

$\log \sin a = 9.99948$
$\log \tan B = 10.07671$
$\log \tan b = 10.07619$
$b = 50°.$

$\log \cos a = 8.68765$
$\log \sin B = 9.88447$
$\log \cos A = 8.57212$
$A = 92° 8' 23''.$

12. Solve the right triangle, given $a = 2° 0' 55''$, $B = 12° 40'$.

$\tan c = \tan a \sec B.$
$\tan b = \sin a \tan B.$
$\cos A = \cos a \sin B.$

$\log \tan a = 8.54639$
$\log \sec B = 0.01070$
$\log \tan c = 8.55709$
$c = 2° 3' 55.7''.$

$\log \sin a = 8.54612$
$\log \tan B = 9.35170$
$\log \tan b = 7.89782$
$b = 0° 27' 10.2''.$

$\log \cos a = 9.99973$
$\log \sin B = 9.34100$
$\log \cos A = 9.34073$
$A = 77° 20' 28.4''.$

13. Solve the right triangle, given $a = 20° 20' 20''$, $B = 38° 10' 10''$.

$\tan c = \tan a \sec B.$
$\tan b = \sin a \tan B.$
$\cos A = \cos a \sin B.$

$\log \tan a = 9.56900$
$\log \sec B = 0.10448$
$\log \tan c = 9.67348$
$c = 25° 14' 38.2''.$

$\log \sin a = 9.54104$
$\log \tan B = 9.89545$
$\log \tan b = 9.43649$
$b = 15° 16' 50.4''.$

$\log \cos a = 9.97204$
$\log \sin B = 9.79098$
$\log \cos A = 9.76302$
$A = 54° 35' 16.7''.$

14. Solve the right triangle, given $a = 54° 30'$, $B = 35° 30'$.

$$\tan c = \tan a \sec B.$$
$$\tan b = \sin a \tan B.$$
$$\cos A = \cos a \sin B.$$

$\log \tan a = 10.14673$
$\log \sec B = \underline{0.08931}$
$\log \tan c = 10.23604$
$c = 59° 51' 20.7''.$

$\log \sin a = 9.91069$
$\log \tan B = \underline{9.85327}$
$\log \tan b = 9.76396$
$b = 30° 8' 39.3''.$

$\log \cos a = 9.76395$
$\log \sin B = \underline{9.76395}$
$\log \cos A = 9.52790$
$A = 70° 17' 35''.$

15. Solve the right triangle, given $c = 69° 25' 11''$, $A = 54° 54' 42''$.

$$\sin a = \sin c \sin A.$$
$$\tan b = \tan c \cos A.$$
$$\cot B = \cos c \tan A.$$

$\log \sin c = 9.97136$
$\log \sin A = \underline{9.91289}$
$\log \sin a = 9.88425$
$a = 50°.$

$\log \tan c = 10.42541$
$\log \cos A = \underline{9.75954}$
$\log \tan b = 10.18495$
$b = 56° 50' 49.3''.$

$\log \cos c = 9.54595$
$\log \tan A = \underline{10.15335}$
$\log \cot B = 9.69930$
$B = 63° 25' 4''.$

16. Solve the right triangle, given $c = 112° 48'$, $A = 56° 11' 56''$.

$$\sin a = \sin c \sin A.$$
$$\tan b = \cos A \tan c.$$
$$\cot B = \cos c \tan A.$$

$\log \sin c = 9.96467$
$\log \sin A = \underline{9.91958}$
$\log \sin a = 9.88425$
$a = 50°.$

$\log \cos A = 9.74532$
$\log \tan c = \underline{10.37638}$
$\log \tan b = 10.12170$
$b = 127° 4' 30''.$

$\log \cos c = 9.58829$
$\log \tan A = \underline{10.17427}$
$\log \cot B = 9.76256$
$B = 120° 3' 50''.$

17. Solve the right triangle, given $c = 46° 40' 12''$, $A = 37° 46' 9''$.

$$\sin a = \sin A \sin c.$$
$$\tan b = \tan c \cos A.$$
$$\cot B = \tan A \cos c.$$

$\log \sin A = 9.78709$
$\log \sin c = \underline{9.86178}$
$\log \sin a = 9.64887$
$a = 26° 27' 24''.$

$\log \tan c = 10.02533$
$\log \cos A = \underline{9.89789}$
$\log \tan b = 9.92322$
$b = 39° 57' 41.5''.$

$\log \cos c = 9.83645$
$\log \tan A = \underline{9.88920}$
$\log \cot B = 9.72565$
$B = 62° 0' 4''.$

18. Solve the right triangle, given $c = 118° 40' 1''$, $A = 128° 0' 4''$.

$$\sin a = \sin c \sin A.$$
$$\tan b = \tan c \cos A.$$
$$\cot B = \cos c \tan A.$$

$\log \sin c = 9.94321$
$\log \sin A = \underline{9.89652}$
$\log \sin a = 9.83973$
$a = 136° 15' 32.3''.$

$$\begin{aligned} \log \tan c &= 10.26222 \\ \log \cos A &= \underline{\ \ 9.78935} \\ \log \tan b &= 10.05157 \\ b &= 48° 23' 38.4''. \end{aligned}$$

$$\begin{aligned} \log \cos c &= \ \ 9.68098 \\ \log \tan A &= \underline{10.10717} \\ \log \cot B &= \ \ 9.78815 \\ B &= 58° 27' 4.3''. \end{aligned}$$

19. Solve the right triangle, given $A = 63° 15' 12''$, $B = 135° 33' 39''$.

$$\begin{aligned} \cos a &= \cos A \csc B. \\ \cos b &= \cos B \csc A. \\ \cos c &= \cot A \cot B. \end{aligned}$$

$$\begin{aligned} \log \cos A &= 9.65326 \\ \text{colog} \sin B &= \underline{0.15480} \\ \log \cos a &= 9.80806 \\ a &= 50° 0' 4''. \end{aligned}$$

$$\begin{aligned} \log \cos B &= 9.85369 \\ \text{colog} \sin A &= \underline{0.04915} \\ \log \cos b &= 9.90284 \\ b &= 143° 5' 12''. \end{aligned}$$

$$\begin{aligned} \log \cot A &= \ \ 9.70241 \\ \log \cot B &= \underline{10.00850} \\ \log \cos c &= \ \ 9.71091 \\ c &= 120° 55' 34.3''. \end{aligned}$$

20. Solve the right triangle, given $A = 116° 43' 12''$, $B = 116° 31' 25''$.

$$\begin{aligned} \cos a &= \cos A \csc B. \\ \cos b &= \cos B \csc A. \\ \log \cos c &= \cot A \cot B. \end{aligned}$$

$$\begin{aligned} \log \cos A &= 9.65286 \\ \log \csc B &= \underline{0.04830} \\ \log \cos a &= 9.70116 \\ a &= 120° 10' 3''. \end{aligned}$$

$$\begin{aligned} \log \cos B &= 9.64988 \\ \log \csc A &= \underline{0.04904} \\ \log \cos b &= 9.69892 \\ b &= 119° 59' 46''. \end{aligned}$$

$$\begin{aligned} \log \cot A &= 9.70190 \\ \log \cot B &= \underline{9.69818} \\ \log \cos c &= 9.40008 \\ c &= 75° 26' 58''. \end{aligned}$$

21. Solve the right triangle, given $A = 46° 59' 42''$, $B = 57° 59' 17''$.

$$\begin{aligned} \cos a &= \cos A \csc B. \\ \cos b &= \cos B \csc A. \\ \cos c &= \cot A \cot B. \end{aligned}$$

$$\begin{aligned} \log \cos A &= 9.83382 \\ \log \csc B &= \underline{0.07164} \\ \log \cos a &= 9.90546 \\ a &= 36° 27'. \end{aligned}$$

$$\begin{aligned} \log \cos B &= 9.72435 \\ \log \csc A &= \underline{0.13591} \\ \log \cos b &= 9.86026 \\ b &= 43° 32' 37''. \end{aligned}$$

$$\begin{aligned} \log \cot A &= 9.96973 \\ \log \cot B &= \underline{9.79599} \\ \log \cos c &= 9.76572 \\ c &= 54° 20'. \end{aligned}$$

22. Solve the right triangle, given $A = 90°$, $B = 88° 24' 35''$.

$$\begin{aligned} \cos a &= \cos A \csc B. \\ \cos b &= \cos B \csc A. \\ \cos c &= \cot A \cot B. \end{aligned}$$

$$\begin{aligned} \cos A &= 0. \\ \therefore \cos a &= 0. \\ \therefore a &= 90°. \end{aligned}$$

$$\begin{aligned} \csc A &= 1. \\ \therefore b &= B. \\ \therefore b &= 88° 24' 35''. \end{aligned}$$

$$\begin{aligned} \cot A &= 0. \\ \therefore \cos c &= 0. \\ \therefore c &= 90°. \end{aligned}$$

23. Define a quadrantal triangle, and show how its solution may be reduced to that of the right triangle.

A quadrantal triangle is a triangle having one or more of its sides equal to a quadrant.

Let $A'B'C'$ be a quadrantal triangle with side $A'B' = 90°$, or a quadrant.

Let ABC be its polar triangle.

Then since

$A'B' + C = 180°$, $C = 90°$.

$\therefore ABC$ is a right triangle.

$\therefore$ all parts of the polar triangle may be found by formulas for right triangle. The parts of $A'B'C'$ may then be found by subtracting proper parts of ABC from 180°.

24. Solve the quadrantal triangle whose sides are:

$$a = 174° 12' 49.1''.$$
$$b = 94° 8' 20''.$$
$$c = 90°.$$

Let A' B', C', a', b' c' represent the corresponding angles and sides of the polar triangle.

Then $A' = 5° 47' 10.9''$.
$B' = 85° 51' 40''$.
$C' = 90°$.

$\tan^2 \frac{1}{2} c'$
$= -\cos(B' + A') \sec(B' - A')$.

$\tan^2 \frac{1}{2} b' = \tan[\frac{1}{2}(B' + A') - 45°]$
$\tan[45° + \frac{1}{2}(B' - A')]$.

$\tan^2 \frac{1}{2} a' = \tan[\frac{1}{2}(B' + A') - 45°]$
$\tan[45° - \frac{1}{2}(B' - A')]$.

$B' + A' = 91° 38' 50.9''$.
$B' - A' = 80° 4' 29.1''$.
$\frac{1}{2}(A' + B') - 45° = 49' 25.5''$.
$45° + \frac{1}{2}(B' - A') = 85° 2' 14.6''$.
$\frac{1}{2}(B' + A') - 45° = 0° 49' 25.5''$.
$45° - \frac{1}{2}(B' - A') = 4° 57' 45.4''$.

$\log\cos(B' + A') = 8.45863$
$\log\sec(B' - A') = \underline{0.76356}$
$2)\,9.22219$
$\log\tan\frac{1}{2}c' = 9.61110$
$\frac{1}{2}c' = 22° 12' 56\frac{1}{2}''$.
$c' = 44° 25' 53''$.
$C = 135° 34' 7''$.

$\log\tan 0° 49' 25.5'' = 8.15770$
$\log\tan 85° 2' 14.6'' = \underline{11.06133}$
$2)\,9.21903$
$\log\tan\frac{1}{2}b' = 9.60952$
$\frac{1}{2}b' = 22° 8' 35''$.
$b' = 44° 17' 10''$.
$B = 135° 42' 50''$.

$\log\tan 0° 49' 25.5'' = 8.15770$
$\log\tan 4° 57' 45.4'' = \underline{8.93867}$
$2)\,7.09637$
$\log\tan\frac{1}{2}a' = 8.54819$
$\frac{1}{2}a' = 2° 1' 25''$.
$a' = 4° 2' 50''$.
$A = 175° 57' 10''$.

25. Solve the quadrantal triangle in which

$$c = 90°.$$
$$A = 110° 47' 50''.$$
$$B = 135° 35' 34.5''.$$

Let A', B', C', a', b', c' represent the corresponding angles and sides of the polar triangle.

Then $a' = 69° 12' 10''$.
$b' = 44° 24' 25.5''$.
$C' = 90°$.

$\tan A' = \tan a' \csc b'$.
$\tan B' = \tan b' \csc a'$.
$\cos c' = \cot A' \cot B'$.

$\log\tan a' = 10.42043$
$\log\csc b' = \underline{0.15505}$
$\log\tan A' = 10.57548$
$A' = 75° 6' 58''$.
$a = 104° 53' 2''$.

$$\log \tan b' = 9.99101$$
$$\log \csc a' = 0.02926$$
$$\log \tan B' = 10.02027$$
$$B' = 46° 20' 12''.$$
$$b = 133° 39' 48''.$$
$$\log \cot A' = 9.42452$$
$$\log \cot B' = 9.97973$$
$$\log \cos c' = 9.40425$$
$$c' = 75° 18' 21''.$$
$$C = 104° 41' 39''.$$

26. Given in a spherical triangle A, C, and $c = 90°$; solve the triangle.

$$\sin a = \sin c \sin A.$$
$$= 1 \times 1.$$
$$\therefore a = 90°.$$

Then B is the pole of b, and $B = b$; but B and b are otherwise indeterminate.

27. Given $A = 60°$, $C = 90°$, and $c = 90°$; solve the triangle.

$$\sin a = \sin c \sin A.$$
$$\tan b = \tan c \cos A.$$
$$\cot B = \cos c \tan A.$$
$$\sin a = \sin A.$$
$$a = A = 60°.$$
$$\tan b = \infty \times \tfrac{1}{2}$$
$$= \infty.$$
$$b = 90°.$$
$$\cot B = 0 \times \sqrt{3}$$
$$= 0.$$
$$B = 90°.$$

28. Given in a right spherical triangle, $A = 42° 24' 9''$, $B = 9° 4' 11''$; solve the triangle.

$$\cos c = \cot A \cot B.$$
$$\cot A > 1.$$
$$\cot B > 1.$$
$$\therefore \cos c > 1,$$

which is impossible.

$\therefore$ triangle is impossible.

29. In a right triangle, given $a = 119° 11'$, $B = 126° 54'$; solve the triangle.

$$\tan b = \sin a \tan B.$$
$$\tan c = \tan a \sec B.$$
$$\cos A = \cos a \sin B.$$
$$\log \sin a = 9.94105$$
$$\log \tan B = 10.12446$$
$$\log \tan b = 10.06551$$
$$b = 130° 41' 42''.$$
$$\log \tan a = 10.25298$$
$$\log \cos B = 0.22154$$
$$\log \tan c = 10.47452$$
$$c = 71° 27' 43''.$$
$$\log \cos a = 9.68807$$
$$\log \sin B = 9.90292$$
$$\log \cos A = 9.59099$$
$$A = 112° 57' 2''.$$

30. In a right triangle, given $c = 50°$, $b = 44° 18' 39''$; solve the triangle.

$$\cos a = \cos c \sec b.$$
$$\sin A = \sin a \csc c.$$
$$\tan B = \tan b \csc a.$$
$$\log \cos c = 9.80807$$
$$\text{colog} \cos b = 0.14535$$
$$\log \cos a = 9.95342$$
$$a = 26° 3' 51''.$$
$$\log \sin a = 9.64284$$
$$\log \csc c = 0.11575$$
$$\log \sin A = 9.75859$$
$$A = 35°.$$
$$\log \tan b = 9.98955$$
$$\log \csc a = 0.35716$$
$$\log \tan B = 10.34671$$
$$B = 65° 46' 7''.$$

31. In a right triangle, given $A = 156° 20' 30''$, $a = 65° 15' 45''$; solve the triangle.

It is impossible, because a and A are unlike in kind.

32. If the legs a and b of a right spherical triangle are equal, prove that $\cos a = \cot A = \sqrt{\cos c}$.

$$\cos c = \cos a \cos b.$$

But $\cos a = \cos b.$

$$\therefore \cos c = \cos^2 a.$$

$$\therefore \cos a = \sqrt{\cos c}.$$

$$\sin b = \tan a \cot A.$$

But $\sin a = \sin b.$

$$\therefore \cos a = \cot A.$$

33. In a right triangle prove that $\cos^2 A \times \sin^2 c = \sin(c - a)\sin(c + a)$.

By [39], $\sin A = \dfrac{\sin a}{\sin c}$.

$$\therefore \cos^2 A = 1 - \frac{\sin^2 a}{\sin^2 c}$$

$$= \frac{\sin^2 c - \sin^2 a}{\sin^2 c}.$$

$$\cos^2 A \sin^2 c = \sin^2 c - \sin^2 a.$$

But

$$\sin(c + a) = \sin c \cos a + \cos c \sin a.$$

$$\sin(c - a) = \sin c \cos a - \cos c \sin a.$$

$$\therefore \sin(c + a)\sin(c - a)$$

$$= \sin^2 c \cos^2 a - \cos^2 c \sin^2 a$$

$$= \sin^2 c(1 - \sin^2 a) - (1 - \sin^2 c)\sin^2 a$$

$$= \sin^2 c - \sin^2 a.$$

$$\therefore \cos^2 A \sin^2 c = \sin(c - a)\sin(c + a).$$

34. In a right triangle prove that $\tan a \cos c = \sin b \cot B$.

$$\sin b = \tan a \cot A.$$

$$\cot A = \frac{\sin b}{\tan a}.$$

$$\cos c = \cot A \cot B.$$

$$\cot A = \frac{\cos c}{\cot B}.$$

$$\therefore \frac{\cos c}{\cot B} = \frac{\sin b}{\tan a}.$$

$$\tan a \cos c = \sin b \cot B.$$

35. In a right triangle prove that $\sin^2 A = \cos^2 B + \sin^2 a \sin^2 B$.

$$\cos B = \cos b \sin A.$$

$$\sin^2 A = \frac{\cos^2 B}{\cos^2 b}$$

$$= \cos^2 B \sec^2 b$$

$$= \cos^2 B(1 + \tan^2 b).$$

$$\sin a = \tan b \cot B.$$

$$\tan b = \sin a \tan B.$$

$$\therefore \sin^2 A = \cos^2 B + \sin^2 a \tan^2 B \cos^2 B$$

$$= \cos^2 B + \sin^2 a \sin^2 B.$$

36. In a right triangle prove that $\sin(b + c) = 2\cos^2 \frac{1}{2} a \cos b \sin c$.

$$\sin(b + c)$$

$$= \sin b \cos c + \cos b \sin c$$

$$= \left(\frac{\sin b \cos c}{\cos b \sin c} + 1\right)\cos b \sin c$$

$$= (\tan b \cot c + 1)\cos b \sin c.$$

But $\tan b \cot c = \cos A.$

$$\therefore \tan b \cot c + 1 = \cos A + 1$$

$$= 2\cos^2 \tfrac{1}{2} A.$$

$$\therefore \sin(b + c) = 2\cos^2 \tfrac{1}{2} A \cos b \sin c.$$

37. In a right triangle prove that $\sin(c - b) = 2\sin^2 \frac{1}{2} a \cos b \sin c$.

$$\sin(c - b)$$

$$= \sin c \cos b - \cos c \sin b$$

$$= \sin c \cos b\left(1 - \frac{\cos c \sin b}{\sin c \cos b}\right)$$

$$= \sin c \cos b(1 - \cot c \tan b)$$

But $\cot c \tan b = \cos A.$

$$\therefore 1 - \cot c \tan b = 1 - \cos A$$

$$= 2\sin^2 \tfrac{1}{2} A.$$

$$\therefore \sin(c - b) = 2\sin^2 \tfrac{1}{2} A \cos b \sin c.$$

38. If, in a right triangle, p denote the arc of the great circle passing through the vertex of the right angle and perpendicular to the hypotenuse, m and n the segments of the hypotenuse made by this arc adjacent to the legs a and b, prove that

(i.) $\tan^2 a = \tan c \tan m.$

(ii.) $\sin^2 p = \tan m \tan n.$

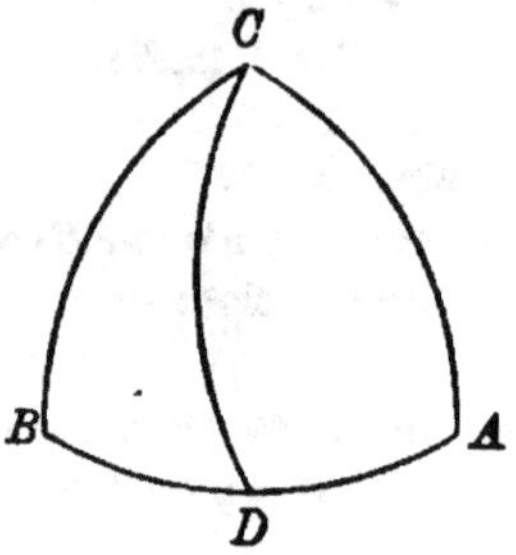

(i.) In triangle BCA

$$\cos B = \tan a \cot c.$$

$$\therefore \tan a = \frac{\cos B}{\cot c}.$$

In right triangle CBD

$$\cos B = \tan BD \cot BC$$
$$= \tan m \cot a.$$

$$\therefore \tan a = \frac{\tan m}{\cos B}.$$

Multiplying the two equations,

$$\tan^2 a = \frac{\tan m}{\cos B} \times \frac{\cos B}{\cot c}$$
$$= \tan m \tan c.$$

(ii.) In triangle CBD

$$\sin p = \tan m \cot BCD;$$

and in triangle CAD

$$\sin p = \tan n \cot DCA.$$

But, since $BCD + DCA = 90°$,

$$\cot BCD \times \cot DCA = 1.$$

$$\therefore \sin^2 p = \tan m \tan n.$$

Exercise XXXIII. Page 149.

1. In an isosceles spherical triangle, given the base b and the side a; find A the angle at the base, B the angle at the vertex, and h the altitude.

Let ABA' be an isosceles triangle, A and A' being the equal angles, a and a' the equal sides.

Let h the arc of a great circle be drawn from B perpendicular to AA', meeting AA' in C.

Then in the right triangle $A'BC$,

$b = \frac{1}{2} b$ in triangle ABA'.

$c = a$ in triangle ABA'.

$B = \frac{1}{2} B$ in triangle ABA'.

$$\cos A = \cot a \tan \tfrac{1}{2} b.$$
$$\sin \tfrac{1}{2} B = \csc a \sin \tfrac{1}{2} b.$$
$$\cos h = \cos a \sec \tfrac{1}{2} b.$$

2. In an equilateral spherical triangle, given the side a; find the angle A.

In the equilateral triangle $AA'A''$ draw arc $AC \perp$ to $A'A''$.

Then in the right triangle $AA'C$,

$$\sin \tfrac{1}{2} a = \sin a \sin \tfrac{1}{2} A.$$

$$\sin \tfrac{1}{2} A = \frac{\sin \tfrac{1}{2} a}{\sin a}$$
$$= \frac{\sin \tfrac{1}{2} a}{2 \sin \tfrac{1}{2} a \cos \tfrac{1}{2} a}$$
$$= \tfrac{1}{2} \sec \tfrac{1}{2} a.$$

3. Given the side a of a regular spherical polygon of n sides; find the angle A of the polygon, the distance R from the centre of the polygon to one of its vertices, and the distance r from the centre to the middle point of one of its sides.

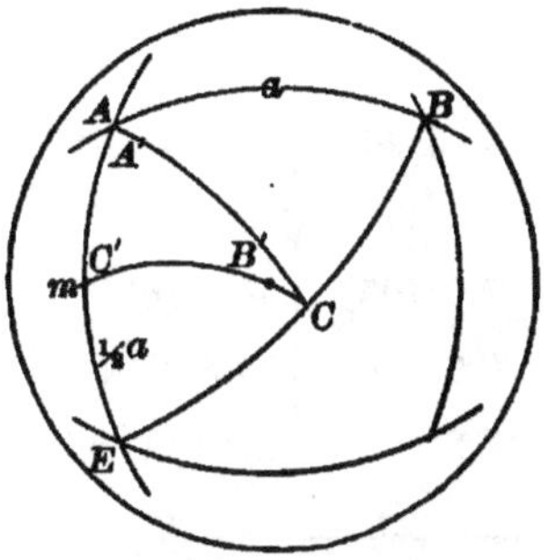

In the regular polygon $ABDE$ draw arcs from the vertices A, B, etc., through the centre C, and from C to M, the middle of one side.

Then $ACB = \frac{360°}{n}$;

$$ACM = \frac{180°}{n},$$
$$CAM = \tfrac{1}{2}A,$$
$$AM = \tfrac{1}{2}a,$$
$$AC = R,$$
$$MC = r.$$
$$\sin \tfrac{1}{2}A = \sec \tfrac{1}{2}a \cos \frac{180°}{n}$$
$$\sin R = \sin \tfrac{1}{2}a \csc \frac{180°}{n}.$$
$$\sin r = \tan \tfrac{1}{2}a \cot \frac{180°}{n}.$$

4. Compute the dihedral angles made by the faces of the five regular polyhedrons.

If a sphere is described about a vertex of the polyhedron as a centre, the adjacent vertices of the polyhedron lie on the sphere and are the vertices of a regular spherical polygon, of which the angles are required.

If a is the length of a side of this polygon, i.e. one of the angles of a face of the polyhedron, and n the number of sides, i.e. the number of faces of the polyhedron which meet at a vertex, we have for the different cases:

POLYHEDRON.	a.	n.
Tetrahedron . . .	60°	3
Cube	90°	3
Octahedron	60°	4
Dodecahedron . . .	108°	3
Icosahedron . . .	60°	5

But, if A is an angle of the spherical polygon, we have from Ex. 3

$$\sin \tfrac{1}{2}A = \sec \tfrac{1}{2}a \cos \frac{180°}{n}.$$

Hence for the different cases:

POLYHEDRON.	SIN ½ A.	LOG SIN ½ A.	A.
Tetrahedron . . .	$\sqrt{\frac{1}{3}}$	9.76144	70° 31′ 46″
Cube	$\sqrt{\frac{1}{2}}$	. . . .	90°
Octahedron	$\sqrt{\frac{2}{3}}$	9.91195	109° 28′ 14″
Dodecahedron . . .	$\frac{1}{2}\sec 54°$	9.92975	116° 33′ 44″
Icosahedron . . .	$\frac{2}{\sqrt{3}}\cos 36°$	9.97043	138° 11′ 36″

5. A spherical square is a regular spherical quadrilateral. Find the angle A of the square, having given the side a.

This is a special case of Ex. 3 for which $n = 4$. Hence

$$\sin \tfrac{1}{2}A = \sec \tfrac{1}{2}a \cos \frac{180}{4}$$

$$= \frac{1}{\sqrt{2}} \sec \tfrac{1}{2}a.$$

Also $\cos \tfrac{1}{2}A = \sqrt{1 - \tfrac{1}{2}\sec^2 \tfrac{1}{2}a}.$

$$\therefore \cot \tfrac{1}{2}A = \frac{\cos \frac{1}{2}A}{\sin \frac{1}{2}A}$$

$$= \sqrt{\frac{1 - \frac{1}{2}\sec^2 \frac{1}{2}a}{\frac{1}{2}\sec^2 \frac{1}{2}a}}$$

$$= \sqrt{2\cos^2 \tfrac{1}{2}a - 1}$$

$$= \sqrt{\cos a}.$$

EXERCISE XXXIV. PAGE 152.

1. What do Formulas [44] become if $A = 90°$? if $B = 90°$? if $C = 90°$? if $a = 90°$? if $A = B = 90°$? if $a = b = 90°$?

If $A = 90°$,

$$\sin a \sin B = \sin b,$$
$$\sin a \sin C = \sin c.$$

If $B = 90°$,

$$\sin a = \sin b \sin A,$$
$$\sin b \sin C = \sin c.$$

If $C = 90°$,

$$\sin a = \sin c \sin A,$$
$$\sin b = \sin c \sin B.$$

If $a = 90°$,

$$\sin B = \sin b \sin A,$$
$$\sin C = \sin c \sin A.$$

If $A = B = 90°$,

$$\sin a = \sin b,$$
$$\sin c = \sin a \sin C = \sin b \sin C.$$

If $a = b = 90°$,

$$\sin B = \sin A,$$
$$\sin C = \sin c \sin A = \sin c \sin B.$$

2. What does the first of [45] become if $A = 0°$? if $A = 90°$? if $A = 180°$?

If $A = 0°$,

$$\cos a = \cos(b - c).$$

If $A = 90°$,

$$\cos a = \cos b \cos c.$$

If $A = 180°$,

$$\cos a = \cos(b + c).$$

3. From Formulas [45] deduce Formulas [46], by means of the relations between polar triangles (§ 48).

Substituting in Formulas [45] for a, b, and c, their equals, $180° - A'$, $180° - B'$, $180° - C'$, we obtain

$$\cos(180° - A') = \cos(180° - B')\cos(180° - C') + \sin(180° - B')\sin(180° - C')\cos(180° - A').$$

$$\therefore -\cos A' = \cos B' \cos C' - \sin B' \sin C' \cos A'.$$

$$\cos A' = -\cos B' \cos C' + \sin B' \sin C' \cos a';$$

and similarly,

$$\cos B' = -\cos A' \cos C' + \sin A' \sin C' \cos b';$$

$$\cos C' = -\cos A' \cos B' + \sin A' \sin B' \cos c'.$$

Exercise XXXV. Page 157.

1. Write formulas for finding, by Napier's Rules, the side a when b, c, and A are given, and for finding the side b when a, c, and B are given.

(i.) In Fig. 46 suppose p drawn from C, dividing c into m and n.

Then the required formulas are obtained by advancing the letters in

$$\tan m = \tan a \cos C.$$
$$\cos c = \cos a \sec m \cos(b - m).$$

They are

$$\tan m = \tan b \cos A.$$
$$\cos a = \cos b \sec m \cos(c - m).$$

(ii.) By drawing p from A, and advancing the letters two steps,

$$\tan m = \tan c \cos B.$$
$$\cos b = \cos c \sec m \cos(a - m).$$

2. Given / find

$a = 88° 12' 20''$, $A = 63° 15' 11''$,
$b = 124° 7' 17''$, $B = 132° 17' 59''$,
$C = 50° 2' 1''$; $c = 59° 4' 18''$.

$\frac{1}{2}(b - a) = 17° 57' 28.5''$.
$\frac{1}{2}(a + b) = 106° 9' 48.5''$.
$\frac{1}{2}C = 25° 1' 0.5''$.

$\log \cos \frac{1}{2}(b - a) = 9.97831$
$\log \sec \frac{1}{2}(a + b) = 0.55536\ (n)$
$\log \cot \frac{1}{2}C = 9.33100$
$\log \tan \frac{1}{2}(A + B) = 0.86467\ (n)$

$\log \sec \frac{1}{2}(A + B) = 0.86868\ (n)$
$\log \cos \frac{1}{2}(a + b) = 9.44464\ (n)$
$\log \sin \frac{1}{2}C = 9.62622$
$\log \cos \frac{1}{2}c = 9.93954$
$\frac{1}{2}c = 29° 32' 9''$.

$\log \sin \frac{1}{2}(b - a) = 9.48900$
$\log \csc \frac{1}{2}(a + b) = 0.01751$
$\log \cot \frac{1}{2}C = 0.33100$
$\log \tan \frac{1}{2}(B - A) = 9.83751$
$\frac{1}{2}(B - A) = 34° 31' 24''$.
$\frac{1}{2}(A + B) = 97° 46' 35''$.
$A = 63° 15' 11''$.
$B = 132° 17' 59''$.
$c = 59° 4' 18''$.

3. Given find

$a = 120° 55' 35''$, $A = 120° 58' 3''$,
$b = 88° 12' 20''$, $B = 63° 15' 9''$,
$C = 47° 42' 1''$; $c = 55° 52' 40''$.

$\frac{1}{2}(a - b) = 16° 21' 37.5''$.
$\frac{1}{2}(a + b) = 104° 33' 57.5''$.
$\frac{1}{2}C = 23° 51' 0.5''$.

$\log \cos \frac{1}{2}(a - b) = 9.98205$
$\log \sec \frac{1}{2}(a + b) = 0.59947\ (n)$
$\log \cot \frac{1}{2}C = 0.35448$
$\log \tan \frac{1}{2}(A + B) = 10.93600$
$\frac{1}{2}(A + B) = 96° 36' 36''\ (n)$

$\log \sin \frac{1}{2}(a - b) = 9.44976$
$\log \csc \frac{1}{2}(a + b) = 0.01419$
$\log \cot \frac{1}{2}C = 0.35448$
$\log \tan \frac{1}{2}(A - B) = 9.81843$
$\frac{1}{2}(A - B) = 33° 21' 27''$.
$\frac{1}{2}(A + B) = 96° 36' 36''$.
$A = 129° 58' 3''$.
$B = 63° 15' 9''$.

$\log \sec \frac{1}{2}(A + B) = 0.93890\ (n)$
$\log \cos \frac{1}{2}(a + b) = 9.40053\ (n)$
$\log \sin \frac{1}{2}C = 9.60675$
$\log \cos \frac{1}{2}c = 9.94618$
$\frac{1}{2}c = 27° 56' 20''$.
$c = 55° 52' 40''$.

4. Given find

$b = 63° 15' 12''$, $B = 88° 12' 24''$,
$c = 47° 42' 1''$, $C = 55° 52' 42''$,
$A = 59° 4' 25''$; $a = 50° 1' 40''$.

$\frac{1}{2}(b + c) = 55° 28' 36.5''$.
$\frac{1}{2}(b - c) = 7° 46' 35.5''$.
$\frac{1}{2}A = 29° 32' 12.5''$.

$\log \cos \frac{1}{2}(b - c) = 9.99599$
$\text{colog} \cos \frac{1}{2}(b + c) = 0.24662$
$\log \cot \frac{1}{2}A = 10.24671$
$\log \tan \frac{1}{2}(B + C) = 10.48932$
$\frac{1}{2}(B + C) = 72° 2' 33''$.

$\log \sin \frac{1}{2}(b - c) = 9.13133$
$\text{colog} \sin \frac{1}{2}(b + c) = 0.08413$
$\log \cot \frac{1}{2}A = 10.24671$
$\log \tan \frac{1}{2}(B - C) = 9.46217$
$\frac{1}{2}(B - C) = 16° 9' 51''$.
$\frac{1}{2}(B + C) = 72° 2' 33''$.
$B = 88° 12' 24''$.
$C = 55° 52' 42''$.

$\log \cos \frac{1}{2}(b + c) = 9.75338$
$\text{colog} \cos \frac{1}{2}(B + C) = 0.51101$
$\log \sin \frac{1}{2}A = 9.69284$
$\log \cos \frac{1}{2}a = 9.95723$
$\frac{1}{2}a = 25° 0' 50''$.
$a = 50° 1' 40''$.

5. Given find

$b = 69° 25' 11''$, $B = 56° 11' 57''$,
$c = 109° 46' 19''$, $C = 123° 21' 12''$,
$A = 54° 54' 42''$; $a = 67° 13'$.

$\frac{1}{2}(c - b) = 20° 10' 34''$.
$\frac{1}{2}(c + b) = 89° 35' 45''$.
$\frac{1}{2}A = 27° 27' 21''$.

$\log \cos \frac{1}{2}(c - b) = 9.97250$
$\text{colog} \cos \frac{1}{2}(c + b) = 2.15157$
$\log \cot \frac{1}{2}A = 10.28434$
$\log \tan \frac{1}{2}(C + B) = 12.40841$
$\frac{1}{2}(C + B) = 89° 46' 34.5''$.

$\log \sin \frac{1}{2}(c - b) = 9.53770$
$\text{colog} \sin \frac{1}{2}(c + b) = 0.0000$
$\log \cot \frac{1}{2}A = 10.28434$
$\log \tan \frac{1}{2}(C - B) = 9.82205$
$\frac{1}{2}(C - B) = 33° 34' 37.8''$.
$C = 123° 21' 12''$.
$B = 56° 11' 57''$.

$\log \cos \frac{1}{2}(c + b) = 7.84843$
$\text{colog} \cos \frac{1}{2}(C + B) = 2.40837$
$\log \sin \frac{1}{2}A = 9.66376$
$\log \cos \frac{1}{2}a = 9.92056$
$\frac{1}{2}a = 33° 36' 30''$.
$a = 67° 13'$.

EXERCISE XXXVI. PAGE 159.

1. What are the formulas for computing A when B, C, and a are given; and for computing B when A, C, and b are given?

(i.) In Fig. 47 suppose p drawn from C. Then advance the letters in

$\cot x = \tan A \csc c,$
$\cos C = \cos A \csc x \sin (B - x).$

The required formulas are

$\cot x = \tan B \csc a,$
$\cos A = \cos B \csc x \sin (C - x).$

(ii.) Suppose p drawn from A, and advance the letters two steps. The required formulas are

$\cot x = \tan C \csc b,$
$\cos B = \cos C \csc x \sin (A - x).$

2. Given find

$A = 26° 58' 46''$, $a = 37° 14' 10''$,
$B = 39° 45' 10''$, $b = 121° 28' 10''$,
$c = 154° 46' 48''$; $C = 161° 22' 11''$.

$\frac{1}{2}(B - A) = 6° 23' 12''.$
$\frac{1}{2}(B + A) = 33° 21' 58''.$
$\frac{1}{2}c = 77° 23' 24''.$

$\log \cos \frac{1}{2}(B - A) = 9.99730$
$\log \sec \frac{1}{2}(B + A) = 0.07823$
$\log \tan \frac{1}{2}c = 10.65032$
$\log \tan \frac{1}{2}(b + a) = 10.72585$

$\log \sin \frac{1}{2}(B + A) = 9.74035$
$\log \sec \frac{1}{2}(b - a) = 0.12972$
$\log \cos \frac{1}{2}c = 9.33908$
$\log \cos \frac{1}{2}C = 9.20915$
$\frac{1}{2}C = 80° 41' 5.4''.$

$\log \sin \frac{1}{2}(B - A) = 9.04625$
$\log \csc \frac{1}{2}(B + A) = 0.25965$
$\log \tan \frac{1}{2}c = 10.65032$
$\log \tan \frac{1}{2}(b - a) = 9.95622$

$\frac{1}{2}(b - a) = 42° 7'.$
$\frac{1}{2}(b + a) = 79° 21' 10''.$
$b = 121° 28' 10''.$
$a = 37° 14' 10''.$
$C = 161° 22' 11''.$

3. Given find

$A = 128° 41' 49''$, $a = 125° 41' 44''$,
$B = 107° 33' 20''$, $b = 82° 47' 34''$,
$c = 124° 12' 31''$; $C = 127° 22'$.

$\frac{1}{2}(A - B) = 10° 34' 14.5''.$
$\frac{1}{2}(A + B) = 118° 7' 34.5''.$
$\frac{1}{2}c = 62° 6' 15.5''.$

$\log \cos \frac{1}{2}(A - B) = 9.99257$
$\text{colog} \cos \frac{1}{2}(A + B) = 0.32660\ (n)$
$\log \tan \frac{1}{2}c = 10.27624$
$\log \tan \frac{1}{2}(a + b) = 10.59541\ (n)$
$\frac{1}{2}(a + b) = 104° 14' 38.5''.$

$\log \sin \frac{1}{2}(A - B) = 9.26351$
$\text{colog} \sin \frac{1}{2}(A + B) = 0.05457$
$\log \tan \frac{1}{2}c = 10.27624$
$\log \tan \frac{1}{2}(a - b) = 9.59432$
$\frac{1}{2}(a - b) = 21° 27' 5''.$
$a = 125° 41' 44''.$
$b = 82° 47' 34''.$

$\log \sin \frac{1}{2}(A + B) = 9.94543$
$\text{colog} \cos \frac{1}{2}(a - b) = 0.03118$
$\log \cos \frac{1}{2}c = 9.67012$
$\log \cos \frac{1}{2}C = 9.64673$
$\frac{1}{2}C = 63° 41'.$
$C = 127° 22'.$

4. Given find

$B = 153° 17' 6''$, $b = 152° 43' 51''$,
$C = 78° 43' 36''$, $c = 88° 12' 21''$,
$a = 86° 15' 15''$; $A = 78° 15' 48''$.

$\frac{1}{2}(B + C) = 116° 0' 21''.$
$\frac{1}{2}(B - C) = 37° 16' 45''.$
$\frac{1}{2}a = 43° 7' 37.5''.$

$\log \cos \tfrac{1}{2}(B - C) = 9.90074$
$\log \sec \tfrac{1}{2}(B + C) = 0.35807 \; (n)$
$\log \tan \tfrac{1}{2}a \qquad = 9.97159$
$\log \tan \tfrac{1}{2}(b + c) \; = \overline{0.23040} \; (n)$
$\qquad \tfrac{1}{2}(b + c) \; = 120°\, 28'\, 6''.$

$\log \sin \tfrac{1}{2}(B - C) = 9.78226$
$\log \csc \tfrac{1}{2}(B + C) = 0.04636$
$\log \tan \tfrac{1}{2}a \qquad = 9.97159$
$\log \tan \tfrac{1}{2}(b - c) \; = \overline{9.80021}$
$\qquad \tfrac{1}{2}(b - c) \; = 32°\, 15'\, 45''.$

$\log \sin \tfrac{1}{2}(B + C) = 9.95364$
$\log \sec \tfrac{1}{2}(b - c) = 0.07283$
$\log \cos \tfrac{1}{2}a \qquad = 9.86322$
$\log \cos \tfrac{1}{2}A \qquad = \overline{9.88969}$
$\qquad\qquad \tfrac{1}{2}A = 39°\, 7'\, 54''.$
$\qquad b = 152°\, 43'\, 51''.$
$\qquad c = \;\; 88°\, 12'\, 21''.$
$\qquad A = \;\; 78°\, 15'\, 48''.$

5. Given   **find**
$A = 125°\, 41'\, 44'', \quad a = 128°\, 31'\, 46'',$
$C = \;\; 82°\, 47'\, 35'', \quad c = 107°\, 33'\, 20'',$
$b = \;\; 52°\, 37'\, 57''; \quad B = \;\; 55°\, 47'\, 40''.$

$\tfrac{1}{2}(A + C) = 104°\, 14'\, 39.5''.$
$\tfrac{1}{2}(A - C) = \;\; 21°\, 27'\, 4.5''.$
$\tfrac{1}{2}b \qquad = \;\; 26°\, 18'\, 58.5''.$

$\log \cos \tfrac{1}{2}(A - C) = 9.96883$
$\log \sec \tfrac{1}{2}(A + C) = 0.60896 \; (n)$
$\log \tan \tfrac{1}{2}b \qquad = \underline{9.69424}$
$\log \tan \tfrac{1}{2}(a + c) \; = 0.27203 \; (n)$
$\qquad \tfrac{1}{2}(a + c) \; = 118°\, 7'\, 33''.$

$\log \sin \tfrac{1}{2}(A + C) = 9.98644$
$\log \sec \tfrac{1}{2}(a - c) = 0.00743$
$\log \cos \tfrac{1}{2}b \qquad = 9.95248$
$\log \cos \tfrac{1}{2}B \qquad = \overline{9.94635}$
$\qquad\qquad \tfrac{1}{2}B = 27°\, 53'\, 50''.$

$\log \sin \tfrac{1}{2}(A - C) = 9.56313$
$\log \csc \tfrac{1}{2}(A + C) = 0.01356$
$\log \tan \tfrac{1}{2}b \qquad = \underline{9.69424}$
$\log \tan \tfrac{1}{2}(a - c) \; = 9.27093$

$\qquad \tfrac{1}{2}(a - c) \; = \;\; 10°\, 34'\, 13''.$
$\qquad\qquad a = 128°\, 41'\, 46''.$
$\qquad\qquad c = 107°\, 33'\, 20''.$
$\qquad\qquad B = \;\; 55°\, 47'\, 40''.$

Exercise XXXVII.   Page 161.

1. Given   **find**
$a = \;\; 73°\, 49'\, 38'', \quad B = 116°\, 42'\, 30'',$
$b = 120°\, 53'\, 35'', \quad c = 120°\, 57'\, 27'',$
$A = \;\; 88°\, 52'\, 42''; \quad C = 116°\, 47'\, 4''.$

$\log \sin A \qquad = 9.99992$
$\log \sin b \qquad = 9.93355$
$\log \csc a \qquad = 0.01753$
$\log \sin B \qquad = 9.95100$
$\qquad B = [180° - (63°\, 17'\, 30'')]$
$\qquad\quad = 116°\, 42'\, 30''.$

(The greater side is opposite the greater angle.)

$\tfrac{1}{2}(B + A) = 102°\, 47'\, 36''.$
$\tfrac{1}{2}(B - A) = \;\; 13°\, 54'\, 54''.$

$\tfrac{1}{2}(b + a) = \;\; 97°\, 21'\, 36.5''.$
$\tfrac{1}{2}(b - a) = \;\; 23°\, 31'\, 58.5''.$

$\log \sin \tfrac{1}{2}(B + A) = \;\; 9.98908$
$\log \csc \tfrac{1}{2}(B - A) = \;\; 0.61892$
$\log \tan \tfrac{1}{2}(b - a) = \;\; 9.63898$
$\log \tan \tfrac{1}{2}c \qquad = 10.24698$
$\qquad\qquad \tfrac{1}{2}c = \;\; 60°\, 28'\, 43.5''.$
$\qquad\qquad\quad c = 120°\, 57'\, 27''.$

$\log \sin \tfrac{1}{2}(b + a) = 9.99641$
$\log \csc \tfrac{1}{2}(b - a) = 0.39873$
$\log \tan \tfrac{1}{2}(B - A) = \underline{9.39401}$
$\log \cot \tfrac{1}{2}C \qquad = \overline{9.78915}$
$\qquad\qquad \tfrac{1}{2}C = \;\; 58°\, 23'\, 32''.$
$\qquad\qquad\quad C = 116°\, 47'\, 4''.$

2. Given $a = 150° 57' 5''$,
$b = 134° 15' 54''$,
$A = 144° 22' 42''$;
find $B_1 = 120° 47' 45''$,
$B_2 = 59° 12' 15''$,
$c_1 = 55° 42' 8''$.
$c_2 = 23° 57' 17.4''$,
$C_1 = 97° 42' 55''$,
$C_2 = 29° 8' 39''$.

$A > 90°$, $(a+b) > 180°$, $a > b$; hence two solutions.

$$\begin{aligned} \log \sin A &= 9.76524 \\ \log \sin b &= 9.85498 \\ \text{colog} \sin a &= 0.31377 \\ \hline \log \sin B &= 9.93399 \end{aligned}$$

$B_1 = 120° 47' 45''$.
$B_2 = 59° 12' 15''$.

$\frac{1}{2}(A + B_1) = 132° 35' 13.5''$.
$\frac{1}{2}(A + B_2) = 101° 47' 28.5''$.
$\frac{1}{2}(A - B_1) = 11° 47' 28.5''$.
$\frac{1}{2}(A - B_2) = 42° 35' 13.5''$.
$\frac{1}{2}(a - b) = 8° 20' 35.5''$.
$\frac{1}{2}(a + b) = 142° 36' 28.5''$.

$$\begin{aligned} \log \sin \tfrac{1}{2}(a+b) &= 9.78338 \\ \log \csc \tfrac{1}{2}(a-b) &= 0.83833 \\ \log \tan \tfrac{1}{2}(A-B_1) &= 9.31963 \\ \hline \log \cot \tfrac{1}{2} C_1 &= 9.94134 \end{aligned}$$

$\frac{1}{2} C_1 = 48° 51' 27.7''$.
$C_1 = 97° 42' 55.4''$.

$$\begin{aligned} \log \sin \tfrac{1}{2}(a+b) &= 9.78338 \\ \log \csc \tfrac{1}{2}(a-b) &= 0.83833 \\ \log \tan \tfrac{1}{2}(A-B_2) &= 9.96338 \\ \hline \log \cot \tfrac{1}{2} C_2 &= 10.58509 \end{aligned}$$

$\frac{1}{2} C_2 = 14° 34' 19.6''$.
$C_2 = 29° 8' 39''$.

$$\begin{aligned} \log \sin \tfrac{1}{2}(A+B_1) &= 9.86703 \\ \text{colog} \sin \tfrac{1}{2}(A-B_1) &= 0.68963 \\ \log \tan \tfrac{1}{2}(a-b) &= 9.16629 \\ \hline \log \tan \tfrac{1}{2} c_1 &= 9.72295 \end{aligned}$$

$\frac{1}{2} c_1 = 27° 51' 4''$.
$c_1 = 55° 42' 8''$.

$$\begin{aligned} \log \sin \tfrac{1}{2}(A+B_2) &= 9.99074 \\ \text{colog} \sin \tfrac{1}{2}(A-B_2) &= 0.16960 \\ \log \tan \tfrac{1}{2}(a-b) &= 9.16629 \\ \hline \log \tan \tfrac{1}{2} c_2 &= 9.32663 \end{aligned}$$

$\frac{1}{2} c_2 = 11° 58' 38.7''$.
$c_2 = 23° 57' 17.4''$.

3. Given | find
$a = 79° 0' 54.5''$, $B = 90°$,
$b = 82° 17' 4''$, $c = 45° 12' 19''$,
$A = 82° 9' 25.8''$; $C = 45° 44'$.

$$\begin{aligned} \log \sin A &= 9.99592 \\ \log \sin b &= 9.99605 \\ \text{colog} \sin a &= 0.00803 \\ \hline \log \sin B &= 0.00000 \end{aligned}$$

$B = 90°$.

$\tan c = \cos A \tan b$.
$\cot C = \tan A \cos b$.

$$\begin{aligned} \log \cos A &= 9.13499 \\ \log \tan b &= 10.86812 \\ \hline \log \tan c &= 10.00311 \end{aligned}$$

$c = 45° 12' 19''$.

$$\begin{aligned} \log \tan A &= 0.86092 \\ \log \cos b &= 9.12793 \\ \hline \log \cot C &= 9.98885 \end{aligned}$$

$C = 45° 44'$.

4. Given $a = 30° 52' 36.6''$, $b = 31° 9' 16''$, $A = 87° 34' 12''$; show that the triangle is impossible.

$\sin B = \sin A \sin b \csc a$.

$$\begin{aligned} \log \sin A &= 9.99961 \\ \log \sin b &= 9.71378 \\ \log \csc a &= 0.28972 \\ \hline \log \sin B &= 9.00311 \end{aligned}$$

$\sin B = 1.009$.

$\therefore$ impossible, since $\sin B > 1$.

EXERCISE XXXVIII. PAGE 162.

1. Given $A = 110^\circ 10'$, $B = 133^\circ 18'$, $a = 147^\circ\ 5' 32''$; find $b = 155^\circ\ 5' 18''$, $c = 33^\circ\ 1' 36''$, $C = 70^\circ 20' 40''$.

$\sin b = \sin a \sin B \csc A.$

$$\begin{aligned}
\log \sin a &= 9.73503\\
\log \sin B &= 9.86200\\
\text{colog} \sin A &= 0.02748\\
\log \sin b &= 9.62451\\
b &= 155^\circ\ 5'\ 18''.
\end{aligned}$$

$$\begin{aligned}
\tfrac{1}{2}(B + A) &= 121^\circ\ 44'.\\
\tfrac{1}{2}(B - A) &= 11^\circ\ 34'.\\
\tfrac{1}{2}(b - a) &= 3^\circ\ 59'\ 53''.\\
\tfrac{1}{2}(b + a) &= 151^\circ\ 5'\ 25''.
\end{aligned}$$

$$\begin{aligned}
\log \sin \tfrac{1}{2}(B + A) &= 9.92968\\
\text{colog} \sin \tfrac{1}{2}(B - A) &= 0.69787\\
\log \tan \tfrac{1}{2}(b - a) &= 8.84443\\
\log \tan \tfrac{1}{2} 3 &= 9.47198\\
\tfrac{1}{2} c &= 16^\circ\ 30'\ 48''.\\
c &= 33^\circ\ 1'\ 36''.
\end{aligned}$$

$$\begin{aligned}
\text{colog} \sin \tfrac{1}{2}(b - a) &= 0.15663\\
\log \sin \tfrac{1}{2}(b + a) &= 9.68433\\
\log \tan \tfrac{1}{2}(B - A) &= 9.31104\\
\log \cot \tfrac{1}{2} C &= 9.15200\\
\tfrac{1}{2} C &= 35^\circ\ 10'\ 20''.\\
C &= 70^\circ\ 20'\ 40''.
\end{aligned}$$

2. Given $A = 113^\circ 39' 21''$, $B = 123^\circ 40' 18''$, $a = 65^\circ 39' 46''$; find $b = 124^\circ\ 7' 20''$, $c = 159^\circ 53'\ 2''$, $C = 159^\circ 43' 35''$.

$$\begin{aligned}
\log \sin a &= 9.95959\\
\log \sin B &= 9.92024\\
\text{colog} \sin A &= 0.03812\\
\log \sin b &= 9.91795\\
b &= 124^\circ\ 7'\ 20''.
\end{aligned}$$

$$\begin{aligned}
\tfrac{1}{2}(B + A) &= 118^\circ\ 39'\ 49.5''.\\
\tfrac{1}{2}(B - A) &= 5^\circ\ 0'\ 28.5''.\\
\tfrac{1}{2}(b - a) &= 29^\circ\ 13'\ 52''.\\
\tfrac{1}{2}(b + a) &= 94^\circ\ 53'\ 33''.
\end{aligned}$$

$$\begin{aligned}
\log \sin \tfrac{1}{2}(B + A) &= 9.94422\\
\text{colog} \sin \tfrac{1}{2}(B - A) &= 1.05901\\
\log \tan \tfrac{1}{2}(b - a) &= 9.74789\\
\log \tan \tfrac{1}{2} c &= 10.75112\\
\tfrac{1}{2} c &= 79^\circ\ 56'\ 51''.\\
c &= 159^\circ\ 53'\ 2''.
\end{aligned}$$

$$\begin{aligned}
\log \sin \tfrac{1}{2}(b + a) &= 9.99842\\
\text{colog} \sin \tfrac{1}{2}(b - a) &= 0.31128\\
\log \tan \tfrac{1}{2}(B - A) &= 8.94264\\
\log \cot \tfrac{1}{2} C &= 9.25234\\
\tfrac{1}{2} C &= 79^\circ\ 51'\ 47.7''.\\
C &= 159^\circ\ 43'\ 35''.
\end{aligned}$$

3. Given $A = 100^\circ\ 2' 11.3''$, $B = 98^\circ 30' 28''$, $a = 95^\circ 20' 38.7''$; find $b = 90^\circ$, $c = 147^\circ 41' 43''$, $C = 148^\circ\ 5' 33''$.

$$\begin{aligned}
\log \sin a &= 9.99811\\
\log \sin B &= 9.99519\\
\log \csc A &= 0.00670\\
\log \sin b &= 0.00000\\
b &= 90^\circ.
\end{aligned}$$

$$\begin{aligned}
\tfrac{1}{2}(A + B) &= 99^\circ\ 16'\ 19.7''.\\
\tfrac{1}{2}(A - B) &= 0^\circ\ 45'\ 51.7''.\\
\tfrac{1}{2}(a - b) &= 2^\circ\ 40'\ 19.4''.\\
\tfrac{1}{2}(a + b) &= 92^\circ\ 40'\ 19.3''.
\end{aligned}$$

$$\begin{aligned}
\log \sin \tfrac{1}{2}(A + B) &= 9.99428\\
\text{colog} \sin \tfrac{1}{2}(A - B) &= 1.87484\\
\log \tan \tfrac{1}{2}(a - b) &= 8.66904\\
\log \tan \tfrac{1}{2} c &= 10.53816\\
\tfrac{1}{2} c &= 73^\circ\ 50'\ 51.7''.\\
c &= 147^\circ\ 41'\ 43''.
\end{aligned}$$

$$\begin{aligned}
\log \sin \tfrac{1}{2}(a + b) &= 9.99953\\
\text{colog} \sin \tfrac{1}{2}(a - b) &= 1.33144\\
\log \tan \tfrac{1}{2}(A - B) &= 8.12520\\
\log \cot \tfrac{1}{2} C &= 9.45617\\
\tfrac{1}{2} C &= 74^\circ\ 2'\ 46.3''.\\
C &= 148^\circ\ 5'\ 33''.
\end{aligned}$$

4. Given $A = 24° 33' 9''$, $B = 38° 0' 12''$, $a = 65° 20' 13''$; show that the triangle is impossible.

$$\begin{aligned} \log \sin a &= 9.95845 \\ \log \sin B &= 9.78937 \\ \log \csc A &= \underline{0.38140} \\ \log \sin b &= 10.12922 \\ \sin b &> 1. \end{aligned}$$

$\therefore$ the triangle is impossible.

EXERCISE XXXIX. PAGE 164.

1. Given $a = 120° 55' 35''$, $b = 59° 4' 25''$, $c = 106° 10' 22''$; find $A = 116° 44' 49''$, $B = 63° 15' 14''$, $C = 91° 7' 21''$.

$$\begin{aligned} a &= 120° 55' 35'' \\ b &= 59° 4' 25'' \\ c &= \underline{106° 10' 22''} \\ 2s &= 286° 10' 22'' \\ s &= 143° 5' 11''. \\ s - a &= 22° 9' 36''. \\ s - b &= 84° 0' 46''. \\ s - c &= 36° 54' 49''. \end{aligned}$$

$$\begin{aligned} \log \sin (s - a) &= 9.57657 \\ \log \sin (s - b) &= 9.99763 \\ \log \sin (s - c) &= 9.77859 \\ \log \csc s &= \underline{0.22141} \\ \log \tan^2 r &= 19.57420 \\ \log \tan r &= 9.78710. \end{aligned}$$

$$\begin{aligned} \log \tan \tfrac{1}{2} A &= 10.21053 \\ \log \tan \tfrac{1}{2} B &= 9.78948 \\ \log \tan \tfrac{1}{2} C &= \underline{10.00851} \\ \tfrac{1}{2} A &= 58° 22' 24.8''. \\ \tfrac{1}{2} B &= 31° 37.2'. \\ \tfrac{1}{2} C &= 45° 33' 40.8''. \end{aligned}$$

$$\begin{aligned} A &= 116° 44' 49''. \\ B &= 63° 15' 14''. \\ C &= 91° 7' 21''. \end{aligned}$$

2. Given $a = 50° 12' 4''$, $b = 116° 44' 48''$, $c = 129° 11' 42''$; find $A = 59° 4' 28''$, $B = 94° 23' 12''$, $C = 120° 4' 52''$.

$$\begin{aligned} a &= 50° 12' 4'' \\ b &= 116° 44' 48'' \\ c &= \underline{129° 11' 42''} \\ 2s &= 296° 8' 34'' \\ s &= 148° 4' 17''. \\ s - a &= 97° 52' 13''. \\ s - b &= 31° 19' 29''. \\ s - c &= 18° 52' 35''. \end{aligned}$$

$$\begin{aligned} \log \sin (s - a) &= 9.99589 \\ \log \sin (s - b) &= 9.71591 \\ \log \sin (s - c) &= 9.50992 \\ \log \csc s &= \underline{0.27666} \\ \log \tan^2 r &= 19.49838 \\ \log \tan r &= 9.74919. \end{aligned}$$

$$\begin{aligned} \log \tan \tfrac{1}{2} A &= 9.75330 \\ \log \tan \tfrac{1}{2} B &= 0.03328 \\ \log \tan \tfrac{1}{2} C &= \underline{0.23927} \\ \tfrac{1}{2} A &= 29° 32' 14''. \\ \tfrac{1}{2} B &= 47° 11' 36''. \\ \tfrac{1}{2} C &= 60° 2' 26''. \end{aligned}$$

$$\begin{aligned} A &= 59° 4' 28''. \\ B &= 94° 23' 12''. \\ C &= 120° 4' 52''. \end{aligned}$$

3. Given $a = 131° 35' 4''$, $b = 108° 30' 14''$, $c = 84° 46' 34''$; find $A = 132° 14' 21''$, $B = 110° 10' 40''$, $C = 99° 42' 24''$.

$$\begin{aligned} a &= 131° 35' 4'' \\ b &= 108° 30' 14'' \\ c &= 84° 46' 34'' \\ 2s &= 324° 51' 52'' \\ s &= 162° 25' 56''. \\ s-a &= 30° 50' 52''. \\ s-b &= 53° 55' 42''. \\ s-c &= 77° 39' 22''. \end{aligned}$$

$$\begin{aligned} \log\sin(s-a) &= 9.70991 \\ \log\sin(s-b) &= 9.90756 \\ \log\sin(s-c) &= 9.98984 \\ \log\csc s &= 0.52023 \\ \log\tan^2 r &= 10.12754 \\ \log\tan r &= 10.06377. \end{aligned}$$

$$\begin{aligned} \log\tan\tfrac{1}{2}A &= 0.35386 \\ \log\tan\tfrac{1}{2}B &= 0.15621 \\ \log\tan\tfrac{1}{2}C &= 0.07393 \\ \tfrac{1}{2}A &= 66° 7' 10.6''. \\ \tfrac{1}{2}B &= 55° 5' 20''. \\ \tfrac{1}{2}C &= 49° 51' 12''. \\ A &= 132° 14' 21''. \\ B &= 110° 10' 40''. \\ C &= 99° 42' 24''. \end{aligned}$$

4. Given $a = 20° 16' 38''$, $b = 56° 19' 40''$, $c = 66° 20' 44''$; find $A = 20° 9' 54''$, $B = 55° 52' 31''$, $C = 114° 20' 17''$.

$$\begin{aligned} a &= 20° 16' 38'' \\ b &= 56° 19' 40'' \\ c &= 66° 20' 44'' \\ 2s &= 142° 57' 2'' \\ s &= 71° 28' 31''. \\ s-a &= 51° 11' 53''. \\ s-b &= 15° 8' 51''. \\ s-c &= 5° 7' 47''. \end{aligned}$$

$$\begin{aligned} \log\sin(s-a) &= 9.89172 \\ \log\sin(s-b) &= 9.41715 \\ \log\sin(s-c) &= 8.95139 \\ \log\csc s &= 0.02311 \\ \log\tan^2 r &= 8.28337 \\ \log\tan r &= 9.14168. \end{aligned}$$

$$\begin{aligned} \log\tan\tfrac{1}{2}A &= 9.24996 \\ \log\tan\tfrac{1}{2}B &= 9.72453 \\ \log\tan\tfrac{1}{2}C &= 10.19029 \\ \tfrac{1}{2}A &= 10° 4' 56.8''. \\ \tfrac{1}{2}B &= 27° 56' 15.5''. \\ \tfrac{1}{2}C &= 57° 10' 8.6''. \\ A &= 20° 9' 54''. \\ B &= 55° 52' 31''. \\ C &= 114° 20' 17''. \end{aligned}$$

Exercise XL. Page 166.

1. Given $A = 130°$, $B = 110°$, $C = 80°$; find $a = 139° 21' 22''$, $b = 126° 57' 52''$, $c = 56° 51' 48''$.

$$\begin{aligned} A &= 130° \\ B &= 110° \\ C &= 80° \\ 2S &= 320° \\ S &= 160°. \end{aligned}$$

$$\begin{aligned} S-A &= 30°. \\ S-B &= 50°. \\ S-C &= 80°. \end{aligned}$$

$$\begin{aligned} \log\cos S &= 9.97299 \\ \log\sec(S-A) &= 0.06247 \\ \log\sec(S-B) &= 0.19193 \\ \log\sec(S-C) &= 0.76033 \\ \log\tan^2 R &= 10.98772 \\ \log\tan R &= 10.49386. \end{aligned}$$

$\log \tan \frac{1}{2} a \quad = 10.43139$
$\log \tan \frac{1}{2} b \quad = 10.30193$
$\log \tan \frac{1}{2} c \quad = 9.73353$

$\frac{1}{2} a = 69° 40' 41''.$
$\frac{1}{2} b = 63° 28' 56''.$
$\frac{1}{2} c = 28° 25' 54''.$
$a = 139° 21' 22''.$
$b = 126° 57' 52''.$
$c = 56° 51' 48''.$

2. Given find

$A = 59° 55' 10'', \quad a = 128° 42' 29'',$
$B = 85° 36' 50'', \quad b = 64° 2' 47'',$
$C = 59° 55' 10''; \quad c = 128° 42' 29'',$

$A = 59° 55' 10''$
$B = 85° 36' 50''$
$C = 59° 55' 10''$
$2S = 205° 27' 10''$
$S = 102° 43' 35''.$
$S - A = 42° 48' 25''.$
$S - B = 17° 6' 45''.$
$S - C = 42° 48' 25''.$

$\log \cos S = 9.34301$
$\log \sec (S - A) = 0.13451$
$\log \sec (S - B) = 0.01967$
$\log \sec (S - C) = 0.13451$
$\log \tan^2 R = 9.63170$
$\log \tan R = 9.81585.$

$\log \tan \frac{1}{2} a = 9.68134$
$\log \tan \frac{1}{2} b = 9.79618$
$\log \tan \frac{1}{2} c = 9.68134$

$\frac{1}{2} a = 25° 38' 45.5''.$
$\frac{1}{2} b = 32° 1' 23.6''.$
$\frac{1}{2} c = 25° 38' 45.5''.$
$a = 51° 17' 31''.$
$b = 64° 2' 47''.$
$c = 51° 17' 31''.$

3. Given find

$A = 102° 14' 12'', \quad a = 104° 25' 9'',$
$B = 54° 32' 24'', \quad b = 53° 49' 25'',$
$C = 89° 5' 46''; \quad c = 97° 44' 24''.$

$A = 102° 14' 12''$
$B = 54° 32' 24''$
$C = 89° 5' 46''$
$2S = 245° 52' 22''$
$S = 122° 56' 11''.$
$S - A = 20° 41' 59''.$
$S - B = 68° 23' 47''.$
$S - C = 33° 50' 25''.$

$\log \cos S = 9.73536$
$\log \sec (S - A) = 0.02898$
$\log \sec (S - B) = 0.43394$
$\log \sec (S - C) = 0.08061$
$\log \tan^2 R = 0.27889$
$\log \tan R = 0.13945.$
$\log \tan \frac{1}{2} a = 0.11047$
$\log \tan \frac{1}{2} b = 9.70551$
$\log \tan \frac{1}{2} c = 0.05885$

$\frac{1}{2} a = 52° 12' 34.6''.$
$\frac{1}{2} b = 26° 54' 42.5''.$
$\frac{1}{2} c = 48° 52' 12''.$
$a = 104° 25' 9''.$
$b = 53° 49' 25''.$
$c = 97° 44' 24''.$

4. Given find

$A = 4° 23' 35'', \quad a = 31° 9' 11'',$
$B = 8° 28' 20'', \quad b = 84° 18' 23'',$
$C = 172° 17' 56''; \quad c = 115° 9' 56''.$

$A = 4° 23' 35''$
$B = 8° 28' 20''$
$C = 172° 17' 56''$
$2S = 185° 9' 51''$
$S = 92° 34' 55.5''.$
$S - A = 88° 11' 20.5''.$
$S - B = 84° 6' 35.5''.$
$S - C = -(79° 43' 0.5'').$

$\log \cos S = 8.65368$
$\log \sec (S - A) = 1.50029$
$\log \sec (S - B) = 0.98876$
$\log \sec (S - C) = 0.74833$
$\log \tan^2 R = 11.89106$
$\log \tan R = 10.94553$

$\log\tan\frac{1}{2}a = 9.44524$
$\log\tan\frac{1}{2}b = 9.95677$
$\log\tan\frac{1}{2}c = 10.19720$
$\frac{1}{2}a = 15° 34' 35.5''.$
$\frac{1}{2}b = 42° 9' 11.5''.$
$\frac{1}{2}c = 57° 34' 58''.$
$a = 31° 9' 11''.$
$b = 84° 18' 23''.$
$c = 115° 9' 56''.$

Exercise XLI. Page 169.

1. Given $A = 84° 20' 19''$, $B = 27° 22' 40''$, $C = 75° 33'$; find $E = 26159''$, $F = 0.12685 R^2$.

$E = A + B + C - 180°$
$A = 84° 20' 19''$
$B = 27° 22' 40''$
$C = 75° 33'$
$187° 15' 59''$
$180°$
$E = 7° 15' 59''$
$= 26159''.$

$\log 26159 = 4.41762$
$\text{colog } 648000 = 4.18842 - 10$
$\log 3.14159 = 0.49715$
$\log F = 9.10319 - 10$
$F = 0.12682 R^2.$

2. Given $a = 69° 15' 6''$, $b = 120° 42' 47''$, $c = 159° 18' 33''$; find $E = 216° 40' 18''$.

$a = 69° 15' 6''$
$b = 120° 42' 47''$
$c = 159° 18' 33''$
$2s = 349° 16' 26''$
$s = 174° 38' 13''.$
$s - a = 105° 23' 7''.$
$s - b = 53° 55' 26''.$
$s - c = 15° 19' 40''.$
$\frac{1}{2}s = 87° 19' 6.5''.$
$\frac{1}{2}(s - a) = 52° 41' 33.5''.$
$\frac{1}{2}(s - b) = 26° 57' 43''.$
$\frac{1}{2}(s - c) = 7° 39' 50''.$

$\log\tan\frac{1}{2}s = 11.32942$
$\log\tan\frac{1}{2}(s - a) = 10.11804$
$\log\tan\frac{1}{2}(s - b) = 9.70645$
$\log\tan\frac{1}{2}(s - c) = 9.12893$
$\log\tan^2\frac{1}{4}E = 10.28284$
$\log\tan\frac{1}{4}E = 10.14142$
$\frac{1}{4}E = 54° 10' 4.6''.$
$E = 216° 40' 18''$

3. Given $a = 33° 1' 45''$, $b = 155° 5' 18''$, $C = 110° 10''$; find $E = 133° 48' 53''$.

$\tan m = \tan a \cos C.$ (§ 50)
$\cos c = \cos a \sec m \cos (b - m).$ (§ 50)

$\log\tan a = 9.81300$
$\log\cos c = 9.53751$
$\log\tan m = 9.35051$
$m = 167° 22''.$
$b - m = -(12° 16' 42'').$

$\log\cos a = 9.92345$
$\log\sec m = 0.01064$
$\log\cos(b - m) = 9.98995$
$\log\cos c = 9.92404$
$c = 147° 5' 30''.$

$a = 33° 1' 45''$
$b = 155° 5' 18''$
$c = 147° 5' 30''$
$2s = 335° 12' 33''$
$s = 167° 36' 16.5''.$
$s - a = 134° 34' 31.5''.$
$s - b = 12° 30' 58.5''.$
$s - c = 20° 30' 46.5''.$

$\frac{1}{2}s = 83° 48' 8.25''.$
$\frac{1}{2}(s-a) = 67° 17' 15.75''.$
$\frac{1}{2}(s-b) = 6° 15' 29.25''.$
$\frac{1}{2}(s-c) = 10° 15' 23.25''.$

$\log \tan \frac{1}{2}s = 0.96419$
$\log \tan \frac{1}{2}(s-a) = 0.37824$
$\log \tan \frac{1}{2}(s-b) = 9.04005$
$\log \tan \frac{1}{2}(s-c) = 9.25755$

$\log \tan^2 \frac{1}{4}E = 9.64003$
$\log \tan \frac{1}{4}E = 9.82002.$

$\frac{1}{4}E = 33° 27' 13\frac{1}{4}''.$
$E = 133° 48' 53''.$

4. Find the area of a triangle on the earth's surface (regarded as spherical), if each side of the triangle is equal to 1°. (Radius of earth = 3958 miles.)

Given a, b, and c each $= 1°$; then

$2s = 3°.$ $\quad \frac{1}{2}s = 45'.$
$s = 1° 30'.$ $\quad \frac{1}{2}(s-a) = 15'.$
$s-a = 30'.$ $\quad \frac{1}{2}(s-b) = 15'.$
$s-b = 30'.$ $\quad \frac{1}{2}(s-c) = 15'.$
$s-c = 30'.$

$\log \tan \frac{1}{2}s = 8.11696$
$\log \tan \frac{1}{2}(s-a) = 7.63982$
$\log \tan \frac{1}{2}(s-b) = 7.63982$
$\log \tan \frac{1}{2}(s-c) = 7.63982$
$\log \tan^2 \frac{1}{4}E = 11.03642$
$\log \tan \frac{1}{4}E = 5.51821.$

$\frac{1}{4}E = 6.802''.$
$E = 27.208.''$

$\log E = 1.43470$
$\log \frac{\pi}{64800} = 4.68557 - 10$
$\log R^2 = 7.19496$
3.31523
$F = 2066.5$ sq. mi.

EXERCISE XLII. PAGE 182.

1. Find the dihedral angle made by adjacent lateral faces of a regular ten-sided pyramid; given the angle $V = 18°$, made at the vertex by two adjacent lateral edges.

About the vertex of the pyramid describe a sphere. It will intersect the lateral surface, forming a regular spherical decagon, of which each side $= 18°$, being measured by the plane angle at the centre.

The required angle is an angle A of this decagon.

By Example 3, Exercise XXXIII

$$\sin \tfrac{1}{2}A = \sec \tfrac{1}{2}a \cos \frac{180°}{10}.$$

$\log \cos 18° = 9.97821$
$\text{colog} \cos 9° = 0.00538$
$\log \sin \frac{1}{2}A = 9.98359$
$\frac{1}{2}A = 74° 21'.$
$A = 148° 42'.$

2. Through the foot of a rod which makes the angle A with a plane, a straight line is drawn in the plane. This line makes the angle B with the projection of the rod upon the plane. What angle does this line make with the rod?

Let CO be a straight line, making the angle A with the plane GH; OI a straight line passing through the foot of CO, making the angle B

with the projection EO of CO upon the plane GH.

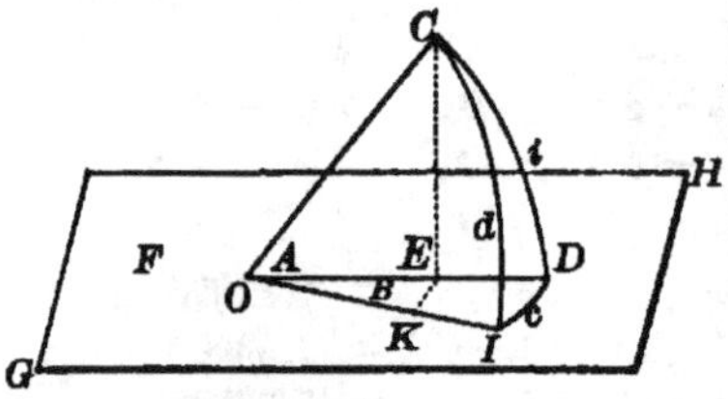

It is required to find the angle $COI = x$.

With a radius equal to unity, from O as a centre, construct the spherical triangle DCI.

Then
$$i = A.$$
$$c = B.$$
$$d = COI = x.$$
$$CDI = \text{rt. angle.}$$

By Formula [38],

$$\cos d = \cos i \cos c.$$
$$\therefore \cos x = \cos A \cos B.$$

3. Find the volume V of an oblique parallelopipedon; given the three unequal edges a, b, c, and the three angles l, m, n, which the edges make with one another.

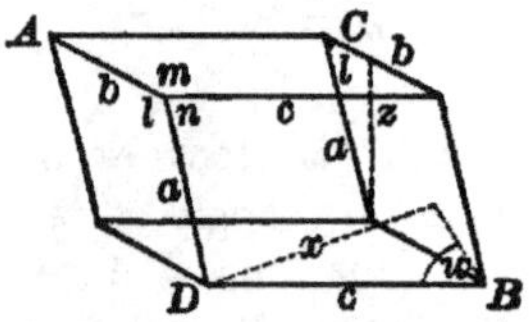

Let AB be a parallelopipedon, and l, m, and n, the angles which the unequal edges a, b, and c make with one another.

Required the volume, V.

Let w = the inclination of the edge c to the plane of a and b.

$$V = \text{area base} \times \text{altitude}.$$
$$\text{Area base} = ab \sin l.$$
$$\text{Altitude} = x = c \sin w.$$
$$\therefore V = abc \sin l \sin w.$$

Suppose a sphere to be described having for its centre the vertex of the trihedral angle whose edges are a, b, and c. The spherical triangle whose vertices are the points where a, b, and c meet the surface has for its sides l, m, n; and w = perpendicular arc from side l to the opposite vertex.

Let L, M, N denote the angles of the triangle opposite l, m, n, respectively.

Then by [39],

$$\sin w = \sin m \sin N,$$
$$= 2 \sin m \sin \tfrac{1}{2} N \cos \tfrac{1}{2} N.$$

Or by [48] if

$$s = \tfrac{1}{2}(l + m + n),$$

$$\sin w = \frac{2}{\sin l}\sqrt{\sin s \sin (s-l) \sin (s-m) \sin (s-n)}$$

$$\therefore V = 2abc\sqrt{\sin s \sin (s-l) \sin (s-m) \sin (s-n)}.$$

4. The continent of Asia has nearly the shape of an equilateral triangle, the vertices being the East Cape, Cape Romania, and the Promontory of Baba. Assuming each side of this triangle to be 4800 geographical miles, and the earth's radius to be 3440 geographical miles, find the area of the triangle:

(i.) regarded as a plane triangle;
(ii.) regarded as a spherical triangle.

(i.) Area $= \tfrac{1}{2}(\text{base} \times \text{altitude})$.

Altitude $= \sqrt{4800^2 - 2400^2}$
$= \sqrt{17280000}.$

$\log \sqrt{17280000} = 3.61877$
$\log 2400 \quad = 3.38021$
$\log \text{area} \quad = 6.99898$
$\text{Area} = 9976500.$

(ii.) $$F = \frac{E}{180°}\pi R^2.$$

$$a, b, \text{ and } c = \frac{4800°}{60} = 80°.$$
$$s = 120°.$$
$$\tfrac{1}{2}(s-a) = 20°.$$
$$\tfrac{1}{2}(s-b) = 20°.$$
$$\tfrac{1}{2}(s-c) = 20°.$$

$\log \tan \frac{1}{2}s \quad = 10.23856$
$\log \tan \frac{1}{2}(s-a) = 9.56107$
$\log \tan \frac{1}{2}(s-b) = 9.56107$
$\log \tan \frac{1}{2}(s-c) = 9.56107$
$\log \tan^2 \frac{1}{4}E \quad = 8.92177$
$\frac{1}{4}E = 16° \ 7' \ 8.1''.$
$E = 64° \ 28' \ 32.5''.$
$= 232112.5''.$

$\log E \quad = 5.36570$
$\log \frac{\pi}{648000} = 4.68557$
$\log R^2 \quad = 7.07312$
$\log F \quad = 7.12439$
$F = 13316560.$

5. A ship sails from a harbor in latitude l, and keeps on the arc of a great circle. Her *course* (or angle between the direction in which she sails and the meridian) at starting is a. Find where she will cross the equator, her course at the equator, and the distance she has sailed.

Let $NESW$ be the earth, WCE the equator, N and S the north and south poles. Let A be the point from which the ship starts, AFD the parallel of latitude the ship

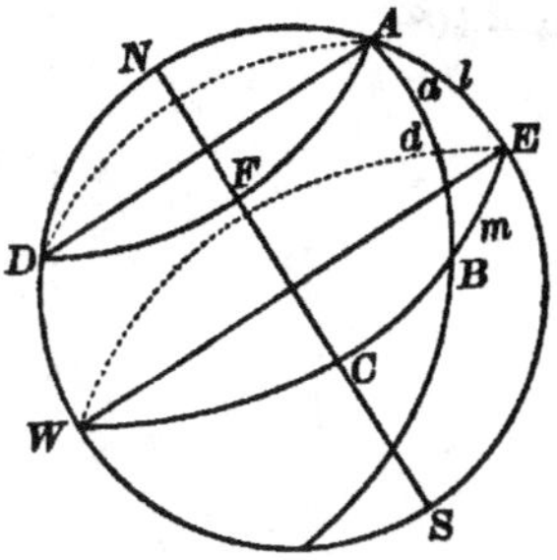

starts from, and AB the great circle of its course.

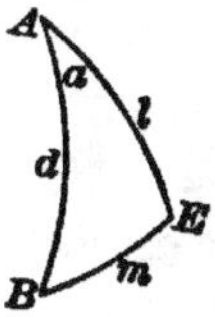

Then

$BAE = a \ =$ course of ship.
$AE = l \ =$ latitude of its starting-place.
$BE = m =$ place of crossing the equator.
$90° - B =$ course at equator.
$AB = d \ =$ distance sailed.

By Napier's Rule,

$$\sin l \ = \tan m \cot a;$$
$$\therefore \tan m = \sin l \tan a.$$
$$\cos B = \cos l \sin a.$$
$$\cot d = \cot l \cos a.$$

6. Two places have the same latitude l, and their distance apart, measured on an arc of a great circle, is d. How much greater is the arc of the parallel of latitude between the places than the arc of the great circle? Compute the results for $l = 45°$, $d = 90°$.

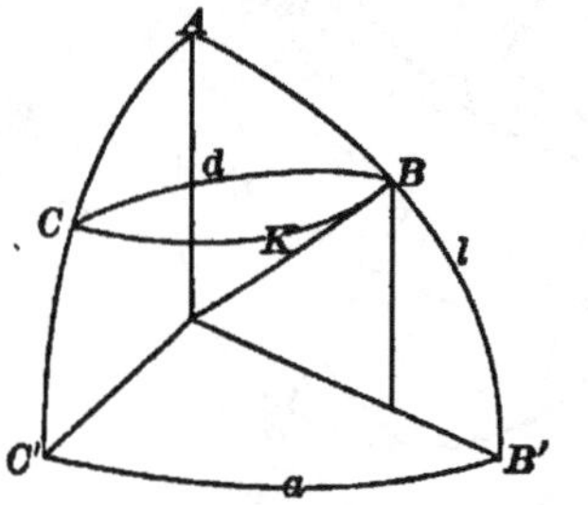

In isosceles spherical triangle ABC

$$\sin \tfrac{1}{2} A = \sin \tfrac{1}{2} d \csc (90° - l)$$
$$= \sin \tfrac{1}{2} d \sec l.$$
$$A = \text{arc } a.$$
$$\text{Arc } k = a \cos l.$$

Let $l = 45°$, $d = 90°$.

$$\log \sin \tfrac{1}{2} d = 9.84949$$
$$\log \sec l = 0.15051$$
$$\log \sin \tfrac{1}{2} A = \overline{10.00000 - 10}$$
$$\tfrac{1}{2} A = 90°.$$
$$A = 180°.$$
$$\text{Arc } a = 180°.$$
$$\text{Arc } K = a \times \cos l$$
$$= \tfrac{1}{2} a \sqrt{2} = 90° \sqrt{2}.$$
$$90° \sqrt{2} - 90° = 90° (\sqrt{2} - 1).$$

7. The shortest distance d between two places and their latitudes l and l' are known. Find the difference between their longitudes.

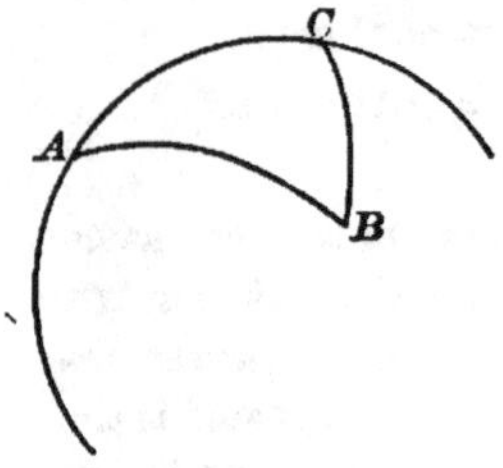

Let C represent the north pole, the position of the one place, B the position of the other, and $AB = d$.

If the latitudes of A and B are l and l',

$$AC = 90° - l,$$
$$BC = 90° - l'.$$

Required C.

By Formula [47],

$$\tan \tfrac{1}{2} C =$$
$$\sqrt{\sec s \sec (s-d) \sin (s-l) \sin (s-l')},$$

where $2s = l + l' + d$.

8. Given the latitude and longitudes of three places on the earth's surface, and also the radius of the earth; show how to find the area of the spherical triangle formed by arcs of great circles passing through the places.

The sides of the triangle are found by § 64; and the area is found from the sides by § 62.

9. The distance between Paris and Berlin (that is, the arc of a great circle between these places) is equal to 472 geographical miles. The latitude of Paris is 48° 50′ 13″; that of Berlin, 52° 30′ 16″. When it is noon at Paris what time is it at Berlin?

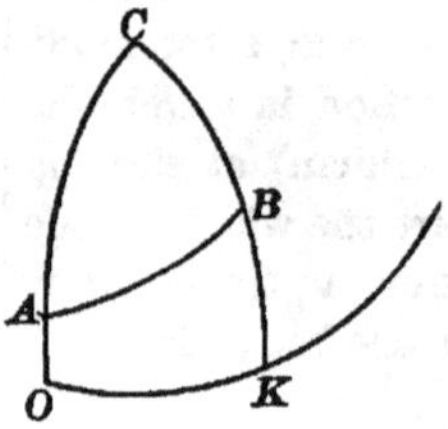

Let AO represent the latitude of Paris, and BK the latitude of Berlin. Then C represents the difference in longitude.

$CA = b = 41° \ 9' \ 47''$
$CB = a = 37° \ 29' \ 44''$
$AB = c = \ \ 7° \ 52' \qquad (472 \div 60)$
$2s = 86° \ 31' \ 31''$
$s = 43° \ 15' \ 45.5''.$
$s - a = \ \ 5° \ 46' \ \ 1.5''.$
$s - b = \ \ 2° \ \ 5' \ 58.5''.$
$s - c = 35° \ 23' \ 45.5''.$

$\tan^2 \frac{1}{2} C = \csc s$
$\sin (s - a) \sin (s - b) \csc (s - c).$

$\log \csc s \qquad = \ \ 0.16409$
$\log \sin (s - a) = \ \ 9.00210$
$\log \sin (s - b) = \ \ 8.56391$
$\log \csc (s - c) = \ \ 0.23716$
$\log \tan^2 \frac{1}{2} C \ = 17.96726$
$\tan \frac{1}{2} C \ = \ \ 8.98363$
$\frac{1}{2} C = \ \ 5° \ 30' \ 2''.$
$C = 11° \ \ 0' \ 4''.$

$1° = 4$ minutes.

$\therefore 11° \ 0' \ 4'' = 44$ min. $\frac{4}{15}$ sec.

Time at Berlin, 12 h. 44 min.

10. The altitude of the pole being 45°, I see a star on the horizon and observe its azimuth to be 45°; find its polar distance.

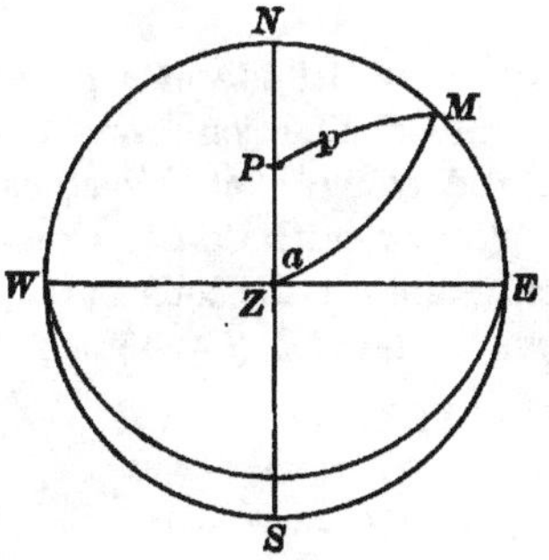

Let Z be the zenith, P the pole, and M the position of the star. In the spherical triangle ZMP

$ZP = 90 - l = 45°,$
$ZM = z = 90°,$
$Z = a = 45°.$

Required p.

By [44],

$\cos p = \sin (90 - l) \cos a$
$= \cos l \cos a$
$= \frac{1}{2}.$
$\therefore p = 60°.$

11. Given the latitude l of the observer, and the declination d of the sun; find the local time (apparent solar time) of sunrise and sunset, and also the azimuth of the sun at these times (refraction being neglected). When and where does the sun rise on the longest day of the year (at which time $d = +23° \ 27'$) in Boston ($l = 42° \ 21'$), and what is the length of the day from sunrise to sunset? Also, find when and where the sun rises in Boston on the shortest day of the year (when $d = -23° \ 27'$), and the length of this day.

(i.) To find the hour angle t when the sun is on the horizon.

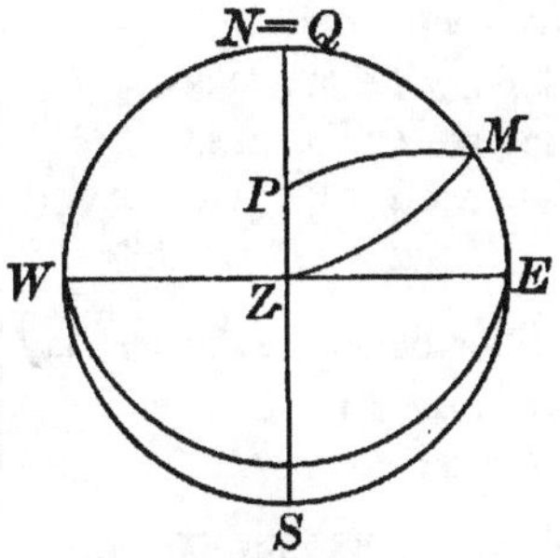

$PM = 90° - d,$
$ZQ = 90°,$
$PQ = l.$

Then in triangle PMQ, by [40],

$$\cos QPM = \tan PQ \cot PM,$$

or, $$\cos t = -\tan l \tan d.$$

Time of sunrise

$$= \left(12 - \frac{t}{15}\right) \text{ o'clock A.M.}$$

Time of sunset

$$= \left(\frac{t}{15}\right) \text{ o'clock P.M.}$$

(ii.) To find azimuth $a = MQ$.

By [38],

$$\cos PM = \cos PQ \cos QM,$$
$$\sin d = \cos l \cos a.$$
$$\therefore \cos a = \sin d \sec l.$$

(iii.) At Boston on the longest day

$$\cos t = -\tan d \tan l.$$

$$\log \tan d = 9.63726$$
$$\log \tan l = 9.95977$$
$$\log \cos t = 9.59703$$
$$t = 113° 17' 34''.$$
$$\frac{t}{15} = 7 \text{ h. } 33 \text{ min. } 10 \text{ sec.}$$
$$12 - \frac{t}{15} = 4 \text{ h. } 26 \text{ min. } 50 \text{ sec.}$$

Length of longest day

$$= 2\frac{t}{15} = 15 \text{ h. } 6 \text{ min. } 20 \text{ sec.}$$

$$\cos a = \sin d \sec l.$$

$$\log \sin d = 9.59983$$
$$\log \sec l = 0.13133$$
$$\log \cos a = 9.73116$$
$$a = 57° 25' 15''.$$

(iv.) At Boston on the shortest day

$$\cos t = \tan d \tan l.$$
$$t = 66° 42' 26''.$$
$$\frac{t}{15} = 4 \text{ h. } 26 \text{ min. } 50 \text{ sec.}$$
$$12 - \frac{t}{15} = 7 \text{ h. } 33 \text{ min. } 10 \text{ sec.}$$

Length of shortest day

$$= 8 \text{ h. } 53 \text{ min. } 40 \text{ sec.}$$
$$\cos a' = -\sin d \sec l.$$
$$\therefore a' = 180 - a$$
$$= 122° 34' 45''.$$

12. When is the solution of the problem in Example 11 impossible, and for what places is the solution impossible?

The solution is impossible if $\cos t > 1$ or < -1 or if $\cos a > 1$, or < -1, i.e., if (for positive declination)

$$\tan l > \cot d,$$

or $$\sin l > \cos d;$$

that is, if $$l > 90° - d.$$

The maximum value of d is 23° 27′; hence the minimum value of l is 66° 33′. The solution is therefore impossible only for places within the Arctic or Antarctic circles. For such places at certain seasons depending on d the sun fails to rise during 24 hours.

13. Given the latitude of a place and the sun's declination; find his altitude and azimuth at 6 o'clock A.M. (neglecting refraction). Compute the results for the longest day of the year at Munich ($l = 48° 9'$).

$$PZM = a.$$
$$PZ = 90° - l.$$
$$PM = 90° - d.$$
$$ZPM = t = 90°.$$
$$ZM = 90° - h.$$
$$l = 48° 9'.$$

Sun's declination on longest day, 23° 27′.

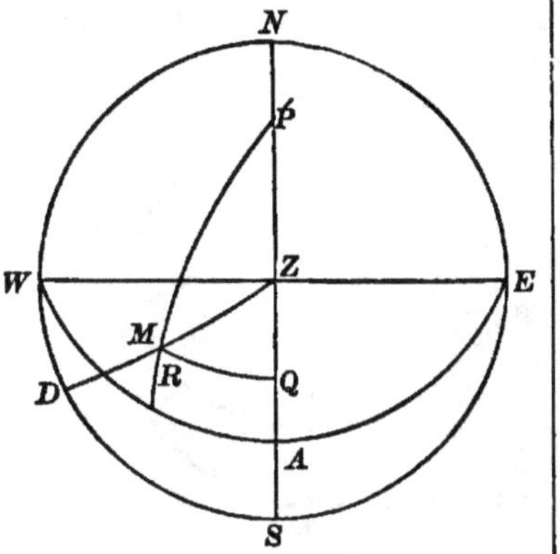

$$\sin h = \sin l \sin d.$$
$$\cot a = \cos l \tan d.$$

$$\log \sin l = 9.87209$$
$$\log \sin d = 9.59983$$
$$\log \sin h = 9.47192$$
$$\text{Altitude} = h = 17° \ 14' \ 35''.$$

$$\log \cos l = 9.82424$$
$$\log \tan d = 9.63726$$
$$\log \cot a = 9.46150$$
$$\text{Azimuth} = a = 73° \ 51' \ 34''.$$

14. How does the altitude of the sun at 6 A.M. *on a given day* change as we go from the equator to the pole? At what time of the year is it a maximum *at a given place?* (Given $\sin h = \sin l \sin d$.)

The farther the place from the equator, the greater the sun's altitude at 6 A.M. in summer. At the equator it is 0°. At the north pole it is equal to the sun's declination. At a given place, the sun's altitude at 6 A.M. is a maximum on the longest day of the year, and then $\sin h = \sin l \sin e$ (where $e = 23° 27'$).

15. Given the latitude of a place north of the equator, and the declination of the sun; find the time of day when the sun bears due east and due west. Compute the results for the longest day at St. Petersburg ($l = 59° 56'$).

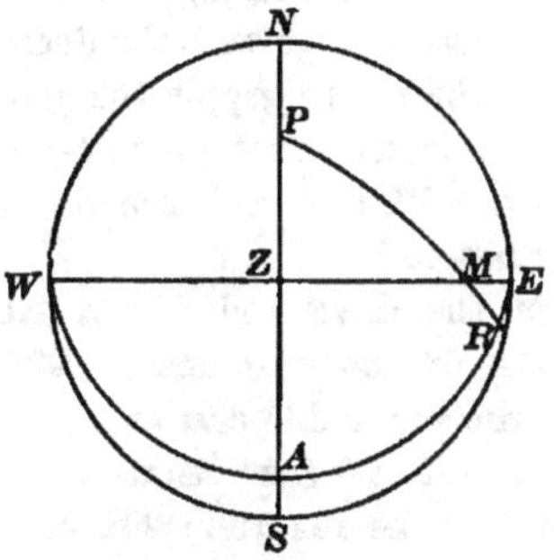

$$PM = 90° - d.$$
$$PZ = 90° - l.$$
$$PZM = 90°.$$
$$\cos t = \tan PZ \cot PM$$
$$= \tan (90° - l) \cot (90° - d).$$
$$\therefore \cos t = \cot l \tan d.$$

The times of bearing due east and west are

$$12 - \frac{t}{15} \text{ A.M. and } \frac{t}{15} \text{ P.M.},$$

respectively.

At St. Petersburg on the longest day $l = 59° 56'$, $d = 23° 27'$.

$$\log \cot l = 9.76361$$
$$\log \tan d = 9.63726$$
$$\log \cos t = 9.39987$$
$$t = 75° \ 27' \ 24''.$$

$$\therefore 12 - \frac{t}{15} = 6 \text{ hrs. } 58 \text{ min. A.M.},$$

$$\text{and} \quad \frac{t}{15} = 5 \text{ hrs. } 2 \text{ min. P.M.}$$

16. Apply the general result in Example 15 ($\cos t = \cot l \tan d$) to the case when the days and nights are equal in length (that is, when $d = 0°$). Why can the sun in summer never be due east before 6 A.M., or due west after 6 P.M.? How does the time of bearing due east and due west change with the declination of the sun? Apply the general result to the cases where $l < d$ and $l = d$. What does it become at the north pole?

When the days and nights are equal, $d = 0°$, $\cos t = 0$, and $t = 90°$; that is, the sun is due east at 6 A.M. and due west at 6 P.M. Since l and d must both be less than 90°, $\cos t$ cannot be negative; therefore t cannot be greater than 90°. As d increases, t decreases; that is, the times of bearing due east and west both approach noon.

If $d = l$, $\cos t = 1$, $t = 0°$, and the times both coincide with noon. If $d > l$, the case is impossible.

The explanation of these results is that, if $d = l$, the sun is in the zenith at noon, and north of the prime vertical at every other time. And if $d > l$, the sun is north of the prime vertical the entire day.

If $l > d$, the diurnal circle of the sun and the prime vertical of the place meet in two points, which separate farther and farther as l increases, the distance between them approaching 180° − 23° as l approaches 90°. At the pole the prime vertical is indeterminate; but near the pole $t = 90°$, and the sun always east at 6 A.M.

17. Given the sun's declination and his altitude when he bears due east; find the latitude of the observer.

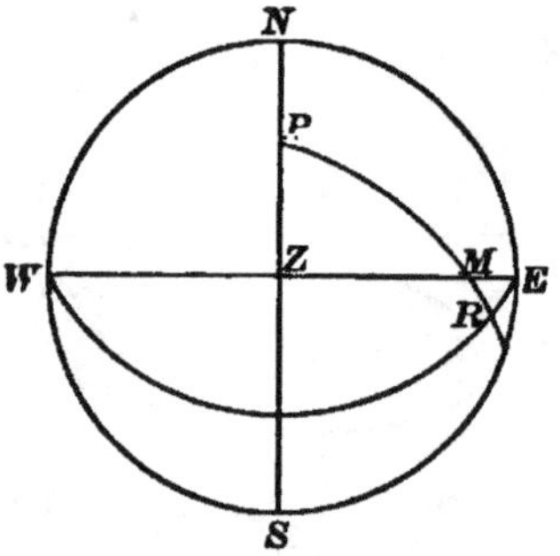

$$ZM = 90° - h.$$
$$PM = 90° - d.$$
$$PZ = 90° - l.$$

Since the sun M bears due east, MZP is a right angle.

$$\cos PM = \cos PZ \cos MZ.$$
$$\therefore \sin d = \sin l \sin h.$$
$$\sin l = \sin d \csc h.$$

18. At a point O in a horizontal plane MN a staff OA is fixed, so that its angle of inclination AOB with the plane is equal to the latitude of the place, 51° 30′ N., and the direction OB is due north. What angle will OB make with the shadow of OA on the plane at 1 P.M.?

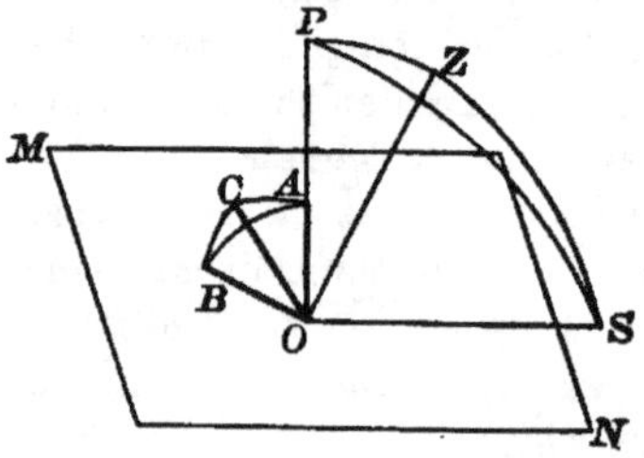

Given direction of OB due north,

$AOB = 51° 30' = l$, and plane MN horizontal; to find BOC.

$SPZ =$ hour angle of sun at 1 P.M. $= 15°$.

$SPZ = CAB$, being vertical angles.

$\therefore CAB = 15°$.

$ABC = 90°$, since OB is the projection of OA on plane MN.

Arc $AB = 51° 30'$, being the measure of plane angle AOB.

Then in right spherical triangle ABC, by [42],

$$\tan BC = \tan BAC \sin AB.$$

$$\begin{aligned} \log \tan 15° &= 9.42805 \\ \log \sin 51° 30' &= \underline{9.89354} \\ \log \tan BC &= 9.32159 \\ \text{Arc } BC &= 11° 50' 35''. \end{aligned}$$

$$\therefore BOC = 11° 50' 35''.$$

19. What is the direction of a wall in latitude 52° 30′ N. which casts no shadow at 6 A.M. on the longest day of the year?

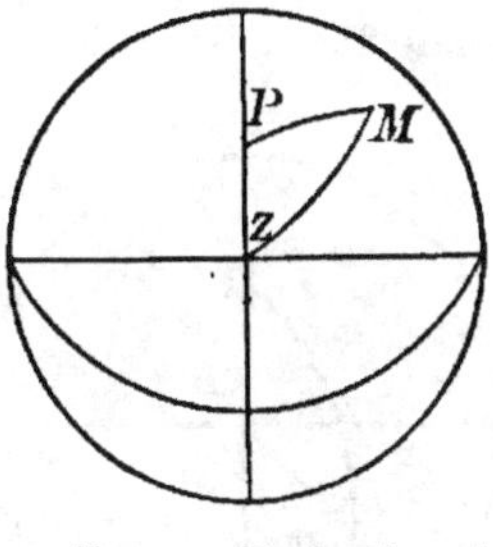

The wall must lie in the plane of ZM in order that it may cast no shadow.

$$PZ = 90 - l,$$
$$PM = 90 - l,$$
$$P = 90°;$$

required $MZP = a$.

By [42],

$$\cos l = \cot e \cot a.$$
$$\therefore \cot a = \cos l \tan e.$$

$$\begin{aligned} \log \cos l &= 9.78445 \\ \log \tan e &= \underline{9.63726} \\ \log \cot a &= 9.42171 \\ a &= 75° 12' 38''. \end{aligned}$$

20. At a certain place the sun is observed to rise exactly in the north-east point on the longest day of the year; find the latitude of the place.

When the sun rises in the north-east on the longest day of the year, $a = 45°$, $d = 23° 27'$.

$$\cos a = \sin d \sec l.$$

$$\begin{aligned} \log \cos 45° &= 0.84949 \\ \log \csc 23° 27' &= \underline{0.40017} \\ \log \sec l &= 0.24966 \\ l &= 55° 45' 6''. \end{aligned}$$

21. Find the latitude of the place at which the sun sets at 10 o'clock on the longest day.

$$\begin{aligned} ZPM &= 15° \times 10 \\ &= 150°, \\ ZM &= 90°, \\ MP &= 90° - l. \end{aligned}$$

$$\cot l = \cos t \cot d.$$
$$t = 150°.$$
$$d = 23° 27'.$$

$$\begin{aligned} \log \cos t &= 9.93753 \\ \log \cot d &= \underline{0.36274} \\ \log \cot l &= 0.30027 \\ l &= 63° 23' 41''. \end{aligned}$$

22. What does the general formula for the hour angle, in § 69, become when (i.) $h = 0°$, (ii.) $l = 0°$ and $d = 0°$, (iii.) l or $d = 90°$?

By § 69,

$$\sin \tfrac{1}{2}t$$
$$= \pm \left[\cos \tfrac{1}{2}(l+p+h) \sin \tfrac{1}{2}(l+p-h) \sec l \csc p\right]^{\frac{1}{2}}$$
$$= \pm \left[\tfrac{1}{2}(\sin\{l+p\}-\sin h)\sec l \csc p\right]^{\frac{1}{2}}.$$

(i.) If $h = 0$,

$$\sin \tfrac{1}{2}t = \pm \left[\tfrac{1}{2}\sin(l+p)\sec l \csc p\right]^{\frac{1}{2}}.$$
$$\cos t = 1 - 2\sin^2 \tfrac{1}{2}t$$
$$= 1 - \sin(l+p)\sec l \csc p$$
$$= 1 - \frac{\sin l \cos p + \cos l \sin p}{\cos l \sin p}$$
$$= -\frac{\sin l \cos p}{\cos l \sin p}$$
$$= -\tan l \cot p.$$

But $\quad p = 90 - d.$

$$\therefore \cos t = -\tan l \tan d.$$

(ii.) If $\quad l = 0$ and $d = 0$,

$$p = 90^\circ - d$$
$$= 90^\circ.$$
$$\sin \tfrac{1}{2}t = \left[\tfrac{1}{2}(1-\sin h)\right]^{\frac{1}{2}}.$$
$$\cos t = 1 - (1 - \sin h)$$
$$= \sin h.$$
$$\therefore t = 90 - h$$
$$= z.$$

(iii.) If l or $d = 90^\circ$, $\sec l$ or $\csc p = \infty$, and the formula is useless. When $d = 90^\circ$, the star is at the pole and its hour angle is indeterminate; and when $l = 90^\circ$, the place of observation is at the terrestrial pole and the meridian is indeterminate.

23. What does the general formula for the azimuth of a celestial body, in § 70, become when $t = 90^\circ = 6$ hours?

From § 70,

$$\tan m = \cot d \cos t.$$
$$\tan a = \sec(l+m)\tan t \sin m.$$

Multiply these two equations together:

$$\tan a \tan m$$
$$= \sec(l+m)\cot d \sin t \sin m.$$
$$\therefore \tan a = \sec(l+m)\cot d \sin t \cos m.$$

Here $t = 90^\circ$, $m = 0$; hence

$$\tan a = \sec l \cot d.$$
$$\cot a = \cos l \tan d.$$

24. Show that the formulas of § 71, if $t = 90^\circ$, lead to the equation $\sin l = \sin h \csc d$; and that if $d = 0^\circ$, they lead to the equation $\cos l = \sin h \sec t$.

From § 71,

$$\tan m = \cot d \cos t. \qquad (1)$$
$$\cos n = \cos m \sin h \csc d. \qquad (2)$$

(i.) If $t = 90^\circ$, $m = 0$ and $n = 90^\circ - l$; hence

$$\cos(90^\circ - l) = \cos 0 \sin h \csc d.$$
$$\sin l = \sin h \csc d.$$

(ii.) If $d = 0$, $m = 90^\circ$, $n = l$.

Divide (2) by (1),

$$\cos n \cot m = \cos m \sin h \sec d \sec t.$$
$$\cos n = \sin m \sin h \sec d \sec t.$$
$$\therefore \cos l = \sin h \sec t.$$

25. Given latitude of place 52° 30′ 16″, declination of star 38°, its hour angle 28° 17′ 15″; find its altitude.

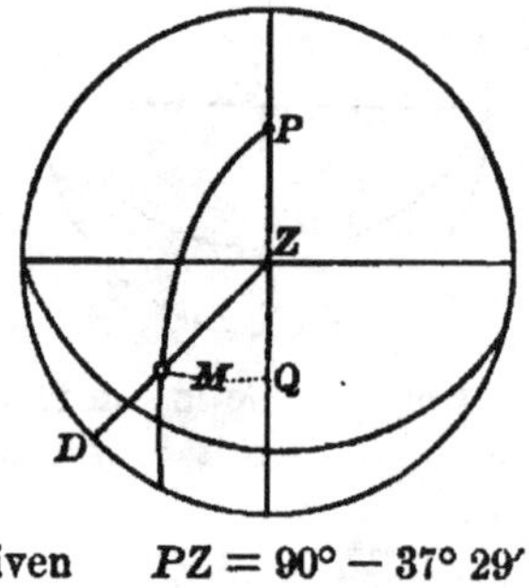

Given $\quad PZ = 90^\circ - 37^\circ\, 29'\, 44''$,

$$PM = 90^\circ - d = 52^\circ,$$
$$ZPM = t = 28^\circ\, 17'\, 15'';$$

required $\quad ZM = 90^\circ - h.$

Let $PQ = m$. Then

$$\tan m = \cot d \cos t.$$
$$\sin h = \sin (l + m) \sin d \sec m.$$

$\log \cot d = 0.10719$
$\log \cos t = 9.94477$
$\log \tan m = 10.05196$
$m = 48° 25' 10''.$

$\log \sin (l+m) = 9.99206$
$\log \sin d = 9.78934$
$\log \sec m = 0.17804$
$\log \sin h = 9.95944 - 10$
$h = 65° 37' 20''.$

26. Given latitude of place 51° 19′ 20″, polar distance of star 67° 59′ 5″, its hour angle 15° 8′ 12″; find its altitude and its azimuth.

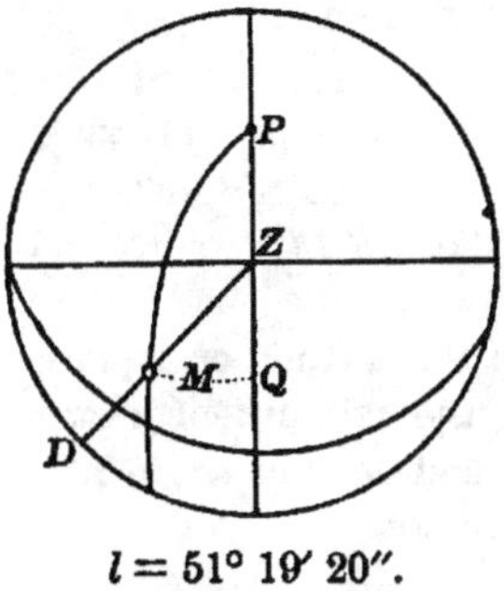

$$l = 51° 19' 20''.$$
$$d = 22° \ 0' 55''.$$
$$t = 15° \ 8' 12''.$$
$$\tan m = \cot d \cos t.$$
$$\sin h = \sin (l+m) \sin d \sec m.$$
$$\tan a = \sec (l+m) \tan t \sin m.$$

$\log \cot d = 10.39326$
$\log \cos t = 9.98466$
$\log \tan m = 10.37792$
$m = 67° 16' 22''.$

$\log \sin (l+m) = 9.94351$
$\log \sin d = 9.57387$
$\log \sec m = 0.41302$
$\log \sin h = 9.93040$
$h = 58° 25' 15''.$

$\log \sec (l+m) = 0.32001$
$\log \tan t = 9.43218$
$\log \sin m = 9.96490$
$\log \tan a = 9.71709$
$a = 152° 28'.$

27. Given the declination of a star 7° 54′, its altitude 22° 45′ 12″, its azimuth 129° 45′ 37″; find its hour angle and the latitude of the observer.

$$\sin t = \sin a \cos h \sec d.$$

$\log \sin a = 9.88577$
$\log \cos h = 9.96482$
$\text{colog} \cos d = 0.00414$
$\log \sin t = 9.85473$
$t = 45° 42'.$

$$\tan m = \cot d \cos t.$$
$$\cos n = \cos m \sin h \csc d.$$
$$l = 90° - (m \pm n).$$

$\log \cot d = 10.85773$
$\log \cos t = 9.84411$
$\log \tan m = 10.70184$
$m = 78° 45' 45''.$

$\log \cos m \quad 9.28976$
$\log \sin h = 9.58745$
$\log \csc d = 0.86187$
$\log \cos n = 9.73908$
$n = 56° 44' 39''.$

$$m - n = 12° \ 1' \ 6''.$$
$$90° - (m - n) = 67° 68' 54''.$$
$$\therefore l = 67° 58' 54''.$$

28. Given the longitude u of the sun, and the obliquity of the ecliptic $e = 23° 27'$; find the declination d, and the right ascension r.

In the figure let P represent the pole of the equinoctial AVB, S the position of the sun, and Q the pole of the ecliptic EVF.

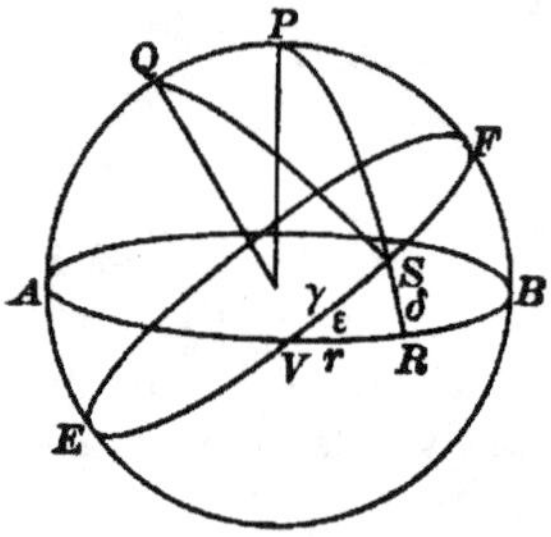

Then $VS = u.$

$VR = r.$

$SR = d.$

$RVS = e.$

Then in the right triangle RVS, by [39],

$$\sin SR = \sin VS \times \sin RVS,$$

or $$\sin d = \sin u \sin e.$$

Also by [40],

$$\cos RVS = \tan RV \cot VS,$$

or $$\cos e = \tan r \cot u.$$

$$\tan r = \tan u \cos e.$$

29. Given the obliquity of the ecliptic $e = 23° 27'$, the latitude of a star 51°, its longitude 315°; find its declination and its right ascension.

In Fig. 51, given

$VT = 315°$ or $-45°$,

$TM = 51°$,

$RVT = 23° 27'$,

to find $VR = r$

and $RM = d$.

In right triangle VTM,

$$\cos VM = \cos VT \cos TM,$$

and $\tan MVT = \tan MT \csc VT.$

log cos 315° = 9.84949
log cos 51° = 9.79887
log cos VM = 9.64836

$VM = 63° 34' 36''$.

log tan 51° = 10.09163
log csc 315° = 0.15051 (n)
log tan MVT = 10.24214 (n)

$MVT = -(60° 12' 14.5'')$.

In right triangle RVM,

$RVM = RVT + TVM$

$= 23° 27' - (60° 12' 14.5'')$

$= -(36° 45' 14.5'')$.

By [39],

$$\sin RM = \sin VM \sin RVM.$$

log sin VM = 9.95208
log sin RVM = 9.77698
log sin RM = 9.72906

$RM = d = 32° 24' 12''$.

Also, by [42],

$$\sin VR = \tan RM \cot RVM.$$

log tan RM = 9.80257
log cot RVM = 0.12677 (n)
log sin VR = 9.92934 (n)

$VR = -(58° 11' 43'')$.

$\therefore VR = 360° - 58° 11' 43''$

$= 301° 48' 17''$.

30. Given the latitude of a place 44° 50′ 14″, the azimuth of a star 138° 58′ 43″, and its hour angle 20°; find its declination.

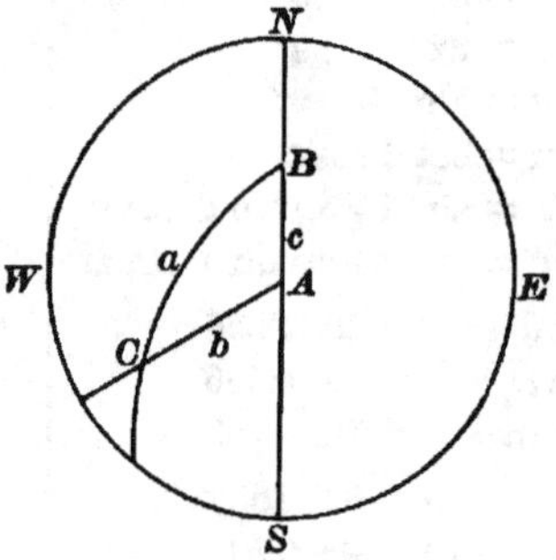

Given $c = 90° - 44° 50' 14''$

$= 45° 9' 46''$.

$A = 138° 58' 43''$.

$B = 20°$.

$\frac{1}{2}(A - B) = 59° 29' 22''.$
$\frac{1}{2}(A + B) = 79° 29' 22''.$
$\frac{1}{2}c \quad = 22° 34' 53''.$

$$\begin{aligned}
\log\cos\tfrac{1}{2}(A-B) &= 9.70560\\
\text{colog}\cos\tfrac{1}{2}(A+B) &= 0.73893\\
\log\tan\tfrac{1}{2}c &= 9.61897\\
\log\tan\tfrac{1}{2}(a+b) &= 0.06350\\
\tfrac{1}{2}(a+b) &= 49° 10' 26''.
\end{aligned}$$

$$\begin{aligned}
\log\sin\tfrac{1}{2}(A-B) &= 9.93528\\
\text{colog}\sin\tfrac{1}{2}(A+B) &= 0.00735\\
\log\tan\tfrac{1}{2}c &= 9.61897\\
\log\tan\tfrac{1}{2}(a-b) &= 9.56160\\
\tfrac{1}{2}(a-b) &= 20° 1' 21.5''.
\end{aligned}$$

$\therefore a = 69° 11' 48''.$
$90° - 69° 11' 48'' = 20° 48' 12''.$

31. Given latitude of place 51° 31′ 48″, altitude of sun west of the meridian 35° 14′ 27″, its declination +21° 27′; find the local apparent time.

By § 69,

$$PZ = 90° - l,$$
$$PM = 90° - d = p,$$
$$ZM = 90° - h;$$

required $t = ZPM.$

$$p = 68° 33'.$$
$$\tfrac{1}{2}(l + h + p) = 77° 39' 37.5''.$$
$$\tfrac{1}{2}(l - h + p) = 42° 25' 10.5''.$$

$$\begin{aligned}
\log\cos\tfrac{1}{2}(l+p+h) &= 9.32982\\
\log\sin\tfrac{1}{2}(l+p-h) &= 9.82901\\
\text{colog}\cos l &= 0.20614\\
\text{colog}\sin p &= 0.03117\\
&2)\,19.39614\\
\log\sin\tfrac{1}{2}t &= 9.69807
\end{aligned}$$

$$\tfrac{1}{2}t = 29° 55' 55.5''.$$
$$t = 59° 51' 51''.$$
$$\frac{t}{15} = 3 \text{ h. } 59 \text{ min. } 27\tfrac{2}{5} \text{ sec. P.M.}$$

32. Given latitude of place l, the polar distance p of a star, and its altitude h; find its azimuth a.

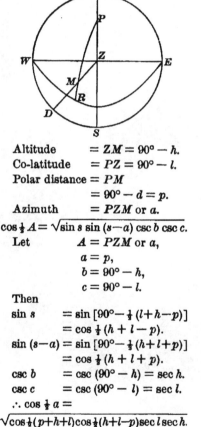

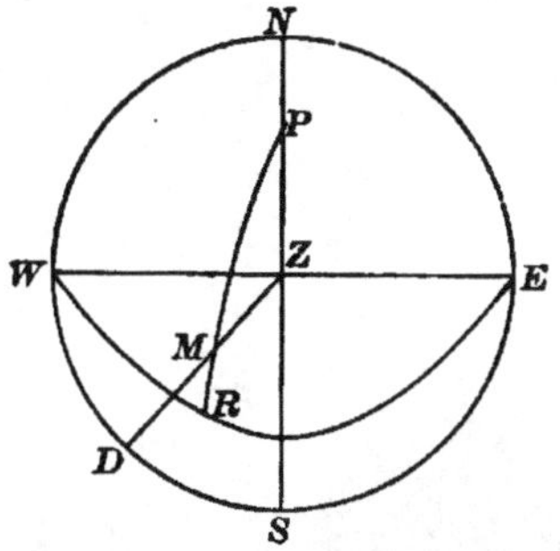

Altitude $= ZM = 90° - h.$
Co-latitude $= PZ = 90° - l.$
Polar distance $= PM$
$= 90° - d = p.$
Azimuth $= PZM$ or a.

$$\cos\tfrac{1}{2}A = \sqrt{\sin s \sin(s-a)\csc b \csc c}.$$

Let $A = PZM$ or a,
$a = p,$
$b = 90° - h,$
$c = 90° - l.$

Then

$$\begin{aligned}
\sin s &= \sin[90° - \tfrac{1}{2}(l+h-p)]\\
&= \cos\tfrac{1}{2}(h+l-p).\\
\sin(s-a) &= \sin[90° - \tfrac{1}{2}(h+l+p)]\\
&= \cos\tfrac{1}{2}(h+l+p).\\
\csc b &= \csc(90° - h) = \sec h.\\
\csc c &= \csc(90° - l) = \sec l.
\end{aligned}$$

$\therefore \cos\tfrac{1}{2}a =$

$$\sqrt{\cos\tfrac{1}{2}(p+h+l)\cos\tfrac{1}{2}(h+l-p)\sec l \sec h}.$$

SURVEYING.

EXERCISE I. PAGE 214.

1. Required the area of a triangular field whose sides are respectively 13, 14, and 15 chains.

$$\text{Area} = \sqrt{s(s-a)(s-b)(s-c)}.$$

$s = \frac{1}{2}(13 + 14 + 15) = 21,$ $\quad s - b = 21 - 14 = 7,$

$s - a = 21 - 13 = 8.$ $\quad s - c = 21 - 15 = 6.$

$$\text{Area} = \sqrt{21 \times 8 \times 7 \times 6} = \sqrt{3^2 \times 7^2 \times 2^4} = 3 \times 7 \times 2^2$$
$$= 84 \text{ sq. ch.} = 8.4 \text{ A.} = 8 \text{ A. } 64 \text{ P.}$$

2. Required the area of a triangular field whose sides are respectively 20, 30, and 40 chains.

$$\text{Area} = \sqrt{45 \times 25 \times 15 \times 5} = \sqrt{3^3 \times 5^5} = 3 \times 5^2\sqrt{3 \times 5}$$
$$= 75\sqrt{15} = 290.4737+.$$

290.4737 sq. ch. = 29.04737 A. = 29 A. 7.579 P. = 29 A. $7\frac{3}{5}$ P., nearly.

3. Required the area of a triangular field whose base is 12.60 chains, and altitude 6.40 chains.

$$\text{Area} = \tfrac{1}{2} \text{ base} \times \text{altitude}.$$
$$\text{Area} = \tfrac{1}{2} \times 12.6 \times 6.4 = 40.32 \text{ sq. ch.} = 4.032 \text{ A.} = 4 \text{ A. } 5\tfrac{3}{25} \text{ P.}$$

4. Required the area of a triangular field which has two sides 4.50 and 3.70 chains, respectively, and the included angle 60°.

$$\text{Area} = \tfrac{1}{2} bc \sin A.$$
$$\text{Area} = \tfrac{1}{2} \times 4.5 \times 3.7 \times 0.866 = 7.20945 \text{ sq. ch.} = 0.7209 \text{ A.}$$
$$= 115\tfrac{7}{20} \text{ P., nearly.}$$

5. Required the area of a field in the form of a trapezium, one of whose diagonals is 9 chains, and the two perpendiculars upon this diagonal from the opposite vertices 4.50 and 3.25 chains.

$$\text{Area} = \tfrac{1}{2} \times 9\,(4.5 + 3.25) = 34.875 \text{ sq. ch.} = 3.4875 \text{ A.}$$
$$= 3 \text{ A. } 78 \text{ P.}$$

6. Required the area of the field $ABCDEF$ (Fig. 19), if $AE = 9.25$ chains, $FF' = 6.40$ chains, $BE = 13.75$ chains, $DD' = 7$ chains, $DB = 10$ chains, $CC' = 4$ chains, and $AA' = 4.75$ chains.

$$
\begin{aligned}
2 \text{ area } AFE &= 6.4 \times 9.25 &&= 59.2\\
2 \text{ area } BDEA &= 13.75\,(4.75 + 7) &&= 161.5625\\
2 \text{ area } BDC &= 10 \times 4 &&= 40\\
\hline
2 \text{ area } ABCDEF &&&= 260.7625\\
\text{area } ABCDEF &&&= 130.38125\\
130.38125 \text{ sq. ch.} &= 13.038125 \text{ A.} &&= 13 \text{ A. } 6\tfrac{1}{10} \text{ P.}
\end{aligned}
$$

7. Required the area of the field $ABCDEF$ (Fig. 20), if $AF' = 4$ chains, $FF' = 6$ chains, $EE' = 6.50$ chains, $AE' = 9$ chains, $AD = 14$ chains, $AC' = 10$ chains, $AB' = 6.50$ chains, $BB' = 7$ chains, $CC' = 6.75$ chains.

$$
\begin{aligned}
2 \text{ area } AFF' &= 4 \times 6 &&= 24\\
2 \text{ area } F'E'EF &= 5\,(6 + 6.5) &&= 62.5\\
2 \text{ area } EE'D &= 6.5 \times 5 &&= 32.5\\
2 \text{ area } ABB' &= 6.5 \times 7 &&= 45.5\\
2 \text{ area } BCC'B' &= 3.5\,(7 + 6.75) &&= 48.125\\
2 \text{ area } CDC' &= 6.75 \times 4 &&= 27\\
\hline
2 \text{ area } ABCDEF &&&= 239.625\\
\text{area } ABCDEF &&&= 119.8125\\
119.8125 \text{ sq. ch.} &= 11.98125 \text{ A.} &&= 11 \text{ A. } 157 \text{ P.}
\end{aligned}
$$

8. Required the area of the field $AGBCD$ (Fig. 15), if the diagonal $AC = 5$, BB' (the perpendicular from B to AC) $= 1$, DD' (the perpendicular from D to AC) $= 1.60$, $EE' = 0.25$, $FF' = 0.25$, $GG' = 0.60$, $HH' = 0.52$, $KK' = 0.54$, $AE' = 0.2$, $E'F' = 0.50$, $F'G' = 0.45$, $G'H' = 0.45$, $H'K' = 0.60$, and $K'B = 0.40$.

$$
\begin{aligned}
2 \text{ area } ADCB &= 5\,(1 + 1.6) &&= 13.\\
2 \text{ area } AEE' &= 0.25 \times 0.2 &&= 0.05\\
2 \text{ area } EE'F'F &= 0.5\,(0.25 + 0.25) &&= 0.25\\
2 \text{ area } FF'G'G &= 0.45\,(0.25 + 0.6) &&= 0.3825\\
2 \text{ area } GG'H'H &= 0.45\,(0.6 + 0.52) &&= 0.504\\
2 \text{ area } HH'K'K &= 0.6\,(0.52 + 0.54) &&= 0.636\\
2 \text{ area } KK'B &= 0.4 \times 0.54 &&= 0.216\\
\hline
2 \text{ area } ADCBKHGFE &&&= 15.0385\\
\text{area } ADCBKHGFE &&&= 7.51925.
\end{aligned}
$$

9. Required the area of the field $AGBCD$ (Fig. 16), if $AD = 3$, $AC = 5$, $AB = 6$, angle $DAC = 45°$, angle $BAC = 30°$, $AE' = 0.75$, $AF' = 2.25$, $AH = 2.53$, $AG' = 3.15$, $EE' = 0.60$, $FF' = 0.40$, and $GG' = 0.75$.

2 area $ADCB$	$= 3 \times 5 \times 0.7071 + 5 \times 6 \times 0.5$	$= 25.6065$
2 area HGB	$= 0.75 \times 3.47$	$= 2.6025$
2 area $ADCBGH$		$= 28.2090$
2 area $AEFH$	$= 0.75 \times 0.6 + 1.5\,(0.6 + 0.4) + 0.4 \times 0.28$	$= 2.062$
2 area $ADCBGHFE$		$= 26.147$
area $ADCBGHFE$		$= 13.0735.$

10. Determine the area of the field $ABCD$ from two interior stations P and P', if $PP' = 1.50$ chains,

angle $PP'C = 89° 35'$,	angle $P'PB = 3° 35'$,
$PP'B = 185° 30'$,	$P'PA = 113° 45'$,
$PP'A = 309° 15'$,	$P'PD = 165° 40'$,
$PP'D = 349° 45'$,	$P'PC = 303° 15'$.

Area $= \triangle PAD + \triangle PCD + \triangle PBC + \triangle PAB.$

$\angle PP'D = 10° 15'$,	$\angle PP'A = 50° 45'$,	$\angle PP'C = 89° 35'$,
$\angle PDP' = 4° 5'$,	$\angle PAP' = 15° 30'$,	$\angle PCP' = 33° 40'$.
$\angle PP'B = 174° 30'$,	$\angle PBP' = 1° 55'$,	

$$PD = \frac{PP' \sin PP'D}{\sin PDP'}.$$

log PP'	$= 0.17609$
log sin $PP'D$	$= 9.25028$
colog sin PDP'	$= 1.14748$
log PD	$= 0.57385$

$$PA = \frac{PP' \sin PP'A}{\sin PAP'}.$$

log PP'	$= 0.17609$
log sin $PP'A$	$= 9.88896$
colog sin PAP'	$= 0.57310$
log PA	$= 0.63815$

$$PC = \frac{PP' \sin PP'C}{\sin PCP'}.$$

log PP'	$= 0.17609$
log sin $PP'C$	$= 9.99999$
colog sin PCP'	$= 0.25621$
log PC	0.43229

$$PB = \frac{PP' \sin PP'B}{\sin PBP'}.$$

log PP'	$= 0.17609$
log sin $PP'B$	$= 8.98157$
colog sin PBP'	$= 1.47566$
log PB	$= 0.63332$

$\angle APD = 51° 55'$, $\angle DPC = 137° 35'$, $\angle BPC = 60° 20'$, $\angle APB = 110° 10'$.

2 area $PAD = PD \times PA \sin APD$.

$$\begin{aligned}\log PD &= 0.57385\\ \log PA &= 0.63815\\ \log \sin APD &= 9.89604\\ \log 2 \text{ area} &= 1.10804\\ 2 \text{ area } PAB &= 12.825.\end{aligned}$$

2 area $PCD = PD \times PC \sin DPC$.

$$\begin{aligned}\log PD &= 0.57385\\ \log PC &= 0.43229\\ \log \sin DPC &= 9.82899\\ \log 2 \text{ area} &= 0.83513\\ 2 \text{ area } PCD &= 6.8412.\end{aligned}$$

2 area $PAB = PA \times PB \sin APB$.

$$\begin{aligned}\log PA &= 0.63815\\ \log PB &= 0.63332\\ \log \sin APB &= 9.97252\\ \log 2 \text{ area} &= 1.24399\\ 2 \text{ area } PAB &= 17.538.\end{aligned}$$

2 area $PBC = PC \times PB \sin BPC$.

$$\begin{aligned}\log PC &= 0.43229\\ \log PB &= 0.63332\\ \log \sin PBC &= 9.93898\\ \log 2 \text{ area} &= 1.00459\\ 2 \text{ area } PBC &= 10.106.\end{aligned}$$

$$\begin{aligned}2\,\triangle PAD &= 12.825\\ 2\,\triangle PCD &= 6.841\\ 2\,\triangle PBC &= 10.106\\ 2\,\triangle PAB &= 17.538\\ 2\,ABCD &= 47.310\\ ABCD &= 23.655 \text{ sq. ch.}\end{aligned}$$

23.655 sq. ch. = 2.3655 A. = 2 A. 58¼ P., nearly.

11. Determine the area of the field $ABCD$ from two exterior stations P and P', if $PP' = 1.50$ chains,

$$\begin{aligned}\text{angle } P'PB &= 41^\circ\ 10', & \text{angle } PP'D &= 66^\circ\ 45',\\ P'PA &= 55^\circ\ 45', & PP'C &= 95^\circ\ 40',\\ P'PC &= 77^\circ\ 20', & PP'B &= 132^\circ\ 15',\\ P'PD &= 104^\circ\ 45', & PP'A &= 103^\circ\ 0'.\end{aligned}$$

Area = $(\triangle P'CB + \triangle P'CD) - (\triangle P'AB + \triangle P'AD)$.

$$\begin{aligned}\angle P'PB &= 41^\circ\ 10', & \angle P'PD &= 104^\circ\ 45', & \angle P'PC &= 77^\circ\ 20',\\ \angle PBP' &= 6^\circ\ 35', & \angle PDP' &= 8^\circ\ 30', & \angle PCP' &= 7^\circ\ 0'.\\ \angle P'PA &= 55^\circ\ 45', & \angle PAP' &= 21^\circ\ 15',\end{aligned}$$

$$P'B = \frac{PP' \sin P'PB}{\sin PBP'}.$$

$$\begin{aligned}\log PP' &= 0.17609\\ \log \sin P'PB &= 9.81839\\ \text{colog} \sin PBP' &= 0.94063\\ \log P'B &= 0.93511\end{aligned}$$

$$P'D = \frac{PP' \sin P'PD}{\sin PDP'}.$$

$$\begin{aligned}\log PP' &= 0.17609\\ \log \sin P'PD &= 9.98545\\ \text{colog} \sin PDP' &= 0.83030\\ \log P'D &= 0.99184\end{aligned}$$

$$P'C = \frac{PP' \sin P'PC}{\sin PCP'}.$$

$\log PP'$	$= 0.17609$
$\log \sin P'PC$	$= 9.98930$
$\text{colog} \sin PCP'$	$= 0.91411$
$\log P'C$	$= 1.07950$

$$P'A = \frac{PP' \sin P'PA}{\sin PAP'}.$$

$\log PP'$	$= 0.17609$
$\log \sin P'PA$	$= 9.91729$
$\text{colog} \sin PAP'$	$= 0.44077$
$\log P'A$	$= 0.53415$

$\angle BP'C = 36° 35'$,

$\angle CP'D = 28° 55'$,

$\angle AP'B = 29° 15'$,

$\angle AP'D = 36° 15'$.

$2 \text{ area } P'CB = P'C \times P'B \sin BP'C.$

$\log P'C$	$= 1.07950$
$\log P'B$	$= 0.93511$
$\log \sin BP'C$	$= 9.77524$
$\log 2 \text{ area}$	$= 1.78985$
$2 \text{ area } P'CB$	$= 61.639.$

$2 \text{ area } P'CD = P'C \times P'D \sin CP'D.$

$\log P'C$	$= 1.07950$
$\log P'D$	$= 0.99184$
$\log \sin CP'D$	$= 9.68443$
$\log 2 \text{ area}$	$= 1.75577$
$2 \text{ area } P'CD$	$= 56.986.$

$2 \text{ area } P'AB = P'B \times P'A \sin AP'B.$

$\log P'B$	$= 0.93511$
$\log P'A$	$= 0.53415$
$\log \sin AP'B$	$= 9.68897$
$\log 2 \text{ area}$	$= 1.15823$
$2 \text{ area } P'AB$	$= 14.396.$

$2 \text{ area } P'AD = P'A \times P'D \sin AP'D.$

$\log P'A$	$= 0.53415$
$\log P'D$	$= 0.99184$
$\log \sin AP'D$	$= 9.77181$
$\log 2 \text{ area}$	$= 1.29780$
$2 \text{ area } P'AD$	$= 19.852.$

$2\triangle P'CB$	$= 61.639$
$2\triangle P'CD$	$= 56.986$
	118.625
	34.248
$2\, ABCD$	$= 84.377$
$ABCD$	$= 42.1885.$

$2\triangle P'AB$	$= 14.396$
$2\triangle P'AD$	$= 19.852$
	34.248

42.1885 sq. ch. = 4.21885 A.
= 4 A. 35 P., nearly.

Exercise II. Page 223.

1.

			N.	S.	E.	W.	M. D.	D.M.D.	N. A.	S. A.
1	S. 75° E.	6.00	. . .	1.55	5.79 ~~5.80~~	. . .	5.79	5.79	. . . . .	8.9745
2	S. 15° E.	4.00	. . .	3.86	1.04	. . .	6.83	12.62	. . . . .	48.7132
3	S. 75° W.	6.93	. . .	1.80 ~~1.79~~	. . .	6.70 ~~6.69~~	0.13	6.96	. . . . .	12.5280
4	N. 45° E.	5.00	3.54	. . .	3.54	. . .	3.67	3.80	13.4520	. . . . .
5	N. 45° W.	5.19½	3.67	. . .	. . .	3.67	0	3.67	13.4689	. . . . .
									26.9209	70.2157
										26.9209
										43.2948
										21.6474

21.647 sq. ch. = 2.1647 A. = 2 A. 26 P., nearly.

2.

			N.	S.	E.	W.	M. D.	D.M.D.	N. A.	S. A.
1	N. 45° E.	10.00	7.07	. . .	7.07	. . .	7.07	7.07	49.9849	. . . . .
2	S. 75° E.	11.55	. . .	2.99	11.16	. . .	18.23	25.30	. . . . .	75.6470
3	S. 15° W.	18.21	. . .	17.59	. . .	4.71	13.52	31.75	. . . . .	558.4825
4	N. 45° W.	19.11	13.51	. . .	. . .	13.52 ~~13.51~~	0	13.52	182.6552	. . . . .
									232.6401	634.1295
										232.6401
										401.4894
										200.7447

200.74 sq. ch. = 20.074 A. = 20 A. 12 P., nearly

3.

			N.	*S.*	*E.*	*W.*	*M. D.*	*D.M.D.*	*N. A.*	*S. A.*
1	N. 15° E.	3.00	2.90	. . .	0.78	. . .	0.78	0.78	2.2620	. . . . .
2	N. 75° E.	6.00	1.55	. . .	5.79 ~~5.80~~	. . .	6.57	7.35	11.3925	. . . . .
3	S. 15° W.	6.00	. . .	5.80	. . .	1.55	5.02	11.59	. . . . .	67.2220
4	N. 75° W.	5.20	1.35	. . .	. . .	5.02	0	5.02	6.7770	. . . . .
									20.4315	67.2220
										20.4315
	23.395 sq. ch. = 2.3395 A. = 2 A. 54 P., nearly.									
										46.7905
										23.3953

4.

			N.	*S.*	*E.*	*W.*	*M. D.*	*D.M.D.*	*N. A.*	*S. A.*
1	N. 89° 45′ E.	4.94	0.00 ~~0.02~~	. . .	4.93 ~~4.94~~	. . .	4.93	4.93	. . . . .	. . . . .
2	S. 7° 00′ W.	2.30	. . .	2.29 ~~2.28~~	. . .	0.29 ~~0.28~~	4.64	9.57	. . . . .	21.9153
3	S. 28° 00′ E.	1.52	. . .	1.34	0.71	. . .	5.35	9.99	. . . . .	13.3866
4	S. 0° 45′ E.	2.57	. . .	2.58 ~~2.57~~	0.02 ~~0.03~~	. . .	5.37	10.72	. . . . .	27.6576
5	N. 84° 45′ W.	5.11	0.45 ~~0.47~~	. . .	. . .	5.10 ~~5.09~~	0.27	5.64	2.5380	. . . . .
6	N. 2° 30′ W.	5.79	5.76 ~~5.78~~	. . .	. . .	0.27 ~~0.25~~	0	0.27	1.5552	. . . . .
									4.0932	62.9595
										4.0932
	2.943 A. = 2 A. 151 P., nearly.									
										58.8663
										29.4332

5.

		N.	S.	E.	W.
N. 51° 45′ W.	2.39	1.48	. . .	. . .	1.88
S. 85° W.	6.47	. . .	0.56	. . .	6.45
S. 55° 10′ W.	1.62	. . .	0.93	. . .	1.33
			1.49	. . .	9.66
			1.48	. . .	. . .
			0.01		

			N.	S.	E.	W.	M. D.	D.M.D.	N. A.	S. A.
2	S. W.	. . .	. . .	0.03 ~~0.01~~	. . .	9.65 ~~9.66~~	9.65	9.65	. . . . .	0.2895
3	N. 3° 45′ E.	6.39	6.36 ~~6.38~~	. . .	0.43 ~~0.42~~	. . .	9.22	18.87	120.0132	. . . . .
4	S. 66° 45′ E.	1.70	. . .	0.67	1.56	. . .	7.63	16.88	. . . . .	11.3096
5	N. 15° E.	4.98	4.80 ~~4.81~~	. . .	1.29	. . .	6.37	14.03	67.3440	. . . . .
6	S. 82° 45′ E.	6.03	. . .	0.77 ~~0.76~~	5.98	. . .	0.39	6.76	. . . . .	5.2052
1	S. 2° 15′ E.	9.68	. . .	9.69 ~~9.67~~	0.39 ~~0.38~~	. . .	0	0.39	. . . . .	3.7791
									187.3572	20.5834
									20.5834	
	8.339 A. = 8 A. 54 P., nearly.									
									166.7738	
									83.3869	

6.

		N.	S.	E.	W.
S. 81° 20′ W.	4.28	. . .	0.65	. . .	4.23
N. 76° 30′ W.	2.67	0.62	. . .	. . .	2.60
			0.65	. . .	6.83
			0.62	. . .	. . .
			0.03		

		N.	S.	E.	W.
S. 7° E.	1.79	. . .	1.78	0.22	. . .
S. 27° E.	1.94	. . .	1.73	0.88	. . .
S. 10° 30′ E.	5.35	. . .	5.26	0.98	. . .
N. 76° 45′ W.	1.70	0.39	. . .	. . .	1.65
			8.77	2.08	. . .
			0.39	1.65	. . .
			8.38	0.43	

			N.	S.	E.	W.	M. D.	D.M.D.	N. A.	S. A.
1	S. W.	. . .	. . .	0.03	. . .	6.80 ~~6.83~~	6.80	6.80	. . . . .	0.2040
2	N. 5° E.	8.68	8.65	. . .	0.79 ~~0.76~~	. . .	6.01	12.81	110.8065	. . . . .
3	S. 87° 30′ E.	5.54	. . .	0.24	5.55 ~~5.53~~	. . .	0.46	6.47	. . . . .	1.5528
4	S. E.	. . .	. . .	8.38	0.46 ~~0.43~~	. . .	0	0.46	. . . . .	3.8548
									110.8065	5.6116
									5.6116	
									105.1949	
									52.597	

5.2597 A. = 5 A. 42 P., nearly.

7.

			N.	*S.*	*E.*	*W.*	*M. D.*	*D.M.D.*	*N.A.*	*S.A.*
3	S. 5° 00′ E.	5.86	. . .	5.83 ~~5.84~~	0.53 ~~0.51~~	. . .	0.53	0.53	. . . . .	3.0899
4	N. 88° 30′ E.	4.12	0.12 ~~0.11~~	. . .	4.14 ~~4.12~~	. . .	4.67	5.20	0.6240	. . . . .
1	N. 6° 15′ W.	6.31	6.28 ~~6.27~~	. . .	. . .	0.67 ~~0.69~~	4.00	8.67	54.4476	. . . . .
2	S. 81° 50′ W.	4.06	. . .	0.57 ~~0.58~~	. . .	4.00 ~~4.02~~	0	4.00	. . . . .	2.2800
									55.0716	5.3699
									5.3699	
	2.485 A. = 2 A. 78 P., nearly.								49.7017	
									24.8508	

8.

			N.	*S.*	*E.*	*W.*	*M. D.*	*D.M.D.*	*N.A.*	*S.A.*
3	S. 3° 00′ E.	5.33	. . .	5.29 ~~5.32~~	0.28	. . .	0.28	0.28	. . . . .	1.4812
4	E.	6.72	0.03 ~~0.00~~	. . .	6.73 ~~6.72~~	. . .	7.01	7.29	0.2187	. . . . .
1	N. 5° 30′ W.	6.08	6.08 ~~6.05~~	. . .	. . .	0.57 ~~0.58~~	6.44	13.45	81.7760	. . . . .
2	S. 82° 30′ W.	6.51	. . .	0.82 ~~0.85~~	. . .	6.44 ~~6.45~~	0	6.44	. . . . .	5.2808
									81.9947	6.7620
									6.7620	
	3.761 A. = 3 A. 122 P., nearly.								75.2327	
									37.6163	

9.

			N.	*S.*	*E.*	*W.*	*M. D.*	*D.M.D.*	*N. A.*	*S. A.*
1	N. 20° 00′ E.	4.62½	4.35	. . .	1.58	. . .	1.58	1.58	6.8730	. . . . .
2	N. 73° 00′ E.	4.16½	1.22	. . .	3.98	. . .	5.56	7.14	8.7108	. . . . .
3	S. 45° 15′ E.	6.18½	. . .	4.35	4.39	. . .	9.95	15.51	. . . . .	67.4685
4	S. 38° 30′ W.	8.00	. . .	6.26	. . .	4.98	4.97	14.92	. . . . .	93.3992
5	Wanting.	. . .	5.04	. . .	. . .	4.97	0	4.97	25.0488	. . . . .
									40.6326	160.8677
										40.6326
	6.012 A. = 6 A. 2 P., nearly.									120.2351
										60.1175

10.

			N.	*S.*	*E.*	*W.*	*M. D.*	*D.M.D.*	*N. A.*	*S. A.*
6	N. 32° 00′ E.	8.68	7.33 ~~7.36~~	. . .	4.61 ~~4.60~~	. . .	4.61	4.61	33.7913	. . . . .
7	S. 75° 50′ E.	6.38	. . .	1.58 ~~1.56~~	6.20 ~~6.19~~	. . .	10.81	15.42	. . . . .	24.3636
8	S. 14° 45′ W.	0.98	. . .	0.95	. . .	0.25	10.56	21.37	. . . . .	20.3015
9	S. 79° 15′ E.	4.52	. . .	0.86 ~~0.84~~	4.44	. . .	15.00	25.56	. . . . .	21.9816
1	S. 3° 00′ E.	4.23	. . .	4.23 ~~4.22~~	0.22	. . .	15.22	30.22	. . . . .	127.8306
2	S. 86° 45′ W.	4.78	. . .	0.29 ~~0.27~~	. . .	4.77	10.45	25.67	. . . . .	7.4443
3	S. 37° 00′ W.	2.00	. . .	1.60	. . .	1.20	9.25	19.70	. . . . .	31.5200
4	N. 81° 00′ W.	7.45	1.14 ~~1.17~~	. . .	. . .	7.35 ~~7.36~~	1.90	11.15	12.7110	. . . . .
5	N. 61° 00′ W.	2.17	1.04 ~~1.05~~	. . .	. . .	1.90	0	1.90	1.9760	. . . . .
									48.4783	233.4416
										48.4783
	9.248 A. = 9 A. 40 P., nearly.									184.9633
										92.48

EXERCISE III. PAGE 224.

1.

			N.	S.	E.	W.	M. D.	D.M.D.	N. A.	S. A.
AB	N.	4.000	4.000	. . .	. . . .	. . .	0	0	. . . .	. . . .
BC	S. 60° E.	4.000	. . .	2.000	3.464	. . .	3.464	3.464	. . . .	6.928
CD	S. 30° E.	6.928	. . .	6.000	3.464	. . .	6.928	10.392	. . . .	62.352
DA	N. 60° W.	8.000	4.000	. . .	. . . .	6.928	0	6.928	27.712	. . . .
									27.712	69.280
										27.712
20.784 sq. ch. = 2.0784 A. = 2 A. 12½ P., nearly.										41.568
										20.784

2.

			N.	S.	E.	W.	M. D.	D.M.D.	N. A.	S. A.
AB	N.	79.86	79.860	. . .	. . . .	. . .	0	0	. . . .	. . . .
BC	N. 80° 20′ E.	121.13	20.338	. . .	119.410	. . .	119.410	119.410	2428.56	. . . .
CD	S. 40° 00′ E.	90.00	. . .	68.943	57.851	. . .	177.261	296.671	. . . .	20453.39
DE	S. 55° 52′ W.	100.65	. . .	59.350	. . . .	81.289	95.972	273.233	. . . .	16216.38
EA	N. 73° 41′ W.	100.00	28.095	. . .	. . . .	95.972	0	95.972	2696.33	. . . .
									5124.89	36669.77
15772.44 P. = 98 A. 92 P., nearly.										5124.89
										31544.88
										15772.44

Exercise IV. Page 231.

1. From the square $ABCD$, containing 6 A. 1 R. 24 P., part off 3 A. by a line EF parallel to AB.

$$6 \text{ A. } 1 \text{ R. } 24 \text{ P.} = 64 \text{ sq. ch.}; \sqrt{64} = 8 \text{ ch.} = AB.$$
$$3 \text{ A.} = 30 \text{ sq. ch.}$$
$$AE = \frac{ABFE}{AB} = \frac{30}{8} = 3.75 \text{ ch.}$$

2. From the rectangle $ABCD$, containing 8 A. 1 R. 24 P., part off 2 A. 1 R. 32 P. by a line EF parallel to $AD = 7$ ch. Then, from the remainder of the rectangle part off 2 A. 3 R. 25 P. by a line GH parallel to EB.

$$8 \text{ A. } 1 \text{ R. } 24 \text{ P.} = 84 \text{ sq. ch.} = ABCD.$$
$$2 \text{ A. } 1 \text{ R. } 32 \text{ P.} = 24.5 \text{ sq. ch.} = AEFD.$$
$$2 \text{ A. } 3 \text{ R. } 25 \text{ P.} = 29.0625 \text{ sq. ch.} = EBHG.$$
$$AE = \frac{AEFD}{AD} = \frac{24.5}{7} = 3.5 \text{ ch.}$$
$$AB = \frac{ABCD}{AD} = \frac{84}{7} = 12 \text{ ch.}$$
$$EB = AB - AE = 12 - 3.5 = 8.5 \text{ ch.}$$
$$EG = \frac{EBHG}{EB} = \frac{29.0625}{8.5} = 3.42 \text{ ch., nearly.}$$

3. Part off 6 A. 3 R. 12 P. from a rectangle $ABCD$, containing 15 A. by a line EF parallel to AB; AD being 10 ch.

$$6 \text{ A. } 3 \text{ R. } 12 \text{ P.} = 68.25 \text{ sq. ch.} = ABFE.$$
$$15 \text{ A.} = 150 \text{ sq. ch.} = ABCD.$$
$$AB = \frac{ABCD}{AD} = \frac{150}{10} = 15 \text{ ch.}$$
$$AE = \frac{ABFE}{AB} = \frac{68.25}{15} = 4.55 \text{ ch.}$$

4. From a square $ABCD$, whose side is 9 ch., part off a triangle which shall contain 2 A. 1 R. 36 P., by a line BE drawn from B to the side AD.

$$2 \text{ A. } 1 \text{ R. } 36 \text{ P.} = 24.75 \text{ sq. ch.}$$
$$AE = \frac{2\,ABE}{AB} = \frac{2 \times 24.75}{9} = 5.50 \text{ ch.}$$

5. From $ABCD$, representing a rectangle, whose length is 12.65 ch., and breadth 7.58 ch., part off a trapezoid which shall contain 7 A. 3 R. 24 P., by a line BE drawn from B to the side DC.

7 A. 3 R. 24 P. = 79 sq. ch.
$ABCD = 12.65 \times 7.58 = 95.887$ sq. ch.
$\triangle BCE = 95.887 - 79 = 16.887$ sq. ch.
$$CE = \frac{2\,BCE}{BC} = \frac{2 \times 16.887}{7.58} = 4.456 \text{ ch., nearly.}$$

6. In the triangle ABC, $AB = 12$ ch., $AC = 10$ ch., and $BC = 8$ ch.; part off 1 A. 2 R. 16 P., by the line DE parallel to AB.

1 A. 2 R. 16 P. = 16 sq. ch.
$CAB = \sqrt{15 \times 3 \times 5 \times 7} = 39.6863$ sq. ch.
$CDE = CAB - ABED = 39.6863 - 16 = 23.6863$ sq. ch.
$CAB : CDE :: \overline{CA}^2 : \overline{CD}^2$
$:: \overline{CB}^2 : \overline{CE}^2.$
$39.6863 : 23.6863 :: 10^2 : \overline{CD}^2. \quad \therefore CD = 7.725$ ch.
$:: 8^2 : \overline{CE}^2. \quad \therefore CE = 6.18$ ch.
$AD = CA - CD = 10 - 7.725 = 2.275$ ch.
$BE = CB - CE = 8 - 6.18 = 1,82$ ch.

7. In the triangle ABC, $AB = 26$ ch., $AC = 20$ ch., and $BC = 16$ ch.; part off 6 A. 1 R. 24 P., by the line DE parallel to AB.

6 A. 1 R. 24 P. = 64 sq. ch.
$CAB = \sqrt{31 \times 5 \times 11 \times 15} = 159.9218$ sq. ch.
$CDE = CAB - ABED = 159.9218 - 64 = 95.9218$ sq. ch.
$CAB : CDE :: \overline{CA}^2 : \overline{CD}^2$
$:: \overline{CB}^2 : \overline{CE}^2.$
$159.9218 : 95.9218 :: 20^2 : \overline{CD}^2. \quad \therefore CD = 15.49$ ch.
$:: 16^2 : \overline{CE}^2. \quad \therefore CE = 12.39$ ch.
$AD = CA - CD = 20 - 15.49 = 4.51$ ch., nearly.
$BE = CB - CE = 16 - 12.39 = 3.61$ ch., nearly.

8. It is required to divide the triangular field ABC among three persons whose claims are as the numbers 2, 3, and 5, so that they may all have the use of a watering-place at C; $AB = 10$ ch., $AC = 6.85$ ch., and $CB = 6.10$ ch.

Since the triangles have the same altitude, they are to each other as their bases. Hence it is only necessary to divide the base 10 into the three parts, 2 ch., 3 ch., 5 ch.

9. Divide the five-sided field $ABCHE$ among three persons, X, Y, and Z, in proportion to their claims, X paying $500, Y paying $750, and Z paying $1000, so that each may have the use of an interior pond, at P, the quality of the land being equal throughout. Given $AB = 8.64$ ch., $BC = 8.27$ ch., $CH = 8.06$ ch., $HE = 6.82$ ch., and $EA = 9.90$ ch. The perpendicular PD upon $AB = 5.60$ ch., PD' upon $BC = 6.08$ ch., PD'' upon $CH = 4.80$ ch., PD''' upon $HE = 5.44$ ch., and PD'''' upon $EA = 5.40$ ch. Assume PH as the divisional fence between X's and Z's shares; it is required to determine the position of the fences PM and PN between X's and Y's shares and Y's and Z's shares, respectively.

If P is joined to the vertices, the field is divided into triangles, whose bases are the sides, and the altitudes the given perpendiculars upon the sides from P.

$$\begin{array}{ll}
APB = 8.64 \times 2.80 = & 24.1920 \text{ sq. ch.} \\
BPC = 8.27 \times 3.04 = & 25.1408 \\
CPH = 8.06 \times 2.40 = & 19.3440 \\
HPE = 6.82 \times 2.72 = & 18.5504 \\
\underline{EPA = 9.90 \times 2.70 =} & \underline{26.7300} \\
ABCHE & = 113.9572
\end{array}$$

The whole area, 113.9572 sq. ch., must be divided as the numbers 500, 750, 1000, or as 2, 3, 4. $2 + 3 + 4 = 9$.

$$\begin{array}{l}
9 : 113.9572 :: 2 : 25.3238 \text{ sq. ch.} = \text{X's share.} \\
\qquad\qquad\qquad :: 3 : 37.9857 \text{ sq. ch.} = \text{Y's share.} \\
\qquad\qquad\qquad :: 4 : 50.6476 \text{ sq. ch.} = \text{Z's share.}
\end{array}$$

PH is assumed as the line between X's and Z's shares. Since the triangle PHE is less than X's share by $25.3238 - 18.5504 = 6.7734$ sq. ch., this difference must be taken from the triangle PEA. The area of PEM is then 6.7734 sq. ch., and the altitude $PD'''' = 5.40$.

$$\therefore EM = \frac{2\,PEM}{PD''''} = \frac{2 \times 6.7734}{5.40} = 2.5087 \text{ ch.}$$

$$PMA = PEA - PEM = 26.7300 - 6.7734 = 19.9566 \text{ sq. ch.}$$

Since Y's share is greater than PMA (19.9566) and less than $PMA + PAB$ (44.1486), the point N is on AB.

Y's share diminished by PMA equals PAN; that is,

$$PAN = 37.9857 - 19.9566 = 18.0291 \text{ sq. ch.}$$

$$AN = \frac{2\,PAN}{PD} = \frac{2 \times 18.0291}{5.60} = 6.439 \text{ ch.}$$

10. Divide the triangular field ABC, whose sides AB, AC, and BC are 15, 12, and 10 ch., respectively, into three equal parts, by fences EG and DF parallel to BC.

$$ABC = \sqrt{18.5 \times 3.5 \times 6.5 \times 8.5} = 59.81169 \text{ sq. ch.}$$
$$ADF = \tfrac{1}{3} \text{ of } 59.81169 = 19.9372 \text{ sq. ch.}$$
$$AEG = \tfrac{2}{3} \text{ of } 59.81169 = 39.8744 \text{ sq. ch.}$$
$$ABC : AEG :: \overline{AB}^2 : \overline{AE}^2$$
$$:: \overline{AC}^2 : \overline{AG}^2.$$
$$59.81169 : 39.8744 :: 15^2 : \overline{AE}^2. \quad \therefore AE = 12.247 \text{ ch.}$$
$$:: 12^2 : \overline{AG}^2. \quad \therefore AG = 9.798 \text{ ch.}$$
$$ABC : ADF :: \overline{AB}^2 : \overline{AD}^2$$
$$:: \overline{AC}^2 : \overline{AF}^2.$$
$$59.81169 : 19.9372 :: 15^2 : \overline{AD}^2. \quad \therefore AD = 8.659 \text{ ch.}$$
$$:: 12^2 : \overline{AF}^2. \quad \therefore AF = 6.928 \text{ ch.}$$

11. Divide the triangular field ABC, whose sides AB, BC, and AC are 22, 17, and 15 ch., respectively, among three persons, A, B, and C, by fences parallel to the base AB, so that A may have 3 A., B 4 A., and C the remainder.

$$CAB = \sqrt{27 \times 5 \times 10 \times 12} = 127.2792 \text{ sq. ch.}$$
$$CDG = CAB - ABGD = 127.2792 - 30 = 97.2792 \text{ sq. ch.}$$
$$CEF = CAB - ABFE = 127.2792 - 70 = 57.2792 \text{ sq. ch.}$$
$$CAB : CDG :: \overline{CB}^2 : \overline{CG}^2$$
$$:: \overline{CA}^2 : \overline{CD}^2.$$
$$127.2792 : 97.2792 :: 17^2 : \overline{CG}^2. \quad \therefore CG = 14.862 \text{ ch.}$$
$$:: 15^2 : \overline{CD}^2. \quad \therefore CD = 13.113 \text{ ch.}$$
$$CAB : CEF :: \overline{CB}^2 : \overline{CF}^2$$
$$:: \overline{CA}^2 : \overline{CE}^2.$$
$$127.2792 : 57.2792 :: 17^2 : \overline{CF}^2. \quad \therefore CF = 11.404 \text{ ch.}$$
$$:: 15^2 : \overline{CE}^2. \quad \therefore CE = 10.062 \text{ ch.}$$

Exercise V. Page 260.

1. Find the difference of level of two places from the following field notes; back-sights, 5.2, 6.8, and 4.0; fore-sights, 8.1, 9.5, and 7.9.

$$\begin{aligned} 8.1 + 9.5 + 7.9 &= 25.5 \\ 5.2 + 6.8 + 4 &= \underline{16} \\ & \quad 9.5 \end{aligned}$$

2. Write the proper numbers in the third and fifth columns of the following table of field notes, and make a profile of the section.

Station.	*+S.*	*H.I.*	*−S.*	*H.S.*	*Remarks.*
B	6.944	....	....	20.	Bench on post 22 feet north of 0.
0	....	26.944	7.4	19.5	
1	....	....	5.6	21.3	
2	....	....	3.9	23.0	
3	....	....	4.6	22.3	
t. p.	3.855	....	5.513	21.431	
4	....	25.286	4.9	20.4	
5	....	....	3.5	21.8	
6	....	....	1.2	24.1	

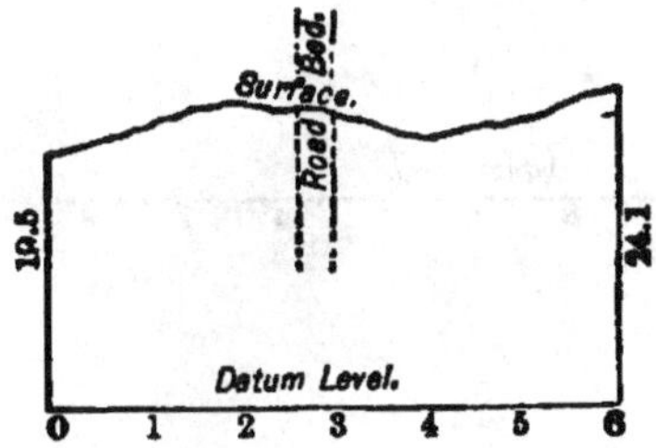

3. Stake 0 of the following notes stands at the lowest point of a pond to be drained into a creek; stake 10 stands at the edge of the bank, and 10.25 at the bottom of the creek. Make a profile, draw the grade line through 0 and 10.25, and fill out the columns *H.G.* and *C.*, the former to show the height of grade line above the datum, and the latter, the depths of cut at the several stakes necessary to construct the drain.

Station.	*+S.*	*H.I.*	*−S.*	*H.S.*	*H.G.*	*C.*	*Remarks.*
B	6.000	....	....	25	....	....	Bench on rock 30 feet west of stake 1.
0	....	....	10.2	....	20.8	0.0	
1	....	....	5.3	....	20.4	5.3	
2	....	....	4.6	....	20.0	6.4	
3	....	....	4.0	....	19.6	7.4	
4	....	....	6.8	....	19.2	5.0	
5	4.572	....	7.090	....	18.8	5.1	
6	....	....	3.9	....	18.4	6.2	
7	....	....	2.0	....	18.0	8.5	
8	....	....	4.9	....	17.6	6.0	
9	....	....	4.3	....	17.2	7.0	
10	....	....	4.5	....	16.8	7.2	
10.25	....	....	11.8	....	16.7	0.0	

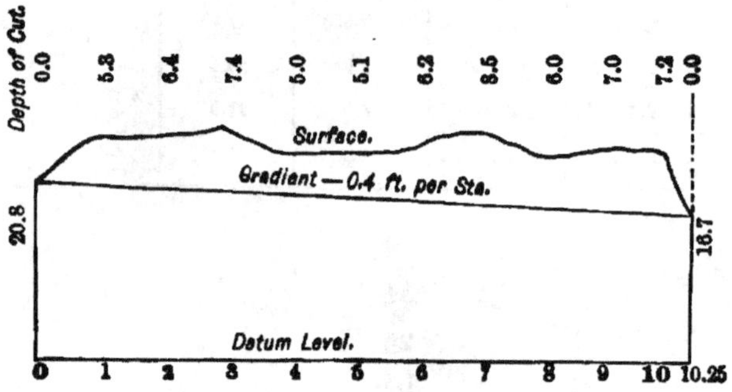

NAVIGATION.

Exercise I. Page 282.

1. Given compass course S., wind E.S.E., leeway 1¼ points, variation 52° 0′ W., deviation 2° 0′ E.; required true course.

Since the wind is E.S.E., and the compass course is S., the ship is on the port tack; hence, leeway is allowed to the right.

Compass course	0 pts. R. of S.
Leeway	1¼ pts. R.
Compass course corrected for leeway .	1¼ pts. R. of S.
	= 14° 3′ 45″ R. of S.
Variation and deviation (52° − 2°) W.	= 50° 0′ 0″ L.
	35° 56′ 15″ L. of S.

True course, S. 35° 56′ E.

2. Given compass course W.N.W., wind N., leeway 3 points, variation 42° 0′ E., deviation 18° 30′ W.; required true course.

Since the wind is N., and the compass course is W.N.W., the ship is on the starboard tack; hence, leeway is allowed to the left.

Compass course	6 pts. L. of N.
Leeway	3 pts. L.
Compass course corrected for leeway .	9 pts. L. of N.
	= 7 pts. R. of S.
	= 78° 45′ R. of S.
Variation and deviation (42° − 18° 30′) E.	= 23° 30′ R.
	102° 15′ R. of S.
	77° 45′ L. of N.

True course, N. 77° 45′ W.

3. Given compass course S.S.E. ½ E., wind S.W. ½ S., leeway 3½ points, variation 2¼ points E., deviation 1¼ points W.; required true course.

Compass course	2½ pts. L. of S.
Leeway (starboard tack)	3½ pts. L.
Variation and deviation	1 pt. R.
	5 pts. L. of S.

True course, S.E. by E.

4. Given true course S. 79° W., wind S. by W., leeway ¾ point, variation 10° 30′ E., deviation 19° 0′ W.; required compass course.

True course	79°	R. of S.
Leeway (port tack)	8° 26′ 15″	L.
Variation and deviation	8° 30′	R.
	79° 3′ 45″	R. of S.

Compass course, S. 79° 4′ W.

5. Given compass course W. ¼ N., wind N.N.W., leeway 1¾ points, variation 8° 30′ E., deviation 15° 35′ E., required true course.

Compass course	7¾ pts.	L. of N.
Leeway (starboard tack)	1¾ pts.	L.
	9½ pts.	L. of N.
	= 6½ pts.	R. of S.
	= 73° 7′ 30″	R. of S.
Variation and deviation	24° 5′	R.
	97° 12′ 30″	R. of S.
	= 82° 47′ 30″	L. of N.

True course, N. 82° 47′ W.

6. Given compass course E. ¼ N., wind N.N.E., leeway 2¼ points, variation 13° 0′ W., deviation 20° 0′ E.; required true course.

Compass course	7¾ pts.	R. of N.
Leeway (port tack)	2¼ pts.	R.
	10 pts.	R. of N.
	= 6 pts.	L. of S.
Variation and deviation	= 67° 30′	L. of S.
	7°	R.
	60° 30′	L. of S.

True course, S. 60° 30′ E.

7. Given true course S. 85° E., wind N. by W., leeway ½ point, variation 14° 0′ E., deviation 19° 0′ E.; required compass course.

True course	85°	L. of S.
Leeway (port tack)	5° 37′ 30″	L.
Variation and deviation	33°	L.
	123° 37′ 30″	L. of S.
	= 56° 22′ 30″	R. of N.

Compass course, N. 56° 22′ E.

8. Given compass course W., wind N.N.W., leeway 1½ points, variation 18° 30′ E., deviation 21° 0′ W.; required true course.

Compass course	8 pts. L. of N.
Leeway (starboard tack)	1¼ pts. L.
	9¼ pts. L. of N.
=	6¾ pts. R. of S.
	= 75° 56′ 15″ R. of S.
Variation and deviation	2° 30′ L.
	73° 26′ 15″ R. of S.

True course, S. 73° 26′ W.

9. Given compass course E. ¼ S., wind N.N.E. ¼ E., leeway 2¾ points, variation 21° 0′ E., deviation 4° 0′ W.; required true course.

Compass course	7¾ pts. L. of S.
Leeway (port tack)	2¾ pts. R.
	5 pts. L. of S.
	= 56° 15′ L. of S.
Variation and deviation	17° R.
	39° 15′ L. of S.

True course, S. 39° 15′ E.

10. Given true course, E. by S. ¼ S., wind N. by W., leeway 2¾ points, variation 2 points W., deviation 3½ points E.; required compass course.

True course	6¾ pts. L. of S.
Leeway (port tack)	2¾ pts. L.
Variation and deviation	1½ pts. L.
	11 pts. L. of S.
	= 5 pts. R. of N.

Compass course, N.E. by E.

11. Given true course N. by W., wind N.E., leeway 3½ points, variation 2¾ points E., deviation 1¾ points E.; required compass course.

True course	1 pt. L. of N.
Leeway (starboard tack)	3½ pts. R.
Variation and deviation	4½ pts. L.
	2 pts. L. of N.

Compass course, N.N.W.

12. Given true course N.N.W., wind S.S.W., leeway 2½ points, variation 2¾ points E., deviation ¾ point E.; required compass course.

True course	2 pts. L. of N.
Leeway (port tack)	2½ pts. L.
Variation and deviation	3½ pts. L.
	8 pts. L. of N.

Compass course, W.

13. Given true course S. 64° E., leeway 0, variation 7° 0′ W., deviation 15° 0′ W.; required compass course.

True course	64° L. of S.
Variation and deviation	22° R.
	42° L. of S.

Compass course, S. 42° E.

14. Given true course N. 44 W., leeway 0, variation 6° 0′ E., deviation 20° 0′ W.; required compass course.

True course	44° L. of N.
Variation and deviation	14° R.
	30° L. of N.

Compass course, N. 30° W.

15. Given compass course N. 65° W., leeway 0, variation 10° 0′ E., deviation 3° 0′ E.; required true course.

Compass course	65° L. of N.
Variation and deviation	13° R.
	52° L. of N.

True course, N. 52° W.

16. Given compass course S. 15° W., leeway 0, variation 6° 0′ W., deviation 18° 0′ E.; required true course.

Compass course	15° R. of S.
Variation and deviation	12° R.
	27° R. of S.

True course, S. 27° W.

17. Given compass course S. 18° E., leeway 0, variation 25° 0′ E., deviation 10° 0′ E., required true course.

Compass course	18° L. of S.
Variation and deviation	35° R.
	17° R. of S.

True course, S. 17° W.

18. Given compass course N. 30° E., wind S. by W., leeway 1½ points, variation 12° 0′ E., deviation 10° 0′ W.; required true course.

Compass course	30°	R. of N.
Leeway (starboard tack)	14° 3′ 45″	L.
Variation and deviation	2°	R.
	17° 56′ 15″	R. of N.

True course, N. 17° 56′ E.

EXERCISE II. PAGE 293.

1. Given L' 49° 57′ N., C S.W. by W., D 488.0; required L'' and p.

$D = 488.0$	$p = D \sin C.$	$L_d = D \cos C.$
$C = 56° 15'$	$\log D = 2.68842$	$\log D = 2.68842$
	$\log \sin C = 9.91985$	$\log \cos C = 9.74474$
	$\log p = 2.60826$	$\log L_d = 2.43316$
	$p = 405.8.$	$L_d = 271'$
		$= 4° 31'$ S.
		$L' = 49° 57'$ N.
		$L'' = 45° 26'$ N.

2. Given L' 1° 45′ N., C S.E. by E., D 487.8; required L'' and p.

$D = 487.8$	$p = D \sin C.$	$L_d = D \cos C.$
$C = 56° 15'$	$\log D = 2.68824$	$\log D = 2.68824$
	$\log \sin C = 9.91985$	$\log \cos C = 9.74474$
	$\log p = 2.60809$	$\log L_d = 2.43298$
	$p = 405.6.$	$L_d = 271'$
		$= 4° 31'$ S.
		$L' = 1° 45'$ N.
		$L'' = 2° 46'$ S.

3. Given L' 3° 15′ S., C N.E. by E. ¾ E., D 449.1; required L'' and p.

$D = 449.1$	$p = D \sin C.$	$L_d = D \cos C.$
$C = 64° 41' 15''$	$\log D = 2.65234$	$\log D = 2.65234$
	$\log \sin C = 9.95616$	$\log \cos C = 9.63099$
	$\log p = 2.60850$	$\log L_d = 2.28333$
	$p = 406.$	$L_d = 192'$
		$= 3° 12'$ N.
		$L' = 3° 15'$ S.
		$L'' = 0° \; 3'$ S.

4. Given L' 2° 10′ S., C N. by E., D 267.0; required L'' and p.

$D = 267.0$	$p = D \sin C.$	$L_d = D \cos C.$
$C = 11° 15'$	$\log D = 2.42651$	$\log D = 2.42651$
	$\log \sin C = 9.29024$	$\log \cos C = 9.99157$
	$\log p = 1.71675$	$\log L_d = 2.41808$
	$p = 52.1$	$L_d = 262'$
		$= 4° 22'$ N.
		$L' = 2° 10'$ S.
		$L'' = 2° 12'$ N.

5. Given L' 41° 30′ N., C S.S.W., D 295.5; required L'' and p.

$D = 295.5$
$C = 22° 30'$

$p = D \sin C.$

$\log D = 2.47056$
$\log \sin C = \underline{9.58284}$
$\log p = 2.05340$
$p = 113.1.$

$L_d = D \cos C.$

$\log D = 2.47056$
$\log \cos C = \underline{9.96562}$
$\log L_d = 2.43618$
$L_d = 273'$
$= 4° 33'$ S.
$L' = \underline{41° 30'}$ N.
$L'' = 36° 57'$ N.

6. Given L' 21° 59′ S., L'' 24° 49′ S., D 360; required C and p.

$D = 360$
$L_d = 170$

$\cos C = \frac{L_d}{D}.$

$\log L_d = 2.23045$
$\log D = \underline{2.55630}$
$\log \cos C = 9.67415$
$C = 61° 49'$
$= 5\frac{1}{2}$ pts., nearly.

$p^2 = (D - L_d)(D + L_d)$
$= 190 \times 530.$

$\log 190 = 2.27875$
$\log 530 = \underline{2.72428}$
$\log p^2 = 5.00303$
$\log p = 2.50151$
$p = 317.3.$

7. Given L' 2° 9′ S., L'' 3° 11′ N., D 354; required C and p.

$D = 354$
$L_d = 320$

$\cos C = \frac{L_d}{D}.$

$\log L_d = 2.50515$
$\log D = \underline{2.54900}$
$\log \cos C = 9.95615$
$C = 25° 19'$
$= 2\frac{1}{4}$ pts.

$p^2 = (D - L_d)(D + L_d)$
$= 34 \times 674.$

$\log 34 = 1.53148$
$\log 674 = \underline{2.82866}$
$\log p^2 = 4.36014$
$\log p = 2.18007$
$p = 151.4.$

8. Given L' 1° 30′ N., L'' 0° 26′ S., C S. by W.; required D and p.

$L_d = 116$
$C = 11° 15'$

$D = L_d \sec C.$

$\log L_d = 2.06446$
$\log \sec C = \underline{0.00843}$
$\log D = 2.07289$
$D = 118.3.$

$p = L_d \tan C.$

$\log L_d = 2.06466$
$\log \tan C = \underline{9.29866}$
$\log p = 1.36312$
$p = 23.1.$

9. Given L' 40° 17′ N., L'' 37° 6′ N., C S. by W. $\frac{1}{2}$ W.; required D and p.

$L_d = 191$
$C = 16° 52' 30''$

$D = L_d \sec C.$

$\log L_d = 2.28103$
$\log \sec C = \underline{0.01911}$
$\log D = 2.30014$
$D = 199.6.$

$p = L_d \tan C.$

$\log L_d = 2.28103$
$\log \tan C = \underline{9.48194}$
$\log p = 1.76297$
$p = 57.9.$

10. Given L' 38° 0′ N., C S.W. by W., p 48.2; required L'' and D.

$p = 48.2$
$C = 56° 15'$

$$D = p \csc C.$$
$$\log p = 1.68305$$
$$\log \csc C = 0.08015$$
$$\log D = 1.76320$$
$$D = 58.0.$$

$$L_d = p \cot C.$$
$$\log p = 1.68305$$
$$\log \cot C = 9.82489$$
$$\log L_d = 1.50794$$
$$L_d = \quad 32' \text{ S.}$$
$$L' = 38° \ 0' \text{ N.}$$
$$L'' = 37° 28' \text{ N.}$$

11. Given L' 18° 25′ N., C S.W. by W. ¾ W., p 65.1; required L'' and D.

$p = 65.1$
$C = 64° 41' 15''$

$$D = p \csc C.$$
$$\log p = 1.81358$$
$$\log \csc C = 0.04384$$
$$\log D = 1.85742$$
$$D = 72.0$$

$$L_d = p \cot C.$$
$$\log p = 1.81358$$
$$\log \cot C = 9.67483$$
$$\log L_d = 1.48841$$
$$L_d = \quad 31' \text{ S.}$$
$$L' = 18° 25' \text{ N.}$$
$$L'' = 17° 54' \text{ N.}$$

12. Given L' 50° 18′ N., L'' 54° 48′ N., D 299.0; required C and p.

$D = 299.0$
$L_d = 270$

$$\cos C = \frac{L_d}{D}.$$
$$\log L_d = 2.43136$$
$$\log D = 2.47567$$
$$\log \cos C = 9.95569$$
$$C = 25° 26' 30''$$
$$= 2\tfrac{1}{4} \text{ pts., nearly.}$$

$$p^2 = (D - L_d)(D + L_d)$$
$$= 29 \times 569.$$
$$\log 29 = 1.46240$$
$$\log 569 = 2.75511$$
$$\log p^2 = 4.21751$$
$$\log p = 2.10875$$
$$p = 128.5.$$

13. Given L' 32° 30′ N., L'' 19° 59′ N., D 812.0; required C and p.

$D = 812.0$
$L_d = 751$

$$\cos C = \frac{L_d}{D}.$$
$$\log L_d = 2.87564$$
$$\log D = 2.90956$$
$$\log \cos C = 9.96608$$
$$C = 22° 21'$$
$$= 2 \text{ pts., nearly.}$$

$$p^2 = (D - L_d)(D + L_d)$$
$$= 61 \times 1563.$$
$$\log 61 = 1.78533$$
$$\log 1563 = 3.19396$$
$$\log p^2 = 4.97929$$
$$\log p = 2.48964$$
$$p = 308.8.$$

14. Given L' 2° 8′ S., C N. 11° E., D 500; required L'' and p.

$$D = 500 \qquad C = 11^\circ$$

$$\begin{aligned} p &= D \sin C. \\ \log D &= 2.69897 \\ \log \sin C &= 9.28060 \\ \log p &= 1.97957 \\ p &= 95.4. \end{aligned}$$

$$\begin{aligned} L_d &= D \cos C. \\ \log D &= 2.69897 \\ \log \cos C &= 9.99195 \\ \log L_d &= 2.69092 \\ L_d &= 491' \\ &= 8^\circ\ 11'\ \text{N.} \\ L' &= 2^\circ\ 8'\ \text{S.} \\ L'' &= 6^\circ\ 3'\ \text{N.} \end{aligned}$$

15. Given L' 20° 21′ S., C N. 20° E., D 402.0; required L'' and p.

$$D = 402.0 \qquad C = 20^\circ$$

$$\begin{aligned} p &= D \sin C. \\ \log D &= 2.60423 \\ \log \sin C &= 9.53405 \\ \log p &= 2.13828 \\ p &= 137.5. \end{aligned}$$

$$\begin{aligned} L_d &= D \cos C. \\ \log D &= 2.60423 \\ \log \cos C &= 9.97299 \\ \log L_d &= 2.57722 \\ L_d &= 378' \\ &= 6^\circ\ 18'\ \text{N.} \\ L' &= 20^\circ\ 21'\ \text{S.} \\ L'' &= 14^\circ\ 3'\ \text{S.} \end{aligned}$$

16. Given L' 40° 25′ S., C N. 87° E., D 240.0; required L'' and p.

$$D = 240.0 \qquad C = 87^\circ$$

$$\begin{aligned} p &= D \sin C. \\ \log D &= 2.38021 \\ \log \sin C &= 9.99940 \\ \log p &= 2.37961 \\ p &= 239.7. \end{aligned}$$

$$\begin{aligned} L_d &= D \cos C. \\ \log D &= 2.38021 \\ \log \cos C &= 8.71880 \\ \log L_d &= 1.09901 \\ L_d &= 13'\ \text{N.} \\ L' &= 40^\circ\ 25'\ \text{S.} \\ L'' &= 40^\circ\ 12'\ \text{S.} \end{aligned}$$

17. Given L' 20° 48′ N., L″ 17° 13′ N., p 289.2 W.; required C and D.

$$p = 289.2 \qquad L_d = 215$$

$$\begin{aligned} \tan C &= \frac{p}{L_d}. \\ \log p &= 2.46120 \\ \log L_d &= 2.33244 \\ \log \tan C &= 10.12876 \\ C &= \text{S. } 53^\circ\ 22'\ 18''\ \text{W.} \\ &= \text{S. } 53^\circ\ 22'\ \text{W.} \end{aligned}$$

$$\begin{aligned} D &= p \csc C. \\ \log p &= 2.46120 \\ \log \csc C &= 0.09554 \\ \log D &= 2.55674 \\ D &= 360.4. \end{aligned}$$

18. Given L' 51° 45′ N., L'' 53° 11′ N., p 128.0 E.; required C and D.

$p = 128.0$
$L_d = 86$

$$\tan C = \frac{p}{L_d}.$$

$\log p = 2.10721$
$\log L_d = 1.93450$
$\log \tan C = 10.17271$
$C =$ N. 56° 6′ 13″ E.
$=$ N.E. by E. nearly.

$D = p \csc C.$
$\log p = 2.10721$
$\log \csc C = 0.08090$
$\log D = 2.18811$
$D = 154.2.$

19. Given L' 0° 20′ S., L'' 0° 18′ N., p 142.7 E.; required C and D.

$p = 142.7$
$L_d = 38$

$$\tan C = \frac{p}{L_d}.$$

$\log p = 2.15442$
$\log L_d = 1.57978$
$\log \tan C = 10.57464$
$C =$ N. 75° 5′ 19″ E.
$=$ N. 75° 5′ E.

$D = p \csc C.$
$\log p = 2.15442$
$\log \csc C = 0.01488$
$\log D = 2.16930$
$D = 147.7.$

20. Given L' 40° 20′ N., L'' 41° 37′ N., p 52.6 W.; required C and D.

$p = 52.6$
$L_d = 77$

$$\tan C = \frac{p}{L_d}.$$

$\log p = 1.72099$
$\log L_d = 1.88649$
$\log \tan C = 9.83450$
$C =$ N. 34° 20′ 17″ W.
$=$ N. 34° 20′ W.

$D = p \csc C.$
$\log p = 1.72099$
$\log \csc C = 0.24867$
$\log D = 1.96966$
$D = 93.3.$

Exercise III. Page 296.

1. Given L 55° 55′, λ' 2° 10′ W., λ'' 12° 52′ E.; required p.

$$p = 15\tfrac{1}{30} \times 33.62 = 505.4 \text{ E.}$$

2. Given L 52° 0′, λ' 0° 59′ W., λ'' 2° 24′ E.; required p.

$$p = 3\tfrac{23}{60} \times 36.94 = 125.0 \text{ E.}$$

3. Given L 61° 25′, λ' 179° 20′ W., λ'' 176° 52′ E.; required p.

$$p = 3\tfrac{48}{60} \times 28.71 = 109.1 \text{ W.}$$

4. Given L 56° 0′, λ' 3° 12′ W., λ'' 4° 8′ E.; required p.

$$p = 7\tfrac{1}{3} \times 33.55 = 246.0 \text{ E.}$$

5. Given L 80° 0′, λ' 10° 0′ W., λ'' 17° 41′ W.; required p.

$$p = 7\tfrac{41}{60} \times 10.42 = 80.1 \text{ W.}$$

6. Given L 60° 0′, p 204.0 E., λ' 160° 2′ E.; required λ''.

$$\lambda_d = 204.0 \div \tfrac{1}{2} = 408' = 6° 48' \text{ E.};$$
$$\lambda'' = 160° 2' + 6° 48' = 166° 50' \text{ E.}$$

7. Given L 51° 28′, p 70.9 E., λ' 32° 7′ W.; required λ''.

$$\lambda_d = 70.9 \div 37.38 = 1.90° = 1° 54' \text{ E.};$$
$$\lambda'' = 32° 7' - 1° 54' = 30° 13' \text{ W.}$$

8. Given L 64° 16′, p. 265.7 W., λ' 170° 0′ W.; required λ''.

$$\lambda_d = 265.7 \div 26.05 = 10.20 = 10° 12' \text{ W.};$$
$$\lambda'' = 170° 0' + 10° 12' = 180° 12' \text{ W.} = 179° 48' \text{ E.}$$

9. Given L 46° 37′, p 352.0 E., λ' 163° 42′ E.; required λ''.

$$\lambda_d = 352.0 \div 41.21 = 8.54° = 8° 33' \text{ E.};$$
$$\lambda'' = 163° 42' + 8° 33' = 172° 15' \text{ E.}$$

10. Given L 39° 57′, p 398.0 W., λ' 4° 8′ W.; required λ''.

$$\lambda_d = 398.0 \div 45.93 = 8.67° = 8° 39' \text{ W.};$$
$$\lambda'' = 4° 8' + 8° 39' = 12° 47' \text{ W.}$$

11. From latitude 32° 3′ S., longitude 179° 45′ W., a ship makes 54 miles west (true). Required the longitude in.

$$\lambda_d = 54 \div 50.85 = 1.06° = 1° 4' \text{ W.};$$
$$\lambda'' = 179° 45' + 1° 4' = 180° 49' \text{ W.} = 179° 11' \text{ E.}$$

12. From latitude 35° 30′ S., longitude 27° 28′ W., a ship sails east (true) 301 miles. Required the longitude in and the compass course; variation $1\tfrac{3}{4}$ points E., leeway $\tfrac{1}{4}$ point to the left, deviation 8° 50′ E.

$$\lambda_d = 301 \div 48.85 = 6.16° = 6° 10';$$
$$\lambda'' = 27° 28' - 6° 10' = 21° 18' \text{ W.}$$

True course	8 pts. R. of N.
Variation and leeway	2 pts. L.
	6 pts. R. of N.
	= 67° 30′ R. of N.
Deviation	8° 50′ L.
Compass course	N. 58° 40′ E.

Exercise IV. Page 301.

1. Given L' 25° 35′ N., L'' 27° 28′ N., λ' 60° 0′ W., λ'' 54° 55′ W.; required C and D.

$L_d = 1° 58'$	$\tan C = \dfrac{\lambda_d \cos L_m}{L_d}$.	$D = L_d \sec C$.
$= 113'$		$\log L_d = 2.05308$
$L_m = 26° 32'$	$\log \lambda_d = 2.48430$	$\log \sec C = 10.41716$
$\lambda_d = 5° 5'$	$\log \cos L_m = 9.95167$	$\log D = 2.47024$
$= 305'$	$\text{colog } L_d = 7.94692$	$D = 295.3$.
	$\log \tan C = 10.38289$	
	$C = \text{N. } 67° 30' \text{ E.} = \text{E.N.E.}$	

2. Given L' 32° 30′ N., L'' 34° 10′ N., λ' 25° 24′ W., λ'' 29° 8′ W.; required C and D.

$L_d = 1° 40'$	$\tan C = \dfrac{\lambda_d \cos L_m}{L_d}$.	$D = L_d \sec C$.
$= 100'$		$\log L_d = 2.00000$
$L_m = 33° 20'$	$\log \lambda_d = 2.35025$	$\log \sec C = 10.32673$
$\lambda_d = 3° 44'$	$\log \cos L_m = 9.92194$	$\log D = 2.32673$
$= 224'$	$\text{colog } L_d = 8.00000$	$D = 212.2$.
	$\log \tan C = 10.27219$	
	$\therefore C = \text{N. } 61° 53' \text{ W.}$	

3. Given L' 39° 30′ S., L'' 41° 0′ S., λ' 74° 20′ E., λ'' 70° 12′ E.; required C and D.

$L_d = 1° 30'$	$\tan = \dfrac{\lambda_d \cos L_m}{L_d}$.	$D = L_d \sec C$.
$= 90'$		$\log L_d = 1.95424$
$L_m = 40° 15'$	$\log \lambda_d = 2.39445$	$\log \sec C = 10.36708$
$\lambda_d = 4° 8'$	$\log \cos L_m = 9.88266$	$\log D = 2.32132$
$= 248'$	$\text{colog } L_d = 8.04576$	$D = 209.6$.
	$\log \tan C = 10.32287$	
	$\therefore C = \text{S. } 64° 34' \text{ W.}$	

4. Given L' 46° 24′ S., λ' 178° 28′ E., C S.E. ¾ E., D 278.0; required L'' and λ''.

$C = 53° 26'$	$L_d = D \cos C$.	$\lambda_d = D \sin C \sec L_m$.
$D = 278.0$	$\log D = 2.44404$	$\log D = 2.44404$
	$\log \cos C = 9.77507$	$\log \sin C = 9.90480$
	$\log L_d = 2.21911$	$\log \sec L_m = 10.17267$
	$L_d = 166'$	$\log \lambda_d = 2.52151$
	$= 2° 46'$	$\lambda_d = 332'$
	$L' = 46° 24'$	$= 5° 32'$
	$L'' = 49° 10' \text{ S.}$	$\lambda' = 178° 28'$
	$L_m = 47° 47'$.	$\lambda'' = 184° \ 0' \text{ E.}$
		$= 176° \ 0' \text{ W.}$

5. Given L' 20° 29′ N., λ' 179° 10′ W., C W. by S. ½ S., D 333.0; required L'' and λ''.

$C = 73° 7'$	$L_d = D \cos C.$	$\lambda_d = D \sin C \sec L_m.$
$D = 333$	$\log D. = 2.52244$	$\log D = 2.52244$
	$\log \cos C = 9.46303$	$\log \sin C = 9.98087$
	$\log L_d = 1.98548$	$\log \sec L_m = 10.02610$
	$L_d = 97'$	$\log \lambda_d = 2.52941$
	$= 1° 37'$	$\lambda_d = 338'$
	$L' = 20° 29'$	$= 5° 38'$
	$L'' = 18° 52'$ N.	$\lambda' = 179° 10'$
	$L_m = 19° 40'.$	$\lambda'' = 184° 48'$ W.
		$= 175° 12'$ E.

6. Given L' 0° 56′ N., λ' 29° 50′ W., C S. 47° E., D 168.0; required L'' and λ''.

$C = 47°$	$L_d = D \cos C.$	$\lambda_d = D \sin C \sec L_m.$
$D = 168$	$\log D = 2.22530$	$\log D = 2.22530$
	$\log \cos C = 9.83378$	$\log \sin C = 9.86413$
	$\log L_d = 2.05908$	$\log \sec L_m = 0.00000$
	$L_d = 115'$	$\log \lambda_d = 2.08943$
	$= 1° 55'$	$\lambda_d = 123'$
	$L' = 0° 56'$	$= 2° 3'$
	$L'' = 0° 59'$ S.	$\lambda' = 29° 50'$
	$L_m = 0° 1'.$	$\lambda'' = 27° 47'$ W.

7. Given L' 42° 25′ N., λ' 66° 14′ W., C S.E. by E., D 25.0; required L'' and λ''.

$C = 56° 15'$	$L_d = D \cos C.$	$\lambda_d = D \sin C \sec L_m.$
$D = 25.0$	$\log D = 1.39794$	$\log D = 1.39794$
	$\log \cos C = 9.74474$	$\log \sin C = 9.91985$
	$\log L_d = 1.14268$	$\log \sec L_m = 0.13098$
	$L_d = 14'$	$\log \lambda_d = 1.44877$
	$L' = 42° 25'$	$\lambda_d = 28'$
	$L'' = 42° 11'$ N.	$\lambda' = 66° 14'$
	$L_m = 42° 18'.$	$\lambda'' = 65° 46'$ W.

8. Given L' 42° 8′ N., λ' 65° 48′ W., C E. ½ S., D 126.0; required L'' and λ''.

$C = 84° 22'$	$L_d = D \cos C.$	$\lambda_d = D \sin C \sec L_m.$
$D = 126.0$	$\log D = 2.10037$	$\log D = 2.10037$
	$\log \cos C = 8.99194$	$\log \sin C = 9.99790$
	$\log L_d = 1.09231$	$\log \sec L_m = 0.12915$
	$L_d = 12'$	$\log L_d = 2.22743$
	$L' = 42° \ 8'$	$\lambda_d = 168'$
	$L'' = 41° 56'$ N.	$= 2° 48'$
	$L_m = 42° 2'.$	$\lambda' = 65° 48'$
		$\lambda'' = 63° 0'$ W.

9. Given L' 41° 52′ N., λ' 62° 47′ W., C E. ½ S., D 161.0; required L'' and λ''.

$C = 84° 22'$	$L_d = D \cos C.$	$\lambda_d = D \sin C \sec L_m.$
$D = 161.0$	$\log D = 2.20683$	$\log D = 2.20683$
	$\log \cos C = 8.99194$	$\log \sin C = 9.99789$
	$\log L_d = 1.19877$	$\log \sec L_m = 0.12712$
	$L_d = 16'$	$\log \lambda_d = 2.33184$
	$L' = 41° 52'$	$\lambda_d = 215'$
	$L'' = 41° 36'$ N.	$= 3° 35'$
	$L_m = 41° 44'$ N.	$\lambda' = 62° 47'$
		$\lambda'' = 59° 12'$ W.

10. Given L' 41° 38′ N., L'' 41° 26′ N., λ' 59° 16′ W., C E. by S.; required λ'' and D.

$L_d = 12$	$D = L_d \sec C.$	$\lambda_d = D \sin C \sec L_m.$
$L_m = 41° 32'$	$\log L_d = 1.07918$	$\log D = 1.78894$
$C = 78° 45'$	$\log \sec C = 0.70976$	$\log \sin C = 9.99157$
	$\log D = 1.78894$	$\log \sec L_m = 0.12577$
	$D = 61.5.$	$\log \lambda_d = 1.90629$
		$\lambda_d = 81'$
		$= 1° 21'$
		$\lambda' = 59° 16'$
		$\lambda'' = 57° 55'$ W.

11. Given L' 41° 19′ N., L'' 41° 11′ N., λ' 57° 47′ W., D 167.0; required λ'' and C.

$L_d = 8$
$L_m = 41° 15'$
$D = 167.0$

$$\cos C = \frac{L_d}{D}.$$

$\log L_d = 0.90309$
$\log D = 2.22272$
$\log \cos C = 8.68037$
$C = 87° 15'.$

$$\lambda_d = D \sin C \sec L_m.$$

$\log D = 2.22272$
$\log \sin C = 9.99950$
$\log \sec L_m = 0.12387$
$\log \lambda_d = 2.34609$
$\lambda_d = 222'$
$= 3° 42'$
$\lambda' = 57° 47'$ W.
$\lambda'' = 61° 29'$ W.
or $54° 5'$ W.

12. Given L' 46° 28′ N., L'' 45° 17′ N., λ' 22° 18′ W., λ'' 19° 39′ W.; required C and D.

$L_d = 71$
$L_m = 45° 52'$
$\lambda_d = 159$

$$\tan C = \frac{\lambda_d \cos L_m}{L_d}.$$

$\log \lambda_d = 2.20140$
$\log \cos L_m = 9.84282$
$\text{colog } L_d = 8.14874$
$\log \tan C = 10.19296$
$C =$ S. 57° 19′ E.

$$D = L_d \sec C.$$

$\log L_d = 1.85126$
$\log \sec C = 0.26761$
$\log D = 2.11887$
$D = 131.5.$

13. Given L' 25° 30′ S., L'' 28° 15′ S., λ' 2° 15′ E., λ'' 11° 17′ E.; required C and D.

$L_d = 165$
$L_m = 26° 52'$
$\lambda_d = 542$

$$\tan C = \frac{\lambda_d \cos L_m}{L_d}.$$

$\log \lambda_d = 2.73399$
$\log \cos L_m = 9.95039$
$\text{colog } L_d = 7.78252$
$\log \tan C = 10.46690$
$C =$ S. 71° 9′ E.

$$D = L_d \sec C.$$

$\log L_d = 2.21748$
$\log \sec C = 0.49067$
$\log D = 2.70815$
$D = 510.7.$

14. Given L' 33° 40′ N., L'' 30° 49′ N., λ' 13° 20′ E., λ'' 17° 56′ E.; required C and D.

$L_d = 171$
$L_m = 31° 44'$
$\lambda_d = 276$

$$\tan C = \frac{\lambda_d \cos L_m}{L_d}.$$

$\log \lambda_d = 2.44090$
$\log \cos L_m = 9.92968$
$\text{colog } L_d = 7.76700$
$\log \tan C = 10.13758$
$C =$ S. 53° 46′ E.

$$D = L_d \sec C.$$

$\log L_d = 2.23300$
$\log \sec C = 0.22836$
$\log D = 2.46136$
$D = 289.3.$

15. Given L' 19° 30′ S., L'' 17° 24′ S., λ' 0° 10′ E., λ'' 1° 28′ W.; required C and D.

$L_d = 126$
$L_m = 18° 27'$
$\lambda_d = 98$

$\tan C = \frac{\lambda_d \cos L_m}{L_d}$.

$\log \lambda_d = 1.99123$
$\log \cos L_m = 9.97708$
$\text{colog } L_d = 7.89963$
$\log \tan C = 9.86794$
$C = \text{N. } 36° 25' \text{ W.}$

$D = L_d \sec C$.

$\log L_d = 2.10037$
$\log \sec C = 0.09435$
$\log D = 2.19472$
$D = 156.6$.

16. A ship sails from Boston light-house, in latitude 42° 20′ N., longitude 71° 4′ W., on a N.N.E. course, 184 miles. Find the latitude and longitude in.

$C = 22° 30'$
$D = 184$

$L_d = D \cos C$.

$\log D = 2.26482$
$\log \cos C = 9.96603$
$\log L_d = 2.23085$
$L_d = 170'$
$= 2° 50'$
$L' = 42° 20'$
$L'' = 45° 10' \text{ N.}$
$L_m = 43° 45'$.

$\lambda_d = D \sin C \sec L_m$.

$\log D = 2.26482$
$\log \sin C = 9.58284$
$\log \sec L_m = 0.14124$
$\log \lambda_d = 1.98890$
$\lambda_d = 97'$
$= 1° 37'$
$\lambda' = 71° 4'$
$\lambda'' = 69° 27' \text{ W.}$

17. A ship sails from Cape May, in latitude 38° 56′ N., longitude 74° 57′ W., on a S.S.E. course, 240 miles. Find the latitude and longitude in.

$C = 22° 30'$
$D = 240$

$L_d = D \cos C$.

$\log D = 2.38021$
$\log \cos C = 9.96603$
$\log L_d = 2.34624$
$L_d = 222'$
$= 3° 42'$
$L' = 38° 56'$
$L'' = 35° 14' \text{ N.}$
$L_m = 37° 5'$.

$\lambda_d = D \sin C \sec L_m$.

$\log D = 2.38021$
$\log \sin C = 9.58284$
$\log \sec L_m = 0.09813$
$\log \lambda_d = 2.06118$
$\lambda_d = 115'$
$= 1° 55'$
$\lambda' = 74° 57'$
$\lambda'' = 73° 2' \text{ W.}$

18. A ship sails from Cape Cod light, in latitude 42° 2′ N., longitude 70° 3′ W., on an E. by N. compass course, 170 miles; wind S.E. by S., leeway ½ point, deviation 17¾° E., variation 11¼° W. Find the latitude and longitude in.

Compass course	7 pts. R. of N.
Leeway	¼ pt. L.
	6¾ pts. R. of N.
	= 75° 56′ R. of N.
Variation and deviation	6° 30′ R.
True course	. N. 82° 26′ E.

$C = 82° 26′$	$L_d = D \cos C.$	$\lambda_d = D \sin C \sec L_m.$
$D = 170$	$\log D = 2.23045$	$\log D = 2.23045$
	$\log \cos C = 9.11951$	$\log \sin C = 9.99620$
	$\log L_d = 1.34996$	$\log \sec L_m = 0.13041$
	$L_d = 22′$	$\log \lambda_d = 2.35706$
	$L′ = 42° 2′$	$\lambda_d = 228′$
	$L″ = 42° 24′$ N.	$= 3° 48′$
	$L_m = 42° 13′.$	$\lambda′ = 70° 3′$
		$\lambda″ = 66° 15′$ W.

19. A ship sails from Cape Cod light on a S.S.E. compass course, 140 miles; deviation 5½° E., variation 11¼° W. Find the latitude and longitude in.

Compass course	22° 30′ L. of S.
Variation and deviation	5° 45′ L.
True course	. S. 28° 15′ E.

$C = 28° 15′$	$L_d = D \cos C.$	$\lambda_d = D \sin C \sec L_m.$
$D = 140$	$\log D = 2.14613$	$\log D = 2.14613$
	$\log \cos C = 9.94492$	$\log \sin C = 9.67516$
	$\log L_d = 2.09105$	$\log \sec L_m = 0.12222$
	$L_d = 123′$	$\log \lambda_d = 1.94351$
	$= 2° 3′$	$\lambda_d = 88′$
	$L′ = 42° 2′$	$= 1° 28′$
	$L′ = 39° 59′$ N.	$\lambda′ = 70° 3′$
	$L_m = 41° 0′.$	$\lambda″ = 68° 35′$ W.

20. A ship sails from latitude 55° 1′ N., longitude 1° 25′ W., on a S.W. compass course, 101 miles; wind W.N.W., leeway 1¼ points, deviation 6° W., variation 24° 56′ W. Find the latitude and longitude in.

Compass course . . .	4 pts. R. of S.	$L_d = D$	$\lambda_d = 0.$
Leeway	1¼ pts. L.	$= 101′$	$\lambda″ = \lambda′$
	2¾ pts. R. of S.	$= 1° 41′$	$= 1° 25′$ W.
	= 30° 56′ R. of S.	$L′ = 55° 1′$	
Variation and deviation	30° 56′ L.	$L″ = 53° 20′$ N.	
True course	S.		

21. A ship sails from the Bermudas, in latitude 32° 18′ N., longitude 64° 50′ W., on a W.S.W. compass course, 190 miles; deviation 1 point W., variation 1 point W. Find the latitude and longitude in.

Compass course	6 pts. R. of S.
Variation and deviation	2 pts. L.
True course	4 pts. R. of S.

$C = 45°$
$D = 190$

$$L_d = D \cos C.$$
$$\log D = 2.27875$$
$$\log \cos C = 9.84949$$
$$\log L_d = 2.12824$$
$$L_d = 134'$$
$$= 2° 14'$$
$$L' = 32° 18'$$
$$L'' = 30° \ 4' \text{ N.}$$
$$L_m = 31° 11'.$$

$$\lambda_d = D \sin C \sec L_m.$$
$$\log D = 2.27875$$
$$\log \sin C = 9.84949$$
$$\log \sec L_m = 0.06777$$
$$\log \lambda_d = 2.19601$$
$$\lambda_d = 157'$$
$$= 2° 37'$$
$$\lambda' = 64° 50'$$
$$\lambda'' = 67° 27' \text{ W.}$$

22. A ship sails from the Bermudas on a W.N.W. compass course, 90 miles; wind S.W., leeway 1 point, deviation 1 point E., variation 1 point W. Find the latitude and longitude in.

Compass course	6 pts. L. of N.
Leeway	1 pt. R.
Variation and deviation	0
True course	5 pts. L. of N.
	= 56° 15′ L. of N.

$C = 56° 15'$
$D = 90$

$$L_d = D \cos C.$$
$$\log D = 1.95124$$
$$\log \cos C = 9.74474$$
$$\log L_d = 1.69898$$
$$L_d = 50'$$
$$L' = 32° 18'$$
$$L'' = 33° \ 8' \text{ N.}$$
$$L_m = 32° 47'.$$

$$\lambda_d = D \sin C \sec L_m.$$
$$\log D = 1.95124$$
$$\log \sin C = 9.91985$$
$$\log \sec L_m = 0.07502$$
$$\log \lambda_d = 1.94911$$
$$\lambda_d = 89'$$
$$= 1° 29'$$
$$\lambda' = 64° 50'$$
$$\lambda'' = 66° 19' \text{ W.}$$

23. A navigator wishes to sail on a rhumb from the Bermudas to Cape Fear, in latitude 33° 52′ N., longitude 78° W.; variation 10° W., deviation 7° W. Find the compass course and distance.

$L_d = 94$
$L_m = 32° 5'$
$\lambda_d = 790$

$$\tan C = \frac{\lambda_d \cos L_m}{L_d}.$$

$\log \lambda_d =$	2.89763
$\log \cos L_m =$	9.92318
$\text{colog } L_d =$	8.02687
$\log \tan C =$	10.84768
$C =$	N. 81° 55′ W.
Var. and dev. =	17° W.
Compass course	N. 64° 55′ W.

$$D = L_d \sec C.$$

$\log L_d =$	1.97313
$\log \sec C =$	0.85197
$\log D =$	2.82510
$D =$	668.5.

24. A ship from latitude 36° 32′ N. sails between south and west until she has made 480 miles of departure, and 9° 22′ of difference of longitude. Required the latitude in, the course steered, and the distance run. [Take $L_m = \frac{1}{2}(L' + L'') + 13'$.]

$p = 480$
$\lambda_d = 562$

$$\cos L_m = \frac{p}{\lambda_d}.$$

$\log p =$	2.68124
$\log \lambda_d =$	2.74974
$\log \cos L_m =$	9.93150
$L_m =$	31° 20′.
$L'' = 2(L_m - 13') - L'$	
$=$	25° 42′ N.
$L_d =$	10° 50′
$=$	650.

$$\tan C = \frac{p}{L_d}.$$

$\log p =$	2.68124
$\log L_d =$	2.81291
$\log \tan C =$	9.86833
$C =$	S. 36° 27′ W.

$$D = p \csc C.$$

$\log p =$	2.68124
$\log \csc C =$	0.22613
$\log D =$	2.90737
$D =$	807.9.

Exercise V. Page 116.

1. Given L' 38° 14′ N., L'' 39° 51′ N., λ' 2° 7′ E., λ'' 4° 18′ E.; required C and D.

39° 51′ N.	Mer. parts = 2596.2	4° 18″ E.
38° 14′ N.	= 2471.8	2° 7″ E.
$L_d =$ 1° 37′ = 97′	Mer. $L_d =$ 124.4	$\lambda_d =$ 2° 11″ = 131′

$$\tan C = \frac{\lambda_d}{\text{Mer. } L_d}$$

$\log \lambda_d$	= 2.11727
colog Mer. L_d	= 7.90518
$\log \tan C$	= 10.02245
$\therefore C$	= N. 46° 29′ E.

$$D = L_d \sec C.$$

$\log L_d$	= 1.98677
$\log \sec C$	= 0.16205
$\log D$	= 2.14882
$\therefore D$	= 140.9.

2. Given L' 49° 53′ N., L'' 48° 28′ N., λ' 6° 19′ W., λ'' 5° 3′ W.; required C and D.

$$\begin{array}{lll} 49^\circ\ 53'\ \text{N.} & \text{Mer. parts} = 3446.0 & 6^\circ\ 19'\ \text{W.} \\ \underline{48^\circ\ 28'\ \text{N.}} & \text{Mer. parts} = \underline{3316.4} & \underline{5^\circ\ \ 3'\ \text{W.}} \\ L_d = 1^\circ\ 25' = 85' & \text{Mer. } L_d = \ 129.6 & \lambda_d = 1^\circ\ 16' = 76' \end{array}$$

$$\tan C = \frac{\lambda_d}{\text{Mer. } L_d}. \qquad D = L_d \sec C.$$

$$\begin{array}{ll} \log \lambda_d = 1.88081 & \log L_d = 1.92941 \\ \text{colog Mer. } L_d = \underline{7.88730} & \log \sec C = \underline{0.06416} \\ \log \tan C = 9.76829 & \log D = 1.99357 \\ \therefore C = \text{S. } 30^\circ\ 23'\ \text{E.} & \therefore D = 98.5. \end{array}$$

3. Given L' 64° 30′ N., L'' 60° 40′ N., λ' 4° 20′ W., λ'' 0° 10′ E.; required C and D.

$$\begin{array}{lll} 64^\circ\ 30'\ \text{N.} & \text{Mer. parts} = 5087.7 & 4^\circ\ 20'\ \text{W.} \\ \underline{60^\circ\ 40'\ \text{N.}} & \text{Mer. parts} = \underline{4582.2} & \underline{0^\circ\ 10'\ \text{E.}} \\ L_d = 3^\circ\ 50' = 230' & \text{Mer. } \lambda_d = \ 505.5 & \lambda_d = 4^\circ\ 30' = 270' \end{array}$$

$$\tan C = \frac{\lambda_d}{\text{Mer. } L_d}. \qquad D = L_d \sec C.$$

$$\begin{array}{ll} \log \lambda_d = 2.43136 & \log L_d = 2.36173 \\ \text{colog Mer. } L_d = \underline{7.29628} & \log \sec C = \underline{0.05569} \\ \log \tan C = 9.72764 & \log D = 2.42742 \\ \therefore C = \text{S. } 28^\circ\ 24'\ \text{E.} & \therefore D = 261.5. \end{array}$$

4. Given L' 54° 54′ S., L'' 34° 22′ S., λ' 60° 28′ W., λ'' 18° 24′ W.; required C and D.

$$\begin{array}{lll} 54^\circ\ 54'\ \text{S.} & \text{Mer. parts} = 3938.7 & 60^\circ\ 28'\ \text{W.} \\ \underline{34^\circ\ 22'\ \text{S.}} & \text{Mer. parts} = \underline{2185.1} & \underline{18^\circ\ 24'\ \text{W.}} \\ L_d = 20^\circ\ 32' = 1232' & \text{Mer. } L_d = 1753.6 & \lambda_d = 42^\circ\ \ 4' = 2524' \end{array}$$

$$\tan C = \frac{\lambda_d}{\text{Mer. } L_d}. \qquad D = L_d \sec C.$$

$$\begin{array}{ll} \log \lambda_d = \ 3.40209 & \log L_d = 3.09061 \\ \text{colog Mer. } L_d = \ \underline{6.75607} & \log \sec C = \underline{0.24376} \\ \log \tan C = 10.15816 & \log D = 3.33437 \\ \therefore C = \text{N. } 55^\circ\ 13'\ \text{E.} & \therefore D = 2160. \end{array}$$

5. Given L' 17° 0′ N., L'' 20° 0′ N., λ' 180° 0′ E., λ'' 177° 0′ E.; required C and D.

$$\begin{array}{lll} 20^\circ\ 0'\ \text{N.} & \text{Mer. parts} = 1217.3 & 180^\circ\ 0'\ \text{E.} \\ \underline{17^\circ\ 0'\ \text{N.}} & \text{Mer. parts} = \underline{1028.6} & \underline{177^\circ\ 0'\ \text{E.}} \\ L_d = 3^\circ\ 0' = 180' & \text{Mer } L_d = \ 188.7 & \lambda_d = 30^\circ\ 0' = 180' \end{array}$$

$$\tan C = \frac{\lambda_d}{\text{Mer. } \lambda_d}.$$

$\log \lambda_d = 2.25527$
$\text{colog Mer. } L_d = 7.72423$
$\log \tan C = 9.97950$
$C = \text{N. } 43^\circ\ 39'\ \text{W.}$

$$D = L_d \sec C.$$

$\log L_d = 2.25527$
$\log \sec C = 0.14052$
$\log D = 2.39579$
$D = 248.8.$

6. Given L' 45° 15′ N., λ' 35° 26′ W., C N. 49° E., D 175; required L'' and λ''.

$$L_d = D \cos C.$$

$\log D = 2.24301$
$\log \cos C = 9.81694$
$\log L_d = 2.05995$
$L_d = 115'$
$= 1^\circ\ 55'$
$L' = 45^\circ\ 15'$
$L'' = 47^\circ\ 10'\ \text{N.}$
$L_m = 46^\circ\ 12'.$

$$\lambda_d = D \sin C \sec L_m.$$

$\log D = 2.24301$
$\log \sin C = 9.87778$
$\log \sec L_m = 10.15980$
$\log \lambda_d = 2.27959$
$\therefore \lambda_d = 190'$
$= 3^\circ\ 11'$
$\lambda' = 35^\circ\ 26'$
$\lambda'' = 32^\circ\ 15'\ \text{W.}$

7. Given L' 55° 1′ N., λ' 1° 25′ E., C N. 10° E., D 246; required L'' and λ''.

$$L_d = D \cos C.$$

$\log D = 2.39094$
$\log \cos C = 9.99313$
$\log D = 2.38407$
$D = 242'$
$= 4^\circ\ 2'$
$L' = 55^\circ\ 1'$
$L'' = 59^\circ\ 3'\ \text{N.}$
$L_m = 57^\circ\ 2'.$

55° 1′ N., Mer. parts = 3950.9
59° 3′ N., Mer. parts = 4395.3
Mer. L_d = 444.4

$$\lambda_d = \text{Mer. } L_d \times \tan C.$$

$\log \text{Mer. } L_d = 2.64777$
$\log \tan C = 9.24632$
$\log \lambda_d = 1.89409$
$\therefore \lambda_d = 78'$
$= 1^\circ\ 18'$
$\lambda' = 1^\circ\ 25'$
$\lambda'' = 2^\circ\ 43'\ \text{E.}$

8. Given L' 50° 48′ N., λ' 9° 10′ W., C S. 41° W., D 275; required L'' and λ''.

$$L_d = D \cos C.$$

$\log D = 2.43993$
$\log \cos C = 9.87778$
$\log L_d = 2.31771$
$\therefore L_d = 208'$
$= 3^\circ\ 28'$
$L' = 50^\circ\ 48'$
$L'' = 47^\circ\ 20'\ \text{N.}$
$L_m = 49^\circ\ 4'.$

50° 48′ N., Mer. parts = 3532.0
47° 20′ N., Mer. parts = 3215.2
Mer. L_d = 316.8

$$\lambda_d = \text{Mer. } L_d \tan C.$$

$\log \text{Mer. } L_d = 2.50079$
$\log \tan C = 9.93916$
$\log \lambda_d = 2.43995$
$\lambda_d = 275'$
$= 4^\circ\ 35'$
$\lambda' = 9^\circ\ 10'$
$\lambda'' = 13^\circ\ 45'\ \text{W.}$

9. Given L' 37° 0′ N., L'' 51° 18′ N., λ' 48° 20′ W., D 1027; required λ'' and C.

$$\begin{aligned} & 51° \; 18' \text{ N.} \\ & \underline{37° \;\; 0' \text{ N.}} \\ L_d = \; & 14° \; 18' \\ = \; & 858'. \end{aligned} \qquad \begin{aligned} \text{Mer. parts} &= 3579.6 \\ \text{Mer. parts} &= \underline{2378.8} \\ \text{Mer. } L_d &= 1200.8 \end{aligned}$$

$$\cos C = \frac{L_d}{D}. \qquad \lambda_d = \text{Mer. } L_d \tan C.$$

$$\begin{aligned} \log L_d &= 2.93349 \\ \log D &= \underline{3.01157} \\ \log \cos C &= 9.92192 \\ C &= \text{N. } 33° \; 20' \text{ W.} \end{aligned} \qquad \begin{aligned} \log \text{Mer. } L_d &= 3.07947 \\ \log \tan C &= \underline{9.81804} \\ \log \lambda_d &= 2.89751 \\ \lambda_d &= 790' \\ &= 13° \; 10' \\ \lambda' &= \underline{48° \; 20'} \\ \lambda'' &= 61° \, 30' \text{ W. or } 35° \, 10' \text{ W.} \end{aligned}$$

10. Given L' 51° 15′ N., L'' 37° 5′ N., λ' 9° 50′ W., C S.W. by S.; required λ'' and D.

$$\begin{aligned} & 51° \; 15' \text{ N.} \\ & \underline{37° \;\; 5' \text{ N.}} \\ L_d = \; & 14° \; 10' \\ = \; & 850'. \end{aligned} \qquad \begin{aligned} \text{Mer. parts} &= 3574.8 \\ \text{Mer. parts} &= \underline{2385.1} \\ \text{Mer. } L_d &= 1189.7 \end{aligned}$$

$$\begin{aligned} C &= 33° \; 45'. \\ D &= L_d \sec C. \\ \log L_d &= 2.92942 \\ \log \sec C &= \underline{0.08015} \\ \log D &= 3.00957 \\ D &= 1022. \end{aligned} \qquad \begin{aligned} \lambda_d &= \text{Mer. } L_d \tan C. \\ \log \text{Mer. } L_d &= 3.07542 \\ \log \tan C &= \underline{9.82489} \\ \log \lambda_d &= 2.90031 \\ \lambda_d &= 795' \\ &= 13° \; 15' \\ \lambda' &= \underline{\;\, 9° \; 50'} \\ \lambda'' &= 23° \;\; 5' \text{ W.} \end{aligned}$$

11. Required the course and distance from Toulon to Valencia, by Mercator's sailing:

$$\text{Toulon} \begin{cases} L = 43° \;\; 8' \text{ N.} \\ \lambda = \;\; 5° \; 56' \text{ E.} \end{cases} \qquad \text{Valencia} \begin{cases} L = 39° \; 27' \text{ N.} \\ \lambda = \;\; 0° \; 19' \text{ W.} \end{cases}$$

$$\begin{aligned} & 43° \;\; 8' \text{ N.} \\ & \underline{39° \; 27' \text{ N.}} \\ L_d = \; & 3° \; 41' = 221' \end{aligned} \qquad \begin{aligned} \text{Mer. parts} &= 2858.3 \\ \text{Mer. parts} &= \underline{2565.2} \\ \text{Mer. } L_d &= \;\; 293.1 \end{aligned} \qquad \begin{aligned} & 5° \; 56' \text{ E.} \\ & \underline{0° \; 19' \text{ W.}} \\ \lambda_d = \; & 6° \; 15' = 375' \end{aligned}$$

$$D = L_d \sec C. \qquad \tan C = \frac{\lambda_d}{\text{Mer. } L_d}.$$

$$\begin{aligned} \log L_d &= 2.34439 \\ \log \sec C &= \underline{0.21050} \\ \log D &= 2.55489 \\ D &= 358.8. \end{aligned} \qquad \begin{aligned} \log \lambda_d &= \;\; 2.57403 \\ \log \text{Mer. } L_d &= \;\; \underline{2.46702} \\ \log \tan C &= 10.10701 \\ C &= \text{S. } 51° \; 59' \text{ W.} \end{aligned}$$

12. Required the compass course and distance from Cape East, New Zealand, to San Francisco; variation 14° 20′ E., and deviation 5° 40′ E.:

Cape East $\begin{cases} L = 37° 40' \text{ S.} \\ \lambda = 178° 36' \text{ E.} \end{cases}$ San Francisco $\begin{cases} L = 37° 48' \text{ N.} \\ \lambda = 122° 24' \text{ W.} \end{cases}$

$$\begin{array}{lll} 37° 48' \text{ N.} & \text{Mer. parts} = 2439.0 & 178° 36' \text{ E.} \\ \underline{37° 40' \text{ S.}} & \text{Mer. parts} = \underline{2428.9} & \underline{122° 24' \text{ W.}} \\ L_d = 75° 28' = 4528' & \text{Mer. } L_d = 4867.9 & \lambda_d = 59° 0' = 3540' \end{array}$$

$$D = L_d \sec C. \qquad \tan C = \frac{\lambda_d}{\text{Mer. } L_d}.$$

$$\begin{array}{ll} \log L_d = 3.65591 & \log \lambda_d = 3.54900 \\ \log \sec C = \underline{0.09222} & \log \text{Mer. } L_d = \underline{3.68734} \\ \log D = 3.74813 & \log \tan C = 9.86166 \\ D = 5599. & C = \text{N. } 36° 2' \text{ E.} \\ & \text{Var. and dev.} = \underline{20° 0' \text{ L.}} \\ & \text{Compass course} = \text{N. } 16° 2' \text{ E.} \end{array}$$

13. Required the course and distance from Cape Lopatka to Callao:

Cape Lopatka $\begin{cases} L = 51° 2' \text{ N.} \\ \lambda = 156° 50' \text{ E.} \end{cases}$ Callao $\begin{cases} L = 12° 4' \text{ S.} \\ \lambda = 77° 14' \text{ W.} \end{cases}$

$$\begin{array}{lll} 51° 2' \text{ N.} & \text{Mer. parts} = 3554.1 & 156° 50' \text{ E.} \\ \underline{12° 4' \text{ S.}} & \text{Mer. parts} = \underline{724.6} & \underline{77° 14' \text{ W.}} \\ L_d = 63° 6' = 3786' & \text{Mer. } L_d = 4278.7 & 234° 4' \end{array}$$

$$360° - 234° 4' = 125° 56' = 7556 \text{ m.} = \lambda_d.$$

$$D = L_d \sec C. \qquad \tan C = \frac{\lambda_d}{\text{Mer. } L_d}.$$

$$\begin{array}{ll} \log L_d = 3.57818 & \log \lambda_d = 3.87829 \\ \log \sec C = \underline{0.30744} & \text{colog Mer. } L_d = \underline{6.36869} \\ \log D = 3.88562 & \log \tan C = 10.24698 \\ D = 7685. & C = \text{S. } 60° 29' \text{ E.} \end{array}$$

14. A ship from latitude 20° 40′ N. sails N.E. by N. until she is in latitude 27° 16′ N. Required the distance and difference of longitude.

$$\begin{array}{ll} 27° 16' \text{ N.} & \text{Mer. parts} = 1691.0 \\ \underline{20° 40' \text{ N.}} & \text{Mer. parts} = \underline{1259.7} \\ L_d = 6° 36' = 396'. & \text{Mer. } L_d = 431.3 \\ C = 33° 45'. & \\ D = L_d \sec C. & \lambda_d = \text{Mer. } L_d \tan C. \\ \log L_d = 2.59770 & \log \text{Mer. } L_d = 2.63478 \\ \log \sec C = \underline{0.08015} & \log \tan C = \underline{9.82489} \\ \log D = 2.67785 & \log \lambda_d = 2.45967 \\ D = 476.3. & \lambda_d = 288' = 4° 48'. \end{array}$$

15. A ship from Cape Clear, in latitude 51° 26′ N. and longitude 9° 29′ W., sails S.W. by S. until the distance run is 1022 miles. Find the latitude and longitude in by Mercator's and Middle Latitude Sailings. Which method is preferable?

By Mercator's Sailing,

Mer. parts of 51° 26′ = 3592.4
Mer. parts of 37° 16′ = 2398.8
Mer. L_d = 1193.6

$$L_d = D \cos C.$$

$D = 1022$
$C = 33° 45'$

$\log D = 3.00945$
$\log \cos C = 9.91985$
$\log L_d = 2.92930$
$L_d = 850'$
$= 14° 10'$
$L' = 51° 26'$
$L'' = 37° 16'$ N.
$L_m = 44° 21'$ N.

$$\lambda_d = \text{Mer. } L_d \tan C.$$

$\log \text{Mer. } L_d = 3.07686$
$\log \tan C = 9.82489$
$\log \lambda_d = 2.90175$
$\lambda_d = 798'$
$= 13° 18'$
$\lambda' = 9° 29'$
$\lambda'' = 22° 47'$ W.

By Mid. Lat. Sailing,

$$\lambda_d = D \sin C \sec L_m.$$

$\log D = 3.00945$
$\log \sin C = 9.74474$
$\log \sec L_m = 0.14564$
$\log \lambda_d = 2.89983$
$\lambda_d = 794'$
$= 13° 14'$
$\lambda' = 9° 29'$
$\lambda'' = 22° 43'$ W.

Mercator's Sailing is preferable, since $C < 45°$.

Exercise VI. Page 312.

1.

C.		D.	N.	S.	E.	W.
S.S.W.	2 pts.	48		44.3		18.4
S.W. by S.	3 pts.	36		29.9		20.
N.E.	4 pts.	24	17		17	
Hence, $L_d = 57.2$ S.				74.2	17	38.4
$= 0° 57'$ S.				17.		17.
$p = 21.4$ W.				57.2		21.4

2.

C.		*D.*	*N.*	*S.*	*E.*	*W.*
S. ½ E.	½ pt.	18		17.9	1.8	
S.W. ½ S.	3½ pts.	37		28.6		23.5
S.S.W. ¼ W.	2¼ pts.	56		50.6		23.9
Hence, $L_d = 97.1$ S.			0	97.1	1.8	47.4
$= 1° 37'$ S.				0.		1.8
$p = 45° 6'$ W.				97.1		45.6

3.

C.		*D.*	*N.*	*S.*	*E.*	*W.*
S.S.W. ¼ W.	2¼ pts.	43		38.9		18.4
S.S.W. ½ W.	2½ pts.	39		34.4		18.4
S. by W. ½ W.	1½ pts.	27		25.8		7.8
Hence, $L_d = 99.1$ S.			0	99.1	0	44.6
$= 1° 39'$ S.				0.		0.
$p = 44.6$ W.				99.1		44.6

4.

C.	*D.*	*N.*	*S.*	*E.*	*W.*
N. 25° W.	16.4	14.9		0.1	6.9
N. 8° E.	7.8	7.7		1.	
N. 19° E.	13.7	13.0		4.5	
N. 76° E.	39.6	9.6		38.4	
Hence, $L_d = 45.2$ N.		45.2	0	44.	6.9
$= 0° 45'$ N.		0.		6.9	
$p = 37.1$ E.		45.2		37.1	

5.

C.		*D.*	*N.*	*S.*	*E.*	*W.*
W.N.W. ¼ W.	6¼ pts.	21	7.1			19.8
N.N.E. ¾ E.	2¾ pts.	9	7.7		4.6	
N. by E. ¾ E.	1¾ pts.	9	8.5		3.0	
S.S.W. ¼ W.	2¼ pts.	30		27.1		12.8
Hence, $L_d = 3.8$ S.			23.3	27.1	7.6	32.6
$= 0° 4'$ S.				23.3		7.6
$p = 25.0$ W.				3.8		25.0

6.

C.	*D.*	*N.*	*S.*	*E.*	*W.*
S. 83° W.	23		2.8		22.8
S. 48° E.	25.2		16.9	18.7	
N. 48° W.	27.1	18.1			20.1
N. 36° W.	2.1	17.			12.3
Hence, $L_d = 15.4$ N.		35.1	19.7	18.7	55.2
$= 0° 15'$ N.		19.7			18.7
$p = 36.5$ W.		15.4			36.5

7.

C.	*D.*	*N.*	*S.*	*E.*	*W.*
S. 17° E.	48		45.9	14.0	
S. 45° W.	19		13.4		13.4
N. 36° W.	18	14.6			10.6
N. 41° W.	50	37.7			32.8
E. (90°).	36	.0	.0	36.0	
		52.3	59.3	50.0	56.8
Hence, $L_d = 0° 7'$ S.			52.3		50.
$p = 6.8$ W.			7.		6.8

8.

C.		D.	N.	S.	E.	W.
N.N.E.	2 pts.	31	28.6		11.9	
E.N.E.	6 pts.	35	13.4		32.3	
E. by S.	7 pts.	36		7	35.3	
S.S.E.	2 pts.	51		47.1	19.5	
S. by E.	1 pt.	60		58.8	11.7	
Hence, $L_d = 70.9$ $= 1° 11' S.$ $p = 110.7$ E.			42.0	112.9 42.0 70.9	110.7	

9.

C.	D.	N.	S.	E.	W.
S. 44° E.	69		49.6	47.9	
S. 85° E.	68		5.9	67.7	
S. 27° E.	25		22.3	11.3	
N. 37° W.	5	4.0			3.0
N. 20° W.	13	12.2			4.4
Hence, $L_d = 61.6$ $= 1° 2' S.$ $p = 119.5$ E.		16.2	77.8 16.2 61.6	126.9 7.4 119.5	7.4

Exercise VII. Page 318.

1. First course: N.N.E. = 2 points R. of N. = N. 22° 30′ E., 31.4 m.
Second course: E.N.E. = 6 points R. of N. = N. 67° 30′ E., 35 m.
Third course: E. by S. = 7 points L. of S. = S. 78° 45′ E., 36.1 m.
Fourth course: S.S.E. = 2 points L. of S. = S. 22° 30′ E., 50.9 m.
Tide course: = 1 point L. of S. = S. 11° 15′ E., 60 m.

THE TRAVERSE.

C.		*D.*	*N.*	*S.*	*E.*	*W.*
N.N.E.	2 pts.	31.4	29.		12.	
E.N.E.	6 pts.	35.	13.4		32.3	
E. by S.	7 pts.	36.1		7.	35.4	
S.S.E.	2 pts.	50.9		47.	19.5	
S. by E.	1 pt.	60.		58.8	11.7	
$L_d = 70.4' = 1° 10'$ S. $L' = 46° 28'$ N. $L'' = 45° 18'$ N.			42.4	112.8 42.4 70.4	110.9	$= p.$

$p = 110.9$ $\qquad \lambda_d = p \sec L_m.$ $\qquad \lambda_d = 159.3'$

$L_m = 45° 53'$ $\qquad \log p = 2.04493$ $\qquad = 2° 39'$ E.

$\log \sec L_m = 0.15731$ $\qquad \lambda' = 22° 18'$ W.

$\log \lambda_d = 2.20224$ $\qquad \lambda'' = 19° 39'$ W.

2. ***First course:***

S. by W. = 1 pt. R. of S. = 11° 15′ R. of S.

Variation . . . . . . 12° 20′ L. of S.

True course . . . . . . 1° 5′ L. of S.

Hence, course and distance S. 1° 5′ E., 40 m.

Second course (starboard tack):

S.W. by S. = 3 pts. R. of S.

Leeway = 1 pt. L.

2 pts. R. of S. = 22° 30′ R. of S.

Variation . . . . . . 12° 20′ L.

True course . . . . . . 10° 10′ R. of S.

Hence, course and distance S. 10° 10′ W., 69.6 m.

Third course:

S.W. by W. = 5 pts. R. of S. = 56° 15′ R. of S.

Variation . . . . . . 12° 20′ L.

True course . . . . . . 43° 55′ R. of S.

Hence, course and distance S. 43° 55′ W., 58.5 m.

Current course:

W.S.W. = 6 pts. R. of S. = 67° 30′ R. of S.

Variation . . . . . . 12° 20′ L.

True course . . . . . . 55° 10′ R. of S.

Hence, course and distance S. 55° 10′ W., 36 m.

THE TRAVERSE.

C.	D.	N.	S.	E.	W.
S. 1° E.	40.		40.	0.7	
S. 10° W.	69.6		68.6		12.1
S. 44° W.	58.5		42.1		40.7
S. 55° W.	36.		20.6		29.5
$L_d = 171.3' = 2° 51'$ S.			171.3	0.7	82.3
$L' = 33° 40'$ N.					0.7
$L'' = 30° 49'$ N.				$p =$	81.6

$p = 81.6$ $\quad$ $\lambda_d = p \sec L_m$. $\quad$ $\lambda_d = 96'$

$L_m = 32° 15'$ $\quad$ $\log p = 1.91169$ $\quad$ $= 1° 36'$ W.

$\log \sec L_m = 0.07277$ $\quad$ $\lambda' = 16° 20'$ W.

$\log \lambda_d = 1.98446$ $\quad$ $\lambda'' = 17° 56'$ W.

3. *First course* (starboard tack):

N. by E. = 1 pt. R. of N.
Leeway = 1 pt. L.
0 = due north = 0°
Variation . . . . . W. 13° 30′ L.
13° 30′ L. of N.

Hence, course and distance N. 13° 30′ W., 37.7 m.

Second course (starboard tack):

N. = 0° pt.
Leeway = 1 pt. L.
1 pt. L. = 11° 15′ L. of N.
Variation . . . . W. 13° 30′ L.
24° 45′ L. of N.

Hence, course and distance N. 24° 45′ W., 38.7 m.

Third course (starboard tack):

N.N.W. = 2 pts. L. of N.
Leeway = 1 pt. L.
3 pts. L. of N. = 33° 45′ L. of N.
Variation . . . . . W. 13° 30′ L.
47° 15′ L. of N.

Hence, course and distance N. 47° 15′ W., 76.5 m.

Current course:

W.N.W. = 6 pts. L. of N. = 67° 30′ L. of N.
Variation . . . . . . 13° 30′ L.
81° 0′ L. of N.

Hence, course and distance N. 81° W., 12 m.

THE TRAVERSE.

C.	*D.*	*N.*	*S.*	*E.*	*W.*
N. 14° W.	37.7	36.6			9.2
N. 25° W.	38.7	35.			16.4
N. 47° W.	76.5	52.1			56.
N. 81° W.	12.	1.9			11.9
$L_d = 125.6' = 2°\ 6'$ N. $L' = 19°\ 30'$ S. $L'' = 17°\ 24'$ S.		125.6		$p =$	93.5

$p = 93.5$
$L_m = 18°\ 27'$

$\lambda_d = p \sec L_m.$
$\log p = 1.97081$
$\log \sec L_m = 0.02296$
$\log \lambda_d = 1.99377$

$\lambda_d = 99'$
$= 1°\ 39'$ W.
$\lambda' = 0°\ 10'$ E.
$\lambda'' = 1°\ 29'$ W.

4. *Departure course* (the opposite of W.S.W.):

E.N.E. The ship's head S.E. by E.; the deviation is the same as for the first course.

E.N.E. = 6 pts. R. of N. = 67° 30′ R. of N.
Variation and deviation . . 17° L.
50° 30′ R. of N.

Hence, course and distance N. 50° 30′ E., 18 m.

First course:

S.E. by E. = 5 pts. L. of S. = 56° 15′ L. of S.
Variation and deviation . . . 17° L.
True course . . . . 73° 15′ L. of S.
Hence, course and distance S. 73° 15′ E., 52 m.

Second course (port tack):

S.E. = 4 pts. L. of S.
Leeway = ½ pt. R.
3½ pts. L. of S. = 39° 22′ L. of S.
Variation and deviation . . . 19° L.
True course . . . . 58° 22′ L. of S.
Hence, course and distance S. 58° 22′ E., 43 m.

Third course (starboard tack):

E. by N. = 7 pts. R. of N.
Leeway = 1 pt. L.

6 pts. R. of N. =	67° 30′ R. of N.
Variation and deviation . .	11° L.
True course	56° 30′ R. of N.

Hence, course and distance N. 56° 30′ E., 36 m.

Fourth course (starboard tack):

E.N.E. = 6 pts. R. of N.
Leeway = $1\frac{1}{2}$ pts. L.

$4\frac{1}{2}$ pts. R. of N. =	50° 37′ R. of N.
Variation and deviation . .	13° L.
True course	37° 37′ R. of N.

Hence, course and distance N. 37° 37′ E., 27 m.

Fifth course (port tack):

S.S.E. = 2 pts. L. of S.
Leeway = 2 pts. R.

0 pts. = due south =	0°
Variation and deviation . .	21° L.
True course	21° L. of S.

Hence, course and distance S. 21° E., 24 m.

Sixth course (port tack):

S.E. by S. = 3 pts. L. of S.
Leeway = $1\frac{1}{4}$ pts. R.

$1\frac{3}{4}$ pts. L. of S. =	19° 41′ L. of S.
Variation and deviation . . .	20° L.
True course	39° 41′ L. of S.

Hence, course and distance S. 39° 41′ E., 29 m.

Current course:

S. by E. = 1 pt. = L. of S. =	11° 15′ L. of S.
Variation	28° L.
	39° 15′ L. of S.

Hence, course and distance S. 39° 15′ E., 12 m.

THE TRAVERSE.

C.	*D.*	*N.*	*S.*	*E.*	*W.*
N. 51° E.	18	11.3		14.0	
S. 73° E.	52		15.2	49.7	
S. 58° E.	43		22.8	36.5	
N. 57° E.	36	19.6		30.2	
N. 38° E.	27	21.3		16.6	
S. 21° E.	24		22.4	8.6	
S. 40° E.	29		22.2	18.6	
S. 39° E.	12		9.3	7.6	
$L_d =$ 40′ S. $L' =$ 47° 31′ N. $L'' =$ 46° 51′ N.		52.2	91.9 52.2 39.7	181.8	$= p$.

$p = 181.8$
$L_m = 47° 11'$

$\lambda_d = p \sec L_m$.
$\log p \quad = 2.25959$
$\log \sec L_m = 0.16771$
$\log \lambda_d \quad = 2.42730$

$\lambda_d = 267'$
$= 4° 27'$ E.
$\lambda' = 52° 33'$ W.
$\lambda'' = 48° \ 6'$ W.

5. *Departure course* (the opposite of W. by S. ¼ S.):

E. by N. ¼ N. = 6¾ pts. R. of N.
= 75° 56′ R. of N.
Variation and deviation . . 34° R.
109° 56′ R. of N.

Hence, course and distance S. 70° E., 17 m.

First course (port tack):

S.S.E. = 2 pts. L. of S.
Leeway = 2¼ pts. R.
¼ pt. R. of S. = 2° 49′ R. of S.
Variation and deviation . . . 34° R.
36° 49′ R. of S.

Hence, course and distance S. 37° W., 21 m.

Second course (starboard tack):

S.S.W. ½ W. = 2½ pts. R. of S.
Leeway = 2¾ pts. L.
¼ pt. L. of S. = 2° 49′ L. of S.
Variation and deviation . . . 27° R.
24° 11′ R. of S.

Hence, course and distance S. 24° W., 20 m.

Third course (port tack):

W.S.W. = 6 pts. R. of S.
Leeway = 2½ pts. R.
8½ pts. R. of S. = 7½ pts. L. of N.
= 84° 22′ L. of N.
Variation and deviation . . . 22° R.
True course . . . . . . 62° 22′ L. of N.

Hence, course and distance N. 62° W., 24 m.

Fourth course (starboard tack):

W. ½ N. = 7½ pts. L. of N. = 84° 22′ L. of N.
Variation and deviation . . . 20° R.
True course . . . . . . 64° 22′ L. of N.

Hence, course and distance N. 64° W., 26 m.

Fifth course (starboard tack):

East = 8 pts. R. of N.
Leeway = 2½ pts. L.
5½ pts. R. of N. = 61° 52′ R. of N.
Variation and deviation . . 41° R.
True course . . . . . . 102° 52′ R. of N.

Hence, course and distance S. 77° E., 19 m.

Sixth course (starboard tack):

E.S.E. = 6 pts. L. of S. = 67° 30′ L. of S.
Variation and deviation . . . 40° R.
True course . . . . . . 27° 30′ L. of S.

Hence, course and distance S. 28° E., 18 m.

Current course:

N.N.E. = 2 pts. R. of N.
= 22° 30′ R. of N.
Variation . . . . 31° R.
True course . . . . . . 53° 30′ R. of N.

Hence, course and distance N. 54° E., 21 m.

THE TRAVERSE.

C.	D.	N.	S.	E.	W.
S. 70° E.	17		5.8	16.0	
S. 37° W.	21		16.8		12.6
S. 24° W.	20		18.3		8.1
N. 62° W.	24	11.3			21.2
N. 64° W.	26	11.4			23.4
S. 77° E.	19		4.3	18.5	
S. 28° E.	18		15.9	8.5	
N. 54° E.	21	12.3		17.	
$L_d =$ 26′ S.		35.0	61.1	60.	65.3
$L' =$ 62° 0′ N.			35.0		60.
$L'' =$ 61° 34′ N.			26.1		5.3

$p = 5.3$

$L_m = 61° 47'$

$\lambda_d = p \sec L_m.$

$\log p = 0.72428$

$\log \sec L_m = 0.32532$

$\log \lambda_d = 1.04960$

$\lambda_d =$ 11′ W.

$\lambda' =$ 150° 0′ E.

$\lambda'' =$ 149° 49′ E.

6. *Departure course* (the opposite of N. ¾ W.):

S. ¾ E. = ¾ of a pt. = 8° 26′ L. of S.

Variation and deviation . . . . 8° R.

0° 26′ L. of S.

Hence, course and distance S., 19 m.

First course (port tack):

S.W. ½ W. = 4½ pts. R. of S. = 50° 37′ R. of S.

Variation and deviation . . . 8° R.

True course . . . . . . 58° 37′ R. of S.

Hence, course and distance S. 59° W., 58 m.

Second course (starboard tack):

N. ¾ E. = ¾ pt. R. of N.

Leeway = 3¼ pts. L.

2½ pts. L. of N. = 28° 7′ L. of N.

Variation and deviation . . . 17° R.

True course . . . . . . 11° 7′ L. of N.

Hence, course and distance N. 11° W., 15 m.

Third course (starboard tack):

S.E. ½ E. = 1½ pts. L. of S.
Leeway = 2¾ pts. L.

4¼ pts. L. of S. = 47° 48′ L. of S.
Variation and deviation . . . 20° R.
True course . . . . . . 27° 48′ L. of S.
Hence, course and distance S. 28° E., 9 m.

Fourth course (port tack):

W. by S. = 7 pts. R. of S.
Leeway = ¼ pt. R.

7¼ pts. R. of S. = 81° 33′ R. of S.
Variation and deviation . . . 0°
True course . . . . . . 81° 33′ R. of S.
Hence, course and distance S. 82° W., 50 m.

Fifth course (starboard tack):

E.N.E. = 6 pts. R. of N.
Leeway = 2½ pts. L.

3½ pts. R. of N. = 39° 22′ R. of N.
Variation and deviation . . . 33° R.
True course . . . . . . 72° 22′ R. of N.
Hence, course and distance N. 72° E., 12 m.

Sixth course (port tack):

S.S.W. ½ W. = 2½ pts. R. of S.
Leeway = 1¾ pts. R.

4¼ pts. R. of S. = 47° 48′ R. of S.
Variation and deviation . . . 10° R.
True course . . . . . . 57° 48′ R. of S.
Hence, course and distance S. 58° W., 22 m.

Current course:

S.W. ¼ W. = 4¼ pts. R. of S. = 47° 48′ R. of S.
Variation . . . . . . 14° R.
True course . . . . . . 61° 48′ R. of S.
Hence, course and distance S. 62° W., 42 m.

THE TRAVERSE.

C.	*D.*	*N.*	*S.*	*E.*	*W.*
S.	19		19.0		
S. 59° W.	58		29.9		49.7
N. 11° W.	15	14.7			2.9
S. 28° E.	9		7.9	4.2	
S. 82° W.	50		7.0		49.5
N. 72° E.	12	3.7		11.4	
S. 58° W.	22		11.7		18.7
S. 62° W.	42		19.7		37.1
		18.4	95.2	15.6	157.9
$L_d = 77' = \ \ 1° \ 17'$ S.			18.4		15.6
$L' = 50° \ 12'$ S.					
$L'' = 51° \ 29'$ S.			76.8		142.3

$p = 142.3$ $\qquad \lambda_d = p \sec L_m. \qquad \lambda_d = 225'$

$L_m = 50° \ 51' \qquad \log p = 2.15320 \qquad = 3° \ 45'$ W.

$\log \sec L_m = 0.19973 \qquad \lambda' = 179° \ 40'$ W.

$\log \lambda_d = 2.35293 \qquad \lambda'' = 176° \ 35'$ E.

7. ***First course:***

S.E. = 4 pts. L. of S.
Variation and deviation . . 1¼ pts. L.

True course . . . . . 5¼ pts. L. of S.
Hence, course and distance S. 5¼ pts. E., 27.8 m.

Second course:

E.S.E. ¼ E. = 6¼ pts. L. of S.
Variation . . . . . 1¾ pts. L.

True course . . . . . 8 pts. L. of S. = due east.
Hence, course and distance E. 75.2 m.

Third course:

E. = 8 pts. R. of N.
Variation and deviation . . 1¼ pts. L.

True course . . . . . 6¾ pts. R. of N.
Hence, course and distance N. 6¾ pts. E., 8.7 m.

THE TRAVERSE.

C.	D.	N.	S.	E.	W.
S. 5¼ pts. E.	27.8		13.1	24.5	
E.	75.2			75.2	
N. 6¾ pts. E.	8.7	2.1		8.4	
$L_d =$ 11′ S. $L' =$ 36° 42′ N. $L'' =$ 36° 31′ N.		2.1	13.1 2.1 11.0	108.1	$= p.$

$p = 108.1$
$L_m = 36° 37'$

$\lambda_d = p \sec L_m.$

$\log p = 2.03383$
$\log \sec L_m = 0.09548$
$\log \lambda_d = 2.12931$

$\lambda_d = 135'$
$= 2° 15'$ E.
$\lambda' = 4° 25'$ W.
$\lambda'' = 2° 10'$ W.

EXERCISE VIII. PAGE 333.

1. Find the elements (initial courses, distance, and latitude and longitude of the vertex) of the great circle track between the Lizard, in latitude 49° 58′ N., longitude 5° 12′ W., and the Bermuda Islands, in latitude 32° 18′ N., longitude 64° 50′ W.

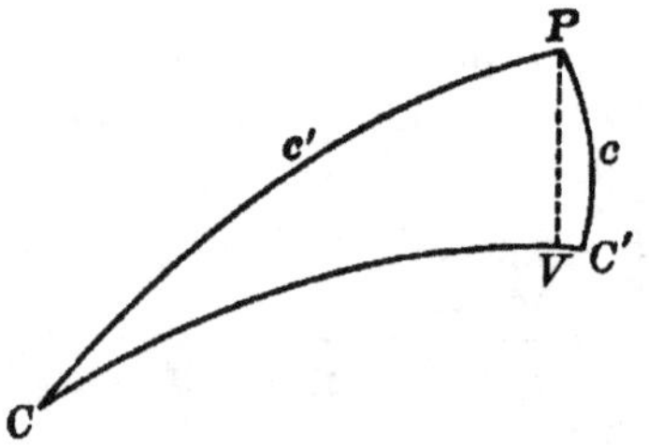

Referring to the triangle CPC',

Lat. $C' = 49° 58'$ N.
Long. $C' = 5° 12'$ W.
Lat. $C = 32° 18'$ N.
Long. $C = 64° 50'$ W.

$c = 90° - 49° 58' = 40° 2'.$
$c' = 90° - 32° 18' = 57° 42'.$
$\lambda_d = 64° 50' - 5° 12' = 59° 38'.$

To find the initial courses:

$$\tan \tfrac{1}{2}(C' + C) = \frac{\cos \frac{1}{2}(c' - c)}{\cos \frac{1}{2}(c' + c)} \cot \tfrac{1}{2}\lambda_d.$$

$$\tan \tfrac{1}{2}(C' - C) = \frac{\sin \frac{1}{2}(c' - c)}{\sin \frac{1}{2}(c' + c)} \cot \tfrac{1}{2}\lambda_d.$$

$\frac{1}{2}(c'-c) = 8° 50'$, log cos $= 9.99482$ log sin $= 9.18628$
$\frac{1}{2}(c'+c) = 48° 52'$, colog cos $= 0.18190$ colog sin $= 0.12310$
$\frac{1}{2}\lambda_d = 29° 48'$, log cot $= \underline{10.24178}$ log cot $= \underline{10.24178}$
log tan $\frac{1}{2}(C'+C) = 10.41850$ log tan $\frac{1}{2}(C'-C) = 9.55116$

$\frac{1}{2}(C'+C) = 69°\ 7'$ $\frac{1}{2}(C'-C) = 19° 35'$.
$\frac{1}{2}(C'-C) = \underline{19° 35'}$
$\therefore C' =$ N. 88° 42′ W. = course from Lizard.
$C =$ N. 49° 32′ E. = course from Bermudas.

To find the distance:

$$\cos \tfrac{1}{2} D = \frac{\cos \frac{1}{2}(c+c')}{\cos \frac{1}{2}(C+C')} \sin \tfrac{1}{2}\lambda_d.$$

$\frac{1}{2}(c+c') = 48° 52'$, log cos $= 9.81810$
$\frac{1}{2}(C+C') = 69°\ 7'$, colog cos $= 0.44798$
$\frac{1}{2}\lambda_d = 29° 49'$, log sin $= \underline{9.69655}$
log cos $\frac{1}{2}D = 9.96263$

$\frac{1}{2}D = 23° 26'$.
$D = 46° 52' = 2812$ m.

To find L of V.

$\sin PV = \sin c' \sin C$.
log sin $c' = 9.92699$
log sin $C = \underline{9.88126}$
log sin $PV = 9.80825$
$PV = 40° 1'$.
L of $V = 90° - 40° 1'$
$= 49° 59'$ N.

To find λ of V.

$\cot CPV = \cos c' \tan C$.
log cos $c' = 9.72783$
log tan $C = \underline{10.06901}$
log cot $CPV = 9.79684$
$CPV = 57° 56'$.
λ of $C = 64° 50'$
$\underline{57° 56'}$
λ of $V = 6° 54'$ W.

2. Find the elements of the great circle track between Boston (Minot's Ledge light-house) in latitude 42° 16′ N., longitude 70° 46′ W., and Cape Clear, in latitude 51° 26′ N., longitude 9° 29′ W. [Take $\frac{1}{2}\lambda_d = 30° 39'$.]

Lat. $C' = 51° 26'$ N.
Long. $C' = 9° 29'$ W.
Lat. $C = 42° 16'$ N.
Long. $C = 70° 46'$ W.

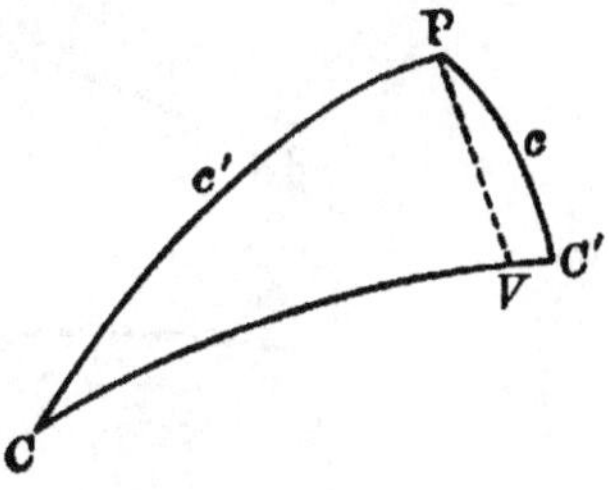

$c = 90° - 51° 26' = 38° 34'$.
$c' = 90° - 42° 16' = 47° 44'$.
$\lambda_d = 70° 46' - 9° 26' = 61° 17'$.

$$\tan \tfrac{1}{2}(C'+C) = \frac{\cos \frac{1}{2}(c'-c)}{\cos \frac{1}{2}(c'+c)} \cot \tfrac{1}{2}\lambda_d.$$

$$\tan \tfrac{1}{2}(C'-C) = \frac{\sin \frac{1}{2}(c'-c)}{\sin \frac{1}{2}(c'+c)} \cot \tfrac{1}{2}\lambda_d.$$

$\frac{1}{2}(c'-c) = 4° 35'$, log cos $= 9.99861$ log sin $= 8.90260$
$\frac{1}{2}(c'+c) = 43° 9'$, colog cos $= 0.13694$ colog sin $= 0.16500$
$\frac{1}{2}\lambda_d = 30° 39'$, log cot $= \underline{10.22726}$ log cot $= \underline{10.22726}$
log tan $\frac{1}{2}(C'+C) = 10.36281$ log tan $\frac{1}{2}(C'-C) = 9.29486$

$\frac{1}{2}(C'+C) = 66° 33'$ $\frac{1}{2}(C'-C) = 11° 9'$.
$\frac{1}{2}(C'-C) = \underline{11° 9'}$

$\therefore C' = $ N. 77° 42′ W. = course from Cape Clear.
$C = $ N. 55° 24′ E. = course from Boston.

To find the distance:

$$\cos\tfrac{1}{2}D = \frac{\cos\frac{1}{2}(c'+c)}{\cos\frac{1}{2}(C'+C)}\sin\tfrac{1}{2}\lambda_d.$$

$\frac{1}{2}(c'+c) = 43° 9'$, log cos $= 9.86306$
$\frac{1}{2}(C'+C) = 66° 33'$, colog cos $= 0.40017$
$\frac{1}{2}\lambda_d = 30° 39'$, log sin $= \underline{9.70739}$
log cos $\frac{1}{2}D = 9.97062$

$\therefore \frac{1}{2}D = 30° 50'$.
$D = 41° 40' = 2500$ m.

To find L of V.

$\sin PV = \sin c' \sin C$.

log sin $c' = 9.86924$
log sin $C = \underline{9.91547}$
log sin $PV = 9.78471$
$\therefore PV = 37° 32'$.
L of $V = 90° - 37° 32'$
$= 52° 28'$ N.

To find λ of V.

$\cot CPV = \cos c' \tan C$.

log cos $c' = 9.82775$
log tan $C = \underline{10.16124}$
log cot $CPV = 9.98899$
$CPV = 45° 44'$
λ of $C = \underline{70° 46'}$
λ of $V = 25° 2'$ W.

3. Find the elements of the great circle track between Vancouver Island, in latitude 50° N., longitude 128° W., and Honolulu, in latitude 21° 18′ N., longitude 157° 52′ W.

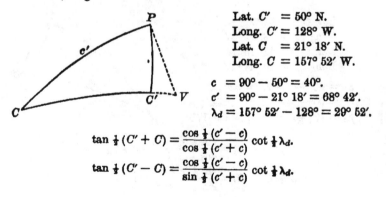

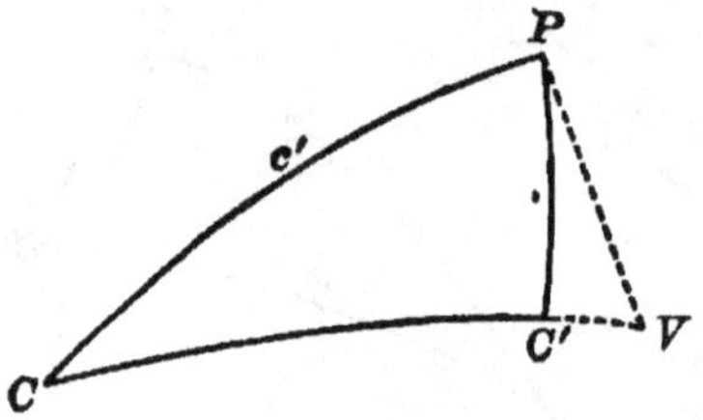

Lat. $C' = 50°$ N.
Long. $C' = 128°$ W.
Lat. $C = 21° 18'$ N.
Long. $C = 157° 52'$ W.

$c = 90° - 50° = 40°$.
$c' = 90° - 21° 18' = 68° 42'$.
$\lambda_d = 157° 52' - 128° = 29° 52'$.

$$\tan\tfrac{1}{2}(C'+C) = \frac{\cos\frac{1}{2}(c'-c)}{\cos\frac{1}{2}(c'+c)}\cot\tfrac{1}{2}\lambda_d.$$

$$\tan\tfrac{1}{2}(C'-C) = \frac{\cos\frac{1}{2}(c'-c)}{\sin\frac{1}{2}(c'+c)}\cot\tfrac{1}{2}\lambda_d.$$

$\frac{1}{2}(c'-c) = 14^\circ\ 21'$, log cos $= 9.98623$ | log sin $= 9.39418$

$\frac{1}{2}(c'+c) = 54^\circ\ 21'$, colog cos $= 0.23446$ | colog sin $= 0.09013$

$\frac{1}{2}\lambda_d = 14^\circ\ 56'$, log cot $= \underline{10.57397}$ | log tan $= \underline{10.57397}$

log tan $\frac{1}{2}(C'+C) = 10.79466$ | log tan $\frac{1}{2}(C'-C) = 10.05828$

$\frac{1}{2}(C'+C) = 80^\circ\ 53'$ | $\frac{1}{2}(C'-C) = 48^\circ\ 50'$.

$\frac{1}{2}(C'-C) = \underline{48^\circ\ 50'}$

$C' = 129^\circ\ 43'$

= S. $50^\circ\ 17'$ W. = course from Vancouver.

$C =$ N. $32^\circ\ 3'$ E. = course from Honolulu.

To find the distance:

$$\cos\tfrac{1}{2}D = \frac{\cos\frac{1}{2}(c'+c)}{\cos\frac{1}{2}(C'+C)}\sin\tfrac{1}{2}\lambda_d.$$

$\frac{1}{2}(c'+c) = 54^\circ\ 21'$, log cos $= 9.76554$

$\frac{1}{2}(C'+C) = 80^\circ\ 53'$, colog cos $= 0.80012$

$\frac{1}{2}\lambda_d = 14^\circ\ 56'$, log sin $= \underline{9.41110}$

log cos $\frac{1}{2}D = 9.97676$

$\frac{1}{2}D = 18^\circ\ 34'$.

$D = 37^\circ\ 8' = 2228$ m.

To find L of V.

sin PV = sin c' sin C.

log sin c' $= 9.96927$

log sin C $= \underline{9.72482}$

log sin $PV = 9.69409$

$PV = 29^\circ\ 38'$.

L of $V = 90^\circ - 29^\circ\ 38'$

$= 60^\circ\ 22'$ N.

To find λ of V.

cot CPV = cos c' tan C.

log cos c' $= 9.56020$

log tan C $= \underline{9.79663}$

log cot $CPV = 9.35683$

$CPV = 77^\circ\ 11'$

λ of $C = \underline{157^\circ\ 52'}$

λ of $V = 80^\circ\ 41'$ W.

4. Find the elements of the great circle track between Cape Clear, in latitude $51^\circ\ 26'$ N., longitude $9^\circ\ 29'$ W., and Sandy Hook, in latitude 40° N., longitude 74° W.

Lat. $C' = 51^\circ\ 26'$ N.

Long. $C' = 9^\circ\ 29'$ W.

Lat. $C = 40^\circ$ N.

Long. $C = 74^\circ$ W.

$c = 90^\circ - 51^\circ\ 26' = 38^\circ\ 34'$.

$c' = 90^\circ - 40^\circ = 50^\circ$.

$\lambda_d = 74^\circ - 9^\circ\ 29' = 64^\circ\ 31'$.

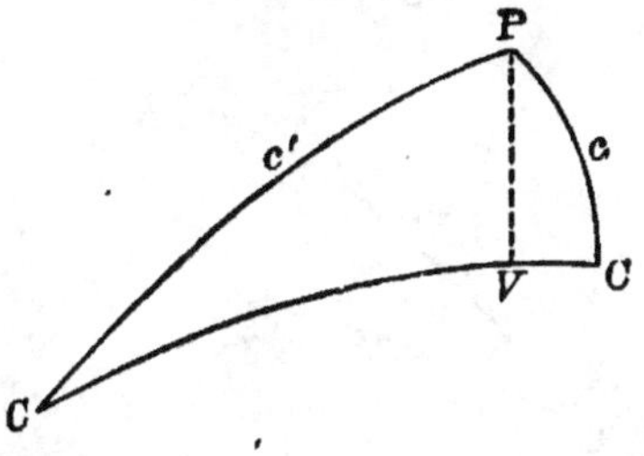

$$\tan\tfrac{1}{2}(C+C') = \frac{\cos\frac{1}{2}(c-c')}{\cos\frac{1}{2}(c+c')}\cot\tfrac{1}{2}\lambda_d.$$

$$\tan\tfrac{1}{2}(C-C') = \frac{\sin\frac{1}{2}(c-c')}{\sin\frac{1}{2}(c+c')}\cot\tfrac{1}{2}\lambda_d.$$

$$\begin{array}{llll}
\tfrac{1}{2}(c'-c) = 5^\circ\ 43', & \log\cos\ = 9.99784 & & \log\sin\ = 8.99830 \\
\tfrac{1}{2}(c+c') = 44^\circ\ 17', & \text{colog}\cos = 0.14515 & & \text{colog}\sin = 0.15602 \\
\tfrac{1}{2}\lambda_d = 32^\circ\ 16', & \log\cot\ = \underline{10.19972} & & \log\cot\ = \underline{10.19972} \\
 & \log\tan\tfrac{1}{2}(C+C') = 10.34271 & & \log\tan\tfrac{1}{2}(C'-C) = 9.35404 \\
 & \tfrac{1}{2}(C+C') = 65^\circ\ 34' & & \tfrac{1}{2}(C'-C) = 12^\circ\ 44'. \\
 & \tfrac{1}{2}(C'-C) = \underline{12^\circ\ 44'} & &
\end{array}$$

$C' =$ N. 78° 18′ W. = course from Cape Clear.

$C\ =$ N. 52° 50′ E. = course from Sandy Hook.

To find the distance:

$$\cos\tfrac{1}{2}D = \frac{\cos\tfrac{1}{2}(c+c')}{\cos\tfrac{1}{2}(C+C')}\sin\tfrac{1}{2}\lambda_d.$$

$$\begin{array}{lll}
\tfrac{1}{2}(c+c') = 44^\circ\ 17', & & \log\cos\ = 9.85485 \\
\tfrac{1}{2}(C+C') = 65^\circ\ 34', & & \text{colog}\cos\ = 0.38338 \\
\tfrac{1}{2}\lambda_d = 32^\circ\ 16', & & \log\sin\ = \underline{9.72743} \\
 & & \log\cos\tfrac{1}{2}D = 9.96566
\end{array}$$

$\tfrac{1}{2}D = 22^\circ\ 29'.$

$D = 44^\circ\ 58' = 2698$ m.

To find L of V.	To find λ of V.
$\sin PV = \sin c \sin C'$.	$\cot CPV = \cos c' \tan C$.
$\log\sin c = 9.79478$	$\log\cos c' = 9.80807$
$\log\sin C' = \underline{9.99088}$	$\log\tan C = \underline{10.12026}$
$\log\sin PV = 9.78566$	$\log\cot CPV = 9.92833$
$PV = 37^\circ\ 37'$.	$CPV = 49^\circ\ 42'$
L of $V = 90^\circ\ 00' - 37^\circ\ 37'$	λ of $C = \underline{74^\circ\ \ 0'}$
$= 52^\circ\ 23'$ N.	λ of $V = 24^\circ\ 18'$ W.

5. Find the elements of the great circle track between Lizard Light, in latitude 49° 58′ N., longitude 5° 12′ W., and Cape Frio, in latitude 23° S., longitude 42° W.

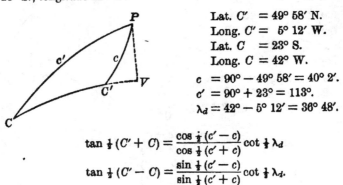

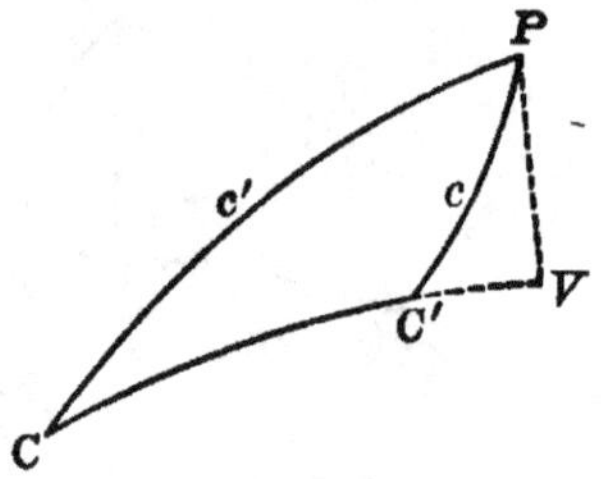

Lat. C' = 49° 58′ N.
Long. C' = 5° 12′ W.
Lat. C = 23° S.
Long. C = 42° W.

$c\ = 90^\circ - 49^\circ\ 58' = 40^\circ\ 2'.$

$c' = 90^\circ + 23^\circ = 113^\circ.$

$\lambda_d = 42^\circ - 5^\circ\ 12' = 36^\circ\ 48'.$

$$\tan\tfrac{1}{2}(C'+C) = \frac{\cos\tfrac{1}{2}(c'-c)}{\cos\tfrac{1}{2}(c'+c)}\cot\tfrac{1}{2}\lambda_d$$

$$\tan\tfrac{1}{2}(C'-C) = \frac{\sin\tfrac{1}{2}(c'-c)}{\sin\tfrac{1}{2}(c'+c)}\cot\tfrac{1}{2}\lambda_d.$$

$\frac{1}{2}(c'-c) = 36° 29'$, log cos $= 9.90527$ log sin $= 9.77421$
$\frac{1}{2}(c'+c) = 76° 31'$, colog cos $= 0.63234$ colog sin $= 0.01214$
$\frac{1}{2}\lambda_d = 18° 24'$, log cot $= \underline{10.47801}$ log cot $= \underline{10.47801}$
log tan $\frac{1}{2}(C'+C) = 11.01562$ log cot $\frac{1}{2}(C'-C) = 10.26436$

$\frac{1}{2}(C'+C) = 84° 29'$ $\frac{1}{2}(C'-C) = 61° 27'$.
$\frac{1}{2}(C'-C) = \underline{61° 27'}$
$C' = $ S. 34° 4′ W. = course from Lizard Light.
$C = $ N. 23° 2′ E. = course from Cape Frio.

To find the distance:

$$\cos\frac{1}{2}D = \frac{\cos\frac{1}{2}(c'+c)}{\cos\frac{1}{2}(C'+C)}\sin\frac{1}{2}\lambda_d.$$

$\frac{1}{2}(c'+c) = 76° 31'$,   log cos $= 9.36766$
$\frac{1}{2}(c'+c) = 84° 29'$,   colog cos $= 1.01712$
$\frac{1}{2}\lambda_d = 18° 24'$,   log sin $= \underline{9.49920}$
log cos $\frac{1}{2}D = 9.88398$

$\therefore \frac{1}{2}D = 40° 3'$.
$D = 80° 6' = 4806$ m.

To find L of V.

$\sin PV = \sin c' \sin C$.

$\sin 180° - 113° = \sin 67° = c'$.
log sin $c' = 9.96403$
log sin $C = \underline{9.59247}$
log sin $PV = 9.55650$
$PV = 21° 7'$ N.
90° 00′ N.
$\underline{21° \; 7' \text{ N.}}$
L of $V = 68° 53'$ N.

To find λ of V.

$\cot CPV = \cos c' \tan C$.

log cos $c' = 9.59188$ (n)
log tan $C = \underline{9.62855}$
log cot $CPV = 9.22043$ (n)
$CPV = 99° 26'$
λ of $C = \underline{42° 00'}$
λ of $V = 57° 26'$ E.

6. Find the elements of the great circle track between Cape Frio and Cape Good Hope, in latitude 34° 20′ S., longitude 18° 30′ E. (Reckon from the nearest pole.)

Lat. $C' = 34° 20'$ S.
Long. $C' = 18° 30'$ E.
Lat. $C = 23°$ S.
Long. $C = 42°$ W.
$c = 90° - 43° 20' = 55° 40'$.
$c' = 90° - 23° = 67°$.
$\lambda_d = 42° + 18° 30' = 60° 30'$.

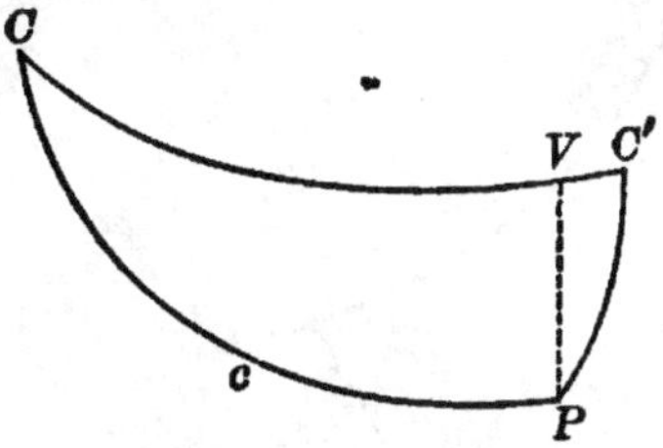

$$\tan\frac{1}{2}(C'+C) = \frac{\cos\frac{1}{2}(c'-c)}{\cos\frac{1}{2}(c'+c)}\cot\frac{1}{2}\lambda_d.$$

$$\tan\frac{1}{2}(C'-C) = \frac{\sin\frac{1}{2}(c'-c)}{\sin\frac{1}{2}(c'+c)}\cot\frac{1}{2}\lambda_d.$$

$\frac{1}{2}(c' - c) = \ \ 5^\circ\ 40'$, $\log\cos = 9.99787$ $\qquad \log\sin = 8.99450$

$\frac{1}{2}(c' + c) = 61^\circ\ 20'$, $\text{colog}\cos = 0.31902$ $\qquad \text{colog}\sin = 0.05679$

$\frac{1}{2}\lambda_d \quad = 30^\circ\ 15'$, $\log\cot = 10.23420$ $\qquad \log\cot = 10.23420$

$\log\tan\frac{1}{2}(C' + C) = 10.55109 \qquad \log\tan(C' - C) = 9.28549$

$\frac{1}{2}(C' + C) = 74^\circ\ 18' \qquad \frac{1}{2}(C' - C) = 10^\circ\ 55'.$

$\frac{1}{2}(C' - C) = 10^\circ\ 55'$

$C' =$ S. $85^\circ\ 13'$ W. = course from C. Good Hope.

$C =$ S. $63^\circ\ 23'$ E. = course from Cape Frio.

To find the distance:

$$\cos\tfrac{1}{2}D = \frac{\cos\frac{1}{2}(c' + c)}{\cos\frac{1}{2}(C' + C)}\sin\tfrac{1}{2}\lambda_d.$$

$\frac{1}{2}(c' + c) = 61^\circ\ 20'$, $\qquad \log\cos = 9.68098$

$\frac{1}{2}(C' + C) = 74^\circ\ 18'$, $\qquad \text{colog}\cos = 0.56767$

$\frac{1}{2}\lambda_d = 30^\circ\ 15'$, $\qquad \log\sin = 9.70224$

$\qquad \log\cos\frac{1}{2}D = 9.95089$

$\frac{1}{2}D = 26^\circ\ 44'.$

$D = 53^\circ\ 28' = 3208$ m.

To find L of V.

$\sin PV = \sin c' \sin C.$

$\log\sin c' = 9.96403$

$\log\sin C = 9.95135$

$\log\sin PV = 9.91538$

$PV = 55^\circ\ 23'$

L of $V = 90^\circ\ \ 0' - 55^\circ\ 23'$

$= 34^\circ\ 37'$ S.

To find λ of V.

$\cot CPV = \cos c' \tan C.$

$\log\cos c' = 9.59188$

$\log\tan C = 10.30005$

$\log\cot CPV = 9.89193$

$CPV = 52^\circ\ 3'$

$42^\circ\ 0'$

λ of $V = 10^\circ\ 3'$ E.

7. Find the elements of the great circle track between Grand Port, Mauritius, in latitude 20° 24′ S., longitude 57° 47′ E., and Perth, in latitude 32° 3′ S., longitude 115° 45′ E.

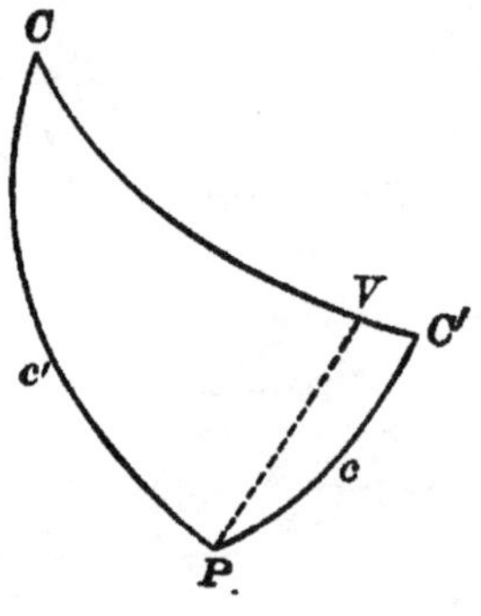

Lat. $C' = 32^\circ\ \ 3'$ S.

Long. $C' = 115^\circ\ 45'$ E.

Lat. $C = 20^\circ\ 24'$ S.

Long. $C = 57^\circ\ 47'$ E.

$c = 90^\circ - 32^\circ\ \ 3' = 57^\circ\ 57'.$

$c' = 90^\circ - 20^\circ\ 24' = 69^\circ\ 36'.$

$\lambda_d = 115^\circ\ 45' - 57^\circ\ 47' = 57^\circ\ 58'.$

$$\tan\tfrac{1}{2}(C' + C) = \frac{\cos\frac{1}{2}(c' - c)}{\cos\frac{1}{2}(c' + c)}\cot\tfrac{1}{2}\lambda_d.$$

$$\tan\tfrac{1}{2}(C' - C) = \frac{\sin\frac{1}{2}(c' - c)}{\sin\frac{1}{2}(c' + c)}\cot\tfrac{1}{2}\lambda_d.$$

$$
\begin{array}{llll}
\tfrac{1}{2}(c'-c)=\ 5^\circ\,50\,, & \log\cos\ =\ 9.99776 & & \log\sin\ =\ 9.00704 \\
\tfrac{1}{2}(c'+c)=63^\circ\,47', & \operatorname{colog}\cos=\ 0.35481 & & \operatorname{colog}\sin=\ 0.04714 \\
\tfrac{1}{2}\lambda_d\ \ =28^\circ\,59', & \log\cot\ =\underline{10.25655} & & \log\cot\ =\underline{10.25655} \\
 & \log\tan\tfrac{1}{2}(C'+C)=10.60912 & & \log\tan\tfrac{1}{2}(C'-C)=\ 9.31073 \\
 & \tfrac{1}{2}(C'+C)=76^\circ\,11' & & \tfrac{1}{2}(C'-C)=11^\circ\,34'. \\
 & \tfrac{1}{2}(C'-C)=\underline{11^\circ\,34'} & & \\
 & C'=\text{S. }87^\circ\,45'\text{ E.}=\text{course from Perth.} & & \\
 & C=\text{S. }64^\circ\,37'\text{ E.}=\text{course from Mauritius.} & &
\end{array}
$$

To find the distance:

$$\cos\tfrac{1}{2}D=\frac{\cos\tfrac{1}{2}(c'+c)}{\cos\tfrac{1}{2}(C'+C)}\sin\tfrac{1}{2}\lambda_d.$$

$$
\begin{array}{ll}
\tfrac{1}{2}(c'+c)\ \ =63^\circ\,47', & \log\cos\ \ =9.64579 \\
\tfrac{1}{2}(C'+C)=76^\circ\,11', & \operatorname{colog}\cos\ =0.62194 \\
\tfrac{1}{2}\lambda_d\ \ \ =28^\circ\,59', & \log\sin\ \ =\underline{9.68534} \\
 & \log\cos\tfrac{1}{2}D=9.95247
\end{array}
$$

$$\tfrac{1}{2}D=26^\circ\,19'.$$
$$D=52^\circ\,38'=3158\text{ m.}$$

To find L of V.		To find λ of V.	
$\sin PV$	$=\sin c'\sin C.$	$\cot CPV$	$=\cos c'\tan C.$
$\log\sin c'$	$=9.97187$	$\log\cos c'$	$=\ 9.54229$
$\log\sin C$	$=\underline{9.95591}$	$\log\tan C$	$=\underline{10.32378}$
$\log\sin PV$	$=9.92778$	$\log\cot CPV$	$=\ 9.86607$
PV	$=57^\circ\,52'.$	CPV	$=\ 53^\circ\,42'$
L of V	$=90^\circ\ 0'-57^\circ\,52'$	λ of C	$=\ \underline{57^\circ\,47'}$
	$=32^\circ\ 8'$ S.	λ of V	$=111^\circ\,29'$ E.

8. Find the elements of the great circle track between A, in latitude 16° 38′ N., longitude 70° 55′ W., and B, in latitude 48° 2′ N., longitude 4° 35′ W.

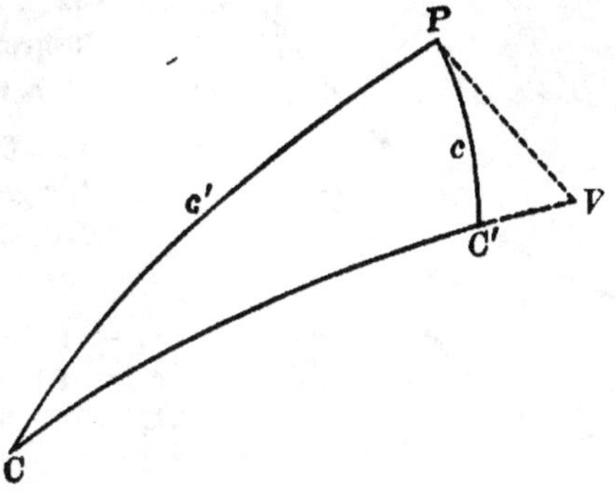

Lat C' $=48^\circ\ 2'$ N.

Long. $C'=\ 4^\circ\,35'$ W.

Lat. C $=16^\circ\,38'$ N.

Long. C $=70^\circ\,55'$ W.

$c\ =90^\circ-48^\circ\ \ 2'=41^\circ\,58'.$

$c'\ =90^\circ-16^\circ\,38'=73^\circ\,22'.$

$\lambda_d=70^\circ\,55'-4^\circ\,35'=66^\circ\,20'.$

$$\tan\tfrac{1}{2}(C'+C)=\frac{\cos\frac{1}{2}(c'-c)}{\cos\frac{1}{2}(c'+c)}\cot\tfrac{1}{2}\lambda_d.$$

$$\tan\tfrac{1}{2}(C'-C)=\frac{\sin\frac{1}{2}(c'-c)}{\sin\frac{1}{2}(c'+c)}\cot\tfrac{1}{2}\lambda_d.$$

$$\begin{array}{llll}
\frac{1}{2}(c'-c)=15^\circ\,42', & \log\cos\ =\ 9.98349 & & \log\sin\ =\ 9.43233\\
\frac{1}{2}(c'+c)=57^\circ\,40', & \text{colog}\cos=\ 0.27177 & & \text{colog}\sin=\ 0.07317\\
\frac{1}{2}\lambda_d\ =33^\circ\,10', & \log\cot\ =\underline{10.18472} & & \log\cot\ =\underline{10.18472}\\
 & \log\tan\frac{1}{2}(C'+C)=10.43998 & & \log\tan\frac{1}{2}(C'-C)=\ 9.69022\\
 & \frac{1}{2}(C'+C)=70^\circ\ \ 3' & & \frac{1}{2}(C'-C)=26^\circ\,6'.\\
 & \frac{1}{2}(C'-C)=\underline{26^\circ\ \ 6'} & &
\end{array}$$

$$\begin{aligned}
C' &= \text{N. } 96^\circ\ 9'\text{ W.}\\
&= \text{S. } 83^\circ\,51'\text{ W.} = \text{course from B.}\\
C &= \text{N. } 43^\circ\,57'\text{ E.} = \text{course from A.}
\end{aligned}$$

To find the distance:

$$\cos\tfrac{1}{2}D=\frac{\cos\frac{1}{2}(c'+c)}{\cos\frac{1}{2}(c'-c)}\sin\tfrac{1}{2}\lambda_d.$$

$$\begin{array}{ll}
\frac{1}{2}(c'+c)\ =57^\circ\,40', & \log\cos\ =9.72823\\
\frac{1}{2}(C'+C)=70^\circ\ \ 3', & \text{colog}\cos\ =0.46699\\
\frac{1}{2}\lambda_d\ =33^\circ\,10', & \log\sin\ =\underline{9.73805}\\
 & \log\cos\frac{1}{2}D=9.93327
\end{array}$$

$$\tfrac{1}{2}D=30^\circ\,57'.$$

$$D=61^\circ\,54'=3714\text{ m.}$$

To find L of V.	To find λ_d of V.
$\sin PV\ =\sin c'\sin C.$	$\cot CPV\ =\cos c'\tan C.$
$\log\sin c\ =9.98144$	$\log\cos c'\ =9.45674$
$\log\sin C'\ =\underline{9.84138}$	$\log\tan C\ =\underline{9.98408}$
$\log\sin PV=9.82282$	$\log\cot CPV=9.44082$
$PV=41^\circ 41'.$	$CPV=74^\circ\,34'$
L of $V=90^\circ\ 0'-41^\circ 41'$	λ of $C=\underline{70^\circ\,55'}$
$=48^\circ\,19'$ N.	λ of $V=\ 3^\circ\,39'$ E.

9. A ship sails from A., in latitude 40° S., longitude 148° 30′ E., to B, in latitude 12° 4′ S., longitude 77° 14′ W. Compare the great circle and the rhumb-line between A and B.

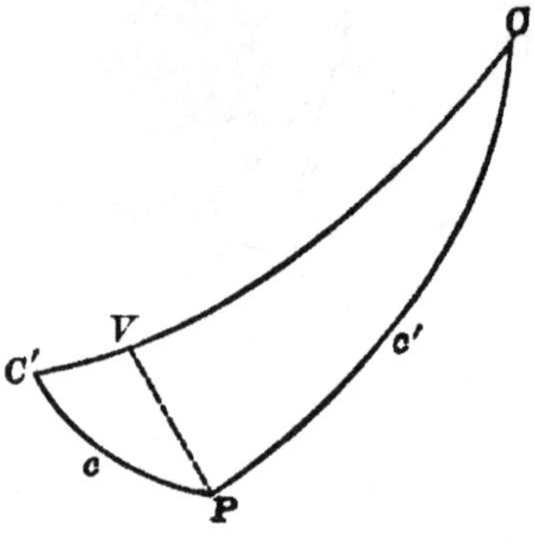

$$\begin{aligned}
\text{Lat. } C' &= 40^\circ\text{ S.}\\
\text{Long. } C' &= 148^\circ\,30'\text{ E.}\\
\text{Lat. } C &= 12^\circ\ \ 4'\text{ S.}\\
\text{Long. } C &= 77^\circ\,14'\text{ W.}
\end{aligned}$$

$$\begin{aligned}
c &= 90^\circ-40^\circ=50^\circ.\\
c' &= 90^\circ-12^\circ\,4'=77^\circ\,56'.\\
\lambda_d &= 148^\circ\,30'+77^\circ\,14'\\
&= 225^\circ\,44',\ \text{or } 134^\circ\,16'.
\end{aligned}$$

$$\tan \tfrac{1}{2}(C' + C) = \frac{\cos \tfrac{1}{2}(c' - c)}{\cos \tfrac{1}{2}(c' + c)} \cot \tfrac{1}{2}\lambda_d.$$

$$\tan \tfrac{1}{2}(C' - C) = \frac{\sin \tfrac{1}{2}(c' - c)}{\sin \tfrac{1}{2}(c' + c)} \cot \tfrac{1}{2}\lambda_d.$$

$\tfrac{1}{2}(c' - c) = 13° 58'$,	log cos = 9.98697	log sin = 9.38266
$\tfrac{1}{2}(c' + c) = 63° 58'$,	colog cos = 0.35764	colog sin = 0.04646
$\tfrac{1}{2}\lambda_d = 67° 8'$,	log cot = 9.62504	log cot = 9.62504
	$\log \tan \tfrac{1}{2}(C' + C) = 9.96965$	$\log \tan \tfrac{1}{2}(C' - C) = 9.05416$
	$\tfrac{1}{2}(C' + C) = 43° 0'$	$\tfrac{1}{2}(C' - C) = 6° 28'$.
	$\tfrac{1}{2}(C' - C) = 6° 28'$	

$C' =$ S. 49° 28′ E. = course from A.
$C =$ S. 36° 32′ W. = course from B.

To find the distance:

$$\cos \tfrac{1}{2}D = \frac{\cos \tfrac{1}{2}(c' + c)}{\cos \tfrac{1}{2}(c' - c)} \sin \tfrac{1}{2}\lambda_d.$$

$\tfrac{1}{2}(c' + c) = 63° 58'$,	log cos = 9.64236
$\tfrac{1}{2}(C' + C) = 43° 0'$,	colog cos = 0.13587
$\tfrac{1}{2}\lambda_d = 67° 8'$,	log sin = 9.96445
	$\log \cos \tfrac{1}{2}D = 9.74268$

$\tfrac{1}{2}D = 56° 26'$.
$D = 112° 52' = 6772$ m.

To find L of V.

$\sin PV = \sin c' \sin C$.

log sin c' = 9.99021
log sin C = 9.77473
log sin PV = 9.76494

$PV = 35° 36'$.
L of $V = 90° 0' - 35° 36'$
$= 54° 24'$ S.

To find λ of V.

$\cot CPV = \cos c' \tan C$.

log cos c' = 9.32025
log tan C = 9.86974
log cot CPV = 9.18999

$CPV = 81° 12'$
λ of $C = 77° 14'$
λ of $V = 158° 26'$ W.

By Rhumb Line:

To find L_d.

$L' = 40°$ S.
$L'' = 12° 4'$ S.
$L_d = 27° 56'$
$= 1676$ m.

To find L_m.

$L' = 40°$ S.
$L'' = 12° 4'$ S.
2)52° 4′
$L_m = 26° 2'$

To find λ_d.

$\lambda' = 148° 30'$ E.
$\lambda'' = 77° 14'$ W.
$\lambda_d = 225° 44'$
$= 360° - 225° 44'$
$= 134° 16' = 8056$ m.

To find the course.

$$\tan C = \frac{\lambda_d \cos L_m}{L_d}.$$

$\log \lambda_d$	=	3.90612
$\log \cos L_m$	=	9.95354
$\text{colog}\, L_d$	=	6.77573
$\log \tan C$	=	10.63539

To find the distance.

$$D = L_d \sec C.$$

$\log L_d$	=	3.22427
$\log \sec C$	=	.64682
$\log D$	=	3.87109
D	=	7432 m.

$C = 76° 58'$. That is, N. 76° 58′ E. from A, or S. 76° 58′ W. from B.

Exercise IX. Page 345.

1.

Observed altitude	25° 6′ 10″	Index correction	+ 1′ 15″
Correction	− 4′ 50″	Dip	− 4′ 2″
		Refraction	− 2′ 3.4″
True altitude	25° 1′ 20″		

2.

Observed altitude	15° 20′ 25″	Index correction	− 2′ 20″
Correction	− 9′ 44″	Dip	− 3′ 55″
		Refraction	− 3′ 29.4″
True altitude	15° 10′ 41″		

3.

Observed altitude	18° 17′ 30″	Index correction	+ 0′ 18″
Correction	9′ 41″	Dip	− 4′ 9″
		Refraction	− 2′ 54″
		Semi-diameter	+ 16′ 18″
		Parallax	+ 8″
True altitude	18° 27′ 11″		

4.

Observed altitude	30° 12′ 40″	Semi-diameter	+ 16′ 4″
Correction	10′ 24″	Parallax	+ 8″
		Index correction	− 0′ 0″
		Dip	− 4′ 16″
		Refraction	− 1′ 39″
True altitude	30° 23′ 4″		

5.

Observed altitude	56° 25′ 20″	Semi-diameter	+ 16′ 3″
Correction	10′ 18″	Parallax	+ 5″
		Index correction	− 1′ 20″
		Dip	− 4′ 2″
		Refraction	− 0′ 38″
True altitude	56° 35′ 28″		

6.

Observed altitude	60° 10′ 10″	Semi-diameter	+ 15′ 48″
Correction	9′ 55″	Parallax	+ 0′ 4″
		Index correction	+ 2′ 15″
		Dip	− 4′ 9″
		Refraction	− 0′ 33″
True altitude	60° 19′ 5″		

7.

Observed altitude	31° 24′ 35″	Semi-diameter	+ 16′ 14″
Correction	10′ 38″	Parallax	+ 8″
		Index correction	− 0′ 0″
		Dip	− 4′ 9″
		Refraction	− 1′ 35″
True altitude	31° 35′ 13″		

8.

Observed altitude	26° 17′ 20″	Semi-diameter	− 16′ 10″
Correction	19′ 53″	Parallax	+ 8″
		Index correction	+ 2′ 15″
		Dip	− 4′ 9″
		Refraction	− 1′ 57″
True altitude	25° 57′ 27″		

9.

Observed altitude	20° 35′ 30″	Semi-diameter	− 15′ 46″
Correction	21′ 49″	Parallax	+ 0′ 8″
		Index correction	+ 0′ 18″
		Dip	− 3′ 55″
		Refraction	− 2′ 34″
True altitude	20° 13′ 41″		

10.

Observed altitude	36° 12′ 10″	Semi-diameter	− 15′ 47″
Correction	20′ 57″	Parallax	+ 0′ 8″
		Index correction	+ 0′ 25″
		Dip	− 4′ 23″
		Refraction	− 1′ 20″
True altitude	35° 51′ 13″		

Exercise X. Page 347.

	Astronomical Time.	d.	h.	m.	s.		Civil Time.	d.	h.	m.	s.
1.	July	8	7	6	10	=	July	8	7	6	10 P.M.
2.	Mar.	7	12	25	30	=	Mar.	8	0	25	30 A.M.
3.	Jan.	1	18	10	10	=	Jan.	2	6	10	10 A.M.
4.	Dec.	31	15	0	0	=	Jan.	1	3	0	0 A.M.
5.	Feb.	2	8	4	30	=	Feb.	2	8	4	30 P.M.

		Civil Time.					Astronomical Time.				
		d.	h.	m.	s.			d.	h.	m.	s.
6.	July	1	11	8	25 A.M.	=	June	30	23	8	25.
7.	Mar.	2	11	56	56 P.M.	=	Mar.	2	11	56	56.
8.	Aug.	3	10	8	20 P.M.	=	Aug.	31	10	8	20.
9.	Sept.	1	0	12	15 A.M.	=	Aug.	31	12	12	15.
10.	Jan.	1	10	41	56 A.M.	=	Dec.	31	22	41	56.

Exercise XI. Page 349.

			d.	h.	m.	s.	
1.	Ship date,	May	4	6	12	15	
	Longitude in time,			+ 11	23	20	15) 170° 50′ 0″
	Greenwich date,	May	4	17	35	35	11 h. 23′ 20″

			d.	h.	m.	s.	
2.	Ship date,	July	30	23	12	30	
	Longitude in time,			+ 2	41	20	15) 40° 20′ 0″
	Greenwich date,	July	31	1	53	50	2 h. 41′ 20″

			d.	h.	m.	s.	
3.	Ship date,	July	31	14	10	15	
	Longitude in time,			5	22	43	15) 80° 40′ 45″
	Greenwich date,	July	31	19	32	58	5 h. 22′ 43″

			d.	h.	m.	s.	
4.	Ship date,	Mar.	2	10	20	0	
	Longitude in time,			3	23	0	15) 50° 45′
	Greenwich date,	Mar.	2	6	57	0	3 h. 23′

			d.	h.	m.	s.	
5.	Ship date,	Mar.	25	11	8	0 P.M.	
	Longitude in time,			6	41	42	15) 100° 25′ 30″
	Greenwich date,	Mar.	25	17	49	42	6 h. 41′ 42″

			d.	h.	m.	s.	
6.	Greenwich date,	Dec.	30	19	47	28	
	Longitude in time,			1	40	28	15) 25° 7′ 0″
		Dec.	30	18	7	0	1 h. 40′ 28″
	Local civil time,	Dec.	31	6	7	0 A.M.	

			d.	h.	m.	s.	
7.	Greenwich date,	July	4	23	51	0	
	Longitude in time,			11	56	0	15) 179° 0′ 0″
	Local civil time,	July	4	11	55	0 P.M.	11 h. 56′

		d.	h.	m.	s.	
8. Greenwich date,	July	3	23	59	0	
Longitude in time,			11	56		
	July	3	35	55	0	
Local civil time,	July	4	11	55	0 P.M.	

		d.	h.	m.	s.	
9. Greenwich date,	May	19	19	40	20	
Longitude in time,				3		15) 45′ 0″
	May	19	19	43	20	3′
Local civil time,	May	20	7	43	20 A.M.	

	1880.	d.	h.	m.	s.	
10. Greenwich date,	Dec.	31	15	8	0	
Longitude in time,				8	40	15) 2° 10′ 0″
	Dec.	31	15	16	40	8′ 40″
Local civil time,	Jan.	1	3	16	40	

Exercise XII. Page 354.

Find the sun's declination and the equation of time corresponding to the following Greenwich dates:

1. 1895 Jan. 7 d. 3 h. apparent time.

			m.	s.
Jan. 7 d. 0 h.	☉'s dec. 22° 22′ 25.8″ S.	Eq. of time	+ 6	29.57
Diff. for 3 h.	− 57.7″		+	3.22
Jan. 7 d. 3 h.	☉'s dec. 22° 21′ 28.1″ S.	Eq. of time	+ 6	32.79

2. 1895 Aug. 1 d. 6 h. 12 m. 20 s. apparent time.

			m.	s.
Aug. 1 d. 0 h.	☉'s dec. 18° 1′ 59.0″ N.	Eq. of time	+ 6	7.87
Diff. for 6 h. 12 m. 20 s.	− 3′ 54.3″		−	0.92
Aug. 1 d. 6 h. 12 m. 20 s.	☉'s dec. 17° 58′ 4.7″ N.	Eq. of time	+ 6	6.95

3. 1895 May 5 d. 10 h. 25 m. apparent time.

			m.	s.
May 5 d. 0 h.	☉'s dec. 16° 15′ 42.5″ N.	Eq. of time	− 3	26.15
Diff. for 10 h. 25 m.	+ 7′ 25.4″		−	2.33
May 5 d. 10 h. 25 m.	☉'s dec. 16° 23′ 7.9″ N.	Eq. of time	− 3	28.48

4. 1895 Aug. 7 d. 15 h. 12 m. apparent time.

			m.	s.
Aug. 8 d. 0 h.	☉'s dec. 16° 9′ 20.0″ N.	Eq. of time	+ 5	28.13
Diff. for 8 h. 48 m.	+ 6′ 15.0″			+ 2.84
Aug. 8 d. 15 h. 12 m.	☉'s dec. 16° 15′ 35.0″ N.	Eq. of time	+ 5	30.97

5. 1895 Dec. 4 d. 6 h. 18 m. apparent time.

			m.	s.
Dec. 4 d. 0 h.	☉'s dec. 22° 15′ 34.0″ S.	Eq. of time	− 9	40.21
Diff. for 6 h. 18 m.	+ 2′ 6.7″			+ 6.38
Dec. 4 d. 6 h. 18 m.	☉'s dec. 22° 17′ 40.7″ S.	Eq. of time	− 9	33.83

6. 1895 July 23 d. 20 h. 16 m. 40 s. apparent time.

			m.	s.
July 24 d. 0 h.	☉'s dec. 19° 53′ 2.9″ N.	Eq. of time	+ 6	16.42
Diff. for 3 h. 43 m. 20 s.	+ 1′ 57.3″			− 2.12
July 23 d. 20 h. 16 m. 40 s.	☉'s dec. 19° 55′ 0.2″ N.	Eq. of time	+ 6	14.30

7. 1895 Nov. 1 d. 3 h. 6 m. apparent time.

			m.	s.
Nov. 1 d. 0 h.	☉'s dec. 14° 26′ 57.9″ S.	Eq. of time	− 16	18.64
Diff. for 3 h. 6 m.	+ 2′ 29.0″			− 0.20
Nov. 1 d. 3 h. 6 m.	☉'s dec. 14° 29′ 26.9″ S.	Eq. of time	− 16	18.84

8. 1895 Oct. 12 d. 5 h. 12 m. apparent time.

			m.	s.
Oct. 12 d. 0 h.	☉'s dec. 7° 24′ 29.3″ S.	Eq. of time	− 13	26.96
Diff. for 5 h. 12 m.	+ 4′ 53.4″			− 3.20
Oct. 12 d. 5 h. 12 m.	☉'s dec. 7° 30′ 22.7″ S.	Eq. of time	− 13	30.16

9. 1895 June 7d. 3 h. 18 m. mean time.

			m.	s.
June 7 d. 0 h.	☉'s dec. 22° 45′ 50.5″ N.	Eq. of time	+ 1	26.59
Diff. for 3 h. 18 m.	+ 47.9″			− 1.51
June 7 d. 3 h. 18 m.	☉'s dec. 22° 46′ 38.4″ N.	Eq. of time	+ 1	25.08

10. 1895 Feb. 3 d. 9 h. 15 m. mean time.

			m.	s.
Feb. 3 d. 0 h.	☉'s dec. 16° 30′ 23.8″ S.	Eq. of time	− 14	2.48
Diff. for 9 h. 15 m.	− 6′ 48.9″			− 2.40
d. 9 h. 15 m.	☉'s dec. 16° 23′ 34.9″ S.	Eq. of time	− 14	4.88

EXERCISE XIII. PAGE 361.

1. Given civil date 1895 Jan. 1, longitude 102° 41′ W., observed meridian altitude of ☉ 59° 59′ 50″ S., index correction + 0′ 50″, eye 15 ft.; find the latitude.

Long. 102° 41′ W. = 6 h. 50 m. 44 s.

☉ 59° 59′ 50″	Index cor.,	+ 0′ 50″	☉'s dec. 23° 0′ 34″ S.	12.49
	Semi-diam.,	+ 16′ 18″	1′ 25″	6.84
+ 12′ 50″	Dip,	− 3′ 48″	d = 22° 59′ 9″ S.	85.43
	Refraction,	− 0′ 34″	z = 29° 47′ 20″ N.	
	Parallax,	+ 0′ 4″	L = 6° 48′ 11″ N.	
60° 12′ 40″				
90°				
z = 29° 47′ 20″ N.				

2. Given civil date 1895 Feb. 1, longitude 78° 14′ E., observed meridian altitude of ☉ 78° 4′ 10″ S., index correction + 0′ 55″, eye 12 ft.; find the latitude.

Long. 78° 14′ = 5 h. 12 m. 56 s.

☉ 78° 4′ 10″	Index cor.,	+ 0′ 55″	☉'s dec. 17° 5′ 1″ S.	42.76
	Semi-diam.,	+ 16′ 16″	3′ 43″	5.22
+ 13′ 37″	Dip,	− 3′ 24″	d = 17° 8′ 44″ S.	223.21
	Refraction,	− 0′ 12″	z = 11° 42′ 13″ N.	
	Parallax,	+ 0′ 2″	L = 5° 26′ 31″ S.	
78° 17′ 47″				
90°				
z = 11° 42′ 13″ N.				

3. Given civil date 1895 Mar. 20, longitude 173° 18′ W., observed meridian altitude of ☉ 89° 37′ 0″ N., index correction + 4′ 32″, eye 18 ft.; find the latitude.

Long. 173° 18′ = 11 h. 33 m. 12 s.

☉ 89° 37′ 0″ N.	Index cor.,	+ 4′ 32″	☉'s dec. 0° 8′ 36″ S.	59.26
	Semi-diam.,	+ 16′ 5″	− 11′ 24″	11.55
+ 16′ 28″	Dip,	− 4′ 9″	d = 0° 2′ 48″ N.	684.45
	Refraction,	− 0′ 0″	z = 0° 6′ 32″ S.	
	Parallax,	+ 0′ 0″	L = 0° 3′ 44″ S.	
89° 53′ 28″				
90°				
z = 0° 6′ 32″ S.				

4. Given civil date 1895 April 1, longitude 87° 42′ W., observed meridian altitude of ⊙ 48° 42′ 30″ S., index correction + 1′ 42″, eye 18 ft.; find the latitude.

Long. 87° 42′ = 5 h. 50 m. 48 s.

⊙ 48° 42′ 30″ S.	Index cor.,	+ 1′ 42″	⊙'s dec. 4° 33′ 23″ N.	57.85
	Semi-diam.,	+ 16′ 2″	+ 5′ 38″	5.85
+ 12′ 50″	Dip,	− 4′ 9″	d = 4° 39′ 1″ N.	338.42
	Refraction,	− 0′ 51″	z = 41° 4′ 40″ N.	
	Parallax,	+ 0′ 6″	L = 45° 43′ 41″ N.	
48° 55′ 20″ S.				
90°				
z = 41° 4′ 40″ N.				

5. Given civil date 1895 Sept. 1, longitude 97° 42′ E., observed meridian altitude of ⊙ 51° 4′ 50″ S., index correction − 6′ 0″, eye 23 ft.; find the latitude.

Long. 97° 42′ = 6 h. 30 m. 48 s.

⊙ 51° 4′ 50″ S.	Index cor.,	− 6′ 0″	⊙'s dec. 8° 17′ 14″ N.	54.43
	Semi-diam.,	+ 15′ 54″	5′ 54″	6.51
+ 4′ 31″	Dip,	− 4′ 42″	d = 8° 23′ 8″ N.	354.34
	Refraction,	− 0′ 47″	z = 38° 50′ 39″ N.	
	Parallax,	+ 0′ 6″	L = 47° 13′ 47″ N.	
51° 9′ 21″ S.				
90°				
z = 38° 50′ 39″ N.				

6. Given civil date 1895 Aug. 26, longitude 92° 3′ E., observed meridian altitude of ⊙ 35° 35′ 20″ N., index correction + 2′ 17″, eye 12 ft.; find the latitude.

Long. 92° 3′ = 6 h. 8 m. 12 s.

⊙ 35° 35′ 20″ N.	Index cor.,	+ 2′ 17″	⊙'s dec. 10° 25′ 18″ N.	52.22
	Semi-diam.,	+ 15′ 52″	5′ 21″	6.14
+ 13′ 31″	Dip,	− 3′ 24″	d = 10° 30′ 39″ N.	320.66
	Refraction,	− 1′ 21″	z = 54° 11′ 9″ S.	
	Parallax,	+ 0′ 7″	L = 43° 40′ 30″ S.	
35° 48′ 51″ N.				
90°				
z = 54° 11′ 9″ S.				

7. Given civil date 1895 May 16, longitude 45° 26′ W., observed meridian altitude of ⊙ 86° 34′ 20″ N., index correction + 4′ 16″, eye 15 ft.; find the latitude.

Long. 45° 26′ = 3 h. 1 m. 44 s.			
⊙ 86° 34′ 20″ N.	Index cor., + 4′ 16″	⊙'s dec. 19° 6′ 29″ N.	34.63
	Semi-diam., + 15′ 51″	1′ 45″	3.03
+ 16′ 16″	Dip, − 3′ 48″	d = 19° 8′ 74″ N.	104.93
	Refraction, − 0′ 4″	z = 3° 9′ 24″ S.	
	Parallax, + 0′ 1″	L = 15° 58′ 50″ N.	
86° 50′ 36″ N.			
90°			
z = 3° 9′ 24″ S.			

8. Given civil date 1895 March 20, longitude 174° 0′ W., observed meridian altitude of ⊙ 89° 56′ 10″ N., index correction − 1′ 15″, eye 15 ft.; find the latitude.

Long. 174° 0′ = 11 h. 36 m.			
⊙ 89° 56′ 10″ N.	Index cor., − 1′ 15″	⊙'s dec. 0° 8′ 36″ S.	59.26
	Semi-diam., + 16′ 5″	11′ 27″	11.60
+ 10′ 52″	Dip, − 3′ 48″	d = 0° 2′ 51″ N.	687.42
	Refraction, − 0′ 0″	z = 0° 7′ 2″ N.	
	Parallax, + 0′ 0″	L = 0° 9′ 53″ N.	
90° 7′ 2″ N.			
90°			
z = 0° 7′ 2″ N.			

9. Given civil date 1895 June 1, longitude 44° 40′ E., observed meridian altitude of ⊙ 72° 14′ 10″ N., index correction + 3′ 45″, eye 22 ft.; find the latitude.

Long. 44° 40′ = 2 h. 58 m. 40 s.			
⊙ 72° 14′ 10″ N.	Index cor., + 3′ 45″	⊙'s dec. 22° 3′ 54″ N.	20.39
	Semi-diam., + 15′ 48″	1′ 1″	2.98
+ 14′ 41″	Dip, − 4′ 36″	d = 22° 2′ 53″ N.	60.76
	Refraction, − 0′ 19″	z = 17° 31′ 9″ S.	
	Parallax, + 0′ 3″	L = 4° 31′ 44″ N.	
72° 28′ 51″ N.			
90°			
z = 17° 31′ 9″ S.			

10. Given civil date 1895 Dec. 1, longitude 67° 56′ E., observed meridian altitude of ⊙ 18° 48′ 10″ S., index correction − 3′ 6″, eye 18 ft.; find the latitude.

Long. 67° 56′ = 4 h. 31 m. 44 s.

⊙ 18° 48′ 10″ S.	Index cor.,	− 3′ 6″	⊙'s dec. 21° 49′ 31″ S.	23.29
	Semi-diam.,	+ 16′ 16″	1′ 45″	4.53
+ 6′ 20″	Dip,	− 4′ 9″	d = 21° 51′ 16″ S.	105.04
	Refraction,	− 2′ 49″	z = 71° 5′ 30″ N.	
	Parallax,	+ 0′ 8″	L = 49° 14′ 14″ N.	
18° 54′ 30″ S.				
90°				
z = 71° 5′ 30″ N.				

11. Given civil date 1895 Sept. 23, longitude 57° 45′ E., observed meridian altitude of ⊙ 84° 10′ 50″ N., index correction − 1′ 36″, eye 16 ft.; find the latitude.

Long. 57° 45′ = 3 h. 51 m.

⊙ 84° 10′ 50″ N.	Index cor.,	− 1′ 36″	⊙'s dec. 0° 4′ 35″ S.	58.49
	Semi-diam.,	+ 15′ 59″	3′ 45″	3.85
+ 10′ 23″	Dip,	− 3′ 55″	d = 0° 0′ 50″ S.	225.19
	Refraction,	− 0′ 6″	z = 5° 38′ 47″ S.	
	Parallax,	+ 0′ 1″	L = 5° 39′ 37″ S.	
84° 21′ 13″ N.				
90°				
z = 5° 38′ 47″ S.				

12. Given civil date 1895 Sept. 23, longitude 119° 54′ E., observed meridian altitude of ⊙ 83° 46′ 0″ S., index correction − 5′ 30″, eye 18 ft.; find the latitude.

Long. 119° 54′ = 7 h. 59 m. 36 s.

⊙ 83° 46′ 0″ S.	Index cor.,	− 5′ 30″	⊙'s dec. 0° 4′ 35″ S.	58.49
	Semi-diam.,	+ 15′ 59″	7′ 48″	7.99
+ 6′ 15″	Dip,	− 4′ 9″	d = 0° 3′ 13″ N.	467.60
	Refraction,	− 0′ 6″	z = 6° 7′ 45″ N.	
	Parallax,	+ 0′ 1″	L = 6° 10′ 58″ N.	
83° 52′ 15″ S.				
90°				
z = 6° 7′ 45″ N.				

13. Given civil date 1895 Nov. 21, longitude 70° 20′ E., observed meridian altitude of ⊙ 80° 20′ 0″ N., index correction − 2′ 50″, eye 20 ft.; the latitude.

Long. 70° 20′ = 4 h. 41 m. 20 s.

⊙ 80° 20′ 0″ N.	Index cor.,	− 2′ 50″	⊙'s dec.	19° 56′ 16″ S.	33.11
	Semi-diam.,	+ 16′ 14″		2′ 35″	4.69
+ 8′ 53″	Dip,	− 4′ 23″	d =	19° 53′ 31″ S.	155.29
	Refraction,	− 0′ 10″	z =	9° 31′ 7″ S.	
	Parallax,	+ 0′ 2″	L =	29° 24′ 38″ S.	

80° 28′ 53″ N.
90°
z = 9° 31′ 7″ S.

14. Given civil date 1895 Dec. 31, longitude 123° 45′ W., observed meridian altitude of ⊙ 67° 8′ 10″ S., index correction + 0′ 9″, eye 13 ft.; find the latitude.

Long. 123° 45′ = 8 h. 15 m.

⊙ 67° 8′ 10″ S.	Index cor.,	+ 0′ 9″	⊙'s dec.	23° 6′ 22″ S.	11.03
	Semi-diam.,	+ 16′ 18″		1′ 31″	8.25
+ 12′ 34″	Dip,	− 3′ 32″	d =	23° 4′ 51″ S.	91.00
	Refraction,	− 0′ 25″	z =	22° 39′ 16″ N.	
	Parallax,	+ 0′ 4″	L =	0° 25′ 35″ S.	

67° 20′ 44″ S.
90°
z = 22° 39′ 16″ N.

15. Given civil date 1895 Oct. 20, longitude 150° 25′ W., observed meridian altitude of ⊙ 49° 58′ 50″ N., index correction + 1′ 10″ eye 19 ft.; find the latitude.

Long. 150° 25′ = 10 h. 1 m. 40 s.

⊙ 49° 58′ 50″ N.	Index cor.,	+ 1′ 10″	⊙'s dec.	10° 21′ 17″ S.	53.89
	Semi-diam.,	+ 16′ 7″		9′ 11″	10.03
+ 12′ 18″	Dip,	− 4′ 16″	d =	10° 30′ 18″ S.	540.52
	Refraction,	− 0′ 49″	z =	39° 48′ 52″ S.	
	Parallax,	+ 0′ 6″	L =	50° 19′ 10″ S.	

50° 11° 8″ N.
90°
z = 39° 48′ 52″ S.

16. Given civil date 1895 June 1, longitude 96° 17′ E., observed meridian altitude of ⊙ 75° 38′ 15″ N., index correction + 0′ 27″, eye 26 ft.; find the latitude.

Long. 96° 17′ = 6 h. 25 m. 8 s.

$\underline{\odot}$ 75° 38′ 15″ N.	Index cor.,	+ 0′ 27″	☉'s dec. 22° 3′ 54″ N.	20.39
	Semi-diam.,	+ 15′ 48″	2′ 11″	6.42
+ 11′ 2″	Dip,	− 5′ 0″	d = 22° 1′ 43″ N.	130.90
	Refraction,	− 0′ 15″	z = 14° 10′ 43″ S.	
	Parallax,	+ 0′ 2″	L = 7° 51′ 0″ N.	
75° 49′ 17″ N.				
90°				
z=14° 10′ 43″ S.				

17. Given civil date 1895 June 25, longitude 59° 15′ E., observed meridian altitude of $\overline{\odot}$ 60° 23′ 15″ N., index correction + 2′ 21″, eye 30 ft.; find the latitude.

Long. 59° 15′ = 3 h. 57 m.

$\overline{\odot}$ 60° 23′ 15″ N.	Index cor.,	+ 2′ 21″	☉'s dec. 23° 24′ 19″ N.	3.93
	Semi-diam.,	− 15′ 46″	15″	3.95
− 19′ 16″	Dip,	− 5′ 22″	d = 23° 24′ 34″ N.	15.52
	Refraction,	− 0′ 33″	z = 29° 56′ 1″ S.	
	Parallax,	+ 0′ 4″	L = 6° 31′ 27″ S.	
60° 3′ 59″ N.				
90°				
z=29° 56′ 1″ S.				

Exercise XIV. Page 362.

1. Given civil date 1895 Jan. 29, observed meridian altitude of Aldebaran 52° 36′ 0″ S., index correction − 0′ 23″, eye 20 ft.; find the latitude.

Obs. alt.	= 52° 36′ 0″ S.		
	− 5′ 31″		
True alt.	= 52° 30′ 29″ S.	Index correction,	− 23″
	90° N.	Dip,	− 4′ 23″
Zenith dis.	= 37° 29′ 31″ N.	Refraction,	− 45″
Dec.	= 16° 18′ 2″ N.		− 5′ 31″
Lat.	= 53° 37′ 33″ N.		

2. Given civil date 1895 Feb. 18, observed meridian altitude of Procyon 77° 18′ 10″ S., index correction + 0′ 19″, eye 16 ft.; find the latitude.

Obs. alt.	= 77° 18′ 10″ S.		
	− 3′ 50″		
True alt.	= 77° 14′ 20″ S.	Index correction,	+ 19″
	90° N.	Dip,	− 3′ 55″
Zenith dis.	= 12° 45′ 40″ N.	Refraction,	− 1′ 13.5″
Dec.	= 5° 29′ 39″ N.		− 3′ 49.5″
Lat.	= 18 15′ 19″ N.		

3. Given civil date 1895 March 20, observed meridian altitude of Arcturus 36° 10′ 20″ N., index correction + 2′ 42″, eye 20 ft.; find the latitude.

Obs. alt.	= 36° 10′ 20″ N.
	− 3′ 1″
True alt.	= 36° 7′ 19″ N.
	90° S.
Zenith dis.	= 53° 52′ 41″ S.
Dec.	= 19° 43′ 23″ N.
Lat.	= 34° 9′ 18″ S.

Index correction,	+ 2′ 42″
Dip,	− 4′ 23″
Refraction,	− 1′ 20″
	− 3′ 1″

4. Given civil date 1895 Aug. 17, observed meridian altitude of Altair 66° 51′ 10″ N., index correction + 0′ 58″, eye 13 ft.; find the latitude.

Obs. alt.	= 66° 51′ 10″ N.
	− 3′ 0″
True alt.	= 66° 48′ 10″ N.
	90° S.
Zenith dis.	= 23° 11′ 50″ S.
Dec.	= 8° 35′ 34″ N.
Lat.	= 14° 36′ 16″ S.

Index correction,	+ 58″
Dip,	− 3′ 32″
Refraction,	− 25.5″
	− 3′ 0″

5. Given civil date 1895 Nov. 4, observed meridian altitude of Fomalhaut 59° 40′ 0″ N., index correction + 1′ 12″, eye 23 ft.; find the latitude.

Obs. alt.	= 59° 40′ 0″ N.
	− 4′ 4″
True alt.	= 59° 35′ 56″ N.
	90° S.
Zenith dis.	= 30° 24′ 4″
Dec.	= 30° 10′ 32″ S.
Lat.	= 60° 34′ 36″ S.

Index correction,	1′ 12″
Dip,	− 4′ 42″
Refraction,	− 34″
	− 4′ 4″

6. Given civil date 1895 Sept. 6, observed meridian altitude of Arcturus 86° 35′ 50″ N., index correction − 1′ 10″, eye 12 ft.; find the latitude.

Obs. alt.	= 86° 35′ 50″
	− 4′ 38″
True alt.	= 86° 31′ 12″
	= 90° S.
Zenith dis.	= 3° 28′ 48″ S.
Dec.	= 19° 43′ 37″ N.
Lat.	= 16° 14′ 49″ N.

Index correction,	− 1′ 10″
Dip,	− 3′ 24″
Refraction,	− 4″
	− 4′ 38″

7. Given civil date 1895 Oct. 6, observed meridian altitude of Markab 54° 10′ 15″ S., index correction 0, eye 13 ft.; find the latitude.

Obs. alt.	= 54° 10′ 15″ S.		
	− 4′ 14″		
True alt.	= 54° 6′ 1″ S.	Index correction,	+ 0′ 0″
	= 90° N.	Dip,	− 3′ 32″
Zenith dis.	= 35° 53′ 59″ N.	Refraction,	− 42″
Dec.	= 14° 38′ 49″ N.		− 4′ 14″
Lat.	= 50° 32′ 48″ N.		

8. Given civil date 1895 Aug. 17, observed meridian altitude of β Centauri 59° 47′ 13″ S., index correction 0, eye 25 ft.; find the latitude.

Obs. alt.	= 59° 47′ 13″ S.		
	− 5′ 28″		
True alt.	= 59° 41′ 45″ S.	Index correction,	+ 0′ 0″
	= 90° N.	Dip,	− 4′ 54″
Zenith dis.	= 30° 18′ 15″ N.	Refraction,	− 34″
Dec.	= 59° 52′ 29″ S.		− 5′ 28″
Lat.	= 29° 34′ 14″ S.		

9. Given civil date 1895 Dec. 4, observed meridian altitude of α Arietis 60° 29′ 50″ S., index correction − 2′ 10″, eye 18 ft.; find the latitude.

Obs. alt.	= 60° 29′ 50″ S.		
	− 6′ 52″		
True alt.	= 60° 22′ 58″ S.	Index correction,	− 2′ 10″
	90° N.	Dip,	− 4′ 9″
Zenith dis.	= 29° 37′ 2″ N.	Refraction,	− 33″
Dec.	= 22° 58′ 26″ N.		− 6′ 52″
Lat.	= 52° 35′ 28″ N.		

10. Given civil date 1895 Feb. 8, observed meridian altitude of Sirius 37° 50′ 20″ S., index correction + 1′ 4″, eye 19 ft.; find the latitude.

Obs. alt.	= 37° 50′ 20″ S.		
	− 4′ 27″		
True alt.	= 37° 45′ 53″ S.	Index correction,	+ 1′ 4″
	90° N.	Dip,	− 4′ 16″
Zenith dis.	= 52° 14′ 7″ N.	Refraction,	− 1′ 15″
Dec.	= 16° 34′ 20″ S.		− 4′ 27″
Lat.	= 35° 41′ 47″ N.		

11. Given civil date 1895 April 9, observed meridian altitude of Sirius 61° 3′ 50″ N., index correction 0, eye 16 ft.; find the latitude.

Obs. alt.	= 61° 3′ 50″ N.		
	4′ 27″	Index correction,	+ 0′ 0″
True alt.	= 60° 59′ 23″ N.	Dip,	− 3′ 55″
	90° S.	Refraction,	− 32″
Zenith dis.	= 29° 0′ 37″ S.		− 4′ 27″
Dec.	16° 34′ 24″ S.		
Lat.	= 45° 35′ 1″ S.		

12. Given civil date 1895 March 30, observed meridian altitude of Spica 52° 14′ 0″ N., index correction 0, eye 19 ft.; find the latitude.

Obs. alt.	= 52° 14′ 0″ N.		
	5′ 1″	Index correction,	+ 0′ 0″
True alt.	= 52° 8′ 59″ N.	Dip,	− 4′ 16″
	90° S.	Refraction,	− 45″
Zenith dis.	= 37° 51′ 1″ S.		− 5′ 1″
Dec.	= 10° 37′ 4″ S.		
Lat.	= 48° 28′ 5″ S.		

13. Given civil date 1895 July 8, observed meridian altitude of Antares 70° 10′ 30″ N., index correction 0, eye 21 ft.; find the latitude.

Obs. alt.	= 70° 10′ 30″ N.		
	4′ 50″	Index correction,	+ 0′ 0″
True alt.	= 70° 5′ 40″ N.	Dip,	− 4′ 29″
	90° S.	Refraction,	− 21″
Zenith dis.	= 19° 54′ 20″ S.		− 4′ 50″
Dec.	= 26° 12′ 12″ S.		
Lat.	= 46° 6′ 32″ S.		

Exercise XV. Page 370.

1. 1895, Oct. 19, A.M., at sea, in latitude 33° 27′ S.; the observed altitude ⨀ 28° 22′ 30″; index correction + 30″; height of eye 18 ft.; Greenwich mean time by chronometer Oct. 18 d. 18 h. 28 m. 38 s. Required the longitude.

⨀ 28° 22′ 30″	Index cor.,	+ 0′ 30″	☉'s dec.	9° 59′ 52.4″ S.	54.27
	Semi-diam.,	+ 16′ 6″		4′ 59.6″	5.52
+ 10′ 47″	Dip,	− 4′ 9″		9° 54′ 52.8″ S.	299.57
	Refraction,	− 1′ 48″		90° S.	
	Parallax,	+ 0′ 8″	p =	80° 5′ 7.2″ S.	
h = 28° 33′ 17″					

$h =$	28° 33′ 17″		
$L =$	33° 27′ 0″	log sec =	0.07864
$p =$	80° 5′ 7″	log csc =	0.00654
$2S =$	142° 5′ 24″		
$S =$	71° 2′ 42″	log cos =	9.51165
$R =$	42° 29′ 25″	log sin =	9.82960
		2)	19.42643
		log sin $\frac{1}{2}t =$	9.71321

Equation of time.

m. s.	
14 56.76	0.447
2.47	5.52
14 54.29	2.47

$\frac{1}{2}t = 31° 6′ 31″$. $t = 62° 13′ 2″ = 4$ h. 8 m. 52 s.

	d.	h.	m.	s.	
Oct.	18	19	51	8	Local apparent astronomical time.
			14	54	Equation of time.
Oct.	18	19	36	14	Local mean astronomical time.
Oct.	18	18	28	38	Greenwich mean time.
		1	7	36	Difference of time.
		16°	54′	0″	E. Longitude.

2. 1895, Oct. 20 A.M., at sea, in latitude 31° 40′ S.; the observed altitude ⨀ 35° 16′ 10″; index correction + 30″; height of eye 18 ft.; Greenwich mean time by chronometer, Oct. 19 d. 19 h. 11 m. 24 s. Required the longitude.

⨀ 35° 16′ 10″	Index cor.,	+ 0′ 30″
	Semi-diam.,	+ 16′ 7″
+ 11′ 14″	Dip,	− 4′ 9″
	Refraction,	− 1′ 22″
	Parallax,	+ 0′ 8″

⊙'s dec. 10° 21′ 30.4″ S.		53.89
4′ 19.2″		4.81
10° 17′ 11″	S.	259.21
90°	S.	
$p =$ 79° 42′ 49″	S.	

$h =$	35° 27′ 24″		
$L =$	31° 40′ 0″	log sec =	0.07001
$p =$	79° 42′ 49″	log csc =	0.00704
$2S =$	146° 50′ 13″		
$S =$	73° 25′ 6″	log cos =	9.45543
$R =$	37° 57′ 42″	log sin =	9.78897
		2)	19.32145
		log sin $\frac{1}{2}t =$	9.66072

Equation of time.

m. s.	
15 7.17	0.421
2.03	4.81
15 5.14	2.03

$\frac{1}{2}t = 27° 14′ 53″$. $t = 54° 29′ 46″ = 3$ h. 37 m. 59 s.

	d.	h.	m.	s.	
Oct.	19	20	22	1	Local apparent astronomical time.
			15	5	Equation of time.
Oct.	19	20	6	56	Local mean astronomical time.
Oct.	19	19	11	24	Greenwich mean time.
			55	32	Difference in time.
		13°	53′	0″	E. Longitude.

3. 1895, Oct. 20, P.M., at sea, in latitude 30° 55′ S.; the observed altitude ⊙ 21° 42′ 30″; index correction + 29″; height of eye 18 ft.; Greenwich mean time by chronometer Oct. 20 d. 3 h. 35 m. 40 s. Required the longitude.

⊙ 21° 42′ 30″	Index cor.,	+ 0′ 29″	⊙ dec.	10° 21′ 30.4″ S.	53.89
	Semi-diam.,	+ 16′ 7″		3′ 13.5″	3.59
+ 10′ 11″	Dip,	− 4′ 9″		10° 24′ 44″ S.	193.47
	Refraction,	− 2′ 24″		90° S.	
	Parallax,	+ 0′ 8″		p = 79° 35′ 16″ S.	

h = 21° 52′ 41″			Equation of time.	
L = 30° 55′ 0″	log sec =	0.06656	m. s.	
p = 79° 35′ 16″	log csc =	0.00721		
$2S$ = 132° 22′ 57″			15 7.17	0.421
S = 66° 11′ 29″	log cos =	9.60604	1.51	3.59
R = 44° 18′ 48″	log sin =	9.84421	15 8.68	1.51
		2)19.52402		

log sin $\frac{1}{2}t$ = 9.76201

$\frac{1}{2}t$ = 35° 19′ 3″. t = 70° 38′ 6″ = 4 h. 42 m. 32 s.

d.	h. m. s.	
Oct. 20	4 42 32	Local apparent astronomical time.
	15 9	Equation of time.
Oct. 20	4 27 23	Local mean astronomical time.
Oct. 20	3 35 40	Greenwich mean time.
	51 43	Difference in time.
	12° 55′ 45″	E. Longitude.

4. 1895, Oct. 21, A.M., at sea, in latitude 29° 35′ S.; the observed altitude ⊙ 24° 26′ 42″; index correction + 29″; height of eye 18 ft.; Greenwich mean time by chronometer Oct. 20 d. 18 h. 30 m. 39 s. Required the longitude.

⊙ 24° 26′ 42″	Index cor.,	+ 0′ 29″	⊙'s dec.	10° 42′ 59.3″ S.	53.50
	Semi-diam.,	+ 16′ 7″		4′ 53.7″	5.49
+ 10′ 27″	Dip,	− 4′ 9″		10° 38′ 6″ S.	293.71
	Refraction,	− 2′ 8″		90° S.	
	Parallax,	+ 0′ 8″		p = 79° 21′ 54″ S.	

h = 24° 37′ 9″			Equation of time.	
L = 29° 35′ 0″	log sec =	0.06066	m. s.	
p = 79° 21′ 54″	log csc =	0.00752		
$2S$ = 133° 34′ 3″			15 16.94	0.394
S = 66° 47′ 1″	log cos =	9.59573	2.16	5.49
R = 42° 9′ 52″	log sin =	9.82691	15 14.78	2.16
		2)19.49082		

log sin $\frac{1}{2}t$ = 9.74541

$\frac{1}{2}t$ = 33° 48′ 33″. t = 67° 37′ 6″ = 4 h. 30 m. 28 s.

	d.	h.	m.	s.	
Oct.	20	19	29	32	Local apparent astronomical time.
			15	15	Equation of time.
Oct.	20	19	14	17	Local mean astronomical time.
Oct.	20	18	30	39	Greenwich mean time.
			43	38	Difference in time.
		10°	54′	30″	E. Longitude.

5. 1895 Jan. 29, P.M., at ship, latitude 42° 26′ N.; observed altitude $\underline{\odot}$ 13° 40′; index error − 1′ 8″; height of eye 16 ft.; time by chronometer 29 d. 6 h. 48 m. 40 s., which was slow 11 m. 22.3 s. for mean noon at Greenwich, Dec. 1, 1894, and on Jan. 1, 1895, was 8 m. 7 s. slow for Greenwich mean noon. Required the longitude.

Chronometer.

	m.	s.
1894 Dec. 1 slow	11	22.3
1895 Jan. 1 slow	8	7.0
31)	3	15.3
		6.3
		28.28
	2	58.2

	d.	h.	m.	s.
Jan.	29	6	48	40
			+8	7
			−2	58
Jan.	29	6	53	49

$\underline{\odot}$ 13° 40′ 0″	Index cor.,	− 1′ 8″	
	Semi-diam.,	+ 16′ 16″	
+ 7′ 27″	Dip,	− 3′ 55″	
	Refraction,	− 3′ 55″	
	Parallax,	+ 0′ 9″	
h = 13° 47′ 27″			

⊙'s dec. 17° 55′ 7.1″ S.		40.43
4′ 39.0″		6.90
17° 50′ 28″ S.		278.97
90° N.		
p = 107° 50′ 28″ N.		

L = 42° 26′ 0″	log sec =	0.13191
p = 107° 50′ 28″	log csc =	0.02141
$2S$ = 164° 3′ 55″		
S = 82° 1′ 57″	log cos =	9.14180
R = 68° 14′ 30″	log sin =	9.96790
		2)19.26302
	log sin $\frac{1}{2}t$ =	9.63151

$\frac{1}{2}t$ = 25° 20′ 42″. t = 50° 41′ 24″ = 3 h. 22 m. 46 s.

Equation of time.

m.	s.	
13	20.81	0.435
	3.10	6.90
13	23.91	3.10

	d.	h.	m.	s.	
Jan.	29	3	22	46	Local apparent astronomical time.
			13	24	Equation of time.
Jan.	29	3	36	10	Local mean astronomical time.
Jan.	29	6	53	49	Greenwich mean time.
		3	17	39	Difference in time.
		49°	24′	45″	W. Longitude.

6. 1895, March 31, A.M., at ship, latitude 26° 9′ N.; observed altitude $\underline{\odot}$ 29° 10′ 20″; height of eye 26 ft.; time by chronometer 31 d. 0 h. 4 m. 50 s., which was 58 m. 58 s. fast for mean noon at Greenwich, Nov. 20, 1894, and on December 31, 1894, was 1 h. 2 m. 55.8 s. fast for mean time at Greenwich. Required the longitude.

Chronometer.

	h.	m.	s.
Nov. 20 fast	0	58	58
Dec. 31 fast	1	2	55.8
41)		3	57.8
			5.8
			90.
		8	42.0

	d.	h.	m.	s.
Mar.	31	0	4	50
		−1	2	56
		−	8	42
Mar.	30	22	53	12

$\underline{\odot}$ 29° 10′ 20″	Index cor.,	+ 0′ 0″
	Semi-diam.,	+ 16′ 2″
+ 9′ 26″	Dip,	− 5′ 0″
	Refraction,	− 1′ 44″
	Parallax,	+ 0′ 8″

⊙'s dec.	4° 10′ 8.3″ N.	58.06
	1′ 5.6″	1.13
	4° 9′ 3″ N.	65.61
	90° N.	
	$p =$ 85° 50′ 57″ N.	

$h =$	29° 19′ 46″		
$L =$	26° 9′ 0″	log sec =	0.04690
$p =$	85° 50′ 57″	log csc =	0.00114
$2S =$	141° 19′ 43″		
$S =$	70° 39′ 51″	log cos =	9.51996
$R =$	41° 20′ 5″	log sin =	9.81984
		2)	19.38784
		log sin $\frac{1}{2}t$ =	9.69392

$\frac{1}{2}t = 29° 37′ 5″$.

$t = 59° 14′ 10″ = 3$ h. 56 m. 57 s.

Equation of time.

m. s.	
4 15.13	0.758
0.86	1.13
4 15.99	0.86

	d.	h.	m.	s.	
Mar.	30	20	3	3	Local apparent astronomical time.
.			4	16	Equation of time.
Mar.	30	20	7	19	Local mean astronomical time.
Mar.	30	22	53	12	Greenwich mean time.
		2	45	53	Difference in time.
		41°	28′	15″	W. Longitude.

7. 1895, May 22, A.M., at ship, latitude 43° 25′ N.; observed altitude $\underline{\odot}$ 32° 8′; index correction + 1′ 28″; height of eye 15 ft.; time by chronometer 21 d. 21 h. 6 m. 10 s., which was slow 12.6 s. for mean noon at Greenwich, Feb. 24, and on April 1 was 2 m. 45 s. fast for mean noon at Greenwich. Required the longitude.

Chronometer.

m. s.		d. h. m. s.
Feb. 24 slow 0 12.6		May 21 21 6 10
Apr. 1 fast 2 45.0		−2 45
36)2 57.6		−4 11
4.93		May 21 20 59 14
51		
4 11.4		

⊙ 32° 8′ 0″	Index cor., + 1′ 28″	⊙'s dec. 20° 23′ 39.9″ N.	29.59
	Semi-diam., + 15′ 50″	1′ 29.1″	3.01
+ 12′ 5″	Dip, − 3′ 48″	20° 22′ 10.8″ N.	89.07
	Refraction, − 1′ 33″	90° N.	
	Parallax, + 0′ 8″	p = 69° 37′ 49″ N.	

h = 32° 20′ 5″		
L = 43° 25′ 0″	log sec = 0.13884	
p = 69° 37′ 49″	log csc = 0.02805	
$2S$ = 145° 22′ 54″		
S = 72° 41′ 27″	log cos = 9.47353	
R = 40° 21′ 22″	log sin = 9.81126	
	2)19.45168	
	log sin $\frac{1}{2}t$ = 9.72584	

Equation of time.

m. s.	
3 34.66	0.182
0.55	3.01
3 35.21	0.55

$\frac{1}{2}t$ = 32° 8′ 6″. t = 64° 16′ 12″ = 4 h. 17 m. 5 s.

d. h. m. s.	
May 21 19 42 55	Local apparent astronomical time.
3 35	Equation of time.
May 21 19 39 20	Local mean astronomical time.
May 21 20 59 14	Greenwich mean time.
1 19 54	Difference in time.
19° 58′ 30″	W. Longitude.

8. 1895, Aug. 24, A.M., at ship, latitude at noon 37° 59′ N.; observed altitude ⊙ 37° 13′ 30″; index correction + 2′ 44″; height of eye 18 ft.; time by chronometer Aug. 23 d. 18 h. 13 m. 24 s., which was 1 m. 5 s. fast for mean noon at Greenwich, August 1, and on August 10 was 0 m. 42 s. slow for mean time at Greenwich; course (true) since observation N.N.W.; distance 22.4 miles. Required the longitude at noon

Chronometer.

m. s.	d. h. m. s.	
Aug. 1 fast 1 5	Aug. 23 18 13 24	L' = 37° 59′ 0″ N.
Aug. 10 slow 0 42	+ 0 42	L_d = 20′ 41″
9)1 47	+ 2 44	L = 37° 38′ 19″ N.
11.89	Aug. 23 18 16 50	
13.76		
2 43.60		

$\odot$ 37° 13′ 30″	Index cor., + 2′ 44″	⊙'s dec. 11° 6′ 46.7″ N.	51.38
	Semi-diam., + 15′ 52″	4′ 53.9″	5.72
+ 13′ 17″	Dip, − 4′ 9″	11° 11′ 40.6″ N.	293.89
	Refraction, − 1′ 17″	90° N.	
	Parallax, + 0′ 7″	$p =$ 78° 48′ 19″ N.	

$h =$ 37° 26′ 47″		Equation of time.		
$L =$ 37° 38′ 19″	log sec = 0.10134	m. s.		
$p =$ 78° 48′ 19″	log csc = 0.00834	2 16.42		0.661
$2S =$ 153° 53′ 25″		3.78		5.72
$S =$ 76° 56′ 42″	log cos = 9.35389	2 20.20		3.78
$R =$ 39° 29′ 55″	log sin = 9.80350			
	2)19.26707			

log sin $\frac{1}{2}t =$ 9.63353

$\frac{1}{2}t =$ 25° 28′ 18″.

$t =$ 50° 56′ 36″ = 3 h. 23 m. 46 s.

		d. h. m. s.	
Mer. L_d=26.1	log=1.41664	Aug. 23 20 36 14	Local appar. ast. time.
C=22° 30′	log tan=9.61722	2 20	Equation of time.
	log λ_d =1.03386	Aug. 23 20 38 34	Local mean ast. time.
	λ_d=10.811	Aug. 23 18 16 50	Greenwich mean time.
	=10′ 49″ W.	2 21 44	Difference in time.
		35° 26′ 0″	E. Long. at sights.
		10′ 49″	W. Long. since observ.
		35° 15′ 11″	E. Long. at noon.

9. 1895, Jan. 29, P.M., at ship, latitude 28° 45′ N.; observed altitude $\odot$ 17° 46′ 30″; index correction − 3′ 18″; height of eye 16 ft., time by chronometer January 28 d. 16 h. 31 m. 30 s., which was 1 m. 16.5 s. fast for Greenwich mean noon, December 17, 1894, and on January 1, 1895, was 1 m. 3 s. slow for Greenwich mean time; course (true) since noon N.W. by W.; distance 20 miles. Required the longitude at the time of observation, and also at noon.

Chronometer.	m. s.	d. h. m. s.
1894, Dec. 17 fast	1 16.5	Jan. 28 16 31 30
1895, Jan. 1 slow	1 3.	+ 1 3
	15)2 19.5	+ 4 21
	9.30	Jan. 28 16 36 54
	28.02	
	4 20.59	

☉̲ 17° 46′ 30″	Index cor., − 3′ 18″	☉̲'s dec. 17° 55′ 7.1″ S.	40.43
	Semi-diam., + 16′ 16″	4′ 58.4″	7.38
+ 6′ 11″	Dip, − 3′ 55″	18° 0′ 5.5″ S.	298.37
	Refraction, − 3′ 0″	90° N.	
	Parallax, + 0′ 8″	$p =$ 108° 0′ 5.5″ N.	
$h =$ 17° 52′ 41″			

$L =$ 28° 45′ 0″	log sec = 0.05714	Equation of time.	
$p =$ 108° 0′ 5″	log csc = 0.02179	m. s.	
$2S =$ 154° 37′ 46″		13 20.81	0.435
$S =$ 77° 18′ 53″	log cos = 9.34163	3.21	7.38
$R =$ 59° 26′ 12″	log sin = 9.93504	13 17.60	3.21
	2)19.35560		
	log sin ½ $t =$ 9.67780		

½ $t =$ 28° 26′ 18″. $t =$ 56° 52′ 36″ = 3 h. 47 m. 30 s.

		d. h. m. s.	
L_d=11′ 7″	log= 1.10037	Jan. 29 3 47 30	Local appar. ast. time.
Mer. L_d=12.6	log tan=10.17511	13 18	Equation of time.
C=56° 15′	log λ_d= 1.27548	Jan. 29 4 0 48	Local mean ast. time.
	λ_d=18.857	Jan. 28 16 36 54	Greenwich mean time.
	=18′ 51″ E.	11 23 54	Difference in time.
		170° 58′ 30″	E. Long. at sights.
		18′ 51″	W. Long. since noon.
		171° 17′ 21″	E. Long. at noon.

10. 1895, Aug. 31, P.M., at ship, latitude 0°; observed altitude ☉̲ 45° 5′ 30″; index correction − 2′ 4″; height of eye 15 ft.; time by chronometer Aug. 31 d. 9 h. 11 m. 28 s., which was 5 m. 20 s. fast for mean noon at Greenwich April 15, and on June 16 was fast 2 m. 43 s. on mean time at Greenwich. Required the longitude.

Chronometer.

m. s.		d. h. m. s.
April 15 fast 5 20		Aug. 31 9 11 28
June 16 fast 2 43		− 2 43
62)2 37		+ 3 12
2.53		Aug. 31 9 11 57
76		
3 12.28		

☉̲ 45° 5′ 30″	Index cor., − 2′ 4″	☉'s dec. 8° 38′ 56.1″ N.	54.11
	Semi-diam., + 15′ 53″	8′ 17.8″	9.20
+ 9′ 9″	Dip, − 3′ 48″	8° 30′ 38.3″ N.	497.81
	Refraction, − 0′ 58″	90° N.	
$h =$ 45° 14′ 39″	Parallax, + 0′ 6″	$p =$ 81° 29′ 22″ N.	

$h =$	$45° 14' 39''$			Equation of time.
$L =$	$0° 0' 0''$	log sec =	0.00000	m. s.
$p =$	$81° 29' 22''$	log csc =	0.00481	0 15.51 | 0.773
$2S =$	$126° 44' 1''$			7.11 | 9.20
$S =$	$63° 22' 0''$	log cos =	9.65155	0 8.40 | 7.11
$R =$	$18° 7' 21''$	log sin =	9.49283	
			2)19.14919	
		log sin $\frac{1}{2}t$ =	9.57459	

$\frac{1}{2}t = 22° 3' 15''$. $t = 44° 6' 30'' = 2$ h. 56 m. 26 s.

d. h. m. s.	
Aug. 31 2 56 26	Local apparent astronomical time.
8	Equation of time.
Aug. 31 2 56 34	Local mean astronomical time.
Aug. 31 9 11 57	Greenwich mean time.
6 15 23	Difference in time.
93° 50′ 45″	W. Longitude.

11. 1895, April 15, A.M., at ship, latitude 48° 52′ N.; observed altitude $\underline{\odot}$ 22° 18′; index correction − 3′ 54″; height of eye 17 ft.; time by chronometer April 14 d. 22 h. 30 m. 42 s., which was 0 m. 4 s. slow for mean noon at Greenwich January 1, and on January 12 was fast 0 m. 2 s. Required the longitude.

Chronometer.

	m. s.		d. h. m. s.
Jan. 1 slow	0 4		Apr. 14 22 30 42
Jan. 12 fast	0 2		−0 2
	11)0 6		− 51
	0.545		Apr. 14 22 29 49
	93		
	50.69		

$\underline{\odot}$ 22° 18′ 0″	Index cor.,	− 3′ 54″	⊙'s dec. 9° 46′ 35.5″ N.	53.58
	Semi-diam.,	+ 15′ 58″	1′ 20.4″	1.50
+ 5′ 49″	Dip,	− 4′ 2″	9° 45′ 15.1″ N.	80.37
	Refraction,	− 2′ 21″	90° N.	
	Parallax,	+ 0′ 8″	$p =$ 80° 14′ 45″ N.	

$h =$	$22° 23' 49''$			Equation of time.
$L =$	$48° 52' 0''$	log sec =	0.18190	m. s.
$p =$	$80° 14' 45''$	log csc =	0.00632	0 1.59 | 0.619
$2S =$	$151° 30' 34''$			0.93 | 1.50
$S =$	$75° 45' 17''$	log cos =	9.39107	0 2.52 | 0.93
$R =$	$53° 21' 28''$	log sin =	9.90438	
			2)19.48367	
		log sin $\frac{1}{2}t$ =	9.74183	

$\frac{1}{2}t = 33° 29' 41''$. $t = 66° 59' 22'' = 4$ h. 27 m. 57 s.

<pre>
       d. h. m.   s.
April 14 19 32   3   Local apparent astronomical time.
                   3   Equation of time.
April 14 19 32   6   Local mean astronomical time.
April 14 22 29 49   Greenwich mean time.
              2 57 43   Difference in time.
          44° 25′ 45″ W. Longitude.
</pre>

12. 1895, Aug. 28, P.M., at ship, latitude 5° S.; observed altitude ☉ 38°; index correction + 5′ 27″; height of eye 21 ft.; time by chronometer Aug. 27 d. 22 h. 20 m. 30 s., which was 10 m. 0 s. slow for mean noon at Greenwich Feb. 19, and on May 30 was 2 m. 20 s. slow on mean noon at Greenwich. Required the longitude.

Chronometer.

<pre>
            m. s.   d. h. m. s.
Feb. 19 slow 10 0   Aug. 27 22 20 30
May 30 slow 2 20   + 2 20
        100)7 40   − 6 54
            4.6   Aug. 27 22 15 56
            90
           6 54
</pre>

<pre>
☉ 38° 0′ 0″ ⎧ Index cor., + 5′ 27″   ☉'s dec. 9° 43′ 13.4″ N. | 53.02
              ⎪ Semi-diam., + 15′ 53″   1′ 31.7″ | 1.73
     + 15′ 44″⎨ Dip,   − 4′ 29″   9° 44′ 45.1″ N. | 91.72
              ⎪ Refraction, − 1′ 14″   90°   S.
              ⎩ Parallax,   + 0′ 7″   p = 99° 44′ 45″ S.
h =   38° 15′ 44″
L =   5° 0′ 0″   log sec = 0.00166   Equation of time.
p =   99° 44′ 45″   log csc = 0.00632
2 S = 143° 0′ 29″   m.   s.
S =   71° 30′ 14″   log cos = 9.50139   1   9.63 | 0.729
R =   33° 14′ 30″   log sin = 9.73891   1.26 | 1.73
                           2)19.24828   1 10.89 | 1.26
              log sin ½t = 9.62414
                     ½t = 24° 53′ 20″.
                      t = 49° 46′ 40″ = 3 h. 19 m. 7 s.
</pre>

<pre>
       d. h. m.   s.
Aug. 28   3 19   7   Local apparent astronomical time.
              1 11   Equation of time.
Aug. 28   3 20 18   Local mean astronomical time.
Aug. 27 22 15 56   Greenwich mean time.
              5   4 22   Difference of time.
          76° 5′ 30″ E. Longitude.
</pre>

13. 1895, Sept. 22, A.M., at ship, on the equator, observed altitude $\overline{\odot}$ 17° 20′ 40″; index correction − 1′ 9″; height of eye 20 ft.; time by chronometer Sept. 22 d. 4 h. 59 m. 16 s., which was 15 s. slow for Greenwich mean noon, April 30, and on June 1 was 10.6 s. fast for mean time at Greenwich. Required the longitude.

Chronometer.

	m.	s.
April 30 slow	0	15
June 1 fast	0	10.6
32)	0	25.6
	0	0.8
		113.2
	1	30.6

	d.	h.	m.	s.
Sept.	22	4	59	16
			−0	10.6
			−1	30.6
Sept.	22	4	57	35

$\overline{\odot}$ 17° 20′ 40″	Index cor.,	− 1′ 19″
	Semi-diam.,	− 15′ 59″
− 24′ 37″	Dip,	− 4′ 23″
	Refraction,	− 3′ 4″
	Parallax,	+ 0′ 8″

⊙'s dec.	0° 18′ 41.2″ N.	58.48
	4′ 50.1″	4.96
	0° 13′ 51.1″ N.	290.06
	90° N.	
$p =$	89° 46′ 9″ N.	

$h =$	16° 56′ 3″		
$L =$	0° 0′ 0″	log sec =	0.00000
$p =$	89° 46′ 9″	log csc =	0.00000
$2S =$	106° 42′ 12″		
$S =$	53° 21′ 6″	log cos =	9.77590
$R =$	36° 25′ 3″	log sin =	9.77354
		2)	19.54944
		log sin $\frac{1}{2}t$ =	9.77472

$\frac{1}{2}t = 36° 31′ 57″$.

$t = 72° 3′ 54″ = 4$ h. 48 m. 16 s.

Equation of time.

m.	s.	
7	15.58	0.807
	4.32	4.96
7	19.90	4.32

d.	h.	m.	s.	
Sept. 21	19	11	44	Local apparent astronomical time.
		7	20	Equation of time.
Sept. 21	19	4	24	Local mean astronomical time.
Sept. 22	4	57	35	Greenwich mean time.
	9	53	11	Difference in time.
	148°	17′	45″	W. Longitude.

14. 1895, Aug. 5, A.M., at ship, latitude at noon 30° 30′ N.; observed altitude $\underline{\odot}$ 35° 6′; height of eye 15 ft.; time by chronometer 5 d. 8 h. 39 m. 22 s., which was fast 29 m. 32.4 s. on Greenwich mean noon, July 8, and on July 20 was fast 30 m. 0 s. on Greenwich mean noon; course (true) till noon W.; distance 48 miles. Required the longitude at noon.

Chronometer.

	m.	s.
July 8 fast	29	32.4
July 20 fast	30	0.0
12)	0	27.6
	0	2.3
		16.4
	0	37.7

	d.	h.	m.	s.
Aug.	5	8	39	22
			− 30	0
			− 0	38
Aug.	5	8	8	44

⊙ 35° 6′ 0″

+ 10′ 45″ { Index cor., + 0′ 0″; Semi-diam., + 15′ 49″; Dip, − 3′ 48″; Refraction, − 1′ 23″; Parallax, + 0′ 7″

⊙'s dec.	16° 59′ 19.9″ N.	40.60
	5′ 30.9″	8.15
	16° 53′ 49.0″ N.	330.89
	90° N.	
$p =$	73° 6′ 11″ N.	

$h =$	35° 16′ 45″		
$L =$	30° 30′ 0″	log sec =	0.06468
$p =$	73° 6′ 11″	log csc =	0.01916
$2S =$	138° 52′ 56″		
$S =$	69° 26′ 28″	log cos =	9.54552
$R =$	34° 9′ 43″	log sin =	9.74938
		2)	19.37874

Equation of time.

m. s.	
5 48.79	0.250
2.04	8.15
5 46.75	2.04

log sin $\frac{1}{2}t =$ 9.68937

$\frac{1}{2}t =$ 29° 16′ 46″.

$t =$ 58° 33′ 32″ = 3 h. 54 m. 14 s.

d. h. m. s.	
Aug. 4 20 5 46	Local apparent astronomical time.
5 47	Equation of time.
Aug. 4 20 11 33	Local mean astronomical time.
Aug. 5 8 8 44	Greenwich mean time.
11 57 11	Difference in time.
179° 17′ 45″	W. Longitude at sight.
48 miles = 55′ 43″	W.
180° 13′ 28″	
= 179° 46′ 32″	E. Longitude at noon.

15. 1895, Nov. 12, A.M., at sea, in latitude 7° 10′ N.; four observed altitudes of the ⊙ were taken at the times (by watch) standing opposite, viz.:

		h. m. s.
Obs. alt. ⊙	21° 8′ 40″	2 55 48
	11′ 50″	56 0
	14′ 50″	56 13
	17′ 30″	56 26.5

Index correction + 31″; height of eye 18 ft.; correction of watch by chronometer − 5 h. 12 m. 2.1 s. Required the longitude.

	h. m. s.
☉ 21° 8′ 40″	2 55 48
11′ 50″	56 0
14′ 50″	56 13
17′ 30″	56 26.5
52′ 50″	224 27.5
21° 13′ 12.5″	2 56 6.9
	−5 12 2.1
	21 44 4.8

☉'s declination.

17° 43′ 11.1″ S.	40.67
1′ 33.3″	2.27
17° 41′ 37.8″ S.	93.32
90° N.	
p = 107° 41′ 38″ N.	

Index correction,	+ 0′ 31″
Semi-diameter,	+ 16′ 12″
Dip,	− 4 9″
Refraction,	− 2′ 29″
Parallax,	− 0′ 8″
	+ 10′ 13″

	+ 10′ 13″	
h =	21° 23′ 25″	
L =	7° 10′ 0″	log sec = 0.00341
p =	107° 41′ 38″	log csc = 0.02104
$2S$ =	136° 15′ 3″	
S =	68° 7′ 31″	log cos = 9.57122
R =	46° 44′ 6″	log sin = 9.86224
		2)19.45791

log sin ½t = 9.72895

½t = 32° 23′ 38″.

t = 64° 47′ 16″ = 4 h. 19 m. 9 s.

Equation of time.

m. s.	
15 44.88	0.324
0.74	2.27
15 45.62	0.74

d. h. m. s.	
Nov. 11 19 40 51	Local apparent astronomical time.
15 46	Equation of time.
Nov. 11 19 25 5	Local mean astronomical time.
Nov. 11 21 44 5	Greenwich mean time.
2 19 0	Difference in time.
34° 45′ 0″	.W. Longitude.

16. 1895, Nov. 13, A.M., at sea, in latitude 9° 30′ N.; five observed altitudes of the ☉ were taken at the times (by watch) standing opposite, viz.:

		h. m. s.
Obs. alt. ☉	18° 58′ 40″	2 59 2
	19° 1′ 20″	13
	3′ 30″	28
	7′ 30″	45
	11′ 0″	57.5

Index correction + 32″; height of eye 18 ft.; correction of watch by chronometer − 5 h. 11 m. 58.4 s. Required the longitude.

	h. m. s.
☉ 18° 58′ 40″	2 59 2
19° 1′ 20″	13
3′ 30″	28
7′ 30″	45
11′ 0″	57.5
22′ 0″	145.5
19° 4′ 24″	2 59 29.1
	−5 11 58.4
	21 47 30.7

☉'s declination.

17° 59′ 18.1″ S.	39.90
1′ 28.2″	2.21
17° 57′ 49.9″ S.	88.18
90° N.	
$p =$ 107° 57′ 50″ N.	

Index correction,	+ 0′ 32″
Semi-diameter,	+ 16′ 13″
Dip,	− 4′ 9″
Refraction,	− 2′ 48″
Parallax,	+ 0′ 8″
	+ 9′ 56″

+ 9′ 56″

$h =$	19° 14′ 20″		
$L =$	9° 30′ 0″	log sec =	0.00600
$p =$	107° 57′ 50″	log csc =	0.02171
$2S =$	136° 42′ 10″		
$S =$	68° 21′ 5″	log cos =	9.56692
$R =$	49° 6′ 45″	log sin =	9.87852
		2)	19.47315
		log sin $\frac{1}{2}t$ =	9.73657

$\frac{1}{2}t = 33° 2' 22''.$

$t = 66° 4' 44'' = 4$ h. 24 m. 19 s.

Equation of time.

m. s.	
15 36.66	0.361
0.80	2.21
15 37.46	0.80

d. h. m. s.	
Nov. 12 19 35 41	Local apparent astronomical time.
15 37	Equation of time.
Nov. 12 19 20 4	Local mean astronomical time.
Nov. 12 21 47 31	Greenwich mean time.
2 27 27	Difference in time.
36° 51′ 45″	W. Longitude.

17. 1895, Nov. 17, A.M., at sea, in latitude 15° 35′ N.; five observed altitudes of the ☉ were taken at the times (by watch) standing opposite, viz.:

	h. m. s.
Obs. alt. ☉ 23° 56′ 0″	4 12 31
24° 0′ 0″	56
4′ 0″	13 2.5
6′ 10″	14
10′ 0″	28.5

Index correction + 31″; height of eye 18 ft.; correction of watch by chronometer − 5 h. 11 m. 43.6 s. Required the longitude.

	h. m. s.
⊙ 23° 56′ 0″	4 12 31
24° 0′ 0″	46
4′ 0″	13 2.5
6′ 10″	14
10′ 0″	28.5
16′ 10″	65 2
24° 3′ 14″	4 13 0.4
	−5 11 43.6
	23 1 16.8

⊙'s declination.

19° 0′ 34.2″ S.	36.63
35.9″	0.98
18° 59′ 58.3″ S.	35.90
90° N.	
$p =$ 108° 59′ 58″ N.	

Index correction,	+ 0′ 31″
Semi-diameter,	+ 16′ 13″
Dip,	− 4′ 9″
Refraction,	− 2′ 10″
Parallax,	+ 0′ 8″
	+ 10′ 33″

+ 10′ 33″

$h =$	24° 13′ 47″	
$L =$	15° 35′ 0″	log sec = 0.01627
$p =$	108° 59′ 58″	log csc = 0.02433
$2S =$	148° 48′ 45″	
$S =$	74° 24′ 22″	log cos = 9.42946
$R =$	50° 10′ 35″	log sin = 9.88537
		2)19.35543

log sin $\frac{1}{2}t$ = 9.67771

$\frac{1}{2}t$ = 28° 25′ 55″.

t = 56° 51′ 50″ = 3 h. 47 m. 27 s.

Equation of time.

m. s.	
14 55.18	0.503
0.49	0.98
14 55.67	0.49

d. h. m. s.	
Nov. 16 20 12 33	Local apparent astronomical time.
14 56	Equation of time.
Nov. 16 19 57 37	Local mean astronomical time.
Nov. 16 23 1 17	Greenwich mean time.
3 3 40	Difference in time.
45° 55′ 0″	W. Longitude.

18. 1895, Nov. 18, A.M., at sea, in latitude 16° 25′ N.; five observed altitudes of the ⊙ were taken at the times (by watch) standing opposite, viz.:

	h. m. s.
Obs. alt. ⊙ 18° 13′ 30″	3 52 42
16′ 10″	53.5
19′ 20″	53 6.5
22′ 30″	23
25′ 30″	38

Index correction + 32″; height of eye 18 ft.; correction of watch by chronometer − 5 h. 11 m. 39.9 s. Required the longitude.

⊙	h. m. s.
⊙ 18° 13′ 30″	3 52 42
16′ 10″	53.5
19′ 20″	53 6.5
22′ 30″	23
25′ 30″	38
97′ 0″	265 43
18° 19′ 24″	3 53 8.6
	−5 11 39.9
	22 41 28.7

⊙'s declination.

19° 15′ 3.2″ S.	35.77
46.9″	1.31
19° 14′ 16.3″ S.	46.86
90° N.	
p = 109° 14′ 16″ N.	

Index correction,	+ 0′ 32″
Semi-diameter,	+ 16′ 14″
Dip,	− 4′ 9″
Refraction,	− 2′ 54″
Parallax,	+ 0′ 8″
	+ 9′ 51″

	+ 9′ 51″	
h =	18° 29′ 15″	
L =	16° 25′ 0″	log sec = 0.01808
p =	109° 14′ 16″	log csc = 0.02493
$2S$ =	144° 8′ 31″	
S =	72° 4′ 15″	log cos = 9.48832
R =	53° 35′ 0″	log sin = 9.90565
		2)19.43698

Equation of time.

m. s.	
14 42.70	0.537
0.70	1.31
14 43.40	0.703

log sin $\frac{1}{2}t$ = 9.71849

$\frac{1}{2}t$ = 31° 31′ 57″.

t = 63° 3′ 54″ = 4 h. 12 m. 16 s.

d. h. m. s.	
Nov. 17 19 47 44	Local apparent astronomical time.
14 43	Equation of time.
Nov. 17 19 33 1	Local mean astronomical time.
Nov. 17 22 41 29	Greenwich mean time.
3 8 28	Difference in time.
47° 7′ 0″	W. Longitude.

19. 1895, Dec. 4, A.M., at sea, in latitude 36° 10′ N.; five observed altitudes of the ⊙ were taken at the times (by watch) standing opposite, viz.:

	h. m. s.
Obs. alt. ⊙ 13° 0′ 30″	6 27 14
3′ 10″	29.5
5′ 40″	49
8′ 50″	28 5
12′ 0″	23

Index correction + 32″; height of eye 18 ft.; correction of watch by chronometer − 5 h. 10 m. 47.1 s. Required the longitude.

	h. m. s.	☉'s declination.	
☉ 13° 0′ 30″	6 27 14	22° 15′ 37.3″ S.	20.10
3′ 10″	29.5	25.7″	1.28
5′ 40″	49	22° 16′ 3.0″ S.	25.73
8′ 50″	28 5	90° N.	
12′ 0″	23	$p =$ 112° 16′ 3″ N.	
30′ 10″	139 0.5		
13° 6′ 2″	6 27 48.1		
	−5 10 47.1		
	1 17 1		

Index correction,	+ 0′ 32″
Semi-diameter,	+ 16′ 16″
Dip,	− 4′ 9″
Refraction,	− 4′ 5″
Parallax,	+ 0′ 9″
	+ 8′ 43″

	+ 8′ 43″	
$h =$	13° 14′ 45″	
$L =$	36° 10′ 0″	log sec = 0.09296
$p =$	112° 16′ 3″	log csc = 0.03366
$2S =$	161° 40′ 48″	
$S =$	80° 50′ 24″	log cos = 9.20192
$R =$	67° 35′ 39″	log sin = 9.96591
		2)19.29445

log sin $\frac{1}{2}t$ = 9.64722

$\frac{1}{2}t$ = 26° 20′ 55″.

t = 52° 41′ 50″ = 3 h. 30 m. 47 s.

Equation of time.

m. s.	
9 40.05	1.013
1.30	1.28
9 38.75	1.30

d. h. m. s.	
Dec. 3 20 29 13	Local apparent astronomical time.
9 39	Equation of time.
Dec. 3 20 19 34	Local mean astronomical time.
Dec. 4 1 17 1	Greenwich mean time.
4 57 27	Difference in time.
74° 21′ 45″	W. Longitude.

20. 1895, Dec. 4, P.M., at sea, in latitude 36° 36′ N.; four observed altitudes of the ☉ were taken at the times (by watch) standing opposite, viz.:

	h. m. s.
Obs. alt. ☉ 16° 31′ 10″	1 7 26.5
30′ 0″	35.5
28′ 30″	46
27′ 20″	56.5

Index correction + 30″; height of eye 18 ft.; correction of watch by chronometer − 5 h. 10 m. 46.1 s. Required the longitude.

	h. m. s.
☉ 16° 31′ 10″	1 7 26.5
30′ 0″	35.5
28′ 30″	46
27′ 20″	56.5
117′ 0″	164.5
16° 29′ 15″	1 7 41.1
	−5 10 46.1
	7 56 55

☉'s declination.

22° 15′ 37.3″ S.	20.10
2′ 39.8″	7.95
22° 18′ 17.1″ S.	159.79
90° N.	
$p=$ 112° 18′ 17″ N.	

Index correction,	+ 0′ 30″
Semi-diameter,	+ 16′ 16″
Dip,	− 4′ 9″
Refraction,	− 3′ 14″
Parallax,	+ 0′ 8″
	+ 9′ 31″

+ 9′ 31″

$h=$ 16° 38′ 46″

$L=$. 36° 36′ 0″   log sec = 0.09538

$p=$ 112° 18′ 17″   log csc = 0.03377

$2\,S=$ 165° 33′ 3″

$S=$ 82° 46′ 31″   log cos = 9.09955

$R=$ 66° 7′ 45″   log sin = 9.96116

2)19.18986

log sin ½ t = 9.59493

½ t = 23° 10′ 18″.

t = 46° 20′ 36″ = 3 h. 5 m. 22 s.

Equation of time.

m. s.	
9 40.05	1.013
8.05	7.95
9 32.00	8.05

d. h. m. s.	
Dec. 4 3 5 22	Local apparent astronomical time.
9 32	Equation of time.
Dec. 4 2 55 50	Local mean astronomical time.
Dec. 4 7 56 55	Greenwich mean time.
5 1 5	Difference in time.
75° 16′ 15″	W. Longitude.

A Treatise on Plane Surveying.

By DANIEL CARHART, C.E., Professor of Civil Engineering in the Western University of Pennsylvania, Allegheny. Illustrated. 8vo. Half leather. xvii + 498 pages. Mailing price, $2.00; for introduction, $1.80.

THIS work covers the whole ground of Plane Surveying. It illustrates and describes the instruments employed, their adjustments and uses; it exemplifies the best methods of solving the ordinary problems occurring in practice, and furnishes solutions for many special cases which not infrequently present themselves. It is the result of twenty years' experience in the field and in technical schools, and the aim has been to make it extremely practical.

W. A. Moody, *Prof. of Mathematics, Bowdoin College:* I consider the book exceptionally fine in execution, subject-matter, and arrangement.

Wm. Hoover, *Prof. of Mathematics, Ohio University:* It is indeed a superior work, and merits the widest adoption.

A Field Book for Civil Engineers.

Department of Special Publication.—By DANIEL CARHART, C.E., Dean and Professor of Civil Engineering, Western University of Pennsylvania. 4½ x 7 inches. Flexible morocco. xii + 282 pages. Retail price, $2.50; for introduction, $2.00.

THIS book shows how to locate a railroad; it gives the organization and describes the outfit of the transit, level, and topographic parties; it indicates the work of the construction corps; tells how slope stakes are set; culverts, trestles, and tunnels staked out; quantities calculated; and frogs, switches, and wyes located. About one hundred diagrams aid in explaining the formulas, and numerous examples of a practical character supplement these. It contains, among many others, tables of all the natural trigonometric functions. It is written for students of civil engineering, and to satisfy the demand of field engineers for a manual convenient in size, containing the desired information, systematically arranged, fully illustrated, and easy of reference.

Engineering News: We are disposed to regard this book on the whole as among the very best field manuals which exist.

Thos. Rodd, *Chief Engineer, Pennsylvania R.R. Co., Pittsburg:* I have gone over Carhart's Field Book with care, and think it a valuable contribution to railroad engineering.

Academic Trigonometry: Plane and Spherical.

By T. M. BLAKSLEE, Ph.D. (Yale), Professor of Mathematics in Des Moines College, Iowa. 12mo. Cloth. 33 pages. Mailing price, 30 cents; for introduction, 25 cents.

THE Plane and Spherical portions are arranged on opposite pages. The memory is aided by analogies, and it is believed that the entire subject can be mastered in less time than is usually given to Plane Trigonometry alone, as the work contains but 29 pages of text.

The Plane portion is compact, and complete in itself.

Examples of Differential Equations.

By GEORGE A. OSBORNE, Professor of Mathematics in the Massachusetts Institute of Technology, Boston. 12mo. Cloth. vii + 50 pages. Mailing Price, 60 cents; for introduction, 50 cents.

A SERIES of nearly three hundred examples with answers, systematically arranged and grouped under the different cases, and accompanied by concise rules for the solution of each case.

Selden J. Coffin, *Prof. of Astronomy, Lafayette College:* Its appearance is most timely, and it supplies a manifest want.

Determinants.

The Theory of Determinants: an Elementary Treatise. By PAUL H. HANUS, B.S., recently Professor of Mathematics in the University of Colorado, now Assistant Professor, Harvard University. 8vo. Cloth. viii + 217 pages. Mailing price, $1.90; for introduction, $1.80.

THIS book is written especially for those who have had no previous knowledge of the subject, and is therefore adapted to self-instruction as well as to the needs of the class-room. The subject is at first presented in a very simple manner. As the reader advances, less and less attention is given to details. Throughout the entire work it is the constant aim to arouse and enliven the reader's interest, by first showing how the various concepts have arisen naturally, and by giving such applications as can be presented without exceeding the limits of the treatise.

William G. Peck, *late Prof. of Mathematics, Columbia College, N.Y.:* A hasty glance convinces me that it is an improvement on Muir.

T. W. Wright, *Prof. of Mathematics, Union Univ., Schenectady, N.Y.:* It fills admirably a vacancy in our mathematical literature, and is a very welcome addition indeed.

Elements of Plane Analytic Geometry.

By JOHN D. RUNKLE, Walker Professor of Mathematics in the Massachusetts Institute of Technology, Boston. 8vo. Cloth. ii + 344 pages. Mailing Price, $2.25; for introduction, $2.00.

IN this work, the author has had particularly in mind the needs of those students who can devote but a limited time to the subject, and yet must become quite familiar with at least its more elementary and fundamental part. For this reason, the earlier chapters are treated with somewhat more fulness than is usual. For some propositions, more than a single proof is given, and particular care has been taken to illustrate and enforce all parts of the subject by a large number of numerical applications. In the matter of problems, only the simpler ones have been selected, and the number has in every case been proportioned to the time that the students will have to devote to them. In general, propositions have been proved first with reference to rectangular axes. The determinant notation has not been used.

Descriptive Geometry.

By LINUS FAUNCE, Assistant Professor of Descriptive Geometry and Drawing in the Massachusetts Institute of Technology. 8vo. Cloth. 54 pages, with 16 lithographic plates, including 88 diagrams. Mailing Price, $1.35; for introduction, $1.25.

IN addition to the ordinary problems of Descriptive Geometry, this work includes a number of practical problems, such as might be met with by the draughtsman at any time, showing the application of the principles of Descriptive Geometry, a feature hitherto omitted in text-books on this subject. All of the problems have been treated clearly and concisely. The author's sole aim has been to present a work of practical value, not only as a text-book for schools and colleges, but also for every draughtsman.

The contents are: Chap. I., Elementary Principles; Notation. Chap. II., Problems relating to the Point, Line, and Plane. Chap. III., Principles and Problems relating to the Cylinder, Cone, and Double Curved Surfaces of Revolution. Chap. IV., Intersection of Planes and Solids, and the Development of Solids; Cylinders; Cones; Double Curved Surfaces of Revolution; Solids bounded by Plane Surfaces. Chap. V., Intersection of Solids. Chap. VI. Miscellaneous Problems.

Wheeler's Plane and Spherical Trigonometry.

By H. N. WHEELER, A.M., formerly of Harvard University. 12mo. Cloth. 211 pages. Mailing price, $1.10; introduction, $1.00. Pierce's Mathematical Tables are included.

THE special aim of the Plane Trigonometry is to give pupils a better idea of the trigonometric functions of obtuse angles than they could obtain from any book heretofore existing.

In the treatment of Spherical Trigonometry special pains has been taken to present applications to Geometry and Astronomy, and problems involving these applications.

Adjustments of the Compass, Transit, and Level.

By A. V. LANE, C.E., Ph.D., formerly Associate Professor of Mathematics, University of Texas, Austin. 12mo. Cloth. v + 43 pages. Mailing price, 33 cents; for introduction, 30 cents.

Principles of Elementary Algebra.

By H. W. KEIGWIN, Teacher of Mathematics, Norwich Free Academy, Norwich, Conn. 12mo. Paper. ii + 41 pages. Mailing and introduction price, 20 cents.

THIS little book is intended as an outline of thorough oral instruction, and is all the "text" which the author has found it necessary to put into his pupils' hands. It should, of course, be accompanied by a good set of exercises and problems.

Metrical Geometry. *An Elementary Treatise on Mensuration.*

By GEORGE BRUCE HALSTED, Ph.D., Professor of Mathematics, University of Texas, Austin. 12mo. Cloth. 246 pages. Mailing price, $1.10; for introduction, $1.00.

THIS work applies new principles and methods to simplify the measurement of lengths, angles, areas, and volumes. It is strictly demonstrative, but uses no Trigonometry, and is adapted to be taken up in connection with or following any elementary Geometry. A hundred illustrative examples are worked out in the course of the book, and at the end are five hundred carefully arranged and indexed exercises, using the metric system.

CPSIA information can be obtained
at www.ICGtesting.com
Printed in the USA
LVHW080727100323
741055LV00064B/361

9 781015 881990